Signals and Communication Technology

More information about this series at http://www.springer.com/series/4748

Shoji Makino
Editor

Audio Source Separation

Springer

Editor
Shoji Makino
University of Tsukuba
Ibaraki
Japan

ISSN 1860-4862 ISSN 1860-4870 (electronic)
Signals and Communication Technology
ISBN 978-3-030-10303-3 ISBN 978-3-319-73031-8 (eBook)
https://doi.org/10.1007/978-3-319-73031-8

Printed on acid-free paper

This Springer imprint is published by Springer Nature
The registered company is Springer International Publishing AG
The registered company address is: Gewerbestrasse 11, 6330 Cham, Switzerland

Preface

We are surrounded by sounds. Such a noisy environment makes it difficult to hear desired speech and to converse comfortably. This makes it important to be able to separate and extract a target speech signal from noisy observations for both human–machine and human–human communication.

Audio source separation is an approach to estimating source signals using information about their mixtures observed in each input channel. The estimation is performed by either spatial filtering based on blind audio source localization or time–frequency filtering based on audio source modeling, or both. The use of audio source separation in the development of suitable acoustic communication channels between humans and machines is widespread.

Some books have been published on audio source separation, independent component analysis (ICA), and related subjects. ICA-based audio source separation has been well studied in the fields of statistics and information theory, for application to a variety of disciplines. In particular, as speech and audio signal mixtures in a real reverberant environment are generally convolutive mixtures and are prevalent in many applications, their separation is a much more challenging task. Recently, nonnegative matrix factorization (NMF) and deep neural networks (DNNs) have been extensively exploited as other means of audio source separation, for which excellent performance has been achieved and useful knowledge about these methods has been acquired.

The goal of this book is to provide a reference to the fascinating topic of audio source separation for convolved speech mixtures. The editor believes that this book is of particular value as it comprises reports on cutting-edge research by internationally recognized scientists and the state of the art. The topic in the individual chapters of this book was selected to be tutorial in nature with specific emphasis on providing an in-depth treatment of recent important results.

This book is organized into three sections that approximately follow the main areas of audio source separation.

Part 1 presents an account of cutting-edge audio source separation based on nonnegative matrix factorization (NMF). Even with a single microphone, we can separate a mixture by using the harmonicity and temporal structure of the sources.

The single-channel NMF approach utilizes frequency diversity to discriminate between desired and undesired components. The multichannel NMF approach utilizes frequency diversity and spatial diversity to discriminate between desired and undesired components.

Part 2 addresses cutting-edge audio source separation based on a deep neural network (DNN). This is a fascinating technique and can be applied to audio source separation problems. Seminal examples of single-channel and multichannel approaches illustrate the bright future of DNNs.

Part 3 describes state-of-the-art audio source separation based on sparse component analysis (SCA). Here, the sparseness of speech sources is very useful and time–frequency diversity plays a key role. In SCA, we can build a probabilistic framework by assuming a source model and separate a mixture by maximizing the a posteriori probability of the sources given the observations.

The authors and the editor hope that this book will serve as a guide for a large audience, inspire many readers, and be a source of new ideas. We hope that it will be a useful resource for readers ranging from students and practicing engineers to advanced researchers.

The editor would like to take this opportunity to express his deep gratitude to all the contributing authors for making this a unique work. Thanks to their cooperation, editing this book has turned out to be a very pleasant experience. Finally, he is very grateful to Tom Spicer and his colleagues from Springer, for their encouragement and kind support.

Tokyo, Japan Shoji Makino

Contents

Chapter 1
Single-Channel Audio Source Separation with NMF: Divergences, Constraints and Algorithms

Cédric Févotte, Emmanuel Vincent and Alexey Ozerov

Abstract Spectral decomposition by nonnegative matrix factorisation (NMF) has become state-of-the-art practice in many audio signal processing tasks, such as source separation, enhancement or transcription. This chapter reviews the fundamentals of NMF-based audio decomposition, in unsupervised and informed settings. We formulate NMF as an optimisation problem and discuss the choice of the measure of fit. We present the standard majorisation-minimisation strategy to address optimisation for NMF with the common β-divergence, a family of measures of fit that takes the quadratic cost, the generalised Kullback-Leibler divergence and the Itakura-Saito divergence as special cases. We discuss the reconstruction of time-domain components from the spectral factorisation and present common variants of NMF-based spectral decomposition: supervised and informed settings, regularised versions, temporal models.

1.1 Introduction

Data is often available in matrix form $\mathbf{V}$, where columns $\mathbf{v}_n$ are data samples and rows are features. Processing such data often entails finding a factorisation of the matrix $\mathbf{V}$ into two unknown matrices $\mathbf{W}$ and $\mathbf{H}$ such that

$$\mathbf{V} \approx \hat{\mathbf{V}} \stackrel{\text{def}}{=} \mathbf{WH}. \tag{1.1}$$

In the approximation (1.1), $\mathbf{W}$ acts as a dictionary of recurring patterns, which is characteristic of the data, and every column $\mathbf{h}_n$ of $\mathbf{H}$ contains the *decomposition* or *activation* coefficients that approximate every $\mathbf{v}_n$ onto the dictionary. In the following

C. Févotte (✉)
CNRS & IRIT, Toulouse, France
e-mail: cedric.fevotte@irit.fr

E. Vincent
Inria, 54600 Villers-lès-Nancy, France

A. Ozerov
Technicolor, Rennes, France

S. Makino (ed.), *Audio Source Separation*, Signals and Communication Technology, https://doi.org/10.1007/978-3-319-73031-8_1

we will refer to $\mathbf{W}$ as the *dictionary* and to $\mathbf{H}$ as the *activation matrix*. The data matrix $\mathbf{V}$ is of dimensions $F \times N$ and the common dimension of $\mathbf{W}$ and $\mathbf{H}$ is denoted K, often referred to as the rank of the factorisation (which might differ from the actual mathematical rank of $\mathbf{V}$).

In the literature, the problem of obtaining the factorisation (1.1) can appear under other domain-specific names such as *dictionary learning*, *low-rank approximation*, *factor analysis* or *latent semantic analysis*. Many forms of factorisation (1.1) have been considered. The most notorious and ancient one is Principal Component Analysis (PCA) [1] which simply minimises the quadratic cost between $\mathbf{V}$ and its approximate $\mathbf{WH}$, where all matrices are treated as real-valued. Independent Component Analysis (ICA) [2] is a major variant of PCA in which the rows of $\mathbf{H}$ are constrained to be mutually independent. Sparse coding [3] and many recent dictionary learning [4] approaches impose some form of sparsity of the activation matrix. Nonnegative matrix factorisation (NMF), the main topic of this chapter, is dedicated to nonnegative data and imposes nonnegativity of the factors $\mathbf{W}$ and $\mathbf{H}$.

Early work on NMF has appeared in applied algebra (under various names) and more notably in chemometrics [5], but it fully came to maturation with the seminal paper of Lee and Seung, published in *Nature* in 1999 [6]. Like PCA, NMF consists of minimising an error of fit between $\mathbf{V}$ and its approximate $\mathbf{WH}$, but subject to nonnegativity of the values of $\mathbf{W}$ and $\mathbf{H}$. The nonnegativity of $\mathbf{W}$ ensures the *interpretability* of the dictionary, in the sense that the extracted patterns $\mathbf{w}_k$ (the columns of $\mathbf{W}$) remain nonnegative, like the data samples. The nonnegativity of $\mathbf{H}$ ensures that $\mathbf{WH}$ is nonnegative, like $\mathbf{V}$, but is also shown to induce a *part-based representation*, in stark contrast with plain PCA that leads to more global or *holistic* representations (where every pattern attempts to generalise as much as possible the whole dataset). Because subtractive combinations are forbidden, the approximate $\mathbf{Wh}_n$ to every sample $\mathbf{v}_n$ can only be formed from building blocks, and thus the estimated patterns tend to be parts of data.

Following the work of Lee and Seung, NMF became an increasingly popular data analysis tool and has been used in many fields. In particular, it has led to important breakthroughs in text retrieval (based on the decomposition of a *bag-of-words* representation [7]), collaborative filtering (completion of missing ratings in users $\times$ items matrices [8]) or spectral unmixing. In the latter case, NMF is for example used in chemical spectroscopy [5], remote sensing (for unmixing of hyperspectral electromagnetic data) [9] and most notably audio signal processing [10]. The seminal work of Smaragdis and Brown [10] has initiated an important thread of NMF-based contributions in music transcription, source separation, speech enhancement, etc. The common principle of all these works is the nonnegative decomposition of the spectrogram of the observed signal onto a dictionary of elementary spectral components, representative of building sound units (notes, chords, percussive sounds, or more complex adaptive structures). This general architecture is detailed in Sect. 1.2. It describes in particular popular NMF models and means of obtaining the factorisation, by optimisation of a cost function. Then it describes how to reconstruct elementary sound components from the nonnegative factorisation of the spectrogram.

This blind decomposition might fail to return adequate and useful results when dealing with complex multi-source signals and the system needs to be "guided" with prior information. Such advanced decompositions for source separation will be covered in Sect. 1.3. Section 1.4 concludes.

1.2 Signal Decomposition by NMF

The general principle of NMF-based audio spectral analysis is depicted in Fig. 1.1. It shows how NMF has the capability of unmixing superimposed spectral components. This is in contrast for example with the Gaussian Mixture Model (GMM), a clustering model that is not designed to handle composite data. In the GMM, each data sample can only be in one among several states. As such, the occurrence of mixed frames in the data represented in Fig. 1.1 (3rd to 5th samples) would count as one state, along the two other states corresponding to pure spectra (red and green). The nonnegativity of **H** encourages so-called *part-based* representations. Because subtractive combinations of dictionary elements are forbidden, the dictionary **W** tends to contain elementary building units. This is a welcome property for analysis tasks such as music transcription or source separation. In contrast, a method such as PCA would instead produce an orthogonal dictionary with a more *holistic* value, that compresses more efficiently the entire dataset. The difference between PCA, NMF and vector quantisation is remarkably illustrated in [6] with comparative experiments using a set of face images. It is shown that where PCA returns *eigenfaces* (sort of template faces), NMF can efficiently capture parts of faces (noise, eyes, etc.). Figure 1.2 dis-

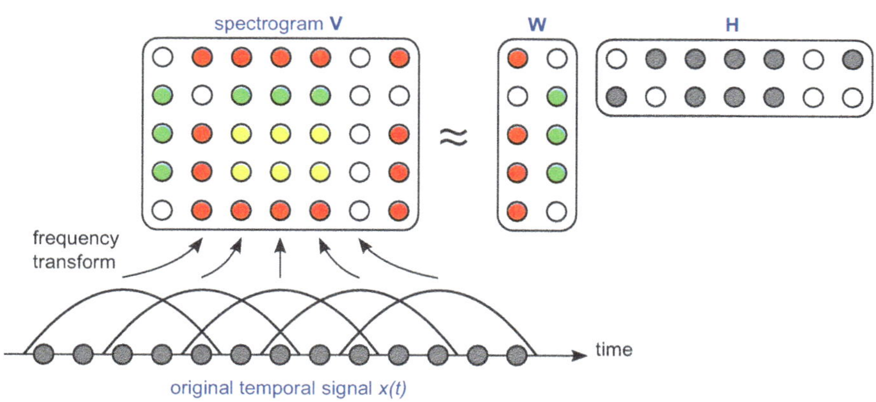

Fig. 1.1 NMF-based audio spectral analysis. A short-time frequency transform, such as the magnitude or power short-time Fourier transform, is applied to the original time-domain signal $x(t)$. The resulting nonnegative matrix is factorised into the nonnegative matrices **W** and **H**. In this schematic example, the red and green elementary spectra are unmixed and extracted into the dictionary matrix **W**. The activation matrix **H** returns the mixing proportions of each time-frame (a column of **W**)

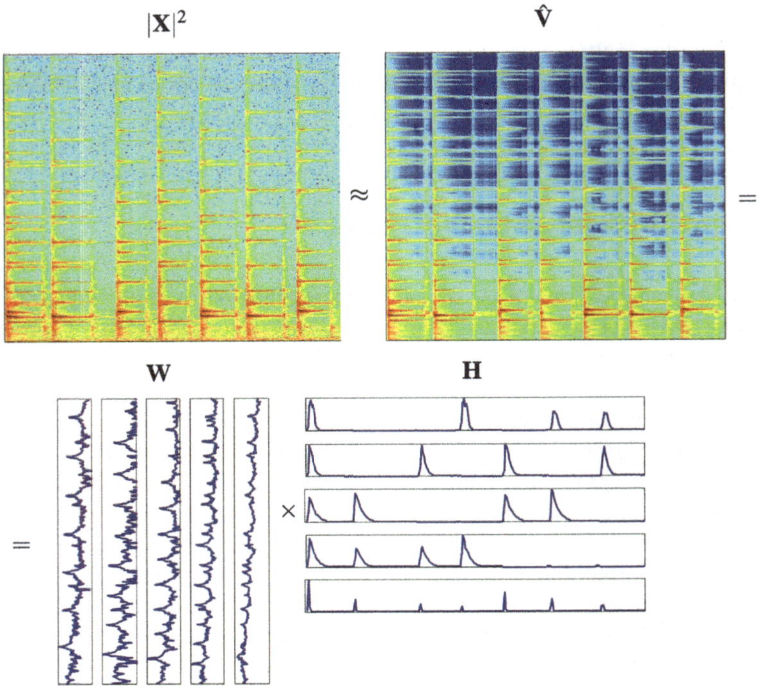

Fig. 1.2 NMF applied to the spectrogram of a short piano sequence composed of four notes. (Data used from [11])

plays the result of NMF applied to the spectrogram of a short piano sequence; see [10, 11] for further illustration on small-scale examples.

1.2.1 NMF by Optimisation

The factorisation (1.1) is usually sought after through the minimisation problem

$$\min_{\mathbf{W},\mathbf{H}} D(\mathbf{V}|\mathbf{WH}) \text{ subject to } \mathbf{W} \geq 0, \mathbf{H} \geq 0 \tag{1.2}$$

where the notation $\mathbf{A} \geq 0$ expresses nonnegativity of the entries of matrix $\mathbf{A}$ (and not semidefinite positiveness), and where $D(\mathbf{V}|\mathbf{WH})$ is a separable measure of fit such that

$$D(\mathbf{V}|\mathbf{WH}) = \sum_{f=1}^{F} \sum_{n=1}^{N} d([\mathbf{V}]_{fn}|[\mathbf{WH}]_{fn}) \tag{1.3}$$

where $d(x|y)$ is a scalar cost function. What we intend by "cost function" is a positive function of $y \in \mathbb{R}_+$ given $x \in \mathbb{R}_+$, with a single minimum for $x = y$.

The quadratic cost function $d_Q(x|y) = \frac{1}{2}(x-y)^2$ is a popular choice when dealing with real numbers. It underlies an additive Gaussian noise model and enjoys convenient mathematical properties for estimation and optimisation problems. For that same reason, it is a less natural choice for nonnegative data because it may generate negative values. Many other choices have been considered in the NMF literature, in particular under the influence of Cichocki et al. Two popular families of NMF cost functions are the α-divergence [12] and the β-divergence [13–15], themselves connected to the wider families of Csiszár or Bregman divergences, see, e.g., [13, 16] in the context of NMF. The β-divergence in particular has enjoyed a certain success in audio signal processing. It can be defined as [17, 18]

$$d_\beta(x|y) \stackrel{\text{def}}{=} \begin{cases} \frac{1}{\beta(\beta-1)}\left(x^\beta + (\beta-1)\, y^\beta - \beta\, x\, y^{\beta-1}\right), & \beta \in \mathbb{R}\backslash\{0, 1\} \\ x \log \frac{x}{y} - x + y = d_{KL}(x|y), & \beta = 1 \\ \frac{x}{y} - \log \frac{x}{y} - 1 = d_{IS}(x|y), & \beta = 0 \end{cases} \tag{1.4}$$

The limit cases $\beta = 0$ and $\beta = 1$ correspond to the Itakura-Saito (IS) and generalised Kullback-Leibler (KL) divergences, respectively. The case $\beta = 2$ corresponds to the quadratic cost $d_Q(x|y)$. The β-divergence forms a continuous family of cost functions that smoothly interpolates between the latter three well-known cases. As noted in [11, 15], a noteworthy property of the β-divergence is its behaviour w.r.t. the scale of the data, as the following equation holds for any value of β:

$$d_\beta(\lambda\, x|\lambda\, y) = \lambda^\beta\, d_\beta(x|y). \tag{1.5}$$

As noted in [11], this implies that factorisations obtained with $\beta > 0$ (such as with the quadratic cost or the KL divergence) will rely more heavily on large data values and less precision is to be expected in the estimation of the low-power components, and conversely factorisations obtained with $\beta < 0$ will rely more heavily on small data values. The IS divergence ($\beta = 0$) is scale-invariant, i.e., $d_{IS}(\lambda\, x|\lambda\, y) = d_{IS}(x|y)$, and is the only one in the family of β-divergences to possess this property. Factorisations with small positive values of β are relevant to decomposition of audio spectra, which typically exhibit exponential power decrease along frequency f and also usually comprise low-power transient components such as note attacks together with higher power components such as tonal parts of sustained notes. For example, [11] presents the results of the decomposition of a piano power spectrogram with IS-NMF and shows that components corresponding to very low residual noise and hammer hits on the strings are extracted with great accuracy, while these components are either ignored or severely degraded when using Euclidean or KL divergences. Similarly, the value $\beta = 0.5$ is advocated by [19, 20] and has been shown to give optimal results in music transcription based on NMF of the magnitude spectrogram by [21].

1.2.2 Composite Models

NMF with the β-divergence as formulated in the previous section fails to give a probabilistic understanding of the modelling assumptions. As a matter of fact, the β-divergence acts as a pseudo-likelihood for the so-called Tweedie distribution, a member of the exponential family, parametrised with respect to its mean, i.e., such that [22]

$$\mathrm{E}\,[\mathbf{V}|\mathbf{WH}] = \mathbf{WH}. \tag{1.6}$$

In particular, the values $\beta = 0, 1, 2$ underlie multiplicative Gamma observation noise ($v_{fn} = [\mathbf{WH}]_{fn}.\varepsilon_{fn}$), Poisson noise ($v_{fn} \sim Po([\mathbf{WH}]_n)$) and Gaussian additive observation noise ($v_{fn} = [\mathbf{WH}]_{fn} + \varepsilon_{fn}$), respectively (see the Appendix for the definitions of the distributions involved).

These probabilistic models characterise the magnitude or power spectrogram $\mathbf{V}$ but do not explicitly characterise the composite structure of sound that is generally looked after in NMF-based decomposition. As such, the *Gaussian Composite Model* (GCM) was introduced in [11] to remedy this limitation. Denoting by x_{fn} the complex-valued coefficients of the short-time Fourier transform (STFT), the GCM is defined by

$$x_{fn} = \sum_k c_{k,fn}, \tag{1.7}$$

$$c_{k,fn} \sim N_c(0, w_{fk}h_{kn}), \tag{1.8}$$

where $N_c(\mu, \lambda)$ refers to the circular complex-valued normal distribution defined in the Appendix. The composite structure of sound (i.e., the superimposition of elementary components) is made explicit by (1.7). Then, (1.8) states that the kth elementary component $c_{k,fn}$ is the expression of the kth the spectral template $\mathbf{w}_k$ amplitude-modulated in time by the activation coefficient h_{kn}. The latent components may also be marginalised from the model to yield more simply

$$x_{fn} \sim N_c(0, [\mathbf{WH}]_{fn}). \tag{1.9}$$

With the uniform phase assumption that defines the circular complex-valued normal distribution, (1.9) itself reduces to

$$v_{fn} = [\mathbf{WH}].\varepsilon_{fn}, \tag{1.10}$$

where $v_{fn} = |x_{fn}|^2$ (the power spectrogram) and ε_{fn} has an exponential distribution with expectation 1 (i.e., using the notations defined in the Appendix, $\varepsilon_{fn} \sim G(1, 1)$). As such, the GCM is tightly connected to the multiplicative Gamma noise model, and we may easily find that

$$- \log p(\mathbf{X}|\mathbf{WH}) = D_{IS}(|\mathbf{X}|^2|\mathbf{WH}) + cst. \tag{1.11}$$

(1.11) shows that factorising the power spectrogram $\mathbf{V} = |\mathbf{X}|^2$ with the IS divergence is equivalent to performing maximum likelihood estimation of $\mathbf{W}$ and $\mathbf{H}$ in the GCM model defined by (1.7) and (1.8). Given estimates of $\mathbf{W}$ and $\mathbf{H}$ (using for example the algorithm presented in the following section), reconstruction of the latent components $c_{k,fn}$ can be done with any estimator. For example, the Minimum Mean Squares Error (MMSE) estimator is given by the so-called Wiener filter

$$\hat{c}_{k,fn} = \mathrm{E}\left[c_{k,fn}|\mathbf{W}, \mathbf{H}\right] = \frac{w_{fk}h_{kn}}{[\mathbf{WH}]_{fn}} x_{fn} \tag{1.12}$$

By construction, the component estimates satisfy $x_{fn} = \sum_k \hat{c}_{k,fn}$. The estimated component STFTs $\hat{\mathbf{C}}_k = \{c_{k,fn}\}_{fn}$ can then be inverse-transformed (using a standard overlap-add procedure) to yield time-domain estimates $\hat{c}_k(t)$ such that $x(t) = \sum_k \hat{c}_k(t)$.

Besides the GCM, other composite interpretations of known NMF models have been proposed in the literature [23]. For example, the Poisson-NMF model

$$v_{fn} \sim Po([\mathbf{WH}]_{fn}) \tag{1.13}$$

is equivalent to

$$x_{fn} = \sum_k c_{k,fn}, \tag{1.14}$$

$$c_{k,fn} \sim Po(w_{fk}h_{kn}). \tag{1.15}$$

It turns out the MMSE estimator of the latent components is again given by (1.12). It is easily shown that

$$- \log p(\mathbf{V}|\mathbf{WH}) = D_{KL}(\mathbf{V}|\mathbf{WH}) + cst \tag{1.16}$$

so that maximum-likelihood estimation of $\mathbf{W}$ and $\mathbf{H}$ in model (1.13) is equivalent to NMF with the generalised KL divergence [11, 24, 25]. A closely related model is PLSA [7] /PLCA [26] which writes

$$\mathbf{v}_n \sim M\left(\sum_f v_{fn}, \mathbf{Wh}_n\right), \tag{1.17}$$

where $M(L, \mathbf{p})$ refers to the multinomial distribution defined in the Appendix and the columns of $\mathbf{W}$ and $\mathbf{H}$ are constrained to sum to 1. PLSA/PLCA can also be shown to be equivalent to a generative model that involves multinomial latent components. PLCA is equivalent to NMF with a weighted KL divergence, such that

$$-\log p(\mathbf{V}|\mathbf{WH}) = \sum_n \|\mathbf{v}_n\|_1 D_{KL}\left(\frac{\mathbf{v}_n}{\|\mathbf{v}_n\|_1}|\mathbf{Wh}_n\right). \tag{1.18}$$

Poisson-NMF and PLCA are also popular models for audio spectrogram decomposition. This is because the KL divergence (used with the magnitude spectrogram $\mathbf{V} = |\mathbf{X}|$) has been experimentally proven to be also a reasonable measure of fit for audio spectral factorisation [27, 28]. However, from a probabilistic generative point of view, the Poisson-NMF and PLCA models are unreasonable because they generate integer values that do not comply with the real-valued nature of spectrograms (as a matter of fact, Poisson-NMF and PLSA/PLCA have been originally designed for count data [7, 24]).

1.2.3 Majorisation-Minimisation

The very large majority of NMF algorithms resort to block-coordinate descent to address problem (1.2). This means the variables $\mathbf{W}$ and $\mathbf{H}$ are updated in turn until a stationary point of $C(\mathbf{W}, \mathbf{H}) = D(\mathbf{V}|\mathbf{WH})$ is reached. Because $C(\mathbf{W}, \mathbf{H})$ is jointly non-convex in $\mathbf{W}$ and $\mathbf{H}$, the stationary point may be not a global minimum (and possibly not even a local minimum). As such, initialisation is an important issue in NMF and running the algorithm from different starting points is usually advised. It is also easy to see that the updates of $\mathbf{W}$ and $\mathbf{H}$ are essentially the same by transposition ($\mathbf{V} \approx \mathbf{WH} \Leftrightarrow \mathbf{V}^T \approx \mathbf{H}^T\mathbf{W}^T$). As such we may restrict our study to the update of $\mathbf{H}$ given $\mathbf{W}$:

$$\min_{\mathbf{W},\mathbf{H}} C(\mathbf{H}) \stackrel{\text{def}}{=} D(\mathbf{V}|\mathbf{WH}) \text{ subject to } \mathbf{H} \geq 0 \tag{1.19}$$

For the divergences considered in Sect. 1.2.1, a standard approach to the conditional updates of $\mathbf{W}$ and $\mathbf{H}$ is Majorisation-minimisation (MM). Generally speaking, MM consists in optimising iteratively an easier-to-minimise tight upper bound of the original objective function $C(\mathbf{H})$ [29].

Denote by $\tilde{\mathbf{H}}$ the estimate of $\mathbf{H}$ at current iteration. The first step of MM consists in building an upper bound $G(\mathbf{H}|\tilde{\mathbf{H}})$ of $C(\mathbf{H})$ which is tight for $\mathbf{H} = \tilde{\mathbf{H}}$, i.e., $C(\mathbf{H}) \leq G(\mathbf{H}|\tilde{\mathbf{H}})$ for all $\mathbf{H}$ and $C(\tilde{\mathbf{H}}) = G(\tilde{\mathbf{H}}|\tilde{\mathbf{H}})$. The second step consists in minimising the bound w.r.t. $\mathbf{H}$, producing a valid descent algorithm. Indeed, at iteration $i + 1$, it holds by construction that $C(\mathbf{H}^{(i+1)}) \leq G(\mathbf{H}^{(i+1)}|\mathbf{H}^{(i)}) \leq G(\mathbf{H}^{(i)}|\mathbf{H}^{(i)}) = C(\mathbf{H}^{(i)})$. The bound $G(\mathbf{H}|\tilde{\mathbf{H}})$ is often referred to as *auxiliary function*. The principle of MM is illustrated in Fig. 1.3.

The question now boils down to whether the construction of such an upper bound, which is amenable to optimisation, is possible. Fortunately, the answer is yes for many divergences, and in particular for the β-divergence discussed in Sect. 1.2.1. The trick is to decompose $C(\mathbf{H})$ into the sum of a convex part and a concave part and to upper-bound each part separately (the concave part is actually inexistent for $1 \leq \beta \leq 2$ where the β-divergence is convex w.r.t. its second argument). The convex part is

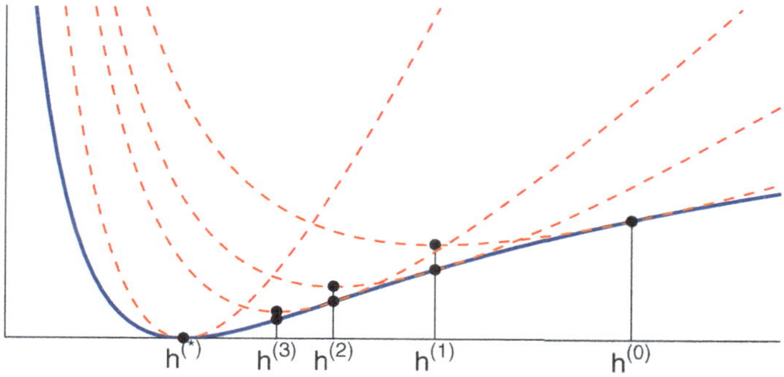

Fig. 1.3 An illustration of the MM principle on a unidimensional problem. Given a current estimate of $\mathbf{W}$, the blue curve acts as the objective function $C(\mathbf{H}) = D(\mathbf{V}|\mathbf{WH})$ to be minimised with respect to $\mathbf{H}$. The MM approach relies on the iterative minimisation of tight upper bounds (dashed red curves). The algorithm is initialised at $\mathbf{H}^{(0)}$, at which the first upper bound is minimised during the first iteration to yield $\mathbf{H}^{(1)}$, and so on until convergence. (Reproduced from [30])

majorised using Jensen's inequality (the definition of convexity) and the concave part is majorised using the tangent inequality. The two separate bounds are summed and the resulting (convex) auxiliary function turns out to have a closed-form minimiser. For illustration, we address the case of NMF with the Itakura-Saito divergence. The more general β-divergence case is addressed in details in [15].

1.2.3.1 A Special Case: NMF with the Itakura-Saito Divergence

Choosing the IS divergence as the measure of fit and addressing the update of $\mathbf{H}$, our goal is to minimise the objective function given by

$$C(\mathbf{H}) = \sum_{fn} \left(\frac{v_{fn}}{[\mathbf{WH}]_{fn}} - \log \frac{v_{fn}}{[\mathbf{WH}]_{fn}} - 1 \right) \tag{1.20}$$

$$= \sum_{fn} \left(\frac{v_{fn}}{[\mathbf{WH}]_{fn}} + \log [\mathbf{WH}]_{fn} \right) + cst \tag{1.21}$$

where cst is a term which is constant w.r.t. $\mathbf{H}$. As such, $C(\mathbf{H})$ can be written as the sum of a convex term $\breve{C}(\mathbf{H}) = \sum_{fn} \frac{v_{fn}}{[\mathbf{WH}]_{fn}}$ and a concave term $\tilde{C}(\mathbf{H}) = \sum_{fn} \log[\mathbf{WH}]_{fn}$. By convexity of $f(x) = 1/x$ for $x \geq 0$ and Jensen's inequality it holds that

$$f(\sum_k \lambda_k x_k) \leq \sum_k \lambda_k f(x_k) \tag{1.22}$$

for any $x_k, \lambda_k \geq 0$ such that $\sum_k \lambda_k = 1$. As such, it holds that

$$\breve{C}(\mathbf{H}) = \sum_{fn} \frac{v_{fn}}{\sum_k \frac{w_{fk}h_{kn}}{\lambda_{fkn}} \lambda_{fkn}} \leq \sum_{fn} v_{fn} \sum_k \frac{\lambda_{fkn}^2}{w_{fk}h_{kn}}, \tag{1.23}$$

for any $\lambda_{fkn} \geq 0$ such that $\sum_k \lambda_{fkn} = 1$. Choosing

$$\lambda_{fkn} = \frac{w_{fk}\tilde{h}_{kn}}{[\mathbf{W}\tilde{\mathbf{H}}]_{fn}} \tag{1.24}$$

and denoting by $\breve{G}(\mathbf{H}|\tilde{\mathbf{H}})$ the right-hand side of (1.23), it can be easily checked that $\breve{G}(\mathbf{H}|\tilde{\mathbf{H}})$ is an auxiliary function for $\breve{C}(\mathbf{H})$.

Now, by concavity of $\hat{C}(\mathbf{H})$ and the tangent inequality applied at $\mathbf{H} = \tilde{\mathbf{H}}$, we may write

$$\hat{C}(\mathbf{H}) \leq \hat{C}(\tilde{\mathbf{H}}) + \sum_{kn} [\nabla \hat{C}(\tilde{\mathbf{H}})]_{kn} (h_{kn} - \tilde{h}_{kn}) \tag{1.25}$$

Using the chain rule, the gradient term is found to be

$$[\nabla \hat{C}(\tilde{\mathbf{H}})]_{kn} = \sum_f \frac{w_{fk}}{[\mathbf{W}\tilde{\mathbf{H}}]_{fn}}. \tag{1.26}$$

By construction, the right hand side of (1.25) defines an auxiliary function $\hat{G}(\mathbf{H}|\tilde{\mathbf{H}})$ of $\hat{C}(\mathbf{H})$. Assembling $\breve{G}(\mathbf{H}|\tilde{\mathbf{H}})$ and $\hat{G}(\mathbf{H}|\tilde{\mathbf{H}})$ defines an auxiliary function $G(\mathbf{H}|\tilde{\mathbf{H}})$ of $C(\mathbf{H})$. The auxiliary function $G(\mathbf{H}|\tilde{\mathbf{H}})$ is convex by construction. Computing and cancelling its gradient leads to

$$h_{kn} = \tilde{h}_{kn} \left(\frac{\sum_f w_{fk} v_{fn} [\mathbf{W}\tilde{\mathbf{H}}]^{-2}}{\sum_f w_{fk} [\mathbf{W}\tilde{\mathbf{H}}]^{-1}} \right)^{\frac{1}{2}}. \tag{1.27}$$

Because the new update is found by multiplying the previous update with a correcting factor, the induced algorithm is coined "multiplicative". Because the correcting factor is nonnegative, nonnegativity of the updates is ensured along the iterations, given positive initialisations. Reference [15] proves that dropping the exponent $\frac{1}{2}$ in (1.27) produces an accelerated descent algorithm. The update (1.27) can then be written in algorithmic form using matrix operations as

$$\mathbf{H} \leftarrow \mathbf{H} \circ \frac{\mathbf{W}^T \left((\mathbf{W}\mathbf{H})^{\circ[-2]} \circ \mathbf{V} \right)}{\mathbf{W}^T (\mathbf{W}\mathbf{H})^{\circ[-1]}} \tag{1.28}$$

where the notation $\circ$ denotes MATLAB-like entry-wise multiplication/exponentiation and the fraction bar denotes entry-wise division. By exchangeability of $\mathbf{W}$ and $\mathbf{H}$ by transposition, the update rule for $\mathbf{W}$ is simply given by

$$\mathbf{W} \leftarrow \mathbf{W} \circ \frac{((\mathbf{WH})^{\circ[-2]} \circ \mathbf{V})\mathbf{H}^T}{(\mathbf{WH})^{\circ[-1]}\,\mathbf{H}^T} \tag{1.29}$$

The two updates (1.28) and (1.29) are applied in turn until a convergence criterion is met. The two updates have linear complexity per iteration, are free of tuning parameters and are very easily implemented.

As detailed in [15], these derivations can easily be extended to the more general case of NMF with the β-divergence. The resulting updates generalise (1.28) and (1.29) and can be written as

$$\mathbf{H} \leftarrow \mathbf{H} \circ \frac{\mathbf{W}^T\,((\mathbf{WH})^{\circ[\beta-2]} \circ \mathbf{V})}{\mathbf{W}^T(\mathbf{WH})^{\circ[\beta-1]}} \tag{1.30}$$

$$\mathbf{W} \leftarrow \mathbf{W} \circ \frac{((\mathbf{WH})^{\circ[\beta-2]} \circ \mathbf{V})\mathbf{H}^T}{(\mathbf{WH})^{\circ[\beta-1]}\,\mathbf{H}^T} \tag{1.31}$$

1.3 Advanced Decompositions for Source Separation

In the previous sections, we described the elementary principles of signal decomposition by NMF. The direct application of these principles leads to so-called *unsupervised NMF*, where both the dictionary and the activation coefficients are estimated from the signal to be separated. This approach yields interesting and useful results on toy data. For real audio signals, however, each sound source rarely consists of a single NMF component. For instance, a music source typically involves several notes with different pitches, while a speech source involves several phonemes. Various techniques have been proposed to classify or to cluster individual NMF components into sources [31, 32]. Nevertheless, several issues remain: the learned components may overfit the test signal, several sources may share similar dictionary elements, and the elegance of NMF is lost. These issues have called for more advanced treatments incorporating prior information about the properties of audio sources in general and/or in a specific signal [33].

1.3.1 Pre-specified Dictionaries

1.3.1.1 Supervised NMF

So-called *supervised NMF* is the simplest such treatment. It assumes that each source is characterised by a fixed source-specific dictionary and only the activation coefficients must be estimated from the signal to be separated [34]. Let us assume that the sources are indexed by $j \in \{1, \dots, J\}$ and denote by $\mathbf{W}_j$ and $\mathbf{H}_j$ the dictionary and the activation matrix associated with source j. The mixture spectrogram $\mathbf{V}$ can then be expressed as in (1.1) where

$$\mathbf{W} = \begin{pmatrix} \mathbf{W}_1 & \cdots & \mathbf{W}_J \end{pmatrix} \tag{1.32}$$

$$\mathbf{H} = \begin{pmatrix} \mathbf{H}_1 \\ \vdots \\ \mathbf{H}_J \end{pmatrix} \tag{1.33}$$

result from the concatenation of the source-specific dictionaries and activation matrices. Given the dictionaries $\mathbf{W}_1, \dots, \mathbf{W}_J$ of all sources, the activation matrices $\mathbf{H}_1, \dots, \mathbf{H}_J$ can be estimated by applying, for instance, the optimisation procedure described in Sect. 1.2. The standard multiplicative update with the β-divergence can be equivalently rewritten in terms of each $\mathbf{H}_j$ as

$$\mathbf{H}_j \leftarrow \mathbf{H}_j \circ \frac{\mathbf{W}_j^T((\mathbf{WH})^{\circ[\beta-2]} \circ \mathbf{V})}{\mathbf{W}_j^T(\mathbf{WH})^{\circ[\beta-1]}}. \tag{1.34}$$

Note that, because $\mathbf{W}$ is here fixed, in the case when the cost function is strictly convex ($1 \leq \beta \leq 2$), the resulting update is guaranteed to converge to a global minimum. Eventually, the complex-valued spectrogram $\mathbf{S}_j$ of each source can be estimated by Wiener filtering as

$$\mathbf{S}_j = \frac{\mathbf{W}_j\mathbf{H}_j}{\mathbf{WH}} \circ \mathbf{X}. \tag{1.35}$$

This is equivalent to extracting the signal corresponding to all NMF components in Sect. 1.2.2 and summing the extracted signals associated with each source. A variant of supervised NMF called *Semi-supervised NMF* assumes that a pre-specified dictionary is available for a subset of sources only and that the remaining sources are jointly represented by an additional dictionary which is estimated from the signal to be separated together with the activation matrices of all sources [35].

In order to apply supervised or semi-supervised NMF, one must design source-specific dictionaries in the first place. This is achieved by learning each dictionary from isolated sounds (e.g., individual notes) or continuous recordings from the desired source. The amount of training data is typically assumed to be large, so that large dictionaries containing hundreds or thousands of components can be trained. Three families of *nonnegative dictionary learning* methods can be found in the literature, which operate by applying NMF or selecting exemplars from the training signals, respectively.

Early dictionary learning methods were based on applying NMF to the training signals [34, 36]. Denoting by $\mathbf{V}_j$ the spectrogram resulting from the concatenation of all training signals for source j, this data can be factorised as

$$\mathbf{V}_j \approx \mathbf{W}_j\mathbf{H}_j. \tag{1.36}$$

The activation matrix $\mathbf{H}_j$ is discarded, while the dictionary $\mathbf{W}_j$ is kept and used together with the dictionaries for the other sources for separation. This method suffers from one major limitation: unless regularisation such as sparsity is enforced (see Sect. 1.3.2), the number of dictionary elements must be smaller than the number of frequency bins. As a consequence, each dictionary element encodes widely different source spectra and it may not account well for the source characteristics. For instance, it has been shown that small dictionaries tend to represent the spectral envelope of the sources but to discard pitch characteristics, which are essential for separation. In order to address this issue, it was recently proposed to construct the dictionary from exemplars, i.e., spectra (columns) selected from the full training set $\mathbf{V}_j$. The number of dictionary elements then becomes unlimited and each element represents a single spectrum at a time, so that all characteristics of the desired source are preserved. If the training set is not too large, $\mathbf{W}_j = \mathbf{V}_j$ itself might be used as the dictionary [37]. Alternatively, the dictionary may be constructed by selecting [38] or clustering [39] the columns of $\mathbf{V}_j$. The selection can be random or exploit prior information about, e.g., the phoneme or the note corresponding to each frame.

1.3.1.2 Convolutive NMF

In [36, 40, 41], the concept of nonnegative dictionary learning was extended to spectrogram patches. The original NMF model in (1.1) can be rewritten in each time frame n as

$$\mathbf{v}_n \approx \mathbf{W}\mathbf{h}_n = \sum_{k=1}^{K} \mathbf{w}_k h_{kn}. \tag{1.37}$$

After replacing each single-frame spectrum $\mathbf{w}_k$ by a spectrogram patch consisting of L consecutive frames

$$\mathbf{W}_k = \left(\mathbf{w}_{k,0} \cdots \mathbf{w}_{k,L-1}\right), \tag{1.38}$$

this model can be extended into

$$\mathbf{v}_n \approx \sum_{k=1}^{K} \sum_{l=0}^{L-1} \mathbf{w}_{k,l} h_{k,n-l}. \tag{1.39}$$

This *convolutive NMF* model assumes that all frames of a given patch are weighted by the same activation coefficient: $\mathbf{w}_{k,0}$ is weighted by h_{kn} in time frame n, $\mathbf{w}_{k,1}$ by the same h_{kn} in time frame $n+1$, $\mathbf{w}_{k,2}$ by the same h_{kn} in time frame $n+2$, and so on. The full spectrogram $\mathbf{V}$ is therefore approximated as a weighted sum of the patches $\mathbf{W}_k$.

The set of patches $\mathbf{W}_k$ can be partitioned into source-specific dictionaries of patches, which can be learned using NMF, exemplar selection, or exemplar clustering similarly to above [36, 38, 39]. The patch length L is typically on the order of 100–300 ms. Figure 1.4 illustrates a subset of exemplars learned on speech.

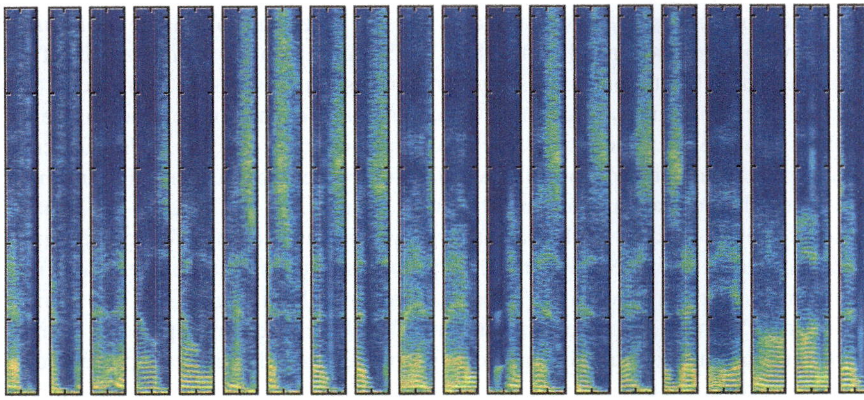

Fig. 1.4 Example convolutive NMF dictionary elements ($\mathbf{W}_k$) learned by random selection of 200 ms exemplars over 500 utterances from a given speaker. Notice how each component represents the spectrogram of a speech phoneme in context

1.3.1.3 Factoring Fine Structure and Envelope

While supervised NMF makes it possible to account for the characteristics of real audio sources, it is rather constrained and may lead to poor separation when the training and test data exhibit some mismatches. This led to the idea of fixing the source characteristics which remain valid in any circumstances and estimating the other characteristics from the signal to be separated.

Harmonic NMF is a first step in this direction. The underlying idea is to decompose each dictionary element $\mathbf{w}_k$ in (1.37) as the sum of narrowband spectral patterns $\mathbf{b}_{km}$ weighted by spectral envelope coefficients e_{km}:

$$\mathbf{w}_k = \sum_{m=1}^{M_k} \mathbf{b}_{km} e_{km}. \tag{1.40}$$

The narrowband patterns $\mathbf{b}_{km}$ represent the fine structure of the spectrum and they can be fixed as either smooth or harmonic spectra. In the former case, the patterns can be fixed as smooth narrowband spectra in order to represent a transient or noisy signal with a locally smooth spectrum. In the latter case, each dictionary index k is associated with a given pitch (fundamental frequency) and the corresponding patterns involve a few successive harmonic partials (i.e., spectral peaks at integer multiples of the given fundamental frequency). This model illustrated in Fig. 1.5 is suitable for voiced speech sounds (e.g., vowels) and pitched musical sounds (e.g., violin). The spectral envelope coefficients e_{km} are not fixed, but estimated from the signal to be separated. In other words, this model does not constrain the dictionary elements to match perfectly the training data, but only to follow a certain fine structure.

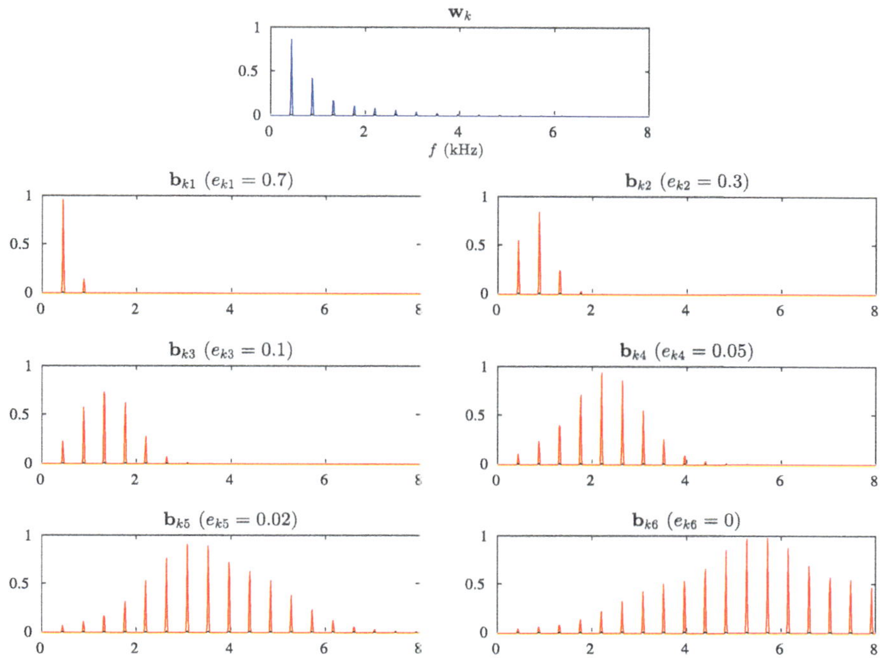

Fig. 1.5 Example narrowband harmonic patterns $\mathbf{b}_{km}$ and resulting dictionary element $\mathbf{w}_k$

An alternative approach is to factor each dictionary element $\mathbf{w}_k$ into the product of an excitation spectrum and a filter [42]. This so-called *excitation-filter* model adheres with the production phenomena of speech and most musical instruments, where an excitation signal is filtered by the vocal tract or the body of the instrument. The latest evolution in this direction is the multilevel NMF framework of [43], embodied in the Flexible Audio Source Separation Toolbox (FASST).[1] This framework represents the observed spectrogram as the product of up to eight matrices, which represent the fine structure or the envelope of the excitation or the filter on the time axis or the frequency axis. It makes it possible to incorporate specific knowledge or constraints in a flexible way and it was shown to outperform conventional NMF in [43].

These extensions of NMF are sometimes grouped under the banner of *nonnegative tensor factorisation* (NTF), a generalisation of NMF to multi-dimensional arrays [44]. Due to the linearity of the models, the NTF parameters can be estimated using multiplicative updates similar to the ones for NMF.

[1] http://bass-db.gforge.inria.fr/fasst/.

1.3.2 Penalised NMF

1.3.2.1 Sparsity

The original NMF model and the above extensions are well suited for the separation of music sources, which typically involve several overlapping notes. Speech sources, however, consist of a single phoneme at a time. NMF can yield disappointing results on mixtures of speech because it can confuse overlapping phonemes from different speakers vs the same speaker. The latter phenomenon cannot occur due to the physical constraints of speech production, but it is possible according to the model. In order to improve the modelling of speech sources, sparsity constraints must be set on the activation matrix **H** [45].

Sparsity signifies that most activation coefficients are very small, and only a small proportion is large. Therefore, it enforces the fact that a single dictionary element predominates in each time frame, and the other dictionary elements are little activated. Sparsity constraints are typically implemented by adding a penalty function to the NMF objective function in Sect. 1.2.1. The ideal penalty function would be the l_0 norm $\|\mathbf{H}\|_0$, that is the number of nonzero entries in **H**. This norm leads to a combinatorial optimisation problem, though, that is difficult to solve. In practice, the l_1 norm $\|\mathbf{H}\|_1 = \sum_{k=1}^{K} \sum_{n=1}^{N} h_{kn}$ is generally used instead:

$$\arg\min_{\mathbf{W},\mathbf{H}} D(\mathbf{V}|\mathbf{W},\mathbf{H}) + \mu\|\mathbf{H}\|_1 \tag{1.41}$$

where $\mu > 0$ is a tradeoff parameter.

The penalised objective function (1.41) can be minimised w.r.t. **H** by adding the constant μ to the denominator of the original multiplicative update [15]:

$$\mathbf{H} \leftarrow \mathbf{H} \circ \frac{\mathbf{W}^T((\mathbf{W}\mathbf{H})^{\circ[\beta-2]} \circ \mathbf{V})}{\mathbf{W}^T(\mathbf{W}\mathbf{H})^{\circ[\beta-1]} + \mu}. \tag{1.42}$$

The greater μ, the sparser the solution. Regarding the dictionary **W**, the classical update in Sect. 1.2.3 cannot be used anymore since **W** must be normalised in some way in order to avoid scaling indeterminacy, e.g., by assuming each $\mathbf{w}_k$ has a unit l_2 norm $\|\mathbf{w}_k\|_2 = 1$. Rescaling **W** a posteriori changes the value of the penalised objective function, so that the **W** resulting from the classical multiplicative update is not optimal anymore. A multiplicative update accounting for this l_2 norm constraint was proposed in [46, 47]. Alternative sparsity promoting penalties were explored in [48, 49].

1.3.2.2 Group Sparsity

Group sparsity is an extension of the concept of sparsity, which enforces simultaneous activation of several dictionary elements. It has been used for two purposes: to automatically group the dictionary elements corresponding to a given phoneme, note or source, in the case when each phoneme, note or source is represented by multiple dictionary elements [50], and to automatically find which sources are active among a pre-specified set of speakers or musical instruments, when the number of sources and the identity of the active sources are unknown [51].

In the latter case, the full dictionary $\mathbf{W}$ can be partitioned into several source-specific dictionaries $\mathbf{W}_j$ as in Sect. 1.3.1.1. Group sparsity means that, if source j is inactive, all entries of the corresponding activation matrix $\mathbf{H}_j$ must be estimated as 0. This behaviour can be enforced by using the mixed $l_{1,2}$ norm as a penalty term:

$$\operatorname*{arg\,min}_{\mathbf{W},\mathbf{H}} D(\mathbf{V}|\mathbf{W},\mathbf{H}) + \mu \sum_{j=1}^{J} \|\mathbf{H}_j\|_2 \tag{1.43}$$

where the l_2 norm is defined by $\|\mathbf{H}_j\|_2 = (\sum_{k=1}^{K}\sum_{n=1}^{N} h_{jkn}^2)^{1/2}$ and $\mu > 0$ is a trade-off parameter. Many variants of this penalty can be designed to favour specific activation patterns. For instance, the penalty $\sum_{j=1}^{J}\sum_{n=1}^{N} \|\mathbf{h}_{jn}\|_2$ favours sparsity both over the sources and over time, but all the dictionary elements corresponding to a given source can be activated at a given time. Alternative group sparsity promoting penalties were explored, for instance in [50].

1.3.2.3 Temporal Dynamics

Another family of NMF models aim to model the dynamics of the activation coefficients over time. The simplest such models account for the temporal smoothness (a.k.a. continuity) of the activation coefficients by constraining the value of h_{kn} given $h_{k,n-1}$ using a suitable penalty function. In [45], the following penalised objective function was proposed:

$$\operatorname*{arg\,min}_{\mathbf{W},\mathbf{H}} D(\mathbf{V}|\mathbf{W},\mathbf{H}) + \sum_{k=1}^{K} \mu_k \sum_{n=2}^{N} (h_{kn} - h_{k,n-1})^2. \tag{1.44}$$

Assuming that μ_k is constant, this penalised objective function can be minimised w.r.t. $\mathbf{H}$ by the following multiplicative update inspired from [45]:

$$\mathbf{H} \leftarrow \mathbf{H} \circ \frac{\mathbf{W}^T((\mathbf{W}\mathbf{H})^{\circ[\beta-2]} \circ \mathbf{V}) + 2\mathbf{M} \circ (\overrightarrow{\mathbf{H}} + \overleftarrow{\mathbf{H}})}{\mathbf{W}^T(\mathbf{W}\mathbf{H})^{\circ[\beta-1]} + 2\mathbf{M} \circ (\mathbf{H} + \overleftrightarrow{\mathbf{H}})} \tag{1.45}$$

where

$$\mathbf{M} = \begin{pmatrix} \mu_1 & \cdots & \mu_1 \\ \mu_2 & \cdots & \mu_2 \\ \vdots & \ddots & \vdots \\ \mu_K & \cdots & \mu_K \end{pmatrix} \tag{1.46}$$

$$\overrightarrow{\mathbf{H}} = \begin{pmatrix} 0 & h_{11} & h_{12} & \cdots & h_{1,N-1} \\ 0 & h_{21} & h_{22} & \cdots & h_{2,N-1} \\ \vdots & \vdots & \vdots & \ddots & \vdots \\ 0 & h_{K1} & h_{K2} & \cdots & h_{K,N-1} \end{pmatrix} \tag{1.47}$$

$$\overleftarrow{\mathbf{H}} = \begin{pmatrix} h_{12} & h_{13} & \cdots & h_{1N} & 0 \\ h_{22} & h_{23} & \cdots & h_{2N} & 0 \\ \vdots & \vdots & \ddots & \vdots & \vdots \\ h_{K2} & h_{K3} & \cdots & h_{KN} & 0 \end{pmatrix} \tag{1.48}$$

$$\overleftrightarrow{\mathbf{H}} = \begin{pmatrix} 0 & h_{12} & \cdots & h_{1,N-1} & 0 \\ 0 & h_{22} & \cdots & h_{2,N-1} & 0 \\ \vdots & \vdots & \ddots & \vdots & \vdots \\ 0 & h_{K2} & \cdots & h_{K,N-1} & 0 \end{pmatrix}. \tag{1.49}$$

The impact of μ_k on the resulting activation coefficients is illustrated in Fig. 1.6. The greater μ_k, the smoother the coefficients. Regarding the dictionary $\mathbf{W}$, once again, a normalisation constraint is required which results in a modified update compared to the one in Sect. 1.2.3. Alternative probabilistically motivated smoothness penalties were proposed in [11, 52].

Building upon this idea, nonnegative continuous-state [53] and discrete-state [34, 54] dynamical models have also been investigated. The latter often limit the number of active dictionary elements at a time and they can be seen as imposing a form of group sparsity. These models account not only for the continuity of the activations, if relevant, but also for typical activation patterns over time due to, e.g., the attack-sustain-decay structure of musical notes or the sequences of phonemes composing common words. For a survey of dynamical NMF models, see [30].

1.3.3 User-guided NMF

While the above methods incorporate general knowledge about speech and music sources, a number of authors have investigated user-guided NMF methods that incorporate specific information about the sources in a given mixture signal. Existing methods can be broadly categorised according to the nature of this information.

A first category of methods exploit information about the activation patterns of the sources. This information is provided by the user based on listening to the original signal or the separated signals and visualising the waveform or the spectrogram.

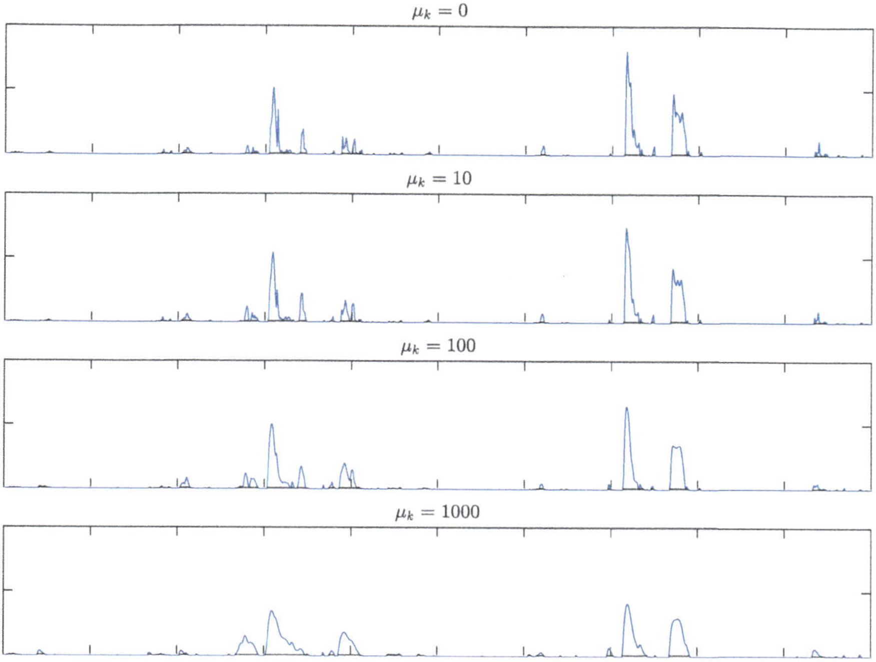

Fig. 1.6 Activation coefficients h_{kn} estimated for one dictionary element k in a music signal for $\beta = 0$ and different values of the smoothness tradeoff parameter μ_k in (1.45)

Given the time intervals when each source is inactive, the corresponding activation coefficients can be fixed to 0, which improves the estimation of the dictionary and the activation coefficients in the other time intervals [55]. In [56], a more advanced method is proposed by which the user can tag a given time-frequency region as active, inactive, or well-separated. The graphical user interface is shown in Fig. 1.7. This information is then iteratively exploited in order to refine the source estimates at each iteration. This method was shown to be effective even without using any isolated training data.

A second category of user-guided methods rely on a (partial) transcription of the signal, that can take the form of a fundamental frequency curve [57], a musical score [58], or the speech transcription. This information can be used to restrict the set of active atoms at a given time, in a similar way as group sparsity except that the set of active atoms is known in advance.

Finally, a third category of methods rely on a reference signal for some or all of the sources to be separated. The user can generate reference signals signal by humming the melody [59] or uttering the same sentence [60]. Reference signals can also be obtained by looking for additional data, e.g., the soundtrack of the same film in a different language, the multitrack cover version of a song, additional data

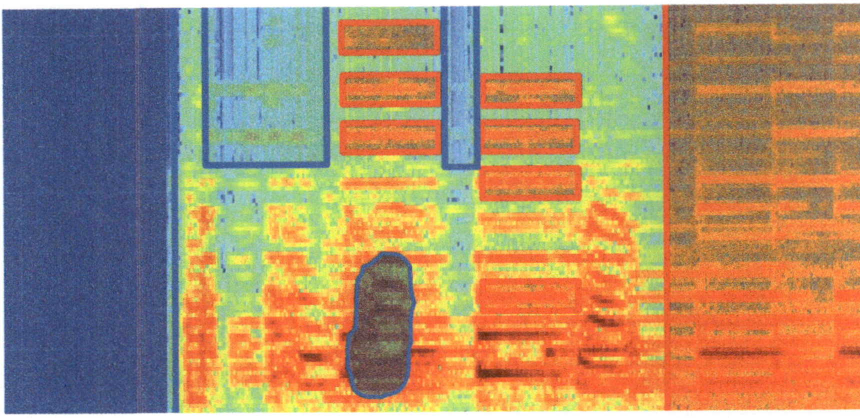

Fig. 1.7 Graphical user interface for user annotation. Piano is labelled as active (resp. inactive) in the red (resp. blue) regions

corresponding to the same speaker or the same musical instrument, or repeated signals (e.g., jingles, background music) in large audio archives [61].

Many user-guided NMF methods can be expressed under the general framework of nonnegative matrix partial co-factorisation (NMPcF), which aims to jointly factor several input matrices into several factor matrices, some of which are shared [62, 63]. For instance, in the case of score-guided or reference-guided separation, the spectrogram to be separated and the score or the reference can be jointly factored using different dictionaries but the same activation matrix.

1.4 Conclusions

In this chapter, we have shown that NMF is a powerful approach for audio source separation. Starting from a simple unsupervised formulation, it makes it possible to incorporate additional information about the sources in a principled optimisation framework. In comparison with deep neural network (DNN) based separation, which has recently attracted a lot of interest, NMF-based separation remains competitive in the situations when the amount of data is medium or small, or user guidance is available. These two situations are hardly handled by DNNs today, due to the need for a large amount of training data and the difficulty of retraining or adapting the DNN at test time based on user feedback. It therefore comes as no surprise that NMF is still the subject of much research today. Most of this research concentrates on overcoming the fundamental limitation of NMF, namely the fact that it models spectro-temporal magnitude or power only, and enabling it to account for phase. For an in-depth discussion of this and other perspectives, see [64].

On a final note, some aspects of NMF for audio signal processing are also covered in other chapters of the present book (Chaps. 2, 3, 4, 5, 6) and in Chaps. 8, 9, and 16 of [64].

Acknowledgements Cédric Févotte acknowledges funding from the European Research Council (ERC) under the European Union's Horizon 2020 research and innovation programme under grant agreement No 681839 (project FACTORY).

Standard Distributions

Poisson

$$Po(x|\lambda) = \exp(-\lambda)\,\frac{\lambda^x}{x!}, \quad x \in \{0, 1, \ldots, \infty\} \tag{1.50}$$

Multinomial

$$M(\mathbf{x}|N, \mathbf{p}) = \frac{N!}{x_1! \ldots x_K!} p_1^{x_1} \ldots p_K^{x_K}, \quad x_k \in \{0, \ldots, N\}, \sum_k x_k = N \tag{1.51}$$

Circular complex normal distribution

$$N_c(x|\mu, \Sigma) = |\pi\,\Sigma|^{-1} \exp -(x-\mu)^H\,\Sigma^{-1}\,(x-\mu), \quad x \in \mathbb{C}^F \tag{1.52}$$

Gamma

$$G(x|\alpha, \beta) = \frac{\beta^\alpha}{\Gamma(\alpha)}\,x^{\alpha-1}\,\exp(-\beta\,x), \quad x \geq 0 \tag{1.53}$$

References

1. C.J.C. Burges, Dimension reduction: a guided tour. Found. Trends Mach. Learn. **2**(4), 275–365 (2009)
2. P. Comon, Independent component analysis, a new concept ? Sig. process. **36**(3), 287–314 (1994)
3. B.A. Olshausen, D.J. Field, Emergence of simple-cell receptive field properties by learning a sparse code for natural images. Nature **381**(6583), 607–609 (1996)
4. M. Aharon, M. Elad, A. Bruckstein, K-SVD: an algorithm for designing overcomplete dictionaries for sparse representation. IEEE Trans. Sig. Process. **54**(11), 4311–4322 (2006)
5. P. Paatero, U. Tapper, Positive matrix factorization: a non-negative factor model with optimal utilization of error estimates of data values. Environmetrics **5**, 111–126 (1994)
6. D.D. Lee, H.S. Seung, Learning the parts of objects with nonnegative matrix factorization. Nature **401**, 788–791 (1999)
7. T. Hofmann, Probabilistic latent semantic indexing, in *Proceedings of the 22nd International Conference on Research and Development in Information Retrieval (SIGIR)* (1999)

8. Y. Koren, R. Bell, C. Volinsky, Matrix factorization techniques for recommender systems. Computers **42**(8), 30–37 (2009)
9. N. Dobigeon, J.-Y. Tourneret, C. Richard, J.C.M. Bermudez, S. McLaughlin, A.O. Hero, Nonlinear unmixing of hyperspectral images: models and algorithms. IEEE Sig. Proccess. Mag. **31**(1), 89–94 (2014)
10. P. Smaragdis, J.C. Brown, Non-negative matrix factorization for polyphonic music transcription, in *Proceedings of the IEEE Workshop on Applications of Signal Processing to Audio and Acoustics (WASPAA)* (2003)
11. C. Févotte, N. Bertin, J.-L. Durrieu, Nonnegative matrix factorization with the Itakura-Saito divergence. With application to music analysis. Neural Comput. **21**(3), 793–830 (2009)
12. A. Cichocki, H. Lee, Y.-D. Kim, S. Choi, Non-negative matrix factorization with α-divergence. Pattern Recognit. Lett. **29**(9), 1433–1440 (2008)
13. A. Cichocki, R. Zdunek, S. Amari, Csiszar's divergences for non-negative matrix factorization: family of new algorithms, in *Proceedings of International Conference on Independent Component Analysis and Blind Signal Separation (ICA), Charleston SC, USA* (2006), pp. 32–39
14. R. Kompass, A generalized divergence measure for nonnegative matrix factorization. Neural Comput. **19**(3), 780–791 (2007)
15. C. Févotte, J. Idier, Algorithms for nonnegative matrix factorization with the beta-divergence. Neural Comput. **23**(9), 2421–2456 (2011)
16. I.S. Dhillon, S. Sra, Generalized nonnegative matrix approximations with Bregman divergences, in *Advances in Neural Information Processing Systems (NIPS)* (2005)
17. A. Basu, I.R. Harris, N.L. Hjort, M.C. Jones, Robust and efficient estimation by minimising a density power divergence. Biometrika **85**(3), 549–559 (1998)
18. S. Eguchi, Y. Kano, Robustifying maximum likelihood estimation, Institute of Statistical Mathematics, Technical report, June 2001, research Memo. 802
19. D. FitzGerald, M. Cranitch, E. Coyle, On the use of the beta divergence for musical source separation, in *Proceedings of the Irish Signals and Systems Conference* (2009)
20. R. Hennequin, R. Badeau, B. David, NMF with time-frequency activations to model non stationary audio events, in *Proceedings of the IEEE International Conference on Acoustics, Speech and Signal Processing (ICASSP)* (2010), pp. 445–448
21. E. Vincent, N. Bertin, R. Badeau, Adaptive harmonic spectral decomposition for multiple pitch estimation. IEEE Trans. Audio Speech Lang. Process. **18**, 528–537 (2010)
22. V.Y.F. Tan, C. Févotte, Automatic relevance determination in nonnegative matrix factorization with the beta-divergence. IEEE Trans. Pattern Anal. Mach. Intell. **35**(7), 1592–1605 (2013)
23. C. Févotte, A.T. Cemgil, Nonnegative matrix factorisations as probabilistic inference in composite models, in *Proceedings of the 17th European Signal Processing Conference (EUSIPCO), Glasgow, Scotland* (2009), pp. 1913–1917
24. J.F. Canny, GaP: a factor model for discrete data, in *Proceedings of the ACM International Conference on Research and Development of Information Retrieval (SIGIR)* (2004), pp. 122–129
25. A.T. Cemgil, Bayesian inference for nonnegative matrix factorisation models. Comput. Intell. Neurosci. **2009**,17 (2009). https://doi.org/10.1155/2009/785152. Article ID 785152
26. P. Smaragdis, B. Raj, M.V. Shashanka, A probabilistic latent variable model for acoustic modeling, in *NIPS Workshop on Advances in Models for Acoustic Processing* (2006)
27. T. Virtanen, Monaural sound source separation by non-negative matrix factorization with temporal continuity and sparseness criteria. IEEE Trans. Audio Speech Lang. Process. **15**(3), 1066–1074 (2007)
28. B. King, C. Févotte, P. Smaragdis, Optimal cost function and magnitude power for NMF-based speech separation and music interpolation, in *Proceedings of the IEEE International Workshop on Machine Learning for Signal Processing (MLSP), Santander, Spain* (2012)
29. D.R. Hunter, K. Lange, A tutorial on MM algorithms. Am. Stat. **58**, 30–37 (2004)
30. P. Smaragdis, C. Févotte, G. Mysore, N. Mohammadiha, M. Hoffman, Static and dynamic source separation using nonnegative factorizations: a unified view. IEEE Sig. Process. Mag. **31**(3), 66–75 (2014)

31. T. Virtanen, Sound source separation using sparse coding with temporal continuity objective, in *Proceedings of the International Computer Music Conference (ICMC)* (2003), pp. 231–234
32. S. Vembu, S. Baumann, Separation of vocals from polyphonic audio recordings, in *Proceedings of the International Conference on Music Information Retrieval (ISMIR)* (2005), pp. 337–344
33. E. Vincent, N. Bertin, R. Gribonval, F. Bimbot, From blind to guided audio source separation: how models and side information can improve the separation of sound. IEEE Sig. Process. Mag. **31**(3), 107–115 (2014)
34. E. Vincent, X. Rodet, Underdetermined source separation with structured source priors, in *Proceedings of the International Conference on Independent Component Analysis and Blind Source Separation (ICA)* (2004), pp. 327–334
35. G.J. Mysore, P. Smaragdis, A non-negative approach to semi- supervised separation of speech from noise with the use of temporal dynamics, in *Proceedings of the IEEE International Conference on Acoustics, Speech, and Signal Processing (ICASSP)* (2011), pp. 17–20
36. P. Smaragdis, Convolutive speech bases and their application to supervised speech separation. IEEE Trans. Audio Speech Lang. Process. **15**(1), 1–12 (2007)
37. P. Smaragdis, M. Shashanka, B. Raj, A sparse non-parametric approach for single channel separation of known sounds, in *Proceedings of the Neural Information Processing Systems (NIPS)* (2009), pp. 1705–1713
38. J.F. Gemmeke, T. Virtanen, A. Hurmalainen, Exemplar-based sparse representations for noise robust automatic speech recognition. IEEE Trans. Audio Speech Lang. Process. **19**(7), 2067–2080 (2011)
39. T. Virtanen, J. Gemmeke, B. Raj, Active-set Newton algorithm for overcomplete non-negative representations of audio. IEEE Trans. Audio Speech Lang. Process. **21**(11), 2277–2289 (2013)
40. P.D. O'Grady, B.A. Pearlmutter, Discovering speech phones using convolutive non-negative matrix factorisation with a sparseness constraint. Neurocomputing **72**(1–3), 88–101 (2008)
41. W. Wang, A. Cichocki, J.A. Chambers, A multiplicative algorithm for convolutive non-negative matrix factorization based on squared Euclidean distance. IEEE Trans. Sig. Process. **57**(7), 2858–2864 (2009)
42. J.-L. Durrieu, G. Richard, B. David, C. Févotte, Source/filter model for unsupervised main melody extraction from polyphonic audio signals. IEEE Trans. Audio Speech Lang. Process. **18**(3), 564–575 (2010)
43. A. Ozerov, E. Vincent, F. Bimbot, A general flexible framework for the handling of prior information in audio source separation. IEEE Trans. Audio Speech Lang. Process. **20**(4), 1118–1133 (2012)
44. D. FitzGerald, M. Cranitch, E. Coyle, Extended nonnegative tensor factorisation models for musical sound source separation. Comput. Intell. Neurosci. **2008** (2008). Article ID 872425
45. T. Virtanen, Monaural sound source separation by nonnegative matrix factorization with temporal continuity and sparseness criteria. IEEE Trans. Audio Speech Lang. Process. **15**(3), 1066–1074 (2007)
46. J. Eggert, E. Körner, Sparse coding and NMF, in *Proceedings of the IEEE International Joint Conference on Neural Networks* (2004), pp. 2529–2533
47. J. Le Roux, F.J. Weninger, J.R. Hershey, Sparse NMF–half-baked or well done? Mitsubishi Electric Research Laboratories (MERL), Technical report TR2015-023, 2015
48. C. Joder, F. Weninger, D. Virette, B. Schuller, A comparative study on sparsity penalties for NMF-based speech separation: beyond Lp-norms, in *Proceedings of the IEEE International Conference on Acoustics, Speech, and Signal Processing (ICASSP)* (2013), pp. 858–862
49. Y. Mitsui, D. Kitamura, S. Takamichi, N. Ono, H. Saruwatari, Blind source separation based on independent low-rank matrix analysis with sparse regularization for time-series activity, in *Proceedings of the IEEE International Conference on Acoustics, Speech, and Signal Processing (ICASSP)* (2017)
50. A. Lefèvre, F. Bach, C. Févotte, Itakura-Saito nonnegative matrix factorization with group sparsity, in *Proceedings of the IEEE International Conference on Acoustics, Speech, and Signal Processing (ICASSP)* (2011), pp. 21–24

51. D.L. Sun, G.J. Mysore, Universal speech models for speaker independent single channel source separation, in *Proceedings of the IEEE International Conference on Acoustics, Speech and Signal Processing (ICASSP)* (2013), pp. 141–145
52. O. Dikmen, A.T. Cemgil, Gamma Markov random fields for audio source modeling. IEEE Trans. Audio Speech Lang. Process. **18**(3), 589–601 (2010)
53. C. Févotte, J. Le Roux, J.R. Hershey, Non-negative dynamical system with application to speech and audio, in *Proceedings of the IEEE International Conference on Acoustics, Speech, and Signal Processing (ICASSP)* (2013), pp. 3158–3162
54. G. Mysore, M. Sahani, Variational inference in non-negative factorial hidden Markov models for efficient audio source separation, in *Proceedings of the International Conference on Machine Learning (ICML)* (2012), pp. 1887–1894
55. A. Ozerov, C. Févotte, R. Blouet, J.-L. Durrieu, Multichannel nonnegative tensor factorization with structured constraints for user-guided audio source separation, in *Proceedings of the IEEE International Conference on Acoustics, Speech and Signal Processing (ICASSP), Prague* (2011), pp. 257–260
56. N.Q.K. Duong, A. Ozerov, L. Chevallier, J. Sirot, An interactive audio source separation framework based on non-negative matrix factorization, in *Proceedings of the IEEE International Conference on Acoustics, Speech, and Signal Processing (ICASSP)* (2014), pp. 1567–1571
57. J.-L. Durrieu, J.-P. Thiran, Musical audio source separation based on user-selected F0 track, in *Proceedings of the International Conference on Latent Variable Analysis and Signal Separation (LVA/ICA)* (2012), pp. 438–445
58. S. Ewert, B. Pardo, M. Müller, M.D. Plumbley, Score-informed source separation for musical audio recordings: an overview. IEEE Sig. Process. Mag. **31**(3), 116–124 (2014)
59. P. Smaragdis, G.J. Mysore, Separation by humming: user-guided sound extraction from monophonic mixtures, in *Proceedings of the IEEE Workshop on Applications of Signal Processing to Audio and Acoustics (WASPAA)* (2009), pp. 69–72
60. L. Le Magoarou, A. Ozerov, N.Q.K. Duong, Text-informed audio source separation using nonnegative matrix partial co-factorization, in *Proceedings of the IEEE International Workshop on Machine Learning for Signal Processing (MLSP)* (2013), pp. 1–6
61. N. Souviraà-Labastie, A. Olivero, E. Vincent, F. Bimbot, Multi-channel audio source separation using multiple deformed references. IEEE/ACM Trans. Audio Speech Lang. Process. **23**(11), 1775–1787 (2015)
62. Y.K. Yılmaz, A.T. Cemgil, U. Şimşekli, Generalized coupled tensor factorization, in *Advances in Neural Information Processing Systems (NIPS)* (2011)
63. N. Seichepine, S. Essid, C. Févotte, O. Cappé, Soft nonnegative matrix co-factorization. IEEE Trans. Sig. Process. **62**(22), 5940–5949 (2014)
64. E. Vincent, T. Virtanen, S. Gannot, *Audio Source Separation and Speech Enhancement* (Wiley, 2017)

Chapter 2
Separation of Known Sources Using Non-negative Spectrogram Factorisation

Tuomas Virtanen and Tom Barker

Abstract This chapter presents non-negative spectrogram factorisation (NMF) techniques which can be used to separate sources in the cases where source-specific training material is available in advance. We first present the basic NMF formulation for sound mixtures and then present criteria and algorithms for estimating the model parameters. We introduce selected methods for training the NMF source models by using either vector quantisation, convexity constraints, archetypal analysis, or discriminative methods. We also explain how the learned dictionaries can be adapted to deal with mismatches between the training data and usage scenario. We present also how semi-supervised learning can be used to deal with unknown noise sources within a mixture and finally we introduce a coupled NMF method which can be used to model large temporal context while retaining low algorithmic latency.

2.1 Introduction

In many source separation cases, we know for a mixture which is subject to separation, what the expected constituent sources will be. For example, in many speech related applications that require source separation, we know that the target source is speech, or even a specific speaker. In music related applications we might be separating particular musical instruments. It is therefore often possible to acquire example material representative of the target sources in order to develop the separation algorithms. Nowadays, many source separation algorithms are based on machine learning methods, which allow learning source models or separation models automatically from some provided training material. Combining this example material of target sources and appropriate machine learning algorithms leads to *supervised methods* for source separation. Supervised learning has long been used for in audio content analysis problems such as automatic speech recognition, but recently it has also become widely used for source separation.

T. Virtanen (✉) · T. Barker
Tampere University of Technology, Korkeakoulunkatu 1,
33720 Tampere, Finland
e-mail: tuomas.virtanen@tut.fi

S. Makino (ed.), *Audio Source Separation*, Signals and Communication Technology, https://doi.org/10.1007/978-3-319-73031-8_2

One popular class of source separation methods is based on *non-negative matrix factorisation*. The term matrix factorisation refers to a model where the magnitude or power spectrum of mixture signals is factored into source components for separation. The descriptor non-negative is used since the factors are purely additive, i.e., subtractive components are not used in the factorisation. The NMF model treats individual sources are the sum of non-negative factors. This makes it suitable for representing various sound sources, since many sounds such as speech, music, and environmental sounds are composed of some elementary units: speech consists of units such as phonemes, syllables and words; music consists of notes played by individual instruments, and environmental sound consists of sound events produced by various sources. Having an additive model for individual sources leads also to an additive model for mixtures, which enables separation algorithms that are relatively simple to implement, use, and extend in various ways.

In this chapter we review the use of NMF for separation of known sources. In Sect. 2.2 we introduce the basic model for representing sound mixtures. In Sect. 2.3 we discuss various types of dictionaries that are used to model individual sources, and how they can be adapted in realistic usage scenarios. Section 2.4 presents the *semi-supervised* extension that allows modeling unknown sources within a mixture. In Sect. 2.5 we present an approach that can be used to achieve low algorithmic latency, allowing to use NMF in real-time applications.

2.2 NMF Model for Separation of Known Sounds

The basic NMF model represents the magnitude (or power) spectrum vector $\mathbf{s}_{j,t}$ of source j in frame $t = 1, \ldots, T$ as the weighted sum of basis vectors $\mathbf{b}_{k,j}$, $k = 1, \ldots, K_j$ (where K_j is the number of basis vectors) as

$$\mathbf{s}_{j,t} = \sum_{k=1}^{K_j} w_{k,t,j} \mathbf{b}_{k,j}. \tag{2.1}$$

Above, $w_{k,t,j}$ is the weight of the kth basis vector in frame t. The length of spectrum vectors and basis vectors is F, the number of frequencies in the spectral representation.

The magnitude (or power) spectrogram matrix $\mathbf{S}_j$ consisting of the frame-wise spectra as $\mathbf{S}_j = \left[\mathbf{s}_{j,1}, \mathbf{s}_{j,2}, \ldots, \mathbf{s}_{j,T}\right]$ (where T is the number of frames) can therefore be expressed as the product of two matrices as

$$\mathbf{S}_j = \mathbf{B}_j \mathbf{W}_j, \tag{2.2}$$

where $\mathbf{B}_j = \left[\mathbf{b}_{1,j}, \mathbf{b}_{2,j}, \ldots, \mathbf{b}_{K,j}\right]$ is the basis matrix and $[\mathbf{W}_j]_{kt} = w_{k,t,j}$ is the weight matrix. In the NMF model, both the basis vectors and the weights, and

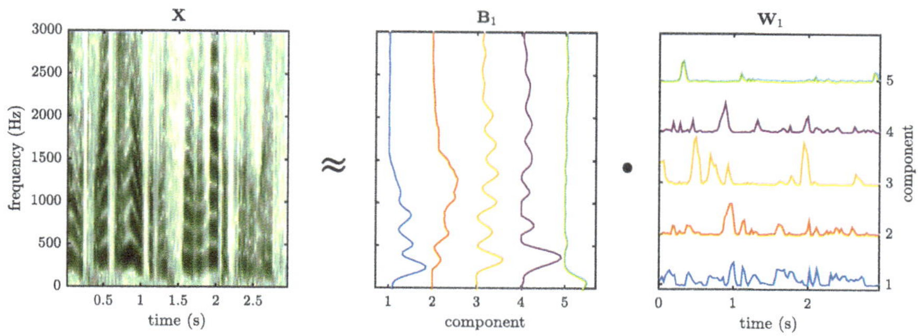

Fig. 2.1 In the NMF model, the magnitude spectrum matrix **X** modeled as the product of basis matrix **B** consisting of basis spectra of components, and weight matrix **W** consisting of the temporal activations of the components

therefore also the basis matrix and weight matrix, are entry-wise non-negative. The model is illustrated in Fig. 2.1.

The magnitude (or power) spectrum of the mixture signal $\mathbf{x}_t$ in frame is modeled as the sum of magnitude spectra of sources as

$$\mathbf{x}_t = \sum_{j=1}^{J} \mathbf{s}_{j,t}, \tag{2.3}$$

where J is the number of sources. Therefore, the model for the mixture magnitude (or power) spectrum is

$$\mathbf{x}_t = \sum_{j=1}^{J} \sum_{k=1}^{K_j} w_{k,t,j} \mathbf{b}_{k,j}, \tag{2.4}$$

and the model for the magnitude (or power) spectrogram

$$\mathbf{X} = \mathbf{B}\mathbf{W}, \tag{2.5}$$

where $\mathbf{B} = [\mathbf{B}_1, \mathbf{B}_2, \ldots, \mathbf{B}_J]$ and $\mathbf{W}^T = [\mathbf{W}_1^T, \mathbf{W}_2^T, \ldots, \mathbf{W}_J^T]$. This NMF model for sound mixtures is illustrated in Fig. 2.2

In the scenario where sources are known, the basis matrix **B** is estimated at the training stage using isolated material from each source, and only weight matrix **W** is estimated based on the mixture. There are various methods for estimating the basis matrix which are discussed in Sect. 2.3, and also also various criteria for estimating the weights that are discussed in Sect. 2.2.1.

Once the parameters in (2.5) have been estimated, we can design a spectrogram mask matrix

$$\mathbf{M}_j = \frac{\mathbf{B}_j \mathbf{W}_j}{\mathbf{B}\mathbf{W}}, \tag{2.6}$$

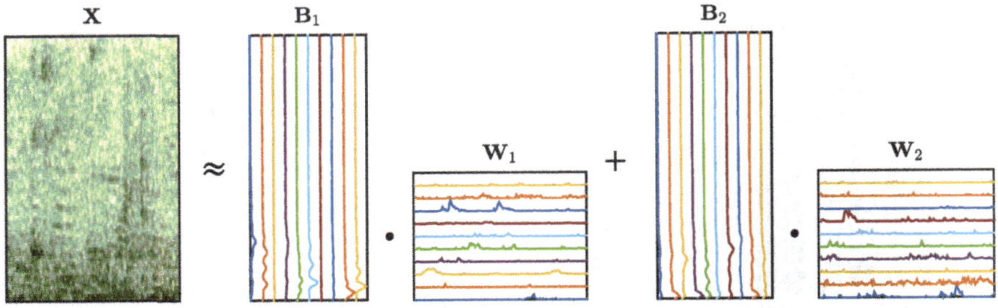

Fig. 2.2 When NMF is used to model a mixture of multiple sources, each source is represented by the NMF model, and the mixture is the sum of the individual source models. This figure illustrates an example mixture of two sources modeled with NMF

which can be used to separate source j by entry-wise multiplying the complex-valued spectrogram of the mixture $\tilde{\mathbf{X}}$ as

$$\tilde{\mathbf{S}}_j = \tilde{\mathbf{X}} \otimes \mathbf{M}_j \tag{2.7}$$

to obtain an estimate of the complex-valued spectrogram $\tilde{\mathbf{S}}_j$ of source j.

All the above processing is done in the short-time spectrum domain. Typical short-time spectrum representations are the short-time Fourier transform (STFT) spectrum and the mel spectrum. The STFT spectrum is obtained by dividing the input signal into windowed frames, and calculating the discrete Fourier transform (DFT) for each frame. The magnitude spectra are obtained by taking the absolute value of each STFT bin. Once complex-valued STFT domain separation has been done using the masking expressed in (2.7), the masked complex spectrum is converted back to the time domain by calculating the inverse discrete Fourier transform, where overlap-add techniques are used to reconstruct audio signal in overlapping frames. NMF can also be used with other representations. For example, we can do NMF in the mel-resolution spectral domain. By interpolating the obtained the mel-resolution mask to linear STFT resolution, source separation on mel-resoultion material can then be done in the STFT domain.

2.2.1 Estimation Criteria and Algorithms

Various criteria can be used for estimating the NMF model parameters . Generally the criteria can be formulated as minimisation of an objective function

$$f(\mathbf{W}) = D(\mathbf{X}|\mathbf{B}\mathbf{W}) + h(\mathbf{W}) \tag{2.8}$$

which consists of divergence D between the observed mixture magnitude spectrogram matrix $\mathbf{X}$ and the model $\mathbf{BW}$, and an optional regularisation function h that can impose constraints on $\mathbf{W}$.

The divergence D is calculated entry wise as

$$D(\mathbf{X}|\mathbf{V}) = \sum_{f=1}^{F}\sum_{t=1}^{T} d([\mathbf{X}]_{f,t}|[\mathbf{V}]_{f,t}), \tag{2.9}$$

where d is the divergence at an individual time-frequency point. The functions typically used for d reach their minimum value zero when $\mathbf{X}_{f,t} = [\mathbf{V}]_{f,t}$. In audio signal processing, commonly used divergences include the generalised Kullback-Leibler divergence

$$d_{KL}(x, v) = x\log(x, v) - x + v \tag{2.10}$$

and Itakura-Saito divergence

$$d_{IS}(x, v) = x/y - \log(x/yv) - 1 \tag{2.11}$$

both of which have been found to produce good results in audio processing [1, 2]. The choice of divergence affects the estimated weights significantly—different divergences produce different weights.

In other applications of NMF, the squared Euclidean distance $d(x, v) = (x - v)^2$ is often used, but it is not optimal for audio applications because of the large dynamic range of sound intensities and non-linearity of perception not mapping well to the properties of Euclidean distance. There is also a broader family of β-divergences [3], and the above-mentioned divergences are specific instances it.

A commonly used regularisation function $h(\mathbf{W})$ is the imposition of sparsity, which means favoring solutions where most of the entries in $\mathbf{W}$ are zero. In order to obtain a function to enforce sparsity which can be easily minimised, the L1-norm calculated as

$$||\mathbf{W}||_1 = \sum_{k=1}^{K}\sum_{t=1}^{T} |[\mathbf{W}]_{k,t}| \tag{2.12}$$

is typically used. Since $\mathbf{W}$ is entry-wise non-negative, the absolute value operator is not needed, but we include it to the above equation to match with the general definition of L1 norm. L_1 is typically weighted by scalar λ, i.e. $h(\mathbf{W}) = \lambda||\mathbf{W}||_1$. This parameter λ needs to be tuned to obtain the desired sparsity.

Within the NMF framework, regulariation functions other than sparsity have also been used. These are more commonly used at the dictionary learning stage though, where the basis vectors are also learned. These methods are discussed in more detail in Sect. 2.3.

There exist various algorithms for minimising the objective function f. The most commonly used algorithms are based on so-called multiplicative updates, where the weights to be estimated are first initialised with positive values, and then iteratively estimated by an update rule where the previous estimates are multiplied with a non-negative correction term.

For example in the case of a function that consists of the generalised Kullback-Leibler divergence and L_1 norm, i.e.

$$f(\mathbf{W}) = D_{KL}(\mathbf{X}|\mathbf{BW}) + \lambda||\mathbf{W}||_1, \tag{2.13}$$

the update rule is given as

$$\mathbf{W} \leftarrow \mathbf{W} \otimes \frac{\mathbf{B}^T \frac{\mathbf{X}}{\mathbf{BW}}}{\mathbf{B}^T \mathbf{1}^{F \times T} + \lambda}, \tag{2.14}$$

where matrix divisions are done entry-wise, $\otimes$ is entry-wise product of matrices, and $\mathbf{1}^{F \times T}$ is a all-one matrix of size F times K. The above update rule requires a large number of iterations (e.g. tens or hundreds) in order to converge, depending on the sizes of the matrices. Its benefit is the simplicity: the above update rule is very easy to implement and extend in various ways. It can also be easily sped up by using parallel computing architectures.

More efficient algorithms for minimising the objective also exist. For example, the active-set Newton method in [4] is based on the property that most of the activation weights will be zero, and then iteratively finds the set of optimal components and estimates their weights using the Newton method. It can produce significantly faster convergence in comparison to the multiplicative updates, especially when a large number of basis vectors are used.

2.3 Sound Dictionary Learning and Adaptation

The performance of NMF-based source separation methods is greatly affected by the basis vectors used to represent each source. There are multiple methods for obtaining the basis vectors. In source separation, basis vectors should be such that they present their target source within the mixture, and do not represent other sources. For example, an identity matrix is not a good basis matrix, since even though it could perfectly represent the target source, it can also represent all other sources in the mixture. In addition to the representation capabilities, there may be also other requirements for the basis matrices. For example, there are typically constraints on the number of basis vectors, since it directly affects the computational complexity of algorithms and their memory usage.

In this section we present selected methods for learning basis matrices, or, *dictionaries*. Sound dictionary learning methods can be roughly divided into two categories, which we here group into generative and discriminative. Generative dictionaries are developed to efficiently model each target source separately, whereas discriminative methods are optimised for their separation capability. We review some of the common generative and discriminative methods in the following sections. All the methods presented assume that the dictionary size, i.e., the number of basis vectors is defined by the developer of the methods. Even though some methods exists for automatically estimating the dictionary size, the sizes are most commonly defined by using some prior knowledge or by developer experimentation with different dictionary sizes.

2.3.1 Generative Dictionaries

Generative sound dictionary learning methods operate by using a set of training instances of spectrum vectors for the target source. Since generative dictionaries are trained separately for each source, and in order to simplify the notation, we will not use the source index j in this section but represent the training spectrum vectors of the target source with vectors $\mathbf{x}_n$, $n = 1, \ldots, N$. The general goal of generative dictionary learning is to estimate a set of basis vectors $\mathbf{b}_k$, $k = 1, \ldots, K$, such that the NMF model

$$\mathbf{x}_n \approx \sum_{k=1}^{K} \mathbf{b}_k w_{k,n}, \quad w_{k,n} \geq 0 \tag{2.15}$$

represents the training instances well. Even when generative dictionaries are used, there are also other requirements for producing the dictionaries, such as minimising the likelihood that each one represents the other sources.

The simplest possible dictionary learning method is based on randomly selecting a subset of training spectrum vectors $\mathbf{x}_n$, and using them directly as basis vectors, or, *dictionary atoms*. Atoms that are instances from the training data are called *exemplars*. Use of exemplars can provide a good separation performance, provided that a large number of atoms are used [5–7]. In addition to simplicity, exemplar-based dictionaries can potentially represent the underlying distribution within training samples more accurately: especially if any non-audio information related to the training data not represented by the training samples is relevant in source separation, any dictionary learning method other than random sampling is likely to lead to a biased estimate of the underlying source distributions. A possible benefit of exemplar-based dictionaries is that each atom corresponds to a real audio spectrum, whereas other dictionary learning methods may produce basis vectors that do not actually correspond to any realistic spectra. Random sampling has the drawback that it typically requires significantly larger dictionaries in comparison to other methods to achieve equivalent source separation performance.

The size of exemplar-based dictionaries can be reduced by applying clustering on the training samples. Clustering operates by finding a set of cluster center vectors $\mathbf{b}_k$ (which correspond to atoms in our cases), and assigning each training instance n to cluster k_n so that the overall distance between training samples and cluster centers

$$\sum_{n=1}^{N} d(\mathbf{x}_n||\mathbf{b}_{k_n}) \tag{2.16}$$

is minimised. Above, $d(\mathbf{x}_n||\mathbf{b}_{k_n})$ is a divergence measure between training sample $\mathbf{x}_n$ and cluster center $\mathbf{b}_{k_n}$. We can for example do clustering with the same criterion as NMF, and use the Kullback-Leibler divergence as the function d. A clustering algorithm minimising the overall Kullback-Leibler divergence between training samples and cluster centers is given in Algorithm 1.

Algorithm 1 Clustering algorithm for minimising the Kullback-Leibler divergence between training samples $\mathbf{x}_n$ and cluster centers $\mathbf{b}_{k_n}$

Require: Input sample vectors $\mathbf{x}_n, n = 1, \ldots, N$ are entry-wise non-negative. The number of clusters K is specified by the developer.

Initialise cluster centers $\mathbf{b}_k, k = 1, \ldots, K$ by choosing a random set of training sample vectors $\mathbf{x}_n$ and assigning each of them to be a cluster center.
repeat
  for $n = 1$ to N **do**
    Find cluster index k_n which minimises the sample-cluster distance $d(\mathbf{x}_n||\mathbf{b}_{k_n})$ in (2.16)
  end for
  for $k = 1$ to K **do**
    Calculate new cluster center $\mathbf{b}_k$ as the mean of training samples $\mathbf{x}_n$ for which $k_n = k$
  end for

until The indexes of the clusters have not changed.
return Cluster centers $\mathbf{b}_k$

A natural extension to the clustering criterion in (2.16) is to use a linear combination of atoms to approximate each training sample, instead of individual cluster centres. This criterion can be written as

$$\sum_{n=1}^{N} d(\mathbf{x}_n||\sum_{k=1}^{K} \mathbf{b}_k w_{k,n}), \quad w_{n,k} \geq 0 \ \forall\, k, n \tag{2.17}$$

where $w_{k,n}$ are the weights of the linear combination. As we can see,this is equivalent to the basic NMF model in (2.5). Similarly to the regular NMF, various constrains can be imposed on the weights w of the linear combination. For example, we can restrict ourselves to convex combinations by constraining the sum of weights to be equal to one as

$$\sum_{n=1}^{N} d(\mathbf{x}_n||\sum_{k=1}^{K} \mathbf{b}_k w_{k,n}), \quad w_{k,n} \geq 0 \ \forall\, n, k; \sum_{k=1}^{K} w_{k,n} = 1 \ \forall\, n. \tag{2.18}$$

Dictionary learning with the pure linear combination criterion in (2.17) can be performed with basic NMF. The standard algorithm for minimising the generalised Kullback-Leibler using multiplicative update rules is given in Algorithm 2, where we use matrix $\mathbf{X} = [\mathbf{x}_1, \ldots, \mathbf{x}_N]$ to denote all the training samples, dictionary matrix $\mathbf{B} = [\mathbf{b}_1, \ldots, \mathbf{b}_K]$ to denote all the basis vectors, and weight matrix $[\mathbf{W}]_{kn} = w_{k,n}$ to denote all the weights. For the case where the generalised Kullback-Leibler

---

Algorithm 2 NMF algorithm for dictionary learning

Require: Input sample matrix $\mathbf{X}$ is entry-wise non-negative. The number of atoms K is specified by the developer.

Initialise basis matrix $\mathbf{B}$ and weight matrix $\mathbf{W}$ with random positive values.

repeat

Update weight matrix as $\mathbf{W} \leftarrow \mathbf{W} \otimes \frac{\mathbf{B}^T \frac{\mathbf{X}}{\mathbf{BW}}}{\mathbf{B}^T \mathbf{1}^{F \times N}}$.

Update dictionary matrix as $\mathbf{B} \leftarrow \mathbf{B} \otimes \frac{\frac{\mathbf{X}}{\mathbf{BW}} \mathbf{W}^T}{\mathbf{1}^{F \times N} \mathbf{W}^T}$.

until The parameters converge.

return Basis matrix $\mathbf{B}$

---

divergence is minimised, and convexity constraints placed on the weights as in (2.18), the update rule should be changed and weights projected to fulfill the convexity constraints, leading to Algorithm 3. When only a small dictionary is used, NMF and

---

Algorithm 3 NMF algorithm for dictionary learning with convexity constraints

---

Require: Input sample matrix $\mathbf{X}$ is entry-wise non-negative. The number of atoms K is specified by the developer.

Initialise basis matrix $\mathbf{B}$ and weight matrix $\mathbf{W}$ with random positive values.

Normalise each column of $\mathbf{W}$ to sum to one by scaling.

repeat

Update weight matrix as $\mathbf{W} \leftarrow \mathbf{W} \otimes \frac{\mathbf{B}^T \frac{\mathbf{X}}{\mathbf{BW}} + \mathbf{1}^{K \times F} \mathbf{BW}}{\mathbf{B}^T \mathbf{1}^{F \times N} + \mathbf{1}^{K \times F} \mathbf{X}}$.

Normalise each column of $\mathbf{W}$ to sum to one by scaling.

Update dictionary matrix as $\mathbf{B} \leftarrow \mathbf{B} \otimes \frac{\frac{\mathbf{X}}{\mathbf{BW}} \mathbf{W}^T}{\mathbf{1}^{F \times N} \mathbf{W}^T}$

until The parameters converge.

return Basis matrix $\mathbf{B}$

NMF with weight convexity constraints can lead to good separation results. However, additional constraints must be used when learning large dictionaries, since without them the approximation error can be minimised with dictionaries such as $\mathbf{B} = \mathbf{I}$ which does not perform any separation.

One of the generic problems with generative dictionary models that are based on linear combinations, is that the learned dictionaries can be too loose, i.e., the convex hull spanned by the basis vectors is unnecessarily large, and may overlap with the subspace of other sources. Applying sparsity constraints or other regularisations can partially deal with this issue, but fundamentally more restrictive models can also be used. Specifically, we can constrain the dictionary atoms to convex combinations of training samples, i.e., by defining them as

$$\mathbf{b}_k = \sum_{n=1}^{N} \mathbf{x}_n c_{n,k}, \quad c_{n,k} \geq 0 \ \forall\, n, k; \quad \sum_{n=1}^{N} c_{n,k} = 1 \ \forall\, n. \tag{2.19}$$

where $c_{n,k}$ are the weights of the convex combination. Together with the model where each training sample is represented as a linear combination of atoms (2.18), this leads to approximation of the training data matrix with the model

$$\mathbf{X} \approx \mathbf{XCW}, \tag{2.20}$$

where matrix $[\mathbf{C}]_{nk} = c_{n,k}$ represents the weights of the convex combination of training samples in (2.19). Both matrices $\mathbf{C}$ and $\mathbf{W}$ are entry-wise non-negative, and the sum of each of their columns is constrained to be one. The dictionary matrix is given simply as $\mathbf{B} = \mathbf{XC}$. This model, *archetypical analysis* originally proposed in [8], allows the learning of a set of basis vectors which produces a tight convex hull, and has been shown to produce better separation results than NMF and VQ [9].

Matrices $\mathbf{C}$ and $\mathbf{W}$ can be again estimated by minimising a criterion such as the Kullback-Leibler divergence between the observations and the model. The standard multiplicative update rules [9] that are derived from the partial derivative of the chosen divergence with respect to each model parameter do not obey the unity-column-sum constraints, and we found out that using the normal multiplicative updates rules followed by a re-scaling to obey to constraint produced suboptimal results. A model that automatically takes into account the unity-column-sum constraint can be written as

$$\mathbf{X} \approx \mathbf{XC} \cdot \mathrm{diag}(\mathbf{1}^{1\times N}\mathbf{C})^{-1}\mathbf{W} \cdot \mathrm{diag}(\mathbf{1}^{1\times K}\mathbf{W})^{-1}, \tag{2.21}$$

where $\mathrm{diag}(\cdot)$ is a diagonal matrix having its input vector values on its diagonal. In this model, matrices $\mathbf{C} \cdot \mathrm{diag}(\mathbf{1}^{1\times N}\mathbf{C})^{-1}$ and $\mathbf{W} \cdot \mathrm{diag}(\mathbf{1}^{1\times K}\mathbf{W})^{-1}$ are now automatically normalised to unity column sum.

By calculating the partial derivative of the Kullback-Leibler divergence with respect to model parameters $\mathbf{C}$ and $\mathbf{W}$ in (2.21), distributing the partial derivative to terms that are either positive or negative as in [1, 10], and by using the identities $\mathbf{1}^{1\times N}\mathbf{C} = \mathbf{1}^{1\times K}$ and $\mathbf{1}^{1\times K}\mathbf{W} = \mathbf{1}^{1\times N}$ (because of the normalisations applied within the algorithm), we obtain Algorithm 4.

Examples of dictionaries learned with different methods discussed above (clustering, NMF with weight convexity constraints, and archetypical analysis) in a simple two-dimensional case are illustrated in Fig. 2.3. It can be seen that the atoms learned by NMF lie outside the subspace of the observations. The convex hull spanned by the NMF basis vectors covers all the observations, but the hull is rather large, and when used to represent a mixture of sounds the NMF dictionary may also represent other sources. The atoms learned by clustering lie within inside the subspace of the

Algorithm 4 Dictionary learning algorithm based on archetypical analysis

Require: Input sample matrix $\mathbf{X}$ is entry-wise non-negative. The number of archetypes K is specified by the developer.
Initialise observation weight $\mathbf{C}$ and atom weight $\mathbf{W}$ matrices with random positive values.
Normalise each column of $\mathbf{C}$ and $\mathbf{W}$ to sum to one by scaling.
repeat
  Calculate basis matrix as $\mathbf{B} = \mathbf{CW}$.
  Update weight matrix as $\mathbf{W} \leftarrow \mathbf{W} \otimes \frac{\mathbf{B}^T \frac{\mathbf{X}}{\mathbf{BW}} + \mathbf{1}^{K \times K} \mathbf{BW}}{\mathbf{B}^T \mathbf{1} + \mathbf{1}^{K \times K} \mathbf{X}}$.
  Normalise each column of $\mathbf{W}$ to sum to one by scaling.
  Calculate basis matrix as $\mathbf{B} = \mathbf{CW}$ and ratio matrix as $\mathbf{R} = \frac{\mathbf{X}}{\mathbf{XCW}}$.
  Update basis weight matrix as $\mathbf{C} \leftarrow \mathbf{C} \otimes \frac{\mathbf{X}^T \mathbf{R} \mathbf{W}^T + \mathbf{1}^{T \times T}((\mathbf{1}^{T \times F} \mathbf{B}) \otimes \mathbf{W}^T)}{\mathbf{X}^T \mathbf{1}^{F \times T} \mathbf{W}^T + \mathbf{1}^{T \times T}((\mathbf{RB}) \otimes \mathbf{W}^T)}$.
  Normalise each column of $\mathbf{C}$ to sum to one by scaling.
until The parameters converge.
return Basis matrix $\mathbf{B} = \mathbf{CW}$.

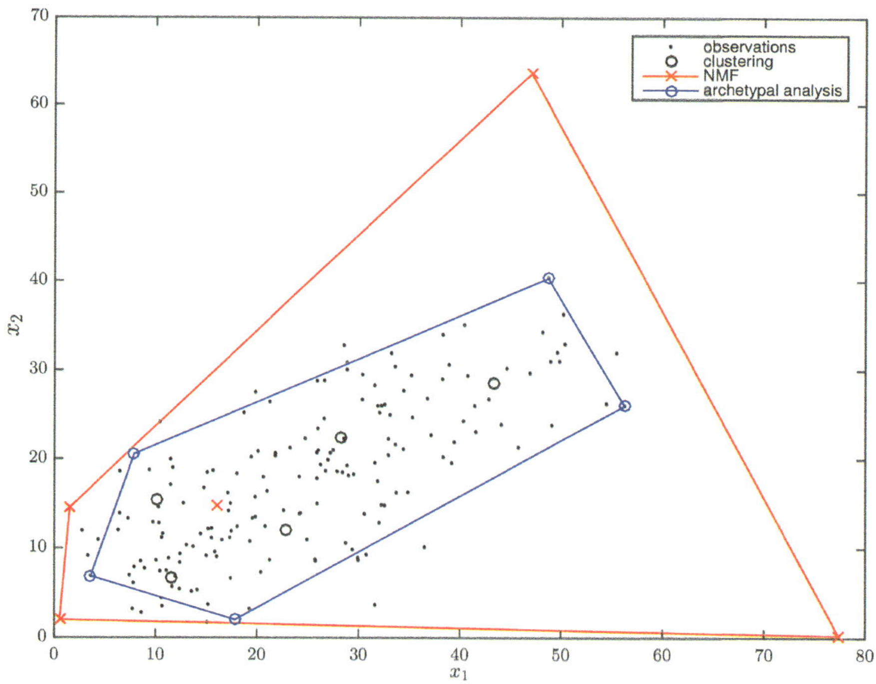

Fig. 2.3 Illustration of dictionaries learned by different methods in a simple two-dimensional case. Black dots represent two-dimensional observations $\mathbf{x}$ used to learn the dictionaries. Circles and crosses represent the two-dimensional basis vectors learned by different algorithms, and they are connected with lines for visualisation. Archetypal analysis learns a convex hull, and almost all the training samples are inside the hull. The convex hull spanned by the NMF basis vectors contains all the training samples, but the hull is not tight. Vector quantisation learns basis that are within the distribution of the training samples

observations, and they are likely to represent the target source in a mixture, provided that the target source observations are close to cluster centers. Observations further away from them may be represented by atoms of another source. The atoms learned by archetypical analysis form a relatively tight convex hull that contains most of the observations.

2.3.2 Discriminative Dictionaries

As we saw from the previous subsection, generative dictionaries provide several alternatives to learn dictionaries to model individual sources, but they are not necessarily optimal for source separation. Discriminative dictionaries offer the possibility to optimise the dictionaries for the separation task. In order to achieve this, the dictionary learning stage needs to be aware of the competing sources, and the learning needs to be either jointly for two sources. Discriminative dictionary learning can be based mixture or isolated sources.

The methods in [11, 12] learned dictionaries using mixtures of sounds. Similarly to other NMF approaches, they estimated the parameters by minimising a divergence, but in their approach the divergence was measured between the ground truth target source (which is known at the training stage), and a reconstructed source. This kind of objective is somewhat more difficult to optimise in comparison to the typical criteria used to estimate NMF model parameters, and therefore, [11] used a two-stage approach, where one set of dictionaries was first used to obtain the weights, and then another set of dictionaries was used to reconstruct the target source. Sprechmann et al. [12] used a single-stage approach where a stochastic gradient algorithm was used to estimate dictionaries.

There are also methods that aim at improving the separation capability of dictionaries by enforcing the dissimilarity of dictionaries of two sources by using additional regularisations such as inter-source dictionary correlation [13]. Here the dictionaries are trained using isolated material of each source, but the dictionary training stage needs to have the knowledge about the other source dictionaries.

2.3.3 Dictionary Adaptation

In practical source separation scenarios, there is a mismatch between the audio material used to develop a system and the target audio at the actual usage scenario. For example, the training material may be recorded in different acoustic environment from the actual usage stage, or different microphones can used. Differences in the impulse responses between the source and the microphone, as well as differences in the impulse responses of microphones can be compensated with linear filters, which are typically assumed to be time-invariant.

Linear filtering in time-domain corresponds to point-wise multiplication in the frequency domain. We can modify the basic NMF model for sound mixtures in (2.4) by adding source-specific compensation matrices $\mathbf{H}_j$ to get the model, introduced in [14],

$$\hat{\mathbf{x}}_t = \sum_{j=1}^{J} \mathbf{H}_j \sum_{k=1}^{K_j} w_{k,t,j} \mathbf{b}_{k,j}. \tag{2.22}$$

Above, $\mathbf{H}_j$, $j = 1, \ldots, J$ are diagonal matrices with the magnitude response of the compensating filter for source j on the diagonal. The compensation filter does not need to be known in advance, but it can be estimated using the same principles with the other NMF models, by minimising a divergence between the model and the observations. An algorithm that minimises the Kullback-Leibler divergence between the observations and the model (2.22) is given in Algorithm 5.

Algorithm 5 Algorithm for estimating basis vector weights and channel compensation filters.

Require: Observation matrix $\mathbf{X}$ and basis vector matrix $\mathbf{B}$ are entry-wise non-negative.
Initialise weight matrices $\mathbf{W}_j$ with random positive values. Initialise channel compensation matrices $\mathbf{H}_j$ with random positive values on the diagonal, and the rest of the entries to zero.
repeat
  Calculate model $\hat{\mathbf{X}}$ according to (2.22)
  for $j = 1$ to J **do**
    Update weight matrix as $\mathbf{W}_j \leftarrow \mathbf{W}_j \otimes \frac{(\mathbf{H}_j\mathbf{B}_j)^T \frac{\mathbf{X}}{\hat{\mathbf{X}}}}{(\mathbf{H}_j\mathbf{B}_j)^T \mathbf{1}^{F\times T}}$.
  end for
  Calculate model $\hat{\mathbf{X}}$ according to (2.22)
  for $j = 1$ to J **do**
    Update compensation matrix as $\mathbf{H}_j \leftarrow \mathbf{H}_j \otimes \frac{\frac{\mathbf{X}}{\hat{\mathbf{X}}}(\mathbf{B}_j\mathbf{W}_j)^T}{\mathbf{1}^{F\times T}(\mathbf{B}_j\mathbf{W}_j)^T}$.
  end for
until The parameters converge.
return Estimated weights $\mathbf{W}$ and source-specific channel compensation filter matrices $\mathbf{H}_j$, $j = 1, \ldots, J$.

Where a high frequency resolution is used, the learned compensation filters have a relatively high number of parameters. This may cause problems when the amount of observations used to do the compensation, or the dictionary size is small since the compensation filter itself may start modeling the source. This problem can be alleviated somewhat by constraining the compensation filter to a linear, low-rank model.

2.4 Semi-supervised Separation

There are often cases where we have information about one source within a mixture, however can not make assumptions about another. For example, in speech enhancement, we can assume that one source will be speech, but the interfering source could be any type of unknown noise, so cannot be modelleled explicitly with a pre-existing dictionary. Here, we can use *semi-supervised* NMF techniques, which make use of a pre-learned dictionary for the known source, but estimate the parameters of the unknown source at the time of separation. We can consider semi-supervised separation from the point of view of speech enhancement, where the known source is speech. We will refer to these unknown sources as *noise* sources from here in, and continue the explanation of semi-supervised NMF from this point of view throughout the rest of this section. These approaches are however much more general, and can of course be used in cases with any type of target signal, not just speech.

In NMF for speech enhancement, as with supervised cases, the aim of the factorisation is that the dictionary for each source effectively models the true contributions to the mixture, from that source. Each time-frequency point in the mixture spectrogram can then be apportioned to the contributing sources in the correct ratio and successful enhancement can occur.

Assuming a well-constructed dictionary for the mixture speech, the noise model should adapt to capture latent structure in only the remaining portion of the mixture, without over-fitting to represent the mixture entire.

Noise can be modelled in the spectrogram as the spectral atoms adapt to fit the residual portion of the mixture which is not approximated by the known source dictionary. The portions of a dictionary relating to speech and noise can then be defined as $\mathbf{B}_s$ and $\mathbf{B}_n$ respectively. The pre-trained atoms in $\mathbf{B}_s$ do not change during factorisation, whilst $\mathbf{B}_n$ is randomly initialised, and updated to minimise cost functions outside of the contributions modelled by $\mathbf{B}_s$ (see Fig. 2.4) .

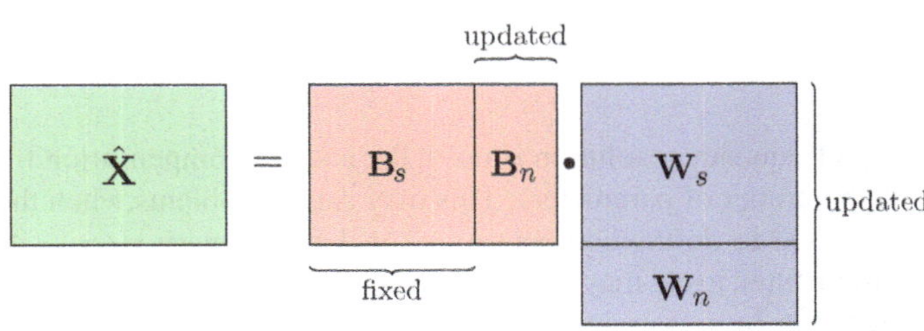

Fig. 2.4 Initialisation and update of weights and parts of the dictionary in semi-supervised NMF. $\mathbf{B}_s$ is initialised from training material, whilst $\mathbf{B}_s$, $\mathbf{W}_s$ and $\mathbf{W}_n$ should be initialised with random non-negative values

The NMF Model for the target matrix

$$\hat{\mathbf{X}} = \mathbf{BW} \tag{2.23}$$

in the semi-supervised case described here becomes

$$\hat{\mathbf{X}} = \left[\mathbf{B}_s \ \mathbf{B}_n\right] \begin{bmatrix} \mathbf{W}_s \\ \mathbf{W}_n \end{bmatrix} \tag{2.24}$$

to denote portions of each matrix pertaining to either speech or nose components in the dictionary. where $\mathbf{B}_s$ remains fixed, whilst $\mathbf{B}_n$ is updated. The weights for both $\mathbf{W}_s$ and $\mathbf{W}_n$ are updated, to reduce an appropriate cost function, for example as in (2.8). Ideally, The factors and activations which effectively reduce the error will reflect the structure present in the sources.

Typically, the number of components used to model noise and speech are quite different. Only a low number of components should be used when constructing the noise model, whereas the speech model will in general benefit from having a larger dictionary. $\mathbf{B}_n$ is kept small in order to prevent overfitting since if too many values are present in $\mathbf{B}_n$ and $\mathbf{W}_n$, then the basis functions simply adapt to model the entire mixture, instead of capturing a low-rank estimation of the noise source structure.

Additionally, sparsity constraints can be imposed on the activations, and the number of iterations for update equations artificially limited, as techniques to reduce overfitting. In this instance, an absolute minimisation of the cost function does not necessarily mean effective source separation, and in fact quite the opposite. We have seen in [15] and independently in [16] that with an increasing number of iterations we do not see increasingly improved performance. In [15], separation increased to a point with a greater number of iterations, then decreased. In [16], for the special case of online noise estimations, only a *single* application of updates to the noise vectors produces greatest separation. In practise, the ideal number of iterations will be a function of the material, imposed sparsity constraints etc. and as such, some manual tuning of the number of iterations used can give significant improvements in the overall enhancement quality. Rather than achieving minimisation of a cost function, in the semi-supervised case, a reduction in cost is generally sufficient. The aim of the semi-supervised factorisation is therefore expressed as reduction of a cost function:

$$\mathbf{B}_n, \mathbf{W} \qquad f(\mathbf{B}_n, \mathbf{W}) = \mathrm{KL}(\mathbf{X} || \left[\mathbf{B}_s \ \mathbf{B}_n\right] \begin{bmatrix} \mathbf{W}_s \\ \mathbf{W}_n \end{bmatrix}) + \lambda ||\mathbf{W}||_1. \tag{2.25}$$

The weights matrix $\mathbf{W} = \begin{bmatrix} \mathbf{W}_s \\ \mathbf{W}_n \end{bmatrix}$ is updated across all values, whilst only the noise portion, $\mathbf{B}_n$ within $\mathbf{B} = \left[\mathbf{B}_s \ \mathbf{B}_n\right]$ is updated with the rule

$$\mathbf{B}_n \leftarrow \mathbf{B}_n \otimes \frac{\frac{\mathbf{X}}{\mathbf{BW}} \mathbf{W}_n^T}{\mathbf{1}^{F \times T} \mathbf{W}_n^T}. \tag{2.26}$$

The weights matrix is updated for all values, and can benefit from the inclusion of sparsity constraints, which can reduce overfitting by keeping the number of activations low, and hence using only the most structurally relevant atoms. To impose a penalty based on the L1 norm, **W** can be updated using the weights matrix update rule (2.14) which was presented in Sect. 2.2.1.

Practical application considering all of the above leads us to the Algorithm 6 for use semi-supervised NMF for speech (or other source) enhancement.

Algorithm 6 Algorithm for estimating basis vector weights and noise model spectra for enhancement filter in semi-supervised speech enhancement

---

Require: Observation matrix $\mathbf{X}$ and speech basis vector matrix $\mathbf{B}_s$ are entry-wise non-negative. $\mathbf{B}_s$ has been obtained from representative speech material.

Select suitable number of atoms for noise model. Initialise weight matrix $\mathbf{W}$ with random positive values. Initialise noise dictionary matrix $\mathbf{B}_n$ with random positive values. Initialise sparsity penalty parameter λ to suitable value.

repeat

  Update weight matrix as $\mathbf{W} \leftarrow \mathbf{W} \otimes \frac{\mathbf{B}^T \frac{\mathbf{X}}{\mathbf{BW}}}{\mathbf{B}^T \mathbf{1}^{F \times T} + \lambda}$, using (2.14)

  Update noise dictionary matrix, $\mathbf{B}_n$ with $\mathbf{B}_n \leftarrow \mathbf{B}_n \otimes \frac{\frac{\mathbf{X}}{\mathbf{BW}} \mathbf{W}_n^T}{\mathbf{1}^{F \times T} \mathbf{W}_n^T}$, using (2.26)

until Sufficient reduction in cost function

Produce speech enhancement mask with $\mathbf{M}_j = \frac{\mathbf{B}_s \mathbf{W}_s}{\mathbf{B}_s \mathbf{W}_s + \mathbf{B}_n \mathbf{W}_n}$, to be applied to mixture spectrogram as in (2.7).

return $\mathbf{M}_j$

2.5 Low-Latency Separation

Many source separation techniques process an entire mixture signal, and perform separation offline, before producing the separated output. There are cases where this is neither practical nor feasible, however, and a low-latency separation algorithm should be used instead. With low-latency separation, as audio samples arrive into the system, they are processed and outputted with a virtually imperceptible delay. Speech enhancement and separation for telecommunication, musical or artistic performance and hearing-aid algorithms are all areas where low-latency separation is an applicable constraint.

The acceptable delay in separation depends somewhat on the context. Since audio delays greater than 20 ms are quite perceivable [17], in order to minimise discomfort for the listener this can be considered the upper-bound for realtime use. Certain scenarios have even stricter latency requirements. In hearing-aid applications for example, even delays lower than 6 ms can even be noticed by the listener, as face-to-face communication produces an uncomfortable mismatch between auditory and visual cues as delays exceed this threshold [18].

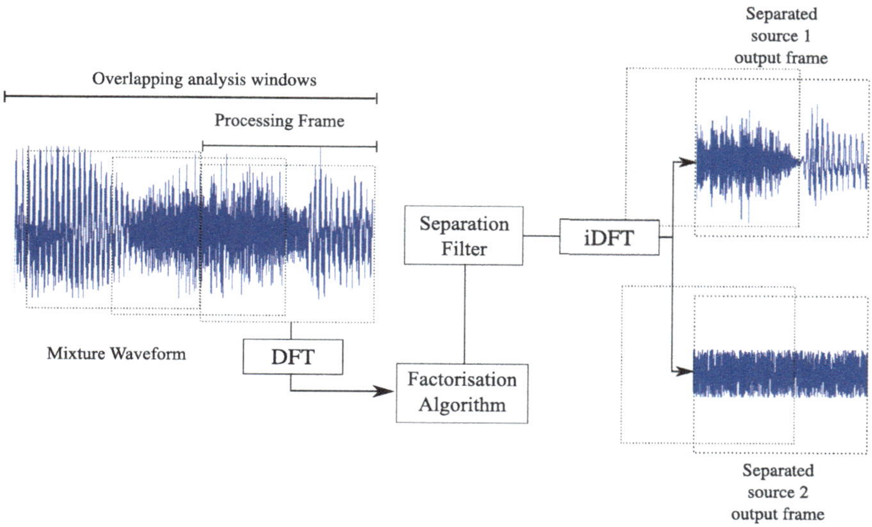

Fig. 2.5 Overview of data handling in sliding-window frame-based separation approach. A single frame of output is produced for each frame of input into the system. The length of this processing frame defines the latency of the system. An entire frame of samples should be buffered for the discrete Fourier transform operation before separation can be performed

NMF algorithms can be used within a frame-based audio framework, as each frame is simply estimated as a sum-of-components for a matrix with only a single column, i.e. a vector. In frame-based spectral decomposition techniques, the lower bound for latency within an algorithm is governed by the length of the processing frame. An entire frame worth of time domain samples needs to be gathered before the discrete Fourier transform (DFT) can be applied. Figure 2.5 shows the basics of how a single frame of separated audio is produced from a stream of time-varying audio data.

2.5.1 *Algorithmic and Processing Latency*

When considering the latency of factorisation techniques, a distinction must be drawn between algorithmic latency and computational latency. Algorithmic latency, T_a, refers to the minimal theoretical latency of the proposed method, and is constrained by the availability, organisation and handling of data within the separation algorithm. Computational latency T_c, on the other hand, refers to the actual time taken to perform the calculations required for the separation estimate. On an infinitely fast processor, this would tend towards zero, whereas the algorithmic latency is a fixed value inherent to the separation approach. The overall latency of any processing approach is a function of the two latencies, and for a frame-based method, is roughly the sum

$$T = T_a + T_c \tag{2.27}$$

since an entire frame of samples is collected before the processing can start to be performed.

In non-theoretical applications, T_c should be less than T_a, so that the results of one frame are ready for output before the next frame is input into the system. With careful planning and design, factorisation algorithms can be made to satisfy this constraint; however, where the chosen separation approach is simply too processing intensive, a compromise may have to be made about the minimal frame length. A longer than ideal processing frame may be required to allow computation to be completed in time.

Several factors can affect the value of T_c, included, but not limited to processor type and architecture, implementation hardware/language etc. As an example, the ASNA factorisation algorithm [4] provides good performance improvements over other NMF algorithms on traditional CPU architectures, but can not be parallelised to take advantage of the increasingly ubiquitous GPU processors. Other matrix factorisation algorithms could benefit greatly from a GPU-based implementation. The rest of the discussion in this section will consider primarily the algorithmic component of low-latency source separation.

2.5.2 Use of Coupled Dictionaries for Very Low Latency Separation

For very low latency applications, the frame length should be kept very short accordingly. However, the lower the frame length, the lower the potential relative entropy which can exist between the two dictionaries becomes, and the greater chance of information distributions within each dictionary overlapping. As the distributions overlap, the potential for assignment of components to incorrect sources increases, decreasing separation quality. There is therefore a trade-off between low latency and the potential for separation performance. To reduce this problem, and produce greater separation in low-latency factorisation, we propose a model in [19] which makes use of a pair of dictionaries, containing two different vector lengths. Factorisation is performed on larger vectors covering extended prior time context, using an *analysis* dictionary. The obtained weights are applied to a dictionary comprised of shorter *reconstruction* vectors, to form the Wiener filter for source separation for each frame. Each atom within a dictionary has a one-to-one relationship with the atom in the other dictionary. In this context, we refer to the dictionaries as *coupled dictionaries*.

Table 2.1 Summary of symbol notations used in this chapter

Symbol	Description
$\mathbf{a}_t$	Time-domain analysis frame
$\mathbf{s}_t$	Time-domain synthesis frame
A	Length in samples of $\mathbf{a}_t$
L	Length in samples of $\mathbf{s}_t$
$\mathbf{y}_t$	Real-valued feature vector formed from $\mathbf{a}_t$
$\mathbf{s}^*$	Complex-valued synthesis vector formed from $\mathbf{s}_t$
$\mathbf{A}$	Analysis dictionary
$\mathbf{R}$	Reconstruction dictionary
$\mathbf{R}_{:,k}$	The k-th column of dictionary $\mathbf{R}$.
$\mathbf{w}$	Weights vector for a single output frame
$\mathbf{s}^j$	The reconstructed frame for the j-th source in a mixture
j	Superscript referring to values associated with the j-th source in dictionaries, weights, or reconstructed frames.

2.5.2.1 System Description and Dictionary Creation

For clarity in the following text, we present here some definitions for frame lengths and symbols. Symbols are also summarised in Table 2.1. Frame data which is processed for the purposes of separated source reconstruction is called the synthesis frame $\mathbf{s}_t$ and is of length L. A buffer $\mathbf{a}_t$ of previous incoming samples, of length A, is maintained (where $A > L$ and A/L is typically an integer). This buffer forms the 'analysis frame'; the temporal context from which filter weights are obtained via factorisation.

Depending on the choice of overlap-add scheme, the update rate of these frames can vary. Updating every $L/2$ samples (50% overlap), though, will achieving an algorithmic latency of L whilst reducing computational costs which would be present with higher overlap values.

The analysis feature vector, $\mathbf{y}_t$, is formed from $\mathbf{a}_t$ by taking the absolute value of the positive frequencies of the discrete Fourier transform (DFT) of analysis subframes length L and concatenating the resulting $(\frac{2A}{L} - 1)$ subframe outputs into a single vector. The overlap ratio and window length can be tailored to the needs of the system, but the point here is that the analysis frame is obtained from greater previous context than the synthesis frame. The complex-valued frequency-domain synthesis vector $\mathbf{s}^*$ is formed by taking absolute values of only positive frequencies from the DFT data in $\mathbf{s}_t$, and so has length $(L/2) + 1$. A Hanning window can be used for overlap add reconstruction.

In the two-source separation case, source-specific dictionary portions are denoted with a superscript e.g. the dictionary contribution for Source 1 is $\mathbf{A}^1$. The analysis dictionary $\mathbf{A}$ is therefore constructed:

$$\mathbf{A} = [\mathbf{A}^1 \ \mathbf{A}^2] \tag{2.28}$$

as is the reconstruction dictionary **R**,

$$\mathbf{R} = [\mathbf{R}^1 \ \mathbf{R}^2]. \tag{2.29}$$

Dictionaries are constructed so that pairs of atoms which are coupled across the dictionaries are derived from very similar temporal context. These coupled atoms are stored at the same column index within the pair of dictionaries so that the k-th atom is coupled:

$$\mathbf{R}_{:,k} \Longleftrightarrow \mathbf{A}_{:,k}. \tag{2.30}$$

Figure 2.6 demonstrates the audio context relationship between coupled atoms. Both share the ultimate time-domain sample, but greater past context is used in the generation of reconstruction dictionary **A**, though, resulting in a increased frame length.

If a reduced size coupled dictionary is required, the techniques described early in this chapter can be applied. However, the pair of dictionaries should be concatenated prior to processing, and then later split, as shown in Fig. 2.7 for a single source dictionary.

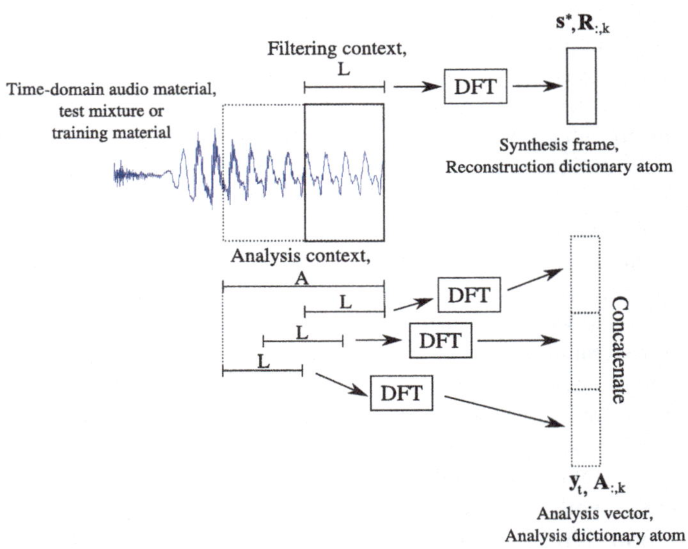

Fig. 2.6 Overview of data handling for vector creation for both dictionary and factorisation/filtering for realtime separation, and the relationship between time contexts for synthesis and analysis vectors/dictionaries

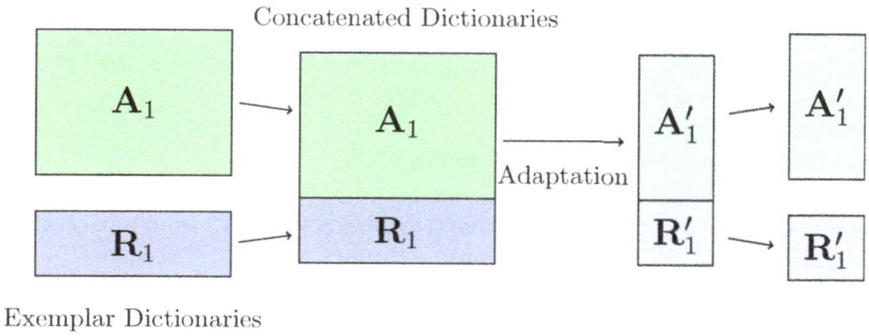

Fig. 2.7 Dictionary adaptation for coupled dictionaries. The coupled dictionaries for a source $\mathbf{A}_1$ and $\mathbf{R}_1$ should first be concatenated so that each atom pair forms one long vector, prior to adaptation. Following processing, the reduced size dictionary is split back into two coupled dictionaries, $\mathbf{A}'_1$ and $\mathbf{R}'_1$ which remain coupled

2.5.3 Factorisation

Similarly to the creation of coupled dictionaries from two contexts, factorisation and filtering are applied across two different contexts, with factorisation vector $\mathbf{y}_t$ being produced in the same way as exemplars for dictionaries in Fig. 2.6.

Analysis is performed by learning the weights $\mathbf{w}$ which minimise KL-divergence between analysis vector $\mathbf{y}_t$ and a weighted sum of atoms from dictionary $\mathbf{A}$

$$\underset{\mathbf{w}}{\text{minimize}} \qquad f(\mathbf{w}) = \text{KL}(\mathbf{y}_t||\mathbf{A}\mathbf{w}) + \lambda||\mathbf{W}||_1 \tag{2.31}$$

The ASNA algorithm [4] is a good candidate for use in realtime applications due to its rapid computation time and guaranteed convergence. Sparsity constraints can also be used.

The learned weights $\mathbf{w}$ are applied to the corresponding coupled dictionary atoms in dictionary $\mathbf{R}$ to form the reconstruction Wiener filters. Filters are applied to the synthesis vector $\mathbf{s}^*$ at each frame processing step so that the positive frequencies of each frame the of separated source 1, $\mathbf{s}^1$ are reconstructed:

$$\mathbf{s}^1 = \mathbf{s}^* \otimes \frac{\mathbf{R}^1\mathbf{w}^1}{\mathbf{R}^1\mathbf{w}^1 + \mathbf{R}^2\mathbf{w}^2}. \tag{2.32}$$

The separated time-domain sources are reconstructed by generating complex conjugates of each $\mathbf{s}^n$ and performing the inverse DFT for each frame to be overlap-add reconstructed into a continuous time output. The various steps of the approach are presented and summarised in Algorithm 7.

This method has been shown to improve separation SDR performance significantly when the latency is 5 or 10 ms and marginally at 20 ms latency [19]. For a 16 kHz samplerate, these values correspond to dictionary vectors of length 80, 160 and 240 samples. Our experiments showed no tangible benefit to use of couple dictionaries once the vector length, and hence minimum factorisation latency exceeds these values.

Algorithm 7 Algorithm for low-latency supervised source separation

Require: Analysis dictionary **A** and reconstruction dictionary **R** are entry-wise non-negative and atoms in each dictionary are coupled to one another.
Select sparsity cost weight, λ.
for each incoming frame of time-domain samples $\mathbf{s}_t$ **do**
Update samples in analysis and synthesis context buffers
Create analysis vector $\mathbf{y}_t$ and complex-valued synthesis vector $\mathbf{s}^*$ from most recent samples.
Initialise weight matrix **W** with random positive values.
Update weight matrix **W** to minimise $\mathrm{KL}(\mathbf{y}||\mathbf{A}\mathbf{w}) + \lambda||\mathbf{W}||_1$ with ASNA or multiplicative update rules.
Multiply source specific weights with source specific reconstruction dictionaries, to produce source-specific filters applied to synthesis vector $\mathbf{s}^*$;
$\mathbf{s}^* \otimes \frac{\mathbf{A}^1\mathbf{W}^1}{\mathbf{A}^1\mathbf{W}^1+\mathbf{A}^2\mathbf{W}^2}$ and $\mathbf{s}^* \otimes \frac{\mathbf{A}^2\mathbf{W}^2}{\mathbf{A}^1\mathbf{W}^1+\mathbf{A}^2\mathbf{W}^2}$
Convert masked synthesis frames to audio with inverse DFT, and overlap-add reconstruct for each separated source.
end for

2.6 Conclusions and Discussion

In this chapter we have presented non-negative matrix factorisation techniques that can be effectively separate known sources in mixtures. NMF is based on a linear model, which can be easily be used to model sound mixtures. The NMF model for each source is obtained from training material where the sources are present in isolation, or using mixtures for which the reference target sources are known. Because of the simplicity of the linear NMF model, the methods used to estimate the model parameters are relatively simple, and the models can be extended in various ways.

We have presented selected dictionary learning algorithms that can be used to model individual sources. Exemplar-based dictionaries are easy to obtain by randomly sampling training data and lead to good separation performance when large dictionaries are used, but they are computationally expensive. Clustering can be used to obtain more compact dictionaries can higher computational efficiency with equal size dictionaries. NMF-based dictionaries can model training samples as the sum of basis vectors, which corresponds to the actual usage of the learned dictionaries. They typically provide the best separation performance with very small dictionaries, but require very tight regularisation to provide meaningful results with larger dictionaries. Archetypical analysis models each dictionary atom as a convex combination of training samples, leading to a compact convex hull in modeling the training data. It has the potential to provide better separation quality in comparison to exemplar-based and clustering-based dictionaries. Discriminative dictionaries are optimised for the separation task and have therefore the potential to achieve the highest accuracy, but are more difficult to estimate. We also present compensation methods that can be used to deal with mismatches between training data and actual test scenario.

Two methods of dictionary-based NMF separation are then described, along with some practical considerations for their implementation. A semi-supervised approach allows one to compensate for parts of a mixture which are not able to be pre-learned, through the use of a dictionary which is adapted to model unknown sources at the point of factorisation. Care must be taken to ensure that the adaptive model does not simply represent the entire mixture, and so we propose the use of limited adaptive dictionary size, sparsity constraints and a user-tuned number of multiplicative update applications to minimise this overfitting. For separation of sources at very close to real time, a low-latency implementation of supervised NMF can be used. In generalised online factorisation, individual incoming frame vectors are factorised as they arrive, and used as a separation filter. At very low latencies though, due to the short frame length, the dictionaries which describe each source are not as discriminative as larger dictionaries would be. Therefore, a coupled-dictionary approach is provided to overcome this problem. Dictionary weights are estimated from data frames covering larger temporal context, and applied to a dictionary of short-context frame, providing desired processing latency.

References

1. T. Virtanen, Monaural sound source separation by non-negative matrix factorization with temporal continuity and sparseness criteria. IEEE Trans. Audio Speech Lang. Process. **15**(3), 1066–1074 (2007)
2. C. Févotte, N. Bertin, J.-L. Durrieu, Nonnegative matrix factorization with the Itakura-Saito divergence: with application to music analysis. Neural Comput. **21**(3), 793–830 (2009)
3. C. Févotte, J. Idier, Algorithms for nonnegative matrix factorization with the beta-divergence. Neural Comput. **23**(9), 2421–2456 (2011)
4. T. Virtanen, B. Raj, J. Gemmeke, H.V. hamme, Active-set Newton algorithm for non-negative sparse coding of audio, in *Proceedings of the International Conference on Acoustics, Speech, and Signal Processing* (2014)
5. J. Gemmeke, T. Virtanen, A. Hurmalainen, Exemplar-based sparse representations for noise robust automatic speech recognition. IEEE Trans. Audio Speech Lang. Process. **19**(7), 2067–2080 (2011)
6. B. Raj, T. Virtanen, S. Chaudhure, R. Singh, Non-negative matrix factorization based compensation of music for automatic speech recognition, in *Proceedings of Interspeech* (2000)
7. P. Smaragdis, M. Shashanka, B. Raj, A sparse non-parametric approach for single channel separation of known sounds, in *Proceedings of Neural Information Processing Systems* (2009)
8. A. Cutler, L. Breiman, Archetypal analysis. Technometrics **36**(4), 338–347 (1996)
9. A. Diment, T. Virtanen, Archetypal analysis for audio dictionary learning, in *Proceedings of IEEE Workshop on Applications of Signal Processing to Audio and Acoustics* (2015)
10. D.D. Lee, H.S. Seung, Algorithms for non-negative matrix factorization, in *Proceedings of Neural Information Processing Systems* (2000), pp. 556–562
11. F. Weninger, J. Le Roux, J.R. Hershey, S. Watanabe, Discriminative NMF and its application to single-channel source separation, in *Proceedings of Interspeech* (2014)
12. P. Sprechmann, A.M. Bronstein, G. Sapiro, Supervised non-euclidean sparse nmf via bilevel optimization with applications to speech enhancement, in *Proceedings of Joint Workshop on Hands-free Speech Communication and Microphone Arrays* (2014)
13. E.M. Grais, H. Erdogan, Discriminative nonnegative dictionary learning using cross-coherence penalties for single channel source separation, in *Proceedings of Interspeech* (2013)
14. J.F. Gemmeke, T. Virtanen, K. Demuynck, Exemplar-based joint channel and noise compensation, in *Proceedings of IEEE International Conference on Audio, Speech and Signal Processing* (2013)

15. T. Barker, T. Virtanen, O. Delhomme, Ultrasound-coupled semi-supervised nonnegative matrix factorisation for speech enhancement, in *IEEE International Conference on Acoustics, Speech and Signal Processing* (2014), pp. 2129–2133
16. C. Joder, F. Weninger, F. Eyben, D. Virette, B. Schuller, Real-time speech separation by semi-supervised nonnegative matrix factorization, in *Proceedings of Latent Variable Analysis and Signal Separation: 10th International Conference*, ed. by F. Theis, A. Cichocki, A. Yeredor, M. Zibulevsky (2012), pp. 322–329
17. S. Laugesen, K. Hansen, J. Hellgren, Acceptable delays in hearing aids and implications for feedback cancellation. J. Acoust. Soc. Am. **105**(2), 1211–1212 (1999)
18. J. Agnew, J. Thornton, Just noticeable and objectionable group delays in digital hearing aids. J. Am. Acad. Audiol. **11**, 330–336 (2000)
19. T. Barker, T. Virtanen, N.H. Pontoppidan, Low-latency sound-source-separation using non-negative matrix factorisation with coupled analysis and synthesis dictionaries, in *Proceedings of the IEEE International Conference on Acoustics, Speech and Signal Processing* (2015)

Chapter 3
Dynamic Non-negative Models for Audio Source Separation

Paris Smaragdis, Gautham Mysore and Nasser Mohammadiha

Abstract As seen so far, non-negative models can be quite powerful when it comes to resolving mixtures of sounds. However, in such models we often ignore temporal information, instead focusing on resolving each incoming spectrum independently. In this chapter we will present some methods that learn to incorporate the temporal aspects of sounds and use that information to perform improved separation. We will show three such models, a conlvolutive model that learns fixed temporal features, a hidden Markov model that learns state transitions and can incorporate language information, and finally a continuous dynamical model that learns how sounds evolve over time and is able to resolve cases where static information is not enough.

3.1 Introduction

Time is of course one of the most important elements of sound. It is the domain over which sounds are defined at multiple levels. At the very lowest level, it is the time ordering of time samples that is essential in representing sound; at a higher level it is temporal structure that helps us to e.g. distinguish "no" from "on", and at the highest level what highlights the difference between "Tom ate a burger" and "a burger ate Tom". Although time is clearly recognizable as a major component in sound, it is often somewhat ignored when performing source separation. As shown in previous chapters, we can perform fairly good separation using only instantaneous spectra. Such a feature encapsulates some of the sample-level temporal structure, but clearly ignores temporal dependencies that take place across spectral frames. In this chapter

P. Smaragdis (✉)
University of Illinois/Adobe Research, Champaign, IL, USA
e-mail: paris@illinois.edu

G. Mysore
Adobe Research, San Francisco, CA, USA
e-mail: gmysore@adobe.com

N. Mohammadiha
Volvo, Gothenburg, Sweden
e-mail: nasser.mohammadiha@volvocars.com

S. Makino (ed.), *Audio Source Separation*, Signals and Communication
Technology, https://doi.org/10.1007/978-3-319-73031-8_3

we will how to incorporate such higher-level time dependencies for improving on source separation.

We will start by defining a probabilistic version of a non-negative sound model, which will then be extended for use in a convolutive model, a Hidden Markov model, and finally a more general dynamical model.

3.2 The PLCA Models

In order to take advantage of standardized dynamical modeling approaches we will examine Nonnegative Matrix Factorization-style models using a probabilistic formulation. A typical NMF model would look like this:

$$\mathbf{X} \approx \mathbf{W} \cdot \mathbf{H} \tag{3.1}$$

where $\mathbf{X}$ would be a matrix representing a magnitude spectrogram, and the two non-negative factors $\mathbf{W}$ and $\mathbf{H}$ will represent a set of components. We will say that the columns of $\mathbf{W}$ will contain a set of spectral bases, and the corresponding rows of $\mathbf{H}$ will contain their respective activations. We will refer to each basis/activation pair as a component, such a model where $\mathbf{W}$ has K columns implies that we have K components (and also implies that $\mathbf{H}$ has K rows).

We will now rewrite this equation as:

$$P(f,t) \approx \sum_z P(f|z)P(z)P(t|z) \tag{3.2}$$

Although it might be a little difficult to see at first, this is the same model as above, only this time we have reinterpreted all the non-negative values as probabilities. The input spectrogram, which was previously the non-negative matrix ($\mathbf{X}$) is now reinterpreted as $P(f,t)$ a distribution of acoustic energy over frequency (f) and time (t). The only difference is that now this quantity has to sum to 1 so that it is a proper distribution (since sounds are scale invariant this is not a complication in our representation).

Now for the approximation part, imagine for a moment that the quantify $P(z)$ did not exist. We would them have:

$$\sum_z P(f|z)P(t|z) \equiv \sum_k W_{i,k}H_{k,j} = \mathbf{W} \cdot \mathbf{H} \tag{3.3}$$

So we effectively implement a matrix multiplication, where the latent variable z is a component index, with $P(f|z)$ being equivalent to the matrix $\mathbf{W}$, and $P(t|z)$ being equivalent to $\mathbf{H}$.

However, since both $P(f|z)$ and $P(t|z)$ are distributions we need to ensure that their bases/activations properly sum to 1. This means that unlike NMF we won't have the ability to scale an entire component by scaling the corresponding column of **W** or row of **H**. In order to maintain that flexibility we add the term $P(z)$ which allows is to manipulate the relative level of each component.

An alternative way to interpret (3.2) would be as a matrix product of three factors:

$$P(f,t) = \sum_z P(f|z)P(z)P(t|z) = \mathbf{F} \cdot \mathrm{diag}(\mathbf{p}) \cdot \mathbf{T} \tag{3.4}$$

where the diag() operator creates a diagonal matrix from a vector. Using this viewpoint, PLCA looks like some sort of a non-negative SVD decomposition, two factors multiplied by a diagonal matrix in the middle.

This being a latent variable model, we can easily estimate its parameters using Expectation-Maximization [1]. The update equations would look as follows. For the E-step we would estimate the posterior distribution for each component:

$$P(f,t|z) = \frac{P(f|z)P(z)P(t|z)}{\sum_z' P(f|z')P(z')P(t|z')} \tag{3.5}$$

And for the M-step we would use that posterior to take a weighted average of the input to estimate the model parameters:

$$P(f|z) = \sum_t P(f,t)P(f,t|z) \tag{3.6}$$

$$P(t|z) = \sum_f P(f,t)P(f,t|z) \tag{3.7}$$

$$P(z) = \sum_{t,f} P(f,t)P(f,t|z) \tag{3.8}$$

More details and a detailed derivation of the model and its updates can be found at [2]. For now we will stop with this intuitive explanation and we will move on to showing how this model can facilitate temporal extensions.

3.3 Convolutional Models

The first model we will show is that takes into account consistent temporal structure by employing convolution. We will do so using the following reformulation of the PLCA model:

$$P(f,t) = \sum_z P(z) \sum_\tau P(f,\tau|z)P(t-\tau|z) \tag{3.9}$$

The difference in this model is that instead of taking a product of two factors we now perform a convolution between a set of two-dimensional bases $P(f,\tau|z)$ and

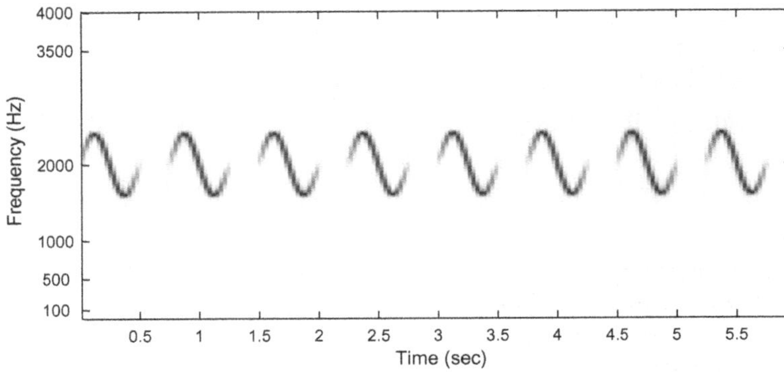

Fig. 3.1 A sound with a specific repetitive temporal structure

one-dimensional activations $P(t|z)$. We can envision $P(f, \tau|z)$ as being brief spectrograms defined over f and τ that get positioned over time using $P(t|z)$. Once again we have $P(z)$ to be able to express a relative scale of all the components.

To better understand this model lets consider the case where we have only one component. This means that the model will look like:

$$P(f, t) = \sum_{\tau} P(f, \tau|z)P(t|z) \tag{3.10}$$

This model would be appropriate for decomposing a spectrogram that has a specific sequence that repeats. As an illustration consider the toy input shown in Fig. 3.1. In it we see a simple sound repeating over time. This sound has a fixed temporal structure that repeats throughout. Suppose that we model that sound using the non-convolutive PLCA model. Asking for one component would result in the estimates shown in Fig. 3.2. Clearly this is a poor approximation of the input since a single spectral basis is not sufficient to approximate the constantly varying input. An 8-component model Fig. 3.3 fares much better, but the temporal structure of the input is lost in the details of this representation.

Applying a convolutive model (Fig. 3.4) produces a much more satisfying representation. The decomposition will be in terms of a time-frequency element that repeats over time. This is of course a better way to approximate the input, and in the results we see that this model has found the repeating element in $P(F, \tau|z)$ and makes use of $P(t|z)$ to position it at the right location.

Naturally, using more components can let us extract more interesting structure. Consider the case in Fig. 3.5. This is a drum loop made out of three sounds; a sweeping bass drum (the L-form pattern), a snare drum (the P-like pattern), and a cowbell (the harmonic sound with the strong midrange peak). During that loop the snare dum sound is never isolated (sounding once in unison with the bass drum, and once more with the cowbell). Despite the mixture, a 3-component PLCA analysis allows us to correctly identify the three sounds in the input.

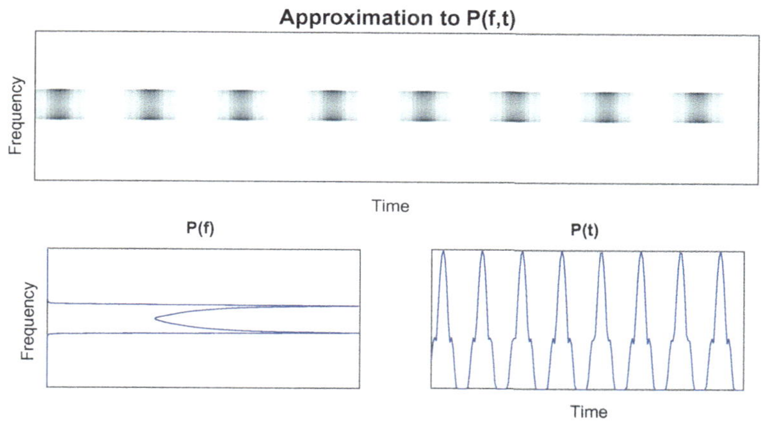

Fig. 3.2 1-component analysis of the input in Fig. 3.1. The single spectral basis in $P(f)$ is not sufficient to model the input well enough, smearing its temporal structure when approximating it

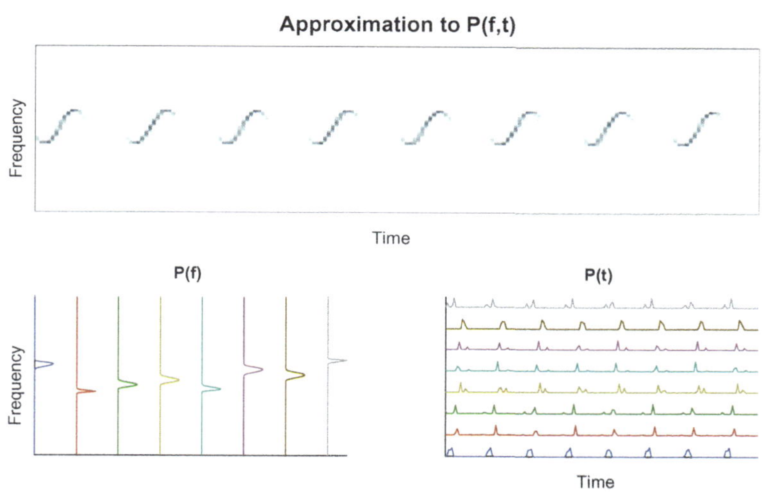

Fig. 3.3 8-component analysis of the input in Fig. 3.1. Using 8 components we can approximate the input much better than with one, but we do not gain any insight on its temporal structure, or the fact that this is the same repeating element

At this point it is worth noting that there is an inherent ambiguity in this model which we need to address. Since convolution is commutative, there is sometimes a possibility that temporal structure from $P(f, \tau|z)$ will be reflected in $P(t|z)$ (e.g. consider for example that $P(f, \tau|z)$ has a pattern that repeats twice, this repetition can also be represented as one instance of the pattern in $P(f, \tau|z)$ and the presence of two peaks in $P(t|z)$). In order to address this problem, we can use sparsity to force either $P(f, \tau|z)$ or $P(t|z)$ to be sparse, therefore allowing us to specify which of the

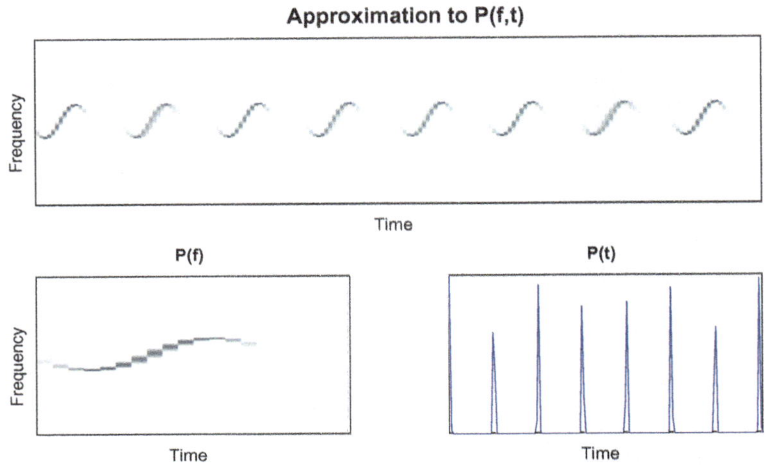

Fig. 3.4 1-component convolutive PLCA analysis of the input in Fig. 3.1. The single spectral basis in $P(f, \tau)$ is sufficient to model the input. Once convolved with $P(t|z)$ it approximates the input well. In addition to that we obtain a very useful representation that easily revelas the repetition in the input and its form

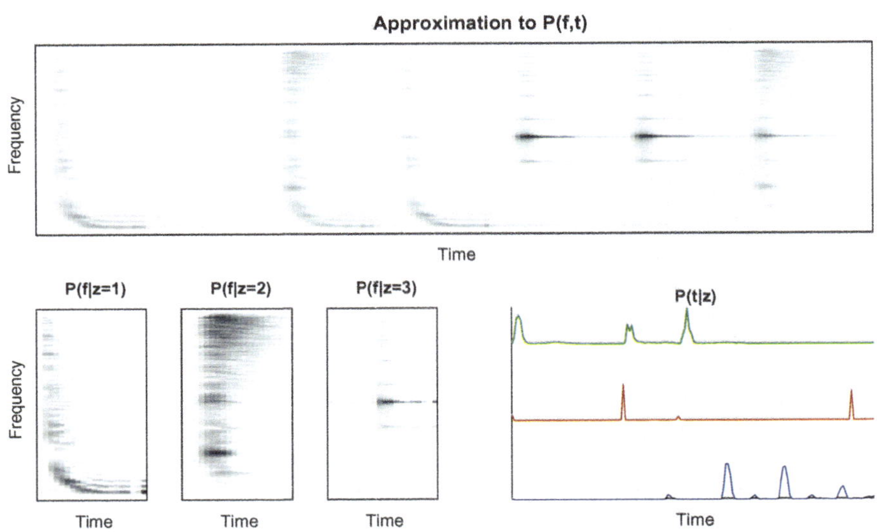

Fig. 3.5 3-component convolutive PLCA analysis of a simple drum loop with three different types of drums. Note that there is often overlap between the three time-frequency patterns that correspond to the three drums. Regardless, this analysis correctly identifies the patterns of the three drum sounds

two factors will have a denser structure. Since this issue is outside the general scope of this paper we refer the interested reader to the discussion in [2]. Likewise, we can use sparsity on $P(z)$ as a method to perform rank selection. For example, when we ask for many components but also request a sparse $P(z)$ the model will return many components with a zero prior, keeping only the components that are needed.

In general, this model is good for extracting fixed time/frequency patterns from sound mixtures, but it isn't as powerful for sounds that might not repeat so strictly. In order to address this case, which is much more realistic, we will develop two new models that allow us to model sounds better.

3.4 Non-negative Hidden Markov Models

A large class of audio signals, such as speech, exhibit a hidden structure in which each time frame corresponds to a discrete hidden state. Moreover, there is typically a relationship between the hidden states at different time frames, in the form of temporal dynamics. For example, each time frame of a speech signal corresponds to a subunit of speech such as a phoneme, which can be modeled as a distinct state. The subunits evolve over time as governed by temporal dynamics. Hidden Markov Models (HMMs) [3] have been used extensively to model such data.

A thread of literature [4–8] combines these ideas with NMF and PLCA to model non-negative data with such structure. In some techniques [4, 7], a state corresponds to a single dictionary element, while in others [5, 6, 8], called non-negative HMMs (N-HMMs), a state corresponds to an entire dictionary. The advantage the of the latter case is that each time frame can be modeled by a linear combination of a number of dictionary elements, which makes it more flexible than using a single dictionary element per state.

Since these models are based on an HMM structure, one can make use of the extensive theory of Markov chains to extend these models in various ways. For example, one can incorporate high level knowledge of a particular class of signals into the model, use higher order Markov chains, or use various natural language processing techniques. Language models were incorporated in this framework [9] as typically done in the speech recognition literature [3]. One could also incorporate other types of temporal structure like music theory rules when dealing with music signals.

The above techniques discuss how to model a single source using an HMM structure. However, in order to perform source separation, we need to model mixtures. This is typically done by combining the individual source models into a non-negative factorial HMM (N-FHMM). [4–6, 8, 10], which allows each source to be governed by a distinct pattern of temporal dynamics. One issue with this strategy is that the computational complexity of inference is exponential in the number of sources. This can be circumvented using approximate inference techniques such as variational inference [11], which makes the complexity linear in the number of sources.

3.4.1 Single Source Models

We start from PLCA and build our model from there. This is illustrated in Fig. 3.6. Consider an example of a piano piece that consists of multiple instances of four distinct notes. We can consider the spectrogram of this signal, to a reasonable level of approximation, to have four distinct patterns. If we use PLCA to learn a dictionary of four dictionary elements, it is likely that each note will closely correspond to a single dictionary element and therefore be primarily represented by a single spectral template. Although, this is a reasonable approximation, it will not be able to capture the variations between multiple instances of a given note.

If we use a larger dictionary, it will be able to better capture the variations and more subtle differences between notes because the dictionary will have more expressive power. However, if the dictionary is too large, it will be less specific to the given sound source and will be able to explain other sound sources as well. This will be problematic for source separation as the model is likely to not be sufficiently discriminative.

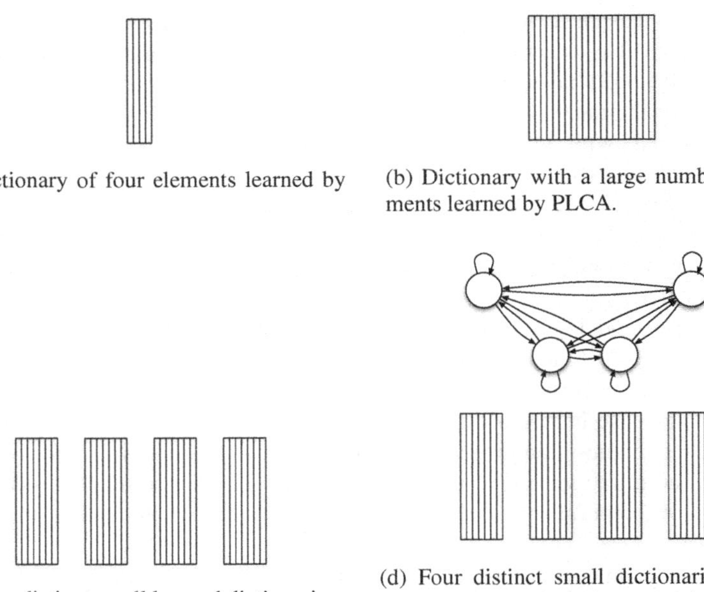

(a) Dictionary of four elements learned by PLCA.

(b) Dictionary with a large number of elements learned by PLCA.

(c) Four distinct small learned dictionaries.

(d) Four distinct small dictionaries and a Markov chain jointly learned using the N-HMM.

Fig. 3.6 Illustration of learned dictionaries. We start with a single small dictionary that is learned using PLCA and work up to several small dictionaries and a Markov chain jointly learned by the N-HMM. In the N-HMM, each dictionary corresponds to a state of the Markov chain. An ergodic (fully-connected) model has been shown for illustration purposes but any kind of Markov chain can be learned

Another approach is to model the spectrogram with multiple dictionaries such that each time frame of the spectrogram is primarily modeled by a single dictionary. We can model our piano example with four dictionaries, each of which has a number of dictionary elements. In this model, each time frame will be primarily explained by a linear combination of dictionary elements from the corresponding dictionary. This effectively creates block sparsity over all of the dictionary elements [12]. The advantage of this over using a single dictionary of four elements is that the variations between multiple instances of a given note can be better modeled since it will be modeled by a linear combination of a number of dictionary elements, as opposed to a single dictionary element (spectral template). The advantage of this over a single large dictionary is that since only a single dictionary is primarily active in a given time frame, each time frame is modeled by a limited number of dictionary elements. This will prevent the dictionary elements from being overly general, which helps during source separation.

With the N-HMM, we also model the temporal dynamics between dictionaries. This is to say that when a given time frame is explained by a given dictionary, we learn a transition matrix or a probability distribution that tells us how likely the next time frame is explained by each of the dictionaries. In this sense each state of the N-HMM has a one to one correspondence to a dictionary.

Speech is a natural candidate to be modeled by the N-HMM since it tends to have distinct spectral patterns (i.e. phonemes), with some amount of variation between multiple instances of a given patterns. It also has distinct temporal patterns that can be reasonably well explained by temporal dynamics. A subset of the dictionaries learned by an N-HMM from training data of a given speaker are shown in Fig. 3.7.

We now briefly describe the model of the N-HMM starting with PLCA. In PLCA, a random variable Z is used denote the dictionary element, which is defined by the multinomial distribution $P(f|z)$. Each dictionary element is analogous to a column of the $\mathbf{W}$ matrix in NMF. Since the N-HMM has multiple dictionaries, we introduce

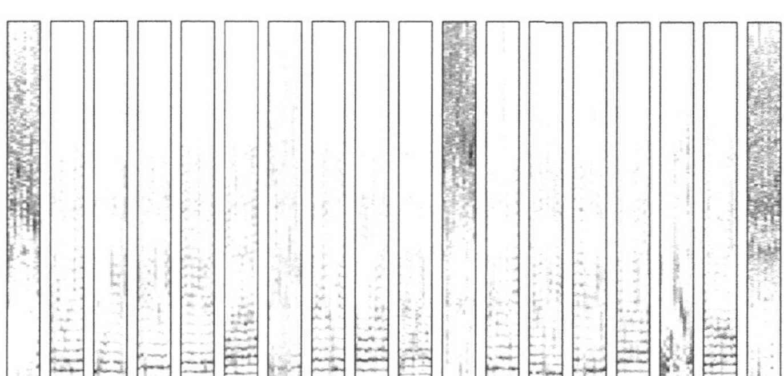

Fig. 3.7 Dictionaries were learned from speech data of a given speaker. Shown are the dictionaries learned for 18 of the 40 states. Each dictionary is comprised of 10 elements that are stacked next to each other. Each of these dictionaries roughly correspond to a subunit of speech, either a voiced or unvoiced phoneme

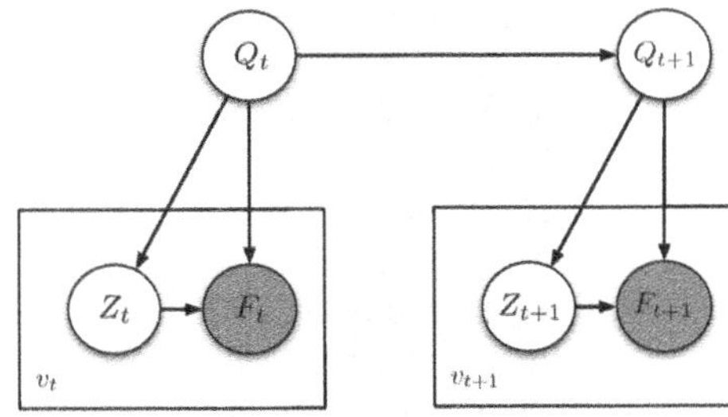

Fig. 3.8 Graphical model of the N-HMM

another random variable Q to denote the dictionary (state). Dictionary element z from dictionary q is therefore represented by a multinomial distribution $P(f|z, q)$.

In a given time frame t, we have a distribution of mixture weights $P(z_t|q_t)$ for each state q_t at that time.

The N-HMM has a transition matrix defined by multinomial distributions $P(q_{t+1}|q_t)$. These distributions together define the temporal dynamics of the model. The prior $P(q_1)$ defines a distribution over states at the first time frame. These distributions are standard HMM distributions [3].

Finally, each state has a Gaussian energy distribution $P(v|q)$. For a given state q, this provides a distribution over the number of counts over all frequency bins when that state is used. This intuitively corresponds to the range of observed loudness of each state.

The graphical model of the N-HMM is shown in Fig. 3.8. The energy distribution is not explicitly shown in the figure, but it is implicit as the number of draws v_t at time t is determined by this distribution.

The N-HMM can be learned from the spectrogram of training data of a given sound source. A detailed derivation of parameter estimation for this model can be found in [13]. The dictionaries, transition matrix, prior probabilities, and energy distributions are characteristic of the source. However, the mixture weights are characteristic of the given instance of the source (i.e. training data), but do not generalize to other instances of that source. Therefore when an N-HMM of a given source is learned, the mixture weights are discarded and all other distributions are retained. These distributions can be used for source separation as shown in the next subsection.

3.4.2 Source Separation

Once N-HMMs for sound sources are learned from training data, they can be combined into a non-negative factorial HMM (N-FHMM) [5] and be used for source separation. The graphical model of the N-FHMM is shown in Fig. 3.9. Comparing this to Fig. 3.8, we can roughly see the graphical models of two individual N-HMMs (one on the top and one of the bottom. The observed variables F_t and the hidden

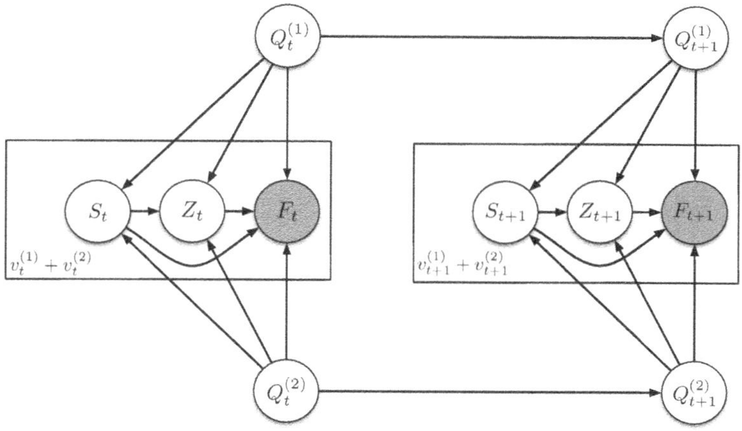

Fig. 3.9 Graphical model of the N-FHMM

variables Z_t are common to both sources. The new hidden variable S_t corresponds to the relative energies of the two sources. The model for two sources is shown to illustrate the concept, but it can be extended to more sources.

Consider an example in which the goal is to separate two sound sources from the mixture spectrogram. We first learn the N-HMMs of each source (all distributions except for the mixture weights) from isolated training data of those sources. We then combine these learned distributions into an N-FHMM. The parameters of all distributions of the N-FHMM except $P(s_t|q_t^{(1)}, q_t^{(2)})$ and $P(z_t|s_t, q_t^{(s)})$ will therefore be fixed. These two distributions can be combined into a single distribution $P(z_t, s_t|q_t^{(1)}, q_t^{(2)})$. Intuitively, this corresponds to distributions of mixture weights over both sources. We perform parameter estimation to estimate these mixture weights. A detailed derivation can be found in [13]. Given the mixture weights and the known parameters, we can reconstruct the spectrogram of each source. We then perform Wiener filtering to obtain the time domain signal of each source. Since we learn N-HMMs from each source in the mixture, this is supervised source separation.

If we have training data for all but one source, we can perform semi-supervised source separation [6]. A common application of this is denoising. If we have training data for speech of a given speaker, we can learn an N-HMM for that speaker. This can be combined into an N-FHMM in which the second source has a single dictionary that is learned during separation. This could be used in different instances in which the second source is for example, different kinds of noise.

Modeling of mixtures using the N-FHMM is illustrated in Fig. 3.10. In this example, we have two sources. In the case of supervised separation, each source has two dictionaries. As shown there are four possible combinations in which they can be combined. Each time frame corresponds to one of these four combinations. Due to the number of possible combinations of dictionaries, the complexity of inference and therefore parameter estimation is exponential in the number of sources. We can

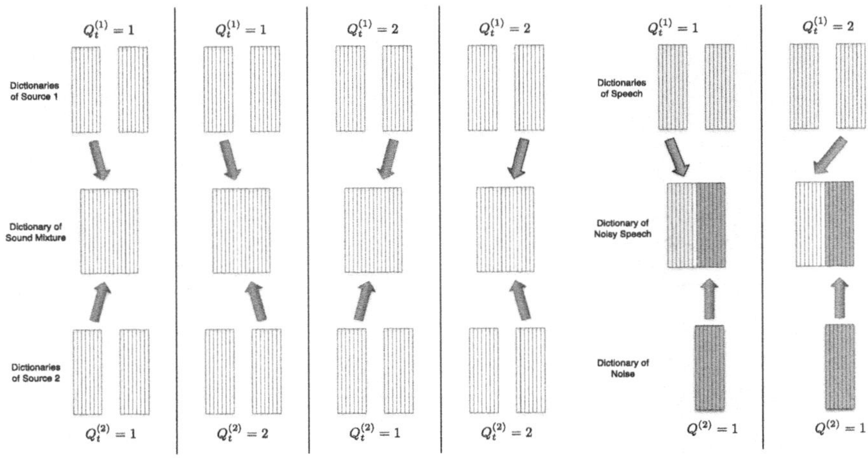

(a) Supervised separation.

(b) Semi-supervised separation.

Fig. 3.10 Illustration of dictionary configurations when using the N-FHMM for source separation

therefore use variational inference to efficiently perform separation [11] with a complexity that is linear in the number of sources. In the case of semi-supervised separation with two sources, the complexity of exact inference is is linear in the number of sources since one source always has a single dictionary (state).

3.4.3 Illustrative Examples

We illustrate source separation using N-FHMMs with a few examples. Our first example, shown in Fig. 3.11, is a toy example to illustrate supervised source separation. In this example, we used synthesized saxophone and the input representation is the Constant-Q transform. Both sources are the exact same notes from the exact same synthesized saxophone. The only difference between the two sources is the sequence of notes. It is therefore not possible to disambiguate the two sources without some sort of source specific temporal information. As shown, this is why PLCA does a poor job of separation and both separated outputs look largely like the input mixture. On the other hand, the N-FHMM has learned the temporal dynamics of each source and is able to preform a fairly clean separation of the two sources.

(a) Original Sources.

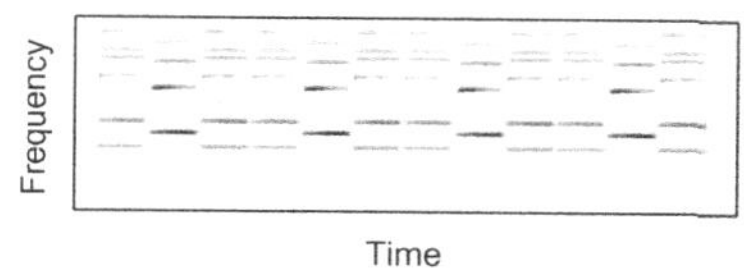

(b) Sound Mixture.

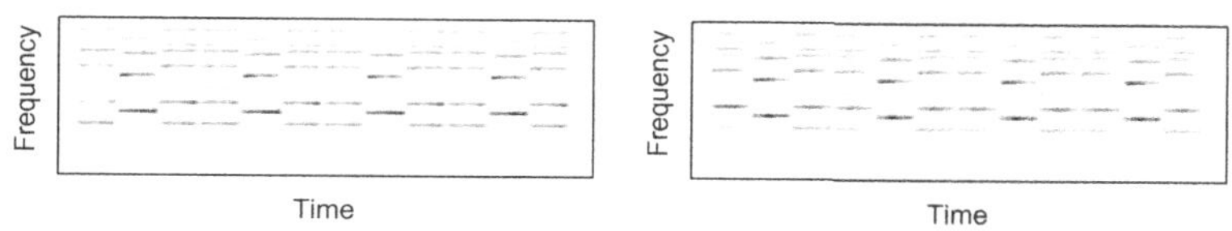

(c) Separated Sources using PLCA.

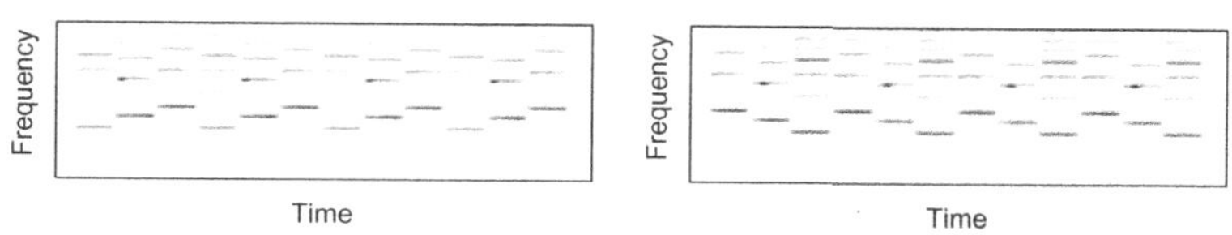

(d) Separated Sources using the N-FHMM..

Fig. 3.11 Comparison of supervised source separation using PLCA and the N-FHMM. The first source is an ascending arpeggio played by a synthesized saxophone. The second source is a descending arpeggio on the same octave played by the same synthesized saxophone

In our second example, shown in Fig. 3.12, we show how temporal dynamics helps perform semi-supervised separation on toy data. The task is to separate synthetic data, which we call the target, from noise using semi-supervised source separation. As shown, the target can be well modeled by two states. In this example, the SNR is low to the point that it is difficult to visually distinguish the target from the noise. The N-FHMM is able to use temporal dynamics to help disambiguate the sources, leading to a significantly higher quality separation than using PLCA.

In Fig. 3.13, we show another example comparing semi-supervised source separation using the N-FHMM and PLCA. In this example, the goal is to separate speech from noise. As shown, the N-FHMM achieves a greater degree of noise suppression.

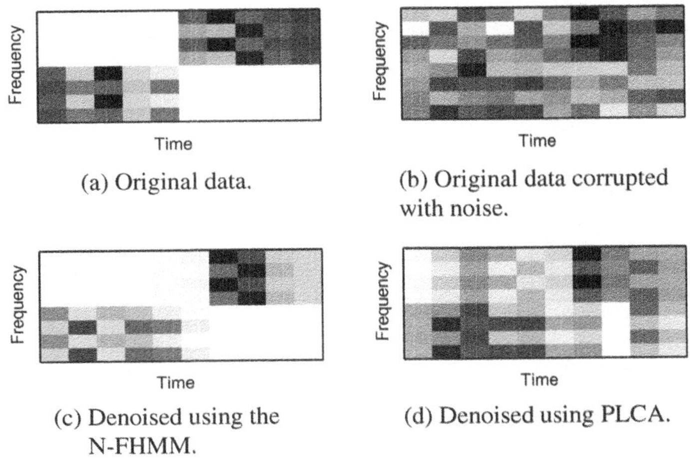

Fig. 3.12 Comparison of semi-supervised source separation using PLCA and the N-FHMM. The noise source is uniformly distributed random noise

Fig. 3.13 Illustration of speech denoising using the N-FHMM and PLCA. The noise source is ambient noise in an airport

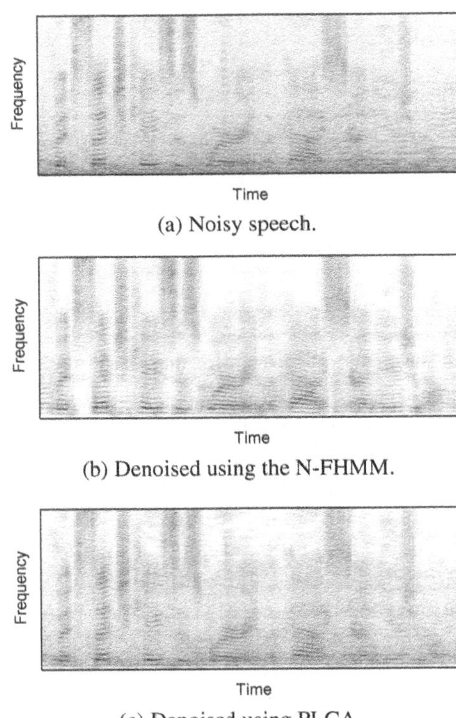

3.5 Dynamic PLCA Using Continuous State-Space Representation

In this section, a method based on continuous state-space representation is presented to use temporal dependencies in NMF [14]. We assume that the NMF coefficients are stochastic processes, and that they evolve through a vector autoregressive (VAR) model over time. Therefore, in addition to the basis matrix, there will be some regression parameters associated with each signal. The proposed method has two steps: firstly, we predict the current NMF coefficients given only the past or both past and future observations, and secondly, the estimates are updated given the current observation. A multiplicative correction of the estimates are used in the second step. The proposed scheme introduces a new way of thinking and although it is quite simple it results in a significant improvement over the baseline PLCA, as will be shown using examples. A more rigorous extension of this idea has also been proposed in [15], where the optimal update rules to estimate the activation vectors as well as the autoregressive coefficient matrices are derived.

3.5.1 Model Definitions

Let us start with representing PLCA using matrix multiplication. PLCA approximates the normalized magnitude spectrogram of the speech $\mathbf{V}$ with elements $v(f,t)$ as: $\mathbf{V} \approx \mathbf{WH}$ where the basis matrix $\mathbf{W}$ and the activation matrix $\mathbf{H}$ are defined as follows:

$$
\begin{aligned}
w_f(z) &= P\left(V(f,t) = v(f,t) \,|H(t) = z\right) \\
h_z(t) &= P\left(H(t) = z\right),
\end{aligned}
$$

where $H(t)$ is the scalar indicator random variable that can take one of the integer values $1, \ldots Z$. Now we have the two equivalent representations for PLCA [14]:

$$
\begin{aligned}
v(f,t) &\approx \sum_{H(t)=1}^{Z} P\left(V(f,t) = v(f,t) \,|H(t) = z\right) P\left(H(t) = z\right) \\
&= \sum_{z=1}^{Z} w_f(z)\, h_z(t). \qquad (3.11)
\end{aligned}
$$

We assume that the activation vectors are modeled by a T' order VAR model as:

$$
\mathbf{h}(t) = \sum_{t'=1}^{T'} \mathbf{D}\left(t'\right)\mathbf{h}\left(t - T'\right) + \boldsymbol{\sigma}(t), \qquad (3.12)
$$

$$
\mathbf{v}(t) = \mathbf{Wh}(t) + \boldsymbol{\varepsilon}(t), \qquad (3.13)
$$

where $D(t')$ is the $Z \times Z$ autoregressive coefficient matrix associated at t'-th lag, $\boldsymbol{\sigma}(t)$ is the process noise, and $\boldsymbol{\varepsilon}(t)$ is the observation noise in the model.

3.5.2 Estimation Methods

In the following, we assume that the basis matrix $\mathbf{W}$ is estimated using some training data and the PLCA update rules and is kept fixed when estimating the activation vectors $\mathbf{h}(t)$. Two approaches are presented in this section to estimate the activation vectors, a filtering and a smoothing approaches. The goal of the filtering approach is to develop an online algorithm to estimate an activation vector $\mathbf{h}(t)$ given all the current and past spectral vectors denoted by $\mathbf{v}_1^t = \{\mathbf{v}(1), \ldots \mathbf{v}(t)\}$. This approach has a prediction step and an update step, similar to a Kalman filtering. The prediction of the activation vector $\mathbf{h}(t)$, given $\mathbf{v}_1^{t-1}$, is denoted by $\hat{\mathbf{h}}(t|t-1)$ and is simply obtained as:

$$\hat{\mathbf{h}}\left(t|t-1\right) = \sum_{t'=1}^{T'} \mathbf{D}\left(t'\right) \hat{\mathbf{h}}\left(t-t'|t-t'\right), \tag{3.14}$$

where $\hat{\mathbf{h}}\left(t-t'|t-t'\right)$ is the updated estimate of $\mathbf{h}(t-t')$ given $\mathbf{v}_1^{t-t'}$. This estimate is corrected using a correction term; to obtain this correction term, the PLCA update rule is applied on $\mathbf{v}(t)$ to find $\tilde{\mathbf{h}}_t$ while initializing the iterative rule at $\hat{\mathbf{h}}\left(t|t-1\right)$). Then, the updated estimate of $\mathbf{h}(t)$ is obtained as:

$$\hat{\mathbf{h}}\left(t|t\right) = \frac{\left(\hat{\mathbf{h}}\left(t|t-1\right)\right)^{\beta} \tilde{\mathbf{h}}\left(t\right)}{\sum \left(\hat{\mathbf{h}}\left(t|t-1\right)\right)^{\beta} \tilde{\mathbf{h}}\left(t\right)}, \tag{3.15}$$

where $(\cdot)^{\beta}$ is an element-wise power operator, β is the prior strength and might be taken different than one, and the normalization is performed to ensure that $\hat{\mathbf{h}}\left(t|t\right)$ is a proper probability vector. The multiplicative update in (3.15) is similar to the forward algorithm in a hidden Markov model (HMM) where the observation likelihood is replaced with $\tilde{\mathbf{h}}(t)$. Therefore, $\hat{\mathbf{h}}\left(t|t\right)$ can be also seen as the posterior probability of the latent variables (hidden states in HMM). Note that there are three different estimations for the activation vector: (1) the instantaneous estimate $\tilde{\mathbf{h}}\left(t\right)$, obtained using the baseline PLCA, (2) forward-predicted estimate $\hat{\mathbf{h}}\left(t|t-1\right)$, and (3) final estimate $\hat{\mathbf{h}}\left(t|t\right)$, obtained by combining both 1 and 2, which will be used in the real applications when we have access to only past data.

The filtering method explained above does not use any future data to refine the estimate of $\mathbf{h}(t)$. The smoothing problem arises when we have observed both past and future data, and we want to estimate the activation vectors $\mathbf{h}(t)$, which are denoted by $\hat{\mathbf{h}}(t|T)$. For this purpose, first the PLCA algorithm is applied on the magnitude spectrogram $\mathbf{V}$ to obtain the activation matrix $\tilde{\mathbf{H}}$, i.e. $\mathbf{V} \approx \mathbf{W}\tilde{\mathbf{H}}$. Then, a forward prediction matrix with columns given by $\hat{\mathbf{h}}(t|t-1)$, and a backward prediction matrix with columns given by $\overline{\mathbf{h}}(t|T)$ are obtained as:

$$\hat{\mathbf{h}}_{t|t-1} = \sum_{t'=1}^{T'} \mathbf{D}\left(t'\right) \tilde{\mathbf{h}}\left(t-t'\right), \tag{3.16}$$

$$\overline{\mathbf{h}}(t|T) = \sum_{t'=1}^{T'} \mathbf{D}^{\top}\left(t'\right) \tilde{\mathbf{h}}\left(t+t'\right). \tag{3.17}$$

In principle, to evaluate (3.16) and (3.17) it suffices to have access to data from $t - T'$ through $t + T'$. Therefore, the algorithm will introduce a delay of T' short time frames. Now we can obtain the final estimate of $\mathbf{h}(t)$ as below:

$$\hat{\mathbf{h}}\left(t|T\right) = \frac{\left(\overline{\mathbf{h}}(t|T)\hat{\mathbf{h}}\left(t|t-1\right)\right)^{\beta} \tilde{\mathbf{h}}\left(t\right)}{\sum \left(\overline{\mathbf{h}}(t|T)\hat{\mathbf{h}}\left(t|t-1\right)\right)^{\beta} \tilde{\mathbf{h}}\left(t\right)}. \tag{3.18}$$

Note that there are four different estimations for the activation matrix: (1) the instantaneous estimate $\tilde{\mathbf{h}}\left(t\right)$, obtained using the baseline static PLCA, (2) forward-predicted estimate $\hat{\mathbf{h}}\left(t|t-1\right)$, (3) backward-predicted estimate $\overline{\mathbf{h}}(t|T)$, and (4) final estimate $\hat{\mathbf{h}}\left(t|T\right)$ combining 1, 2, and 3, which will be used in the enhancement or separation applications when we have access to both past and future data.

The VAR coefficients $\mathbf{D}(t'), t' = 1, \ldots T'$, can be estimated in different ways, e.g. [16, ch. 11]. Using a sub-optimal approach, as explained here, was also found sufficiently good in practice. Let $\mathbf{H}^{(t')}$ denote the matrix $\mathbf{H}$, in which the columns are shifted by t', i.e. $h_z^{(t')}(t) = h_z(t+t')$. Then, $\mathbf{D}(t')$ is estimated as: $\mathbf{D}(t') = \mathbf{H}^{(t')}\mathbf{H}^{\top}$ where $\top$ represents the matrix transpose. The columns of $\mathbf{D}(t')$ are finally normalized to sum to one.

Finally, to separate unknown sources from a given mixture, we can learn the basis matrices and VAR coefficients matrices for all the involved sources offline, similar to the static PLCA, and then concatenate them properly to explain the mixed signal. Then, we can use (3.15) or (3.18) to estimate the activation vector $\mathbf{h}(t)$ and consequently use a Wiener-type filter to estimate the source signals. Let $\mathbf{x}(t) = \sum_k \mathbf{s}^k(t)$ be the observed mixture, where $\mathbf{s}^k(t)$ represents the t-th column of the k-th source. Each source is estimated using a Wiener-type filter as:

$$\hat{\mathbf{s}}^k\left(t\right) = \frac{\mathbf{W}^k\mathbf{h}^k\left(t\right)}{\sum_k \mathbf{W}^k\mathbf{h}^k\left(t\right)}\mathbf{x}\left(t\right), \tag{3.19}$$

where $\mathbf{x}(t)$ is the magnitude spectrogram of the mixture and $\mathbf{W}^k$ is the basis matrix of the k-th source, and $\mathbf{h}^k(t)$ is a part of $\mathbf{h}$(t) that is associated with this source. The separated/enhanced time-domain signals are obtained using the phase of the mixed input signal.

3.5.3 Illustrative Examples

As the first example, we consider a signal separation task with artificially generated signals, see Fig. 3.14. We generated one second of a two-tone sinusoidal signal with incremental frequencies over time as the waveform of the first source. The second source was generated as the time-reversed version of the first signal. Two sources were summed to obtain the mixture signal. Discrete Fourier transform (DFT) with a frame length of 128 ms and 75% overlapped windows using a Hann window was applied to obtain the magnitude spectrogram of the signals as the input to the PLCA algorithms. We learned 20 basis vectors for each source, which were kept fixed during the separation. Since the basis matrices for both of the sources are identical, a static PLCA algorithm can not segregate the input sources, and two extracted sources will be very similar to the mixture. The smoothing algorithm (3.18) with $T' = 4$ and $\beta = 1$ was applied to extract the sources, and as Fig. 3.14 shows a good separation is achieved, and the estimated spectrograms are very similar to the original ones.

As the second example, we applied the smoothing algorithm (3.18) to a mixed signal where the mixture was obtained as the sum of a temporally structured speech signal (see Fig. 3.15) and its time-reverse version. For this example, the sampling rate was 8 kHz, and a frame length of 128 ms with 75% overlap was used in computing the DFT. As mentioned before, since the basis matrices for two source signals are very similar, the static PLCA algorithm can not separate the sources. Bottom panels of Fig. 3.15 show the separated spectrograms, which are obtained using 60 trained basis vectors for each source with parameters set to $T' = 4$, $\beta = 1$. This experiment verifies the benefit of temporal modeling in a difficult separation task. The separation performance in this case is around 11 dB improvement in source to distortion ratio (SDR) [17], while the baseline PLCA fails to separate the sources.

Now we consider a more real problem of noise reduction application where the desired speech signal is corrupted by an additive noise at 0 dB input SNR. A speaker-dependent approach is followed here in which a separate basis matrix is trained for each speaker and each noise type beforehand. The experiment was done for 100 randomly chosen speakers with different genders from the TIMIT database, where 9 out of the 10 available sentences were used for training speech model, and the other sentence was used for test purposes. The denoising algorithms were evaluated for babble and factory noises taken from NOISEX-92 database. All the signals were down-sampled to 16 kHz. The frame length and overlap length in the DFT analysis were set to 64 and 60 ms, respectively. We trained 60 basis vectors for speech while

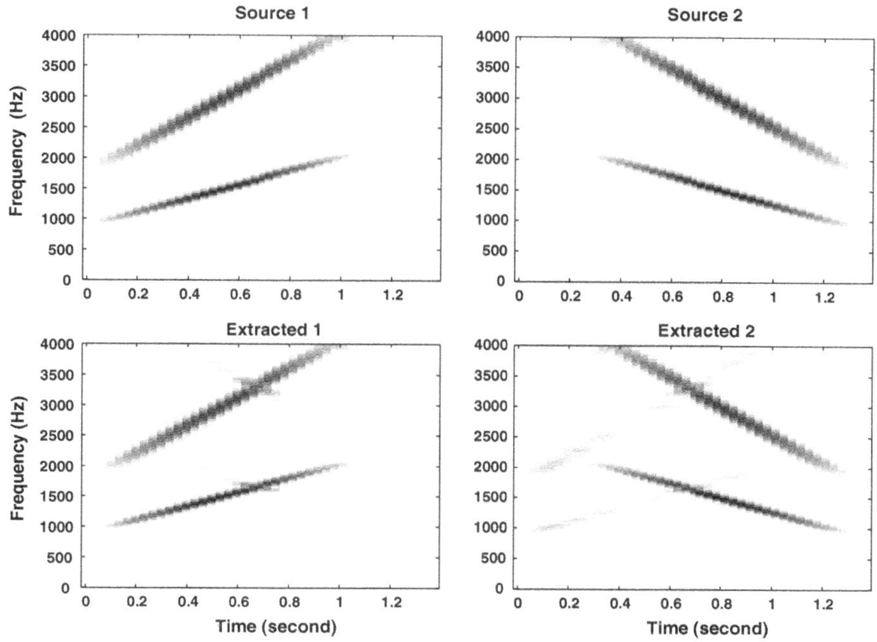

Fig. 3.14 An artificial example: the magnitude spectrograms of the original (top) and extracted signals using the smoothing algorithm (bottom)

for babble and factory noises 20 and 30 basis vectors were learned, respectively. Since speech and noise signals have different temporal characteristics, it is preferred to use different powers (β) in (3.18) for speech (β_{speech}) and noise (β_{noise}) coefficients, which are set experimentally.[1] The performance is measured using SDR, SIR and SAR [17]. We also evaluated the perceptual quality of the enhanced speech using PESQ [18].

Table 3.1 shows the results, where it can be seen that applying the temporal dynamics has increased SIR while reducing SAR compared to the baseline PLCA. Nevertheless, the SDR that gives an indication of the overall quality of the speech has increased significantly for both noise types. In fact, the algorithms have led to a fair trade off between the removing noise and introducing artifacts in the enhanced signal. The PESQ values also confirm a very good quality improvement using the proposed algorithms. Specifically in the case of the factory noise and using the smoothing algorithm, PESQ is improved by 0.46 MOS compared to the baseline. Additionally, the evaluation shows that the smoothing algorithm has produced slightly better SDR and PESQ values than the filtering approach.

[1] In this experiment, we have used $M = 1$, $\beta_{speech} = 0.5$, $\beta_{noise} = 0.2$ for filtering, and $\beta_{speech} = 0.9$, $\beta_{noise} = 0.6$ for smoothing.

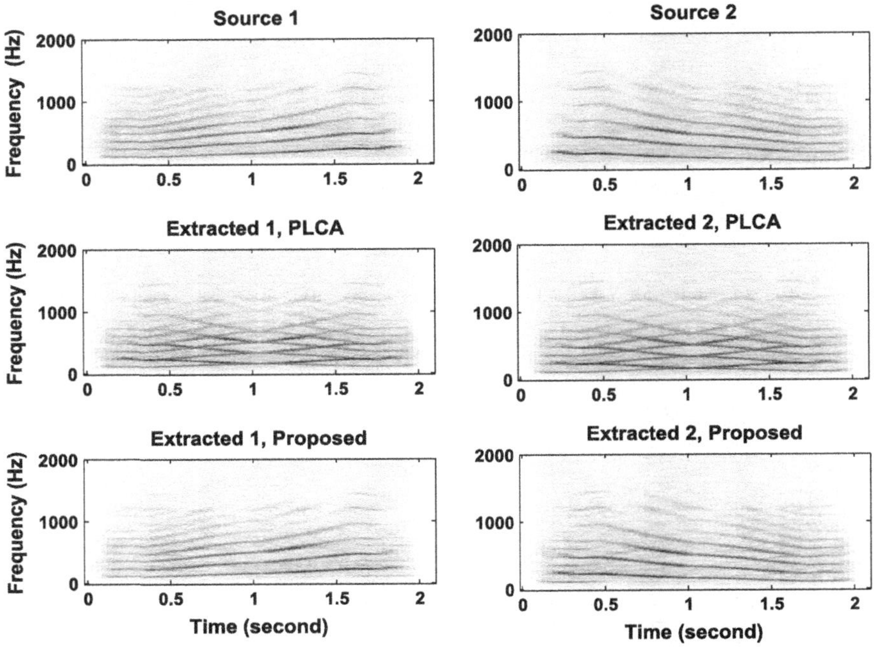

Fig. 3.15 Magnitude spectrogram of the input (top) and separated sources using the static PLCA algorithm (middle) and separated sources using the dynamic PLCA (smoothing algorithm, top). For legibility reasons we only show the frequency range 0–2 kHz

Table 3.1 Results for a denoising problem in the presence of added factory noise at 0 dB input SNR

Algorithm	SDR (dB)	SIR (dB)	SAR (dB)	PESQ (MOS)
Baseline PLCA	3.7	5	**11**	1.79
Filtering	6.7	12	8.5	2.15
Smoothing	**6.9**	**14.7**	7.8	**2.25**

Finally, let us consider the smoothing approach (3.18) applied to the babble case, and study the effect of the model order (M) and prior strength (β) on the performance. Figure 3.16 shows three objective measures as functions of the model order ($M = 1, 2, 3, 4$), and noise prior strength (β_{noise}) while $\beta_{\text{speech}} = 0.9$. As it can be seen in the figure, increasing model order from 1 to 4 has not changed the peak performance, however, it has made the algorithm more robust to the value of β_{noise}. Also, the previously used $\beta_{\text{noise}} = 0.6$ falls into the optimal range of β_{noise}.

It has to be mentioned that adaption of the presented method in this section to a other types of NMF formulations is straightforward. As shown using different examples, the presented method is able to effectively capture the temporal dependencies

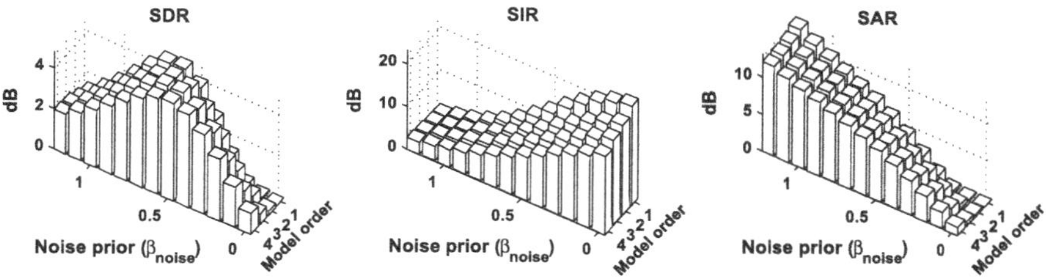

Fig. 3.16 Effect of the VAR model order and noise prior strength on the performance of speech denoising using the smoothing algorithm

and results in substantial improvement in the speech quality in the real applications. Noticeably, it is shown that the method can lead to satisfactory results in source separation even when the basis matrices of two underlying sources are identical. This case is an example where a basic NMF can not separate the sources at all.

3.6 Conclusions

In this chapter we discussed a variety of non-negative models that take advantage of temporal dependencies in order to achieve better source separation performance. We have shown three distinct models. First, we derived a static convolutional model that is well-suited for discovering and extracting sources that exhibit consistent temporal structure, in which non-negative components were defined as having both a frequency and temporal dimension. Secondly, we have shown a hidden Markov non-negative model, which uses a non-negative state model that models states in a manner that supports source separation Using this model, we can incorporate state transition information when extracting a source, and take advantage of temporal consistencies that we can find in, e.g. language, or musical structure. Finally, we have shown a non-negative dynamical model, which models lower-level temporal dependencies than the non-negative HMM, but in a manner that is not as rigid as the convolutional model, thereby being more flexible and applicable for modeling dynamic sources.

It is important to note, that none of these models is better than the others. They are all designed to model different temporal attributed of sounds, and depending on the deployment situation either might be the best choice. As a rule of thumb, the convolutional model is best at extracting sources with consistent temporal structure (e.g. drums, or synthetic sounds that repeat verbatim), the Markov model is better when we have higher-level temporal structure (e.g. in the form of a language—such as in speech or music), and the dynamical model is best at describing sounds that exhibit a more stochastic temporal structure.

Using these models as a starting point, one can also derive more elaborate versions, e.g. convolutional models that employ two-dimensional convolution, thereby learning invariant structures over the frequency space, more general Markov field methods that generalize the non-negative HMM, or even models that combine the above approaches, e.g. a non-negative HMM model that employs a convolutional state model. The space of dynamical models can certainly be very broad, and we expect that the material in this chapter can help you get started with this exciting area of sound modeling.

References

1. A.P. Dempster, N.M. Laird, D.B. Rubin, Maximum likelihood from incomplete data via the EM algorithm. J. Roy. Stat. Soc. **39**(1), 1–38 (1977)
2. P. Smaragdis, B. Raj, Shift-invariant probabilistic latent component analysis. Technical Report TR2007-009 (Mitsubishi Electric Research Labs, 2007)
3. L.R. Rabiner, A tutorial on hidden Markov models and selected applications in speech recognition. Proc. IEEE **77**(2), 257–286 (1989)
4. A. Ozerov, C. Févotte, M. Charbit, Factorial scaled hidden Markov model for polyphonic audio representation and source separation in *Proceedings of IEEE Workshop Applications of Signal Processings Audio Acoustics (WASPAA)* (2009) pp. 121–124
5. G.J. Mysore, P. Smaragdis, B. Raj, Non-negative hidden Markov modeling of audio with application to source separation in *Proceedings of the International Conference Latent Variable Analysis and Signal Separation (LVA/ICA)* (2010) pp. 140–148
6. G.J. Mysore P. Smaragdis, A non-negative approach to semi-supervised separation of speech from noise with the use of temporal dynamics in *Proceedings of the IEEE International Conference Acoustics, Speech, and Signal Processing (ICASSP)* (2011)
7. M. Nakano, J.L. Roux, H. Kameoka, Y. Kitano, N. Ono, S. Sagayama, Nonnegative matrix factorization with Markov-chained bases for modeling time-varying patterns in music spectrograms in *Proceedings of the International Conference, Latent Variable Analysis and Signal Separation (LVA/ICA)* (2010)
8. N. Mohammadiha, A. Leijon, Nonnegative hmm for babble noise derived from speech hmm: application to speech enhancement. IEEE Trans. Audio Speech Lang. Process. **21**(5), 998–1011 (2013)
9. G.J. Mysore, P. Smaragdis, A non-negative approach to language informed speech separation in *Proceedings of the International Conference Latent Variable Analysis and Signal Separation (LVA/ICA)* (2012)
10. M. Nakano, J. Le Roux, H. Kameoka, T. Nakamura, N. Ono, S. Sagayama, Bayesian nonparametric spectrogram modeling based on infinite factorial infinite hidden Markov model in *Proceedings of IEEE Workshop Applications of Signal Processing Audio Acoustics (WASPAA)* (2011)
11. G.J. Mysore, M. Sahani, Variational inference in non-negative factorial hidden Markov models for efficient audio source separation in *Proceedings of the International Conference, Machine Learning (ICML)* (2012)
12. G.J. Mysore, A block sparsity approach to multiple dictionary learning for audio modeling in *Proceedings of the International Conference, Machine Learning (ICML)* (2012)
13. G.J. Mysore, A non-negative framework for joint modeling of spectral structure and temporal dynamics in sound mixtures. Ph.D. Dissertation, Stanford University, 2010
14. N. Mohammadiha, P. Smaragdis, A. Leijon, Prediction based filtering and smoothing to exploit temporal dependencies in NMF in *Proceedings of IEEE International Conference Acoustics, Speech, and Signal Processing (ICASSP)* (2013) pp. 873–877
15. N. Mohammadiha, P. Smaragdis, G. Panahandeh, S. Doclo, A state-space approach to dynamic nonnegative matrix factorization. IEEE Trans. Signal Process. **63**(4), 949–959 (2015)
16. J.D. Hamilton, *Time Series Analysis* (Princeton University Press, New Jersey, 1994)

17. E. Vincent, R. Gribonval, C. Févotte, Performance measurement in blind audio source separation. IEEE Trans. Audio Speech Lang. Process. **14**(4), 1462–1469 (2006)
18. ITU-T. P.862, Perceptual evaluation of speech quality (PESQ), and objective method for end-to-end speech quality assesment of narrowband telephone networks and speech codecs. Technical Report (2000)

Chapter 4
An Introduction to Multichannel NMF for Audio Source Separation

Alexey Ozerov, Cédric Févotte and Emmanuel Vincent

Abstract This chapter introduces multichannel nonnegative matrix factorization (NMF) methods for audio source separation. All the methods and some of their extensions are introduced within a more general local Gaussian modeling (LGM) framework. These methods are very attractive since allow combining spatial and spectral cues in a joint and principal way, but also are natural extensions and generalizations of many single-channel NMF-based methods to the multichannel case. The chapter introduces the spectral (NMF-based) and spatial models, as well as the way to combine them within the LGM framework. Model estimation criteria and algorithms are described as well, while going deeper into details of some of them.

4.1 Introduction

Nonnegative matrix factorisation (NMF) [1] is a dimensionality reduction technique that consists in approximating a nonnegative data matrix (a matrix with nonnegative entries) as a product of two nonnegative matrices of lower rank than the initial data matrix. This also can be viewed as an approximation of data matrix as a sum of few rank-1 nonnegative matrices. It was first successfully applied for single-channel source separation [2], where the nonnegative matrix of magnitude or power spectrogram is decomposed, and became a state of the art reference. The success of this method is mainly due to universality of this quite simple modeling (it is applicable to various types of audio sources including speech [3, 4], music [2, 5],

A. Ozerov (✉)
Technicolor, Rennes, France
e-mail: Alexey.Ozerov@technicolor.com

C. Févotte
CNRS & IRIT, Toulouse, France
e-mail: cedric.fevotte@irit.fr

E. Vincent
Inria, 54600 Villers-lès-Nancy, France

S. Makino (ed.), *Audio Source Separation*, Signals and Communication Technology, https://doi.org/10.1007/978-3-319-73031-8_4

environmental sounds [6], etc.) and due to the flexibility of this modeling allowing adding various constraints to it, such as for example harmonicity of spectral patterns [7], smoothness of their activation coefficients [2, 5], pre-trained spectral patterns [8, 9], etc.

Given the success of the NMF for single-channel source separation, there were several attempts to extend it to the case of multichannel source separation. Earlier ideas were relying on stacking magnitude or power spectrograms of all channels into a 3-valence nonnegative tensor and decomposing it with nonnegative tensor factorisation (NTF) methods [10] or other NTF-like nonnegative structured approximations [11, 12]. This gave some interesting results. However, since only nonnegative power spectrograms are involved, such approaches rely only on the amplitude information, while completely discarding the phases of the short time Fourier transforms (STFTs). In other words, these approaches do not allow exploiting the interchannel phase differences (IPDs), but only the interchannel level differences (ILDs). However, the IPDs may be very important for multichannel source separation, and they are indeed exploited by several clustering-based methods [13, 14]. Using IPDs becomes even more critical for the far-field case (i.e., when the distances between the microphones are much smaller than the distances between the sources and microphones), where the information carried by the ILDs becomes almost non-discriminating.

It is clear that a fully nonnegative (e.g., NTF-like) modeling is unable to model jointly source power spectrograms, ILDs and IPDs, since the phase information is discarded in the nonnegative tensor of multichannel mixture power spectrograms. As such, it was proposed to resort to a semi-nonnegative modeling [8, 12, 15–17], where the latent source power spectrograms are modeled with NMF [8, 12] or NTF [15–17], while the mixing system is modeled differently, not with a nonnegative model. This modeling, often referred to as *multichannel NMF* [12] or *multichannel NTF* [15][1] depending on the model of the source power spectrograms, is usually achieved via a Gaussian probabilistic modeling applied directly to the complex-valued STFTs of all channels.

The multichannel NMF modeling treats the complex-valued STFT coefficients as realizations of zero-mean circular complex-valued Gaussian random variables with structured variances (via NMF) and covariances. This leads to the fact that this modeling reduces to Itakura Saito (IS) NMF in the single channel case (see Chap. 1), thus being its natural extension to the multichannel case. Moreover, it allows integrating many other NMF-like models (see Chap. 1 and [8]) in an easy and flexible manner. Finally, it combines both spectral and spatial (including ILDs and IPDs) cues within a unified framework. When one of these two cues does not allow separating the sources efficiently, the algorithm relies on the other cue, and vice versa. In our opinion the multichannel NMF is one of the first attempts of combining these two cues in a systematic and principal way.

[1] Throughout the chapter we will generally refer to all these methods as multichannel NMF, while precising when we are speaking about multichannel NTF.

4.2 Local Gaussian Model

Multichannel NMF can be formulated as based on a so-called *local Gaussian model (LGM)* that is more general itself (than the multichannel NMF) and allows modeling and combining spatial and spectral cues in a systematic way. In a most general manner the LGM may be formulated as follows. Let us first assume that we deal with a multichannel (I-channel) mixture of J sources to be separated. Assuming all the signals are converted into the STFT domain, this can be written as

$$\mathbf{x}_{fn} = \sum_{j=1}^{J} \mathbf{y}_{jfn}, \tag{4.1}$$

where $\mathbf{x}_{fn} = \left[x_{1,fn}, \ldots, x_{I,fn}\right]^T \in \mathbb{C}^I$ and $\mathbf{y}_{jfn} = \left[y_{1,jfn}, \ldots, x_{I,jfn}\right]^T \in \mathbb{C}^I$ ($j = 1, \ldots, J$) are the channel-wise vectors of STFT coefficients of the mixture and of the j-th source *spatial image*,[2] respectively; and $f = 1, \ldots, F$ and $n = 1, \ldots, N$ are the frequency and time indices, respectively. Given the above-introduced notations, the LGM modeling [18] assumes that each source image (I-length complex-valued vector $\mathbf{y}_{jfn}$) is modeled as a zero-mean circular complex Gaussian random vector as follows

$$\mathbf{y}_{jfn} \sim \mathcal{N}_c\left(0, \mathbf{R}_{jfn} v_{jfn}\right), \tag{4.2}$$

where the complex-valued covariance matrix is positive definite Hermitian, and it is composed of two factors:

- a *spatial covariance* $\mathbf{R}_{jfn} \in \mathbb{C}^{I \times I}$ representing the spatial characteristics of the j-th source image at the time-frequency (TF) point (f, n), and
- a *spectral variance* $\mathbf{v}_{jfn} \in \mathbb{R}$ representing the spectral characteristics of the j-th source image at the TF point (f, n).

Given the model parameters, i.e., the spatial covariances $\mathbf{R}_{jfn}$ and the spectral variances $\mathbf{v}_{jfn}$, the random vectors $\mathbf{y}_{jfn}$ in (4.2) are also assumed mutually independent in time, frequency and between sources. Note that the LGM modeling was not proposed in [18] for the first time, indeed, its variants were already considered in [19, 20]. However, the formulation from [18] is quite general to cover all the cases, that is why we have chosen here this formulation.

Given the multichannel mixing equation and the above independence assumptions, the mixture STFT coefficients may be shown distributed as

$$\mathbf{x}_{fn} \sim \mathcal{N}_c\left(0, \sum_{j=1}^{J} \mathbf{R}_{jfn} v_{jfn}\right). \tag{4.3}$$

---

[2] The spatial image of a source means not the source signal itself, but its contribution into the I-channel mixture.

The model parameters are usually estimated in the maximum likelihood (ML) sense from the observed mixture $\mathbf{X} = \{x_{ifn}\}_{i,f,n}$. However, a direct ML estimation of parameters under the modeling (4.3) would lead to the data overfitting, since the number of scalar parameters exceeds the number of the mixture STFT coefficients. As such, various constraints are applied to both spectral variances and spatial covariances, as it is presented in detail in Sects. 4.3 and 4.4 respectively. In the case of multichannel NMF we address in this chapter, the spectral variances are usually represented by low-rank nonnegative matrices or tensors. However, other approaches consider different models (e.g., such as composite autoregressive models [21], source-excitation models [8] or hidden Markov models [22]) to structure the spectral variances, that is why the LGM modeling is more general than the multichannel NMF. As it is discussed in Sect. 4.4 below, spectral covariances are usually not modeled with fully nonnegative structures. This is the reason why we are speaking about semi-nonnegative modeling in the introduction.

For the sake of better understanding, we now give an interpretation to the spatial covariance matrix $\mathbf{R}_{jfn}$, and relate it to the methods used for multichannel audio compression. For the sake of simplicity and also since most of audio recording are stereo (i.e., two channel mixtures), we consider the case of $I = 2$. The spatial covariance matrix $\mathbf{R}_{jfn}$ is in general a full-rank positive definite Hermitian complex-valued matrix. An example of a spatial covariance matrix is represented on Fig. 4.1. Note that this is a rather "fake" (or incomplete) representation, since it is difficult to represent a 2-dimensional complex-valued covariance matrix on a 2-dimentional real plane.

Since the spatial covariance matrix $\mathbf{R}_{jfn}$ is complex-valued Hermitian, it can be easily shown that in the 2-dimensional case we consider here it is uniquely encoded

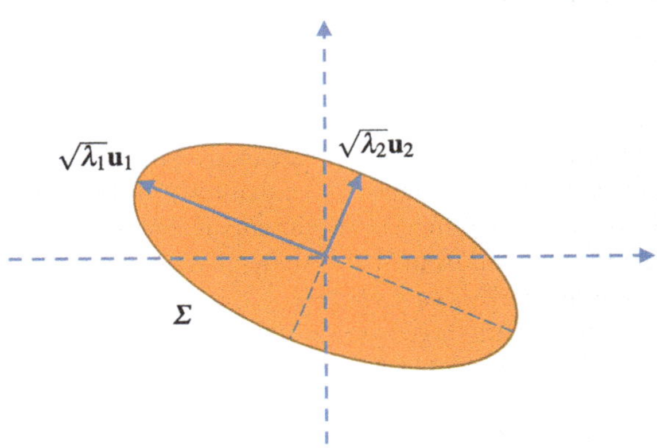

Fig. 4.1 An illustration of a spatial covariance matrix $\mathbf{R}_{jfn}$ in the 2-channel case ($I = 2$). While dropping the indices j, f and n, the covariance matrix eigendecomposition may be written as $\mathbf{R} = \mathbf{U}\boldsymbol{\Lambda}\mathbf{U}^H$, with $\mathbf{U} = [\mathbf{u}_1, \mathbf{u}_2]$, $\mathbf{u}_1, \mathbf{u}_2 \in \mathbb{C}^2$ being the eigenvectors and $\boldsymbol{\Lambda} = \mathrm{diag}([\lambda_1, \lambda_2])$, $\lambda_1, \lambda_2 \in \mathbb{R}_+$ being the eigenvalues. This illustration is not fully complete, since a 2D complex-valued covariance matrix is represented on a 2D real plane

by only four real scalars. Indeed, its 2 diagonal entries are real and the 2 complex-valued off-diagonal entries are conjugate. These four real-valued parameters may be uniquely converted into the following, in a sense more meaningful, real-valued parameters:

- Loudness,[3]
- ILD,
- IPD,
- Diffuseness that can be also replaced by interchannel coherence (IC) [23].

It is worth to note that the last three spatial parameters (ILD, IPD and IC) are also used for parametric coding of stereo audio [23]. This is somehow expected, indeed, the models that are suitable for compression should be also suitable for sources separation, since in both cases the models tend to reduce the redundancy in the signal.

Finally, let us also stress that the LGM modeling seems more general (and thanks to Gaussian formulation more principal) than blind source separation (BSS) approaches based on ILD/IPD clustering [13, 24]. Indeed, the diffuseness or IC is not taken at all into account within the latter approaches.

4.3 Spectral Models

In this section we present and discuss spectral models used within various multichannel NMF approaches. These models include NMF models, NTF models and their extensions.

4.3.1 NMF Modeling of Each Source

NMF modeling of each source, which is usually referred to as mutichannel NMF, consists in structuring the source variances v_{jfn} in (4.2) with NMF structure as in the single-channel NMF case (see Chap. 1):

$$v_{jfn} = \sum_{k=1}^{K_j} w_{jfk} h_{jkn}, \tag{4.4}$$

where the source-dependent K_j is usually smaller than both F and N, and w_{jfk} and h_{jkn} are all nonnegative. By introducing nonnegative matrices (i.e., matrices

[3]Due to the scale ambiguity between $\mathbf{R}_{jfn}$ and v_{jfn} in (4.2), the loudness can be fully attributed to v_{jfn}.

with nonnegative entries) $\mathbf{V}_j = [v_{jfn}]_{f,n} \in \mathbb{R}_+^{F\times N}$, $\mathbf{W}_j = [w_{jfk}]_{f,k} \in \mathbb{R}_+^{F\times K_j}$, and $\mathbf{H}_j = [h_{jkn}]_{k,n} \in \mathbb{R}_+^{K_j\times N}$, (4.4) may be rewritten in a matrix form as:

$$\mathbf{V}_j = \mathbf{W}_j \mathbf{H}_j. \tag{4.5}$$

A visualization of these NMF spectral models is shown on Fig. 4.2.

This kind of spectral models in the case of mutichannel source separation were first introduced in [25, 26], though with more sophisticated NMF-like structures suitable for harmonic music instruments and with different optimization criteria than those we discuss in this chapter. Spectral models based on usual NMF, exactly as in (4.5), were proposed in [12], and then extended/re-considered in many other works [8, 15–17, 27].

A very attractive property of this modeling is that any NMF or NMF-like structure based on the IS divergence, such as for example harmonic NMF [7], smooth NMF [2, 5] or excitation-filter NMF [28] (see also Chap. 1) may be incorporated easily and in a systematic manner within the framework. This was remarked and addressed in [8], where a general source separation framework allowing specifying various spectral and spatial models for each individual source is proposed. The latter research work is supplied with a software called Flexible Audio Source Separation Toolbox (FASST) that implements all these possible model variants in a flexible way. Finally, let us note that many informed or user-assisted/guided audio source separation approaches were extended to the multichannel case within the same paradigm [15, 29].

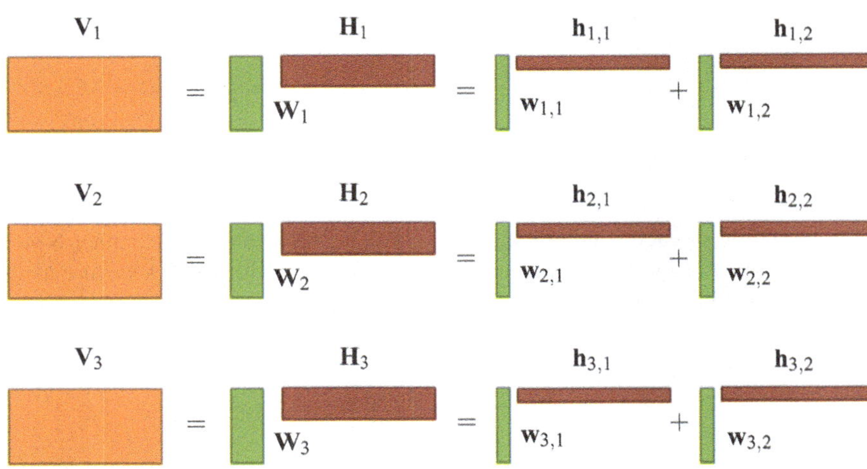

Fig. 4.2 A visualization of spectral models of multichannel NMF. Source variances $\mathbf{V}_j$ of each of J (here $J = 3$) sources are modeled with NMF with K_j (here $K_j = 2$) components, which can be decomposed as a sum of K_j rank-1 matrices ($\mathbf{w}_{j,k}$ and $\mathbf{h}_{j,k}$ are the columns and the lines of matrices **W** and **H**, respectively)

4.3.2 Joint NTF Modeling of All Sources

One of the shortcomings of the multichannel NMF modeling presented in Sect. 4.3.1 is the following. While for single-channel NMF one needs fixing an appropriate number of components K or determining this number automatically, which is not always easy (see, e.g., [30]), in the multichannel NMF, as presented in Sect. 4.3.1, one needs determining not only the total number of components $K = \sum_{j=1}^{J} K_j$, but also the number of components K_j for each source, which may vary from one source to another. To overcome this problem the following idea was introduced in [15], and then extended in other works [16, 17]. It is now assumed that instead of representing each source with an individual NMF $\{\mathbf{W}_j, \mathbf{H}_j\}$ all the sources share the components of the same NMF $\{\mathbf{W}, \mathbf{H}\}$, where $\mathbf{W} = [w_{fk}]_{f,k} \in \mathbb{R}_+^{F\times K}$, and $\mathbf{H} = [h_{kn}]_{k,n} \in \mathbb{R}_+^{K\times N}$. Moreover, in order to specify associations between K NMF components and J sources, a new $(J \times K)$ nonnegative matrix $\mathbf{Q} = [q_{jk}]_{j,k} \in \mathbb{R}_+^{J\times K}$ is introduced, and the source variances v_{jfn} are now structured as:

$$v_{jfn} = \sum_{k=1}^{K} w_{fk} h_{kn} q_{jk}. \tag{4.6}$$

Assuming the columns of $\mathbf{Q}$ are normalized to sum to one (i.e., $\sum_{j=1}^{J} q_{jk} = 1$), which is always possible to achieve thanks to scale ambiguity between the columns of $\mathbf{Q}$ and that of say $\mathbf{W}$ in (4.6), each q_{jk} represents the proportion of association of the component k to the source j.

By denoting with $\mathbf{V} = \{v_{jfn}\}_{j,f,n}$ a 3-valence tensor of source variances, (4.6) may be also rewritten in a tensor/vector form as a sum of K rank-1 tensors:

$$\mathbf{V} = \sum_{k=1}^{K} \mathbf{w}_k \circ \mathbf{h}_k^T \circ \mathbf{q}_k, \tag{4.7}$$

where "$\circ$" denotes the tensor outer product, $\mathbf{w}_k$ and $\mathbf{q}_k$ are the k-th columns of matrices $\mathbf{W}$ and $\mathbf{Q}$ respectively, and $\mathbf{h}_k$ is the k-th line of matrix $\mathbf{H}$. The tensor decomposition as in (4.6) and (4.7) is called parallel factor (PARAFAC) or canonical decomposition (CANDECOMP) [31]. A visualization of these NTF spectral models is shown on Fig. 4.3.

We here call this model multichannel NTF, as introduced in [15], though some authors [16, 17] continue calling it multichannel NMF. Note also that a fully nonnegative NTF modeling [10–12] was applied for multichannel audio source separation as well. Those approaches apply an NTF decomposition directly to the nonnegative tensor of power spectrograms of the multichannel mixture, while here it is applied to the latent nonnegative tensor of power spectrograms of the sources, and the overall modeling is not fully nonnegative, as mentioned in the introduction.

One can easily note that the NTF decomposition (4.6) generalizes that of (4.4). Indeed, (4.6) can be reduced to (4.4) by setting for each column of $\mathbf{Q}$ all the values to

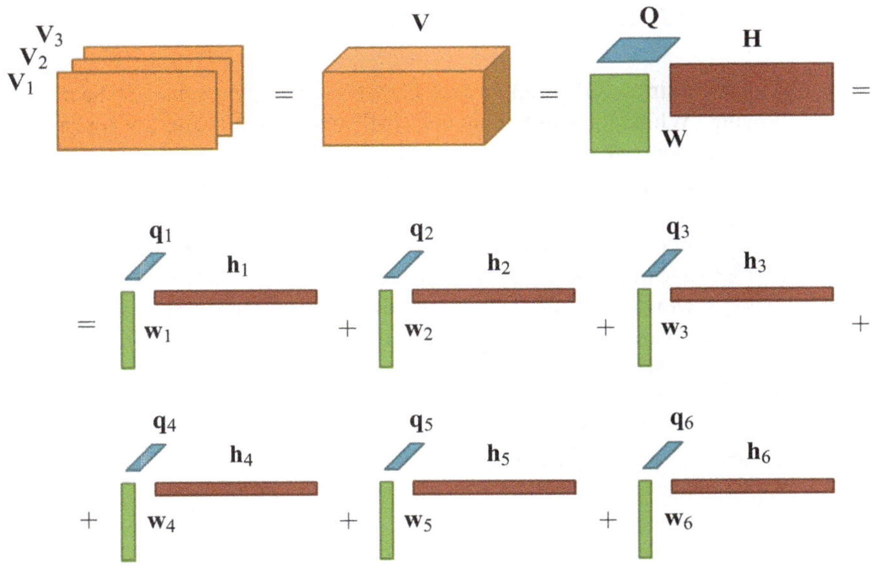

Fig. 4.3 A visualization of spectral models of multichannel NTF. Source variances $\mathbf{V}_j$ are stuck in a common 3-valence tensor $\mathbf{V}$ modeled with PARAFAC model [31] with K (here $K = 6$) components, which can be decomposed as a sum of K rank-1 3-valence tensors

0 except one that is set to 1, and by fixing the values of $\mathbf{Q}$. Finally, the multichannel NTF modeling has the following potential advantages over the multichannel NMF modeling:

- One does not need specifying in advance the number of components K_j for each source, but only the total number of components K. The components are then allocated automatically via the matrix $\mathbf{Q}$, which may be also more optimal than a manual user-specified allocation.
- Some components may be shared between different sources, which means that the modeling is more compact. This happens when there are more than one non-zero entry in one column of matrix $\mathbf{Q}$.

It should be noted however that it is desirable that the matrix $\mathbf{Q}$ is quite sparse, i.e., that there are few components for which there are more than one non-zero entry in the corresponding column of matrix $\mathbf{Q}$. Otherwise, the components are not well allocated between sources, and this may not lead to a good separation result. Thus, it is possibly desirable to add some sparsity-inducing penalty on $\mathbf{Q}$ to the corresponding optimization criterion.

4.4 Spatial Models and Constraints

Spatial covariance $\mathbf{R}_{jfn}$ might be assumed fully unconstrained, though in that case, as already mentioned in Sect. 4.2, the parameter estimation would certainly lead to data overfitting, since there are more parameters than observations, i.e., the STFT coefficients in the multichannel mixture. In order to cope with that it is necessary to introduce some constraints on spatial covariances.

First of all, when the sources are static, it is reasonable to assume that the spatial covariances are time-invariant, i.e., $\mathbf{R}_{jfn} = \mathbf{R}_{jf}$ are independent of n. This assumption is made in many approaches [8, 12, 16–18] and it allows highly reducing the number of free parameters to be estimated. We assume the time-invariant case within this section and the time-varying case will be briefly discussed at the end.

On top of the time-invariance, additional constraints may be introduced as well, and most often it is achieved either by imposing some particular structure or via probabilistic priors.

The early works [12, 19, 20] constraint the spatial covariance $\mathbf{R}_{jf}$ further and assume that the rank of the matrix is one, which is refereed to as *rank-1 spatial covariance*. This was introduced based on the following reasoning. Let us assume that the mixture (4.1) is a convolutive mixture of J point sources. In that case the spatial images $\mathbf{y}_{jfn}$ in (4.1) may be approximated as [32]

$$\mathbf{y}_{jfn} = \mathbf{a}_{jf} s_{jfn}, \tag{4.8}$$

where $s_{jfn} \in \mathbb{C}$ are the STFT coefficients of the point sources and $\mathbf{a}_{jf} = \left[a_{1jf}, \ldots, a_{Ijf}\right]^T \in \mathbb{C}^I$ are the channel-wise vectors of discrete Fourier transforms (DFTs) of the impulse responses of the convolutive mixing filters. The equality in (4.8) holds indeed only approximately and becomes more and more accurate when the sizes of the mixing filters impulse responses are comparable or smaller than the length of the STFT analysis window [32]. This approximation is referred to as *narrowband approximation*. Assuming now that each source STFT coefficient s_{jfn} follows a zero-mean Gaussian distribution with variance v_{jfn}, one can easily show that source images $\mathbf{y}_{jfn}$ are distributed as in (4.2) with

$$\mathbf{R}_{jf} = \mathbf{a}_{jf} \mathbf{a}_{jf}^H. \tag{4.9}$$

We see that the spatial covariance $\mathbf{R}_{jf}$ in (4.9) is indeed a rank-1 matrix.

It was proposed in [18] not to constraint the spatial covariance $\mathbf{R}_{jf}$ or to parametrize it in a different way (see [18] for details), but in both cases so as the matrix remains full rank. This modeling, refereed to as *full rank spatial covariance*, allows to go beyond the limits of the narrowband approximation (4.8), thus it is more suitable than the rank-1 model in case of long reverberation times. It may be also more suitable in case when the point sources assumption is not fully verified. Indeed, as explained in Sect. 4.7.2 below, modeling a source image with a full rank model

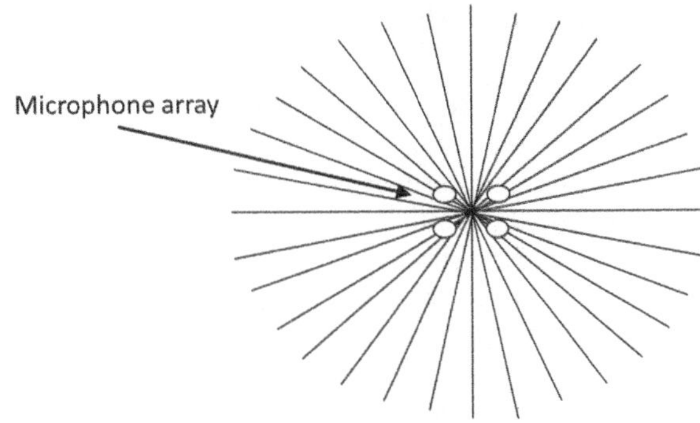

Fig. 4.4 Example of a set of predefined directions in 2D plane for a given microphone array

can be recast as a sum of I point sources with different rank-1 spatial covariances and shared spectral variance.

Another approach [17] consists in assuming that the spatial covariance is a weighted sum of so-called *direction of arrival (DOA) kernels* that are rank-1 spatial covariances modeling plane waves coming from several predefined directions. These directions may be specified in 2D plane or in 3D space (see Fig. 4.4 for a 2D example). Rank-1 DOA kernels corresponding to these directions θ_l ($l = 1, \dots, L$) are then defined as

$$\mathbf{K}_{fl} = \mathbf{d}(f, \theta_l)\mathbf{d}(f, \theta_l)^H \tag{4.10}$$

with $\mathbf{d}(f, \theta_l)$ being a *relative steering vector* for the direction θ_l defined as

$$\mathbf{d}(f, \theta_l) = \left[1, e^{-2\pi\tau_{2,1}(\theta_l)\nu_f/c}, \dots, e^{-2\pi\tau_{I,1}(\theta_l)\nu_f/c}\right]^T, \tag{4.11}$$

where c is the speed of the sound (343 m/s), ν_f is the frequency (in Hz) corresponding to the frequency bin f, and $\tau_{i,i'}(\theta_l)$ is the time difference of arrival (TDOA) (in seconds) between microphones i and i' from the direction θ_l. Note that this relative steering vector is defined without taking into account the ILDs, but only IPDs (see [33] for a definition taking as well into account ILDs). Finally, the spatial covariance is defined as a weighted sum of DOA kernels $\mathbf{K}_{fl}$ from (4.10) as

$$\mathbf{R}_{jf} = \sum_{l=1}^{L} z_{jl}\mathbf{K}_{fl}, \tag{4.12}$$

with z_{jl} being nonnegative weights.

If the DOAs of all or of some sources are known to some extend, it is possible to introduce this information for example via prior distributions on the spatial covariances. In [34] those priors are defined via inverse Wishart distributions as follows

$$p\left(\mathbf{R}_{jf}|\boldsymbol{\Psi}_{jf}, m\right) = \frac{|\boldsymbol{\Psi}_{jf}|^{m}|\mathbf{R}_{jf}|^{-(m+I)}e^{-\mathrm{tr}\left[\boldsymbol{\Psi}_{jf}\mathbf{R}_{jf}^{-1}\right]}}{\pi^{I(I-1)/2}\prod_{i=1}^{I}\Gamma(m-i+1)}, \tag{4.13}$$

with

$$\boldsymbol{\Psi}_{jf} = (m - I)\left(\mathbf{d}(f,\theta_l)\mathbf{d}(f,\theta_l)^{H} + \sigma_{\mathrm{rev}}^{2}\boldsymbol{\Omega}_f\right), \tag{4.14}$$

where $\mathbf{d}(f,\theta_l)$ is a steering vector which may be defined as in (4.11), $\boldsymbol{\Omega}_f = \left[\sin(2\pi\nu_f q_{ii'}/c)/(2\pi\nu_f q_{ii'}/c)\right]_{ii'}$ is a matrix modeling reverberation part (i.e., non-direct part) of the impulse response, and $\sigma_{\mathrm{rev}}^{2}$ is a positive constant depending on the amount of reverberation as compared to the direct part of impulse response.

There are also other models that do not fall into the LGM framework as formulated here. These models include for example multichannel high-resolution NMF (HR-NMF) [35] or a method where the source variance prior parametrization is factorized by NMF [36].

Finally, several approaches [37–39] address time-varying case, where $\mathbf{R}_{jfn}$ is not independent any more on n, though still constrained in different ways.

4.5 Main Steps and Sources Estimation

Let us denote by $\boldsymbol{\theta} = \{\mathbf{R}_{jfn}, v_{jfn}\}_{j,f,n}$ the whole set of model parameters, assuming some constraints from those overviewed in Sects. 4.3 and 4.4 hold. Given a model $\boldsymbol{\theta}$ specified and an estimation criterion (see Sect. 4.6 below) chosen, most of LGM-based approaches are based on the following main steps:

1. The STFT $\mathbf{X}$ of the multichannel mixture signal is computed.
2. The model is estimated with an algorithm (see Sect. 4.7 below) optimizing the chosen criterion.
3. The source images are estimated in the STFT domain via Wiener filtering as:

$$\hat{\mathbf{y}}_{jfn} = \mathbf{R}_{jfn}v_{jfn}\left[\sum_{j=1}^{J}\mathbf{R}_{jfn}v_{jfn}\right]^{-1}\mathbf{x}_{fn}, \tag{4.15}$$

   where $\mathbf{R}_{jfn}$ and v_{jfn} are the spatial covariances and spectral variances as specified in (4.2).
4. The source images in time domain are then reconstructed by applying the inverse STFT to $\widehat{\mathbf{Y}} = \{\hat{\mathbf{y}}_{jfn}\}_{j,f,n}$.

In the online approaches [40, 41], where the separation must be performed for every new frame, the same steps are repeated for each frame and the model estimation algorithm is modified so as to update the model parameters in an incremental and causal (i.e., only the passed and current frames are used) manner.

4.6 Model Estimation Criteria

In order to estimate the model parameters $\boldsymbol{\theta}$ from the observed data, i.e., from the STFT of the multichannel mixture signal $\mathbf{X}$, one needs specifying a model estimation criterion.

4.6.1 Maximum Likelihood

One of the most popular choices for model estimation is the maximum likelihood (ML) criterion that writes

$$\boldsymbol{\theta} = \arg\max_{\boldsymbol{\theta}'} p(\mathbf{X}|\boldsymbol{\theta}'). \tag{4.16}$$

In the case of LGM modeling (4.2) this criterion can be shown [16] equivalent to minimizing the following cost function:

$$C_{\text{IS}}(\boldsymbol{\theta}) = \sum_{f,n=1}^{F,N} \text{tr}\left(\widehat{\boldsymbol{\Sigma}}_{\mathbf{x},fn}\boldsymbol{\Sigma}_{\mathbf{x},fn}^{-1}\right) - \log\det\left(\widehat{\boldsymbol{\Sigma}}_{\mathbf{x},fn}\boldsymbol{\Sigma}_{\mathbf{x},fn}^{-1}\right) - I, \tag{4.17}$$

where

$$\widehat{\boldsymbol{\Sigma}}_{\mathbf{x},fn} = \mathbf{x}_{fn}\mathbf{x}_{fn}^H \quad \text{and} \quad \boldsymbol{\Sigma}_{\mathbf{x},fn} = \mathbf{R}_{jfn}v_{jfn}. \tag{4.18}$$

Note that the cost (4.17) is not well defined (i.e., its value is infinite) when $I > 1$ and matrices $\widehat{\boldsymbol{\Sigma}}_{\mathbf{x},fn}$ are not full rank, which is the case in definition (4.18). However, this is not a problem per se. Indeed, the infinite term $-\log\det\left(\widehat{\boldsymbol{\Sigma}}_{\mathbf{x},fn}\right)$ is independent on $\boldsymbol{\theta}$ and can be simply removed from the cost (4.17), since it has no influence on the optimization over $\boldsymbol{\theta}$. Otherwise, a small regularization term may be added to $\widehat{\boldsymbol{\Sigma}}_{\mathbf{x},fn}$, which would make it full rank. Also, there exist alternative definitions of $\widehat{\boldsymbol{\Sigma}}_{\mathbf{x},fn}$ [8, 42], where it might be full rank by construction.

Formulation with the cost (4.17) is interesting, since, as one can note, it is a generalization of the IS-NMF cost in the single channel case (see Chap. 1). Indeed, $C_{\text{IS}}(\boldsymbol{\theta})$ becomes the single channel IS divergence when $I = 1$.

4.6.2 Maximum a Posteriori

When a prior distribution $p(\boldsymbol{\theta})$ on model parameters is specified, like for example the spatial covariance prior in (4.13), the maximum a posteriori (MAP) criterion is usually used instead of the ML criterion. It writes

$$\boldsymbol{\theta} = \arg\max_{\boldsymbol{\theta}'} p(\boldsymbol{\theta}'|\mathbf{X}) = \arg\max_{\boldsymbol{\theta}'} p(\mathbf{X}|\boldsymbol{\theta}')p(\boldsymbol{\theta}'). \tag{4.19}$$

Note that in case of prior in (4.13) we have $p(\boldsymbol{\theta}) = \prod_{f=1}^{F} p\left(\mathbf{R}_{jf}|\boldsymbol{\Psi}_{jf}, m\right)^N$, since the prior is applied to each time-frequency bin.

If one tries rewriting (4.19) in a form similar to (4.17), it would result in simply adding $-\log p(\boldsymbol{\theta}')$ term to (4.17).

4.6.3 Other Criteria

Several other criteria were proposed as well. For example, we have seen that the ML criterion formulated as in (4.17) generalizes the single channel IS NMF to the multichannel case, as such it was proposed in [16] to generalize the single-channel NMF with Euclidean distance (EUC NMF) to the multichannel case. This is achieved by replacing the cost function (4.17) with the following one

$$C_{\text{FRB}}(\boldsymbol{\theta}) = \sum_{f,n=1}^{F,N} \left\| \widehat{\boldsymbol{\Sigma}}_{\mathbf{x},fn} - \boldsymbol{\Sigma}_{\mathbf{x},fn} \right\|_F^2, \tag{4.20}$$

where $\|\mathbf{A}\|_F$ denotes the Frobenius norm of a matrix $\mathbf{A}$, and the data covariance matrix $\widehat{\boldsymbol{\Sigma}}_{\mathbf{x},fn}$ is defined slightly differently than in (4.18). Notably, it is defined as [16, 17]

$$\widehat{\boldsymbol{\Sigma}}_{\mathbf{x},fn} = \sqrt{\left|\mathbf{x}_{fn}\mathbf{x}_{fn}^H\right|} \times \text{sign}\left(\mathbf{x}_{fn}\mathbf{x}_{fn}^H\right), \tag{4.21}$$

where all the operation, i.e., the absolute value $|\cdot|$, the square root $\sqrt{\cdot}$, the multiplication $\times$ and the sign ($\text{sign}(a) = a/|a|$), are applied element-wise to the corresponding matrices.

There is also the variational Bayes (VB) criterion [43], which consists in computing directly the posterior distribution of the source STFT coefficients while marginalizing over all possible model parameters.

4.7 Model Estimation Algorithms

There exist several model parameter estimation algorithms [8, 16]. Though, due to the probabilistic formulation of the LGM model (4.2), the expectation-maximization (EM) algorithm [44] is one of the most popular choices. As we will see below, the use of the EM algorithm results not in just one algorithm, but it leads to a family of algorithms. Indeed, each particular implementation of the EM algorithm depends on several choices, as will be explained below. Because of the EM popularity we will mostly concentrate here on the different variants of EM and will only mention briefly other algorithms.

To present the variants of EM algorithm we consider the LGM model (4.2) with time-invariant unconstrained full rank spatial covariances $\mathbf{R}_{jf}$ and spatial variances v_{jfn} structured with NTF model (4.6). This is in fact a variant of multichannel NTF similar to the one described in [15], but with full rank covariances instead of rank-1 covariances as in [15]. Since no probabilistic priors on parameters are assumed, the variants of EM algorithm presented below are for the optimization of the ML criterion (4.16).

4.7.1 Variants of EM Algorithm

In one of its general formulations the EM algorithm [44] to optimize the ML criterion (4.16) consists first in specifying

- so-called *observed data* $\mathbf{X}$ that are usually the multichannel mixture STFT coefficients in the case of multichannel source separation, as considered here, and
- so-called *latent data* $\mathbf{Z}$. The choice of latent data may be quite different and different choices would lead to different EM variants.

Assuming that a probabilistic model parametrized by $\boldsymbol{\theta}$ is specified, the EM algorithm is usually applied in the following case. It is applied when it is difficult to optimize in a closed form the ML criterion (4.16) maximizing $\log p(\mathbf{X}|\boldsymbol{\theta})$, while it is easy to maximize in a closed form or via some simplified iterative procedure the log-likelihood $\log p(\mathbf{X}, \mathbf{Z}|\boldsymbol{\theta})$ of so-called *complete data* $\{\mathbf{X}, \mathbf{Z}\}$. The choice of latent data $\mathbf{Z}$ is usually done accordingly.

The EM algorithm consists then in iterating the following two steps:

- **E-step**: Compute an auxiliary function as follows:

$$Q(\boldsymbol{\theta}, \boldsymbol{\theta}^{(\ell)}) = \mathbb{E}_{\mathbf{X}|\mathbf{Z},\boldsymbol{\theta}^{(\ell)}} \log p(\mathbf{X}, \mathbf{Z}|\boldsymbol{\theta}). \tag{4.22}$$

- **M-step**: Optimize the auxiliary function to update model parameters according to the following criterion:

$$\boldsymbol{\theta}^{(\ell+1)} = \arg\max_{\boldsymbol{\theta}} Q(\boldsymbol{\theta}, \boldsymbol{\theta}^{(\ell)}), \tag{4.23}$$

  where $\boldsymbol{\theta}^{(\ell)}$ denotes the model parameters estimated at the ℓ-th iteration.

It is often possible to optimize the criterion (4.23) in a closed form. However, sometimes, depending on the choice of latent data $\mathbf{Z}$, it is not possible. In that case either another iterative optimization algorithm may be applied or any algorithm can be used provided that it assures at each iteration of EM the following non-decreasing of the auxiliary function:

$$Q(\boldsymbol{\theta}^{(\ell+1)}, \boldsymbol{\theta}^{(\ell)}) \geq Q(\boldsymbol{\theta}^{(\ell)}, \boldsymbol{\theta}^{(\ell)}). \tag{4.24}$$

In the latter case the algorithm is called generalized EM (GEM) [44], and the ways the optimization (4.24) is performed lead again to different variants of the algorithm.

To summarize let us list various choices that lead to different EM algorithm variants and thus different model parameters estimation results. These choices include:

1. Choice of latent data $\mathbf{Z}$, for example:
   - Latent data consist of NMF/NTF components [12] defined as

     $$c_{kjfn} \sim \mathcal{N}_c(0, w_{jfk} h_{jkn}), \qquad k = 1, \ldots, K_j \tag{4.25}$$

     in case of NMF spectral model (4.4), or as

     $$c_{kjfn} \sim \mathcal{N}_c(0, w_{fk} h_{kn} q_{jk}), \qquad k = 1, \ldots, K \tag{4.26}$$

     in case of NTF spectral model (4.6).
   - Latent data consist of so-called *sub-sources* [8] (see Sect. 4.7.2 below).
   - Latent data consist of point sources [15] s_{jfn} as in the narrowband approximation (4.8).
   - Latent data consist of spatial source images [27] $\mathbf{y}_{jfn}$ as in (4.2).
   - Latent data consist of binary TF activations of the predominant source (see, e.g., [45] for details).
2. Choice of maximization step updates in case of GEM algorithm, for example:
   - Closed-form updates in case of EM algorithm.
   - Alternating closed-form updates over subsets of parameters [27] (each subset of parameters is updated by a closed-form update, while the other parameters are fixed).
   - Multiplicative update (MU) rules [5] to update NMF/NTF spectral model parameters [8].
3. Choice of initial parameters $\boldsymbol{\theta}^{(0)}$, for example:
   - Random parameters initialization [8].
   - Parameters initialization using the source separation results obtained by a different algorithm [12].
4. Choice of number of EM algorithm iterations, for example:
   - Fixed number of iterations (the most common choice).
   - Iterating till some stopping criterion depending on the likelihood value is satisfied.

A so-called spatial image EM (SIEM) algorithm, where the latent data are the spatial source images, is given in details in the Chap. 7. In the following section we present in details a so-called sub-source EM algorithm based on MU rules

(SSEM/MU) [8], where the latent data are the sub-sources and MU rules are used for the NTF spectral model parameters updates within the M-step. Other variants of the EM and GEM algorithms may be found in the corresponding papers.

4.7.2 Detailed Presentation of SSEM/MU Algorithm

Recall that our model consists of time-invariant unconstrained full rank spatial covariances $\mathbf{R}_{jf}$ and spatial variances v_{jfn} structured with NTF model (4.6). Thus, it can be parametrized as

$$\boldsymbol{\theta} = \left\{\{\mathbf{R}_{jf}\}_{j,f}, \mathbf{Q}, \mathbf{W}, \mathbf{H}\right\}, \tag{4.27}$$

with nonnegative matrices **Q**, **W** and **H** specified in Sect. 4.3.2.

The SSEM/MU algorithm presented below is a partial case of a more general algorithm from [8], though applied to a slightly different model (here the spectral variances are structured with NTF model, while in [8] they are structured with NMF model).

Each spatial $I \times I$ covariance $\mathbf{R}_{jf}$ being full rank, its rank equals to I. For each source j we introduce I so-called point *sub-sources* $s_{ji,fn} \in \mathbb{C}$ $(i = 1, \dots, I)$ that share the same spectral variance v_{jfn}, in other words they are distributed as

$$s_{ji,fn} \sim \mathcal{N}_c(0, v_{jfn}). \tag{4.28}$$

Moreover, each spatial covariance $\mathbf{R}_{jf}$ can be non-uniquely represented as

$$\mathbf{R}_{jf} = \mathbf{A}_{jf}\mathbf{A}_{jf}^H, \tag{4.29}$$

where $\mathbf{A}_{jf}$ is an $I \times I$ complex-valued matrix. By introducing a $J\,I$-length vector

$$\mathbf{s}_{fn} = \left[s_{11,fn}, \dots, s_{1I,fn}, s_{21,fn}, \dots, s_{2I,fn}, \cdots, s_{J1,fn}, \dots, s_{JI,fn}\right]^T, \tag{4.30}$$

and an $I \times J\,I$ matrix

$$\mathbf{A}_f = \left[\mathbf{A}_{1f}, \mathbf{A}_{2f}, \dots, \mathbf{A}_{Jf}\right], \tag{4.31}$$

one can show [8] that the LGM modeling (4.3) is equivalent (up to the noise term $\mathbf{b}_{fn}$) to

$$\mathbf{x}_{fn} = \mathbf{A}_{fn}\mathbf{s}_{fn} + \mathbf{b}_{fn}, \tag{4.32}$$

with $s_{ji,fn}$ (components of $\mathbf{s}_{fn}$) being mutually independent and distributed as in (4.28), the noise term $\mathbf{b}_{fn}$ being distributed as

$$\mathbf{b}_{fn} \sim \mathcal{N}_c(0, \boldsymbol{\Sigma}_{\mathbf{b},fn}), \tag{4.33}$$

with an anisotropic covariance matrix $\boldsymbol{\Sigma}_{\mathbf{b},fn} = \sigma^2_{\mathbf{b},f}\mathbf{I}_I$. The noise term $\mathbf{b}_{fn}$ is needed for a so-called *simulated annealing* procedure that is necessary in this case (see [12] for details), where the noise variance $\sigma^2_{\mathbf{b},f}$ is usually decreased over the algorithm iterations.

Let us now compute the auxiliary function $Q(\boldsymbol{\theta}, \boldsymbol{\theta}^{(\ell)})$ defined in (4.22). Below we will omit sometimes the indexing of parameters with (ℓ), and it will be clear from the context what are the parameters estimated on previous step and what are the parameters to be updated on the current step. The log-likelihood of the complete data $\{\mathbf{X}, \mathbf{Z}\}$ writes[4]

$$\begin{aligned} \log p(\mathbf{X}, \mathbf{Z}|\boldsymbol{\theta}) &= \log p(\mathbf{X}|\mathbf{Z}, \boldsymbol{\theta}) + \log p(\mathbf{Z}|\boldsymbol{\theta}) \\ &\stackrel{c}{=} -\sum_{f,n} \mathbf{tr}\left[\boldsymbol{\Sigma}^{-1}_{\mathbf{b},fn}\left(\boldsymbol{\Sigma}_{\mathbf{x},fn} - \mathbf{A}_{fn}\boldsymbol{\Sigma}^H_{\mathbf{xs},fn} - \boldsymbol{\Sigma}_{\mathbf{xs},fn}\mathbf{A}^H_{fn} + \mathbf{A}_{fn}\boldsymbol{\Sigma}_{\mathbf{s},fn}\mathbf{A}^H_{fn}\right)\right] \\ &\qquad - \sum_{f,n} \log\left|\boldsymbol{\Sigma}_{\mathbf{b},fn}\right| - I\sum_{j,f,n} d_{IS}(\xi_{jfn}|v_{jfn}), \end{aligned} \tag{4.34}$$

where

$$\boldsymbol{\Sigma}_{\mathbf{x},fn} = \widehat{\boldsymbol{\Sigma}}_{\mathbf{x},fn} = \mathbf{x}_{fn}\mathbf{x}^H_{fn} \tag{4.35}$$

is computed as in (4.18),

$$\boldsymbol{\Sigma}_{\mathbf{xs},fn} = \mathbf{x}_{fn}\mathbf{s}^H_{fn}, \tag{4.36}$$

$$\boldsymbol{\Sigma}_{\mathbf{s},fn} = \mathbf{s}_{fn}\mathbf{s}^H_{fn}, \tag{4.37}$$

$$\xi_{j,fn} = \frac{1}{I}\sum_{i=1}^{I} |s_{ji,fn}|^2, \tag{4.38}$$

and $d_{IS}(x|y) = \frac{x}{y} - \log\frac{x}{y} - 1$ is the scalar IS divergence (see Chap. 1).

By applying the conditional expectation operator $\mathbb{E}_{\mathbf{X}|\mathbf{S},\boldsymbol{\theta}^{(\ell)}}[\cdot]$ the auxiliary function $Q(\boldsymbol{\theta}, \boldsymbol{\theta}^{(\ell)})$ writes then

$$\begin{aligned} Q(\boldsymbol{\theta}, \boldsymbol{\theta}^{(\ell)}) &\stackrel{c}{=} -\sum_{f,n} \mathbf{tr}\left[\boldsymbol{\Sigma}^{-1}_{\mathbf{b},fn}\left(\widehat{\boldsymbol{\Sigma}}_{\mathbf{x},fn} - \mathbf{A}_{fn}\widehat{\boldsymbol{\Sigma}}^H_{\mathbf{xs},fn} - \widehat{\boldsymbol{\Sigma}}_{\mathbf{xs},fn}\mathbf{A}^H_{fn} + \mathbf{A}_{fn}\widehat{\boldsymbol{\Sigma}}_{\mathbf{s},fn}\mathbf{A}^H_{fn}\right)\right] \\ &\qquad - \sum_{f,n} \log\left|\boldsymbol{\Sigma}_{\mathbf{b},fn}\right| - I\sum_{j,f,n} d_{IS}(\hat{\xi}_{jfn}|v_{jfn}), \end{aligned} \tag{4.39}$$

with $\widehat{\boldsymbol{\Sigma}}_{\mathbf{xs},fn}$, $\widehat{\boldsymbol{\Sigma}}_{\mathbf{s},fn}$ and $\hat{\xi}_{jfn}$ defined as

---

[4]When we write $\stackrel{c}{=}$, that means that the equality is up to some constant that is independent on model parameters $\boldsymbol{\theta}$, and thus has no influence on the optimization over parameters in (4.23).

$$\widehat{\boldsymbol{\Sigma}}_{\mathbf{xs},fn} = \mathbb{E}_{\mathbf{X}|\mathbf{S},\boldsymbol{\theta}^{(\ell)}}\left[\boldsymbol{\Sigma}_{\mathbf{xs},fn}\right], \tag{4.40}$$

$$\widehat{\boldsymbol{\Sigma}}_{\mathbf{s},fn} = \mathbb{E}_{\mathbf{X}|\mathbf{S},\boldsymbol{\theta}^{(\ell)}}\left[\boldsymbol{\Sigma}_{\mathbf{s},fn}\right], \tag{4.41}$$

$$\hat{\xi}_{jfn} = \mathbb{E}_{\mathbf{X}|\mathbf{S},\boldsymbol{\theta}^{(\ell)}}\left[\xi_{jfn}\right], \tag{4.42}$$

and computed as follows:

$$\widehat{\boldsymbol{\Sigma}}_{\mathbf{xs},fn} = \widehat{\boldsymbol{\Sigma}}_{\mathbf{x},fn}\boldsymbol{\Omega}^H_{\mathbf{s},fn}, \tag{4.43}$$

$$\widehat{\boldsymbol{\Sigma}}_{\mathbf{s},fn} = \boldsymbol{\Omega}_{\mathbf{s},fn}\widehat{\boldsymbol{\Sigma}}_{\mathbf{x},fn}\boldsymbol{\Omega}^H_{\mathbf{s},fn} + (\mathbf{I}_{JI} - \boldsymbol{\Omega}_{\mathbf{s},fn}\mathbf{A}_f)\boldsymbol{\Sigma}_{\mathbf{s},fn}, \tag{4.44}$$

$$\hat{\xi}_{jfn} = \frac{1}{I}\sum_{i=(j-1)I+1}^{jI} \widehat{\boldsymbol{\Sigma}}_{\mathbf{s},fn}(i,i), \tag{4.45}$$

where

$$\boldsymbol{\Omega}_{\mathbf{s},fn} = \boldsymbol{\Sigma}_{\mathbf{s},fn}\mathbf{A}_f^H\boldsymbol{\Sigma}^{-1}_{\mathbf{x},fn}, \tag{4.46}$$

$$\boldsymbol{\Sigma}_{\mathbf{x},fn} = \mathbf{A}_f\boldsymbol{\Sigma}_{\mathbf{s},fn}\mathbf{A}_f^H + \boldsymbol{\Sigma}_{\mathbf{b},fn}, \tag{4.47}$$

$$\boldsymbol{\Sigma}_{\mathbf{s},fn} = \operatorname{diag}\left(\left[\underbrace{v_{1,fn},\dots,v_{1,fn}}_{I\text{ times}},\underbrace{v_{2,fn},\dots,v_{2,fn}}_{I\text{ times}},\cdots,\underbrace{v_{J,fn},\dots,v_{J,fn}}_{I\text{ times}}\right]\right) \tag{4.48}$$

We now proceed with the M-step (4.23). Maximizing the auxiliary function (4.39) over $\mathbf{A}_f$ leads to the following closed-form solution[5]:

$$\mathbf{A}_f = \widehat{\boldsymbol{\Sigma}}_{\mathbf{xs},fn}\widehat{\boldsymbol{\Sigma}}^{-1}_{\mathbf{s},fn}. \tag{4.49}$$

Maximization of the auxiliary function (4.39) over **Q**, **W** and **H**, i.e., the minimization of $\sum_{j,f,n} d_{IS}(\hat{\xi}_{jfn}|v_{jfn})$ with v_{jfn} computed as in (4.6), does not allow a closed-form solution. As such, to update **Q**, **W** and **H**, several iterations of the following MU rules [15] are applied:

$$q_{jk} \leftarrow q_{jk}\left(\frac{\sum_{f,n} w_{fk}h_{kn}\hat{\xi}_{jfn}v^{-2}_{jfn}}{\sum_{f,n} w_{fk}h_{kn}v^{-1}_{jfn}}\right), \tag{4.50}$$

$$w_{fk} \leftarrow w_{fk}\left(\frac{\sum_{j,n} h_{kn}q_{jk}\hat{\xi}_{jfn}v^{-2}_{jfn}}{\sum_{j,n} h_{kn}q_{jk}v^{-1}_{jfn}}\right), \tag{4.51}$$

$$h_{kn} \leftarrow h_{kn}\left(\frac{\sum_{j,f} w_{fk}q_{jk}\hat{\xi}_{jfn}v^{-2}_{jfn}}{\sum_{j,f} w_{fk}q_{jk}v^{-1}_{jfn}}\right). \tag{4.52}$$

---

[5]Note that if the spatial covariances $\mathbf{R}_{jf}$ are needed, they can be always computed with (4.29).

Applying these MU rules does not guarantee auxiliary function minimization as in (4.23), but only its non-decreasing as in (4.24). As such, this is in fact a GEM algorithm.

Algorithm 1 summarizes one iteration of the SSEM/MU algorithm derived above.

Algorithm 1 One iteration of SSEM/MU algorithm

- **E-step**: Compute statistics $\widehat{\boldsymbol{\Sigma}}_{\mathbf{x},fn}$, $\widehat{\boldsymbol{\Sigma}}_{\mathbf{xs},fn}$, $\widehat{\boldsymbol{\Sigma}}_{\mathbf{s},fn}$ and $\hat{\xi}_{jfn}$ as in (4.35), (4.40), (4.41) and (4.42).
- **M-step**:
  - Update $\mathbf{A}_f$ as in (4.49).
  - Update $\mathbf{Q}$, $\mathbf{W}$ and $\mathbf{H}$ iterating (4.50), (4.51) and (4.52) several times.
  - Renormalize $\mathbf{A}_f$, $\mathbf{Q}$, $\mathbf{W}$ and $\mathbf{H}$ to remove scale ambiguity (see [12]).

4.7.3 Other Algorithms

Another very popular choice for multichannel NMF model parameters estimation is the majorization-minimization (MM) algorithm [46], which is used for example in [16, 17]. Note that the EM algorithm is interpretable as a partial case of the MM algorithm.

4.8 Conclusion

In this chapter we have introduced multichannel NMF methods for audio source separation. Potential advantages and disadvantages of these methods are discussed. Despite a quickly growing popularity of deep learning that is now of a great interest for audio source separation, multichannel NMF methods remain still an important area of research and in our opinion cannot be completely replaced by deep learning-based methods in all situations. Indeed, especially in fully blind settings, where no training data are available, deep learning is not a suitable path any more, while multichannel NMF is still applicable.

As for the further research on multichannel NMF we would like highlighting the following possible paths which have been already started to be explored. One research direction consists in proposing more sophisticated spatial and spectral models adapted to the mixing conditions and sources of interest, as well as in proposing new models going beyond the limitations of the LGM modeling. Another direction consists in combining some aspects of multichannel NMF with deep learning.

Acknowledgements Cédric Févotte acknowledges funding from the European Research Council (ERC) under the European Union's Horizon 2020 research and innovation programme under grant agreement No 681839 (project FACTORY).

References

1. D.D. Lee, H.S. Seung, Learning the parts of objects with nonnegative matrix factorization. Nature **401**, 788–791 (1999)
2. T. Virtanen, Monaural sound source separation by non-negative matrix factorization with temporal continuity and sparseness criteria. IEEE Trans. Audio Speech Lang. Process. **15**(3),1066–1074 (2007)
3. M.N. Schmidt, R.K. Olsson, Single-channel speech separation using sparse non-negative matrix factorization, in *Spoken Language Proceesing, ISCA International Conference on (INTERSPEECH)* (2006)
4. L. Le Magoarou, A. Ozerov, N.Q. Duong, Text-informed audio source separation. Example-based approach using non-negative matrix partial co-factorization. J. Signal Process. Syst. **79**(2), 117–131 (2015)
5. C.Févotte, N. Bertin, J.-L. Durrieu, Nonnegative matrix factorization with the Itakura-Saito divergence: With application to music analysis. Neural Comput. **21**(3), 793–830 (2009)
6. D. El Badawy, N.Q. Duong, A. Ozerov, On-the-fly audio source separation—a novel user-friendly framework. IEEE/ACM Trans. Audio Speech Lang. Process. **25**(2), 261–272 (2017)
7. E. Vincent, N. Bertin, R. Badeau, Adaptive harmonic spectral decomposition for multiple pitch estimation. IEEE Trans. Audio Speech Lang. Process. **18**, 528–537 (2010)
8. A. Ozerov, E. Vincent, F. Bimbot, A general flexible framework for the handling of prior information in audio source separation. IEEE Trans. Audio Speech Lang. Process. **20**(4), 1118–1133 (2012)
9. N. Mohammadiha, P. Smaragdis, A. Leijon, Supervised and unsupervised speech enhancement using nonnegative matrix factorization. IEEE Trans. Audio Speech Lang. Process. **21**(10), 2140–2151 (2013)
10. D. FitzGerald, M. Cranitch, E. Coyle, Non-negative tensor factorisation for sound source separation, in *Proceeding of the Irish Signals and Systems Conference*, Dublin, Ireland, Sept 2005
11. D. FitzGerald, M. Cranitch, E. Coyle, Extended nonnegative tensor factorisation models for musical sound source separation. Comput. Intell. Neurosci. **2008**(872425),15 (2008)
12. A. Ozerov, C. Févotte, Multichannel nonnegative matrix factorization in convolutive mixtures for audio source separation. IEEE Trans. Audio Speech Lang. Process. **18**(3), 550–563 (2010)
13. H. Sawada, R. Mukai, S. Araki, S. Makino, A robust and precise method for solving the permutation problem of frequency-domain blind source separation. IEEE Trans. Speech Audio Process. **12**(5), 530–538 (2004)
14. M.I. Mandel, D.P. Ellis, T. Jebara, An EM algorithm for localizing multiple sound sources in reverberant environments. NIPS. **19** (2006)
15. A. Ozerov, C. Févotte, R. Blouet, J.-L. Durrieu, Multichannel nonnegative tensor factorization with structured constraints for user-guided audio source separation, in *Proceedings of the IEEE International Conference on Acoustics, Speech and Signal Processing (ICASSP)*, Prague, (May 2011), pp. 257–260
16. H. Sawada, H. Kameoka, S. Araki, N. Ueda, Multichannel extensions of non-negative matrix factorization with complex-valued data. IEEE Trans. Audio Speech Lang. Process. **21**(5), 971–982 (2013)
17. J. Nikunen, T. Virtanen, Direction of arrival based spatial covariance model for blind sound source separation. IEEE/ACM Trans. Audio Speech Lang. Process. **22**(3), 727–739 (2014)
18. N.Q. Duong, E. Vincent, R. Gribonval, Under-determined reverberant audio source separation using a full-rank spatial covariance model. IEEE Trans. Audio Speech Lang. Process. **18**(7), 1830–1840 (2010)

19. C.Févotte, J.-F. Cardoso, Maximum likelihood approach for blind audio source separation using time-frequency gaussian source models, in *IEEE Workshop on Applications of Signal Processing to Audio and Acoustics*, (IEEE, 2005), pp. 78–81
20. E. Vincent, S. Arberet, R. Gribonval, Underdetermined instantaneous audio source separation via local gaussian modeling, in *International Conference on Independent Component Analysis and Signal Separation*. (Springer, 2009), pp. 775–782
21. H. Kameoka, T. Yoshioka, M. Hamamura, J. Le Roux, K. Kashino, Statistical model of speech signals based on composite autoregressive system with application to blind source separation, in *International Conference on Latent Variable Analysis and Signal Separation*, (Springer, 2010), pp. 245–253
22. T. Higuchi, H. Takeda, T. Nakamura, H. Kameoka, A unified approach for underdetermined blind signal separation and source activity detection by multichannel factorial hidden markov models, in *INTERSPEECH*, (2014), pp. 850–854
23. J. Breebaart, S. van de Par, A. Kohlrausch, E. Schuijers, Parametric coding of stereo audio. EURASIP J. Appl. Signal Process. **2005**, 1305–1322 (2005)
24. M.I. Mandel, R.J. Weiss, D.P. Ellis, Model-based expectation-maximization source separation and localization. IEEE Trans. Audio Speech Lang. Process. **18**(2), 382–394 (2010)
25. E. Vincent, X. Rodet, Underdetermined source separation with structured source priors, in *International Conference on Independent Component Analysis and Signal Separation*, (Springer, 2004), pp. 327–334
26. E. Vincent, Musical source separation using time-frequency source priors. IEEE Trans. Audio Speech Lang. Process. **14**(1), 91–98 (2006)
27. S. Arberet, A. Ozerov, N.Q. Duong, E. Vincent, R. Gribonval, F. Bimbot, P. Vandergheynst, Nonnegative matrix factorization and spatial covariance model for under-determined reverberant audio source separation, in *10th International Conference on Information Sciences Signal Processing and their Applications (ISSPA), 2010*, (IEEE, 2010), pp. 1–4
28. T. Virtanen, A. Klapuri, Analysis of polyphonic audio using source-filter model and non-negative matrix factorization, in *Advances in Models for Acoustic Processing, Neural Information Processing Systems Workshop*, (Citeseer, 2006)
29. N. Souviraà-Labastie, A. Olivero, E. Vincent, F. Bimbot, Multi-channel audio source separation using multiple deformed references. IEEE/ACM Trans. Audio Speech Lang. Process. (TASLP) **23**(11), 1775–1787 (2015)
30. V.Y.F. Tan, C. Févotte, Automatic relevance determination in nonnegative matrix factorization with the beta-divergence. IEEE Trans. Pattern Anal. Mach. Intell. **35**(7), 1592–1605 (2013)
31. R. Bro, Parafac. tutorial and applications. Chemom. Intell. Lab. Syst. **38**(2), 149–171 (1997)
32. L. Parra, C. Spence, Convolutive blind separation of non-stationary sources. IEEE Trans. Speech Audio Process. **8**(3), 320–327 (2000)
33. S. Gannot, E. Vincent, S. Markovich-Golan, A. Ozerov, A consolidated perspective on multimicrophone speech enhancement and source separation. IEEE/ACM Trans. Audio Speech Lang. Process. **25**(4), 692–730 (2017)
34. N.Q. Duong, E. Vincent, R. Gribonval, Spatial location priors for gaussian model based reverberant audio source separation. EURASIP J. Adv. Signal Process. **2013**(1), 149 (2013)
35. R. Badeau, M.D. Plumbley, Multichannel high-resolution nmf for modeling convolutive mixtures of non-stationary signals in the time-frequency domain. IEEE/ACM Trans. Audio Speech Lang. Process. (TASLP) **22**(11), 1670–1680 (2014)
36. D. Kounades-Bastian, L. Girin, X. Alameda-Pineda, S. Gannot, R. Horaud, An inverse-gamma source variance prior with factorized parameterization for audio source separation, in *IEEE International Conference on Acoustics, Speech and Signal Processing (ICASSP)*, (IEEE, 2016), pp. 136–140
37. N.Q. Duong, H. Tachibana, E. Vincent, N. Ono, R. Gribonval, S. Sagayama, Multichannel harmonic and percussive component separation by joint modeling of spatial and spectral continuity, in *IEEE International Conference on Acoustics, Speech and Signal Processing (ICASSP)*, (IEEE, 2011), pp. 205–208

38. T. Higuchi, N. Takamune, T. Nakamura, H. Kameoka, Underdetermined blind separation and tracking of moving sources based on DOA-HMM, in *IEEE International Conference on Acoustics, Speech and Signal Processing (ICASSP)*, (IEEE, 2014), pp. 3191–3195
39. D. Kounades-Bastian, L. Girin, X. Alameda-Pineda, S. Gannot, R. Horaud, A variational EM algorithm for the separation of time-varying convolutive audio mixtures. IEEE/ACM Trans. Audio Speech Lang. Process. **24**(8), 1408–1423 (2016)
40. M. Togami, Online speech source separation based on maximum likelihood of local gaussian modeling, in *IEEE International Conference on Acoustics, Speech and Signal Processing (ICASSP)*. (IEEE, 2011), pp. 213–216
41. L.S. Simon, E. Vincent, A general framework for online audio source separation, in *International conference on Latent Variable Analysis and Signal Separation*, (Springer, 2012), pp. 397–404
42. N.Q. Duong, E. Vincent, R. Gribonval, Under-determined reverberant audio source separation using local observed covariance and auditory-motivated time-frequency representation, in *International Conference on Latent Variable Analysis and Signal Separation*, (Springer, 2010), pp. 73–80
43. K. Adiloğlu, E. Vincent, Variational bayesian inference for source separation and robust feature extraction. IEEE/ACM Trans. Audio Speech Lang. Process. **24**(10), 1746–1758 (2016)
44. A.P. Dempster, N.M. Laird, D.B. Rubin, Maximum likelihood from incomplete data via the EM algorithm. J. R. Stat.Soc. Ser. B (Statistical Methodology) **39**, 1–38 (1977)
45. J. Thiemann, E. Vincent, A fast EM algorithm for Gaussian model-based source separation, in *Proceedings of the 21st European Signal Processing Conference (EUSIPCO)*, (IEEE, 2013), pp. 1–5
46. D.R. Hunter, K. Lange, A tutorial on mm algorithms. Am. Stat. **58**(1), 30–37 (2004)

Chapter 5
General Formulation of Multichannel Extensions of NMF Variants

Hirokazu Kameoka, Hiroshi Sawada and Takuya Higuchi

Abstract Blind source separation (BSS) is generally a mathematically ill-posed problem that involves separating out individual source signals from microphone array inputs. The frequency domain BSS approach is particularly notable in that it provides the flexibility needed to exploit various models for the time-frequency representations of source signals and/or array responses. Many frequency domain BSS approaches can be categorized according to the way in which the source power spectrograms and/or the mixing process are modeled. For source power spectrogram modeling, the non-negative matrix factorization (NMF) model and its variants have recently proved very powerful. For mixing process modeling, one reasonable way involves introducing a plane wave assumption so that the spatial covariances of each source can be described explicitly using the direction of arrival (DOA). This chapter provides a general formulation of the frequency domain BSS that makes it possible to incorporate the models for the source power spectrogram and the source spatial covariance matrix. Through this formulation, we reveal the relationship between the state-of-the-art BSS approaches. We further show that combining these models allows us to solve the problems of source separation, DOA estimation, dereverberation, and voice activity detection in a unified manner.

5.1 Introduction

Blind source separation (BSS) is a technique for separating out individual source signals from microphone array inputs when the transfer characteristics between the sources and microphones are unknown. Since this problem is mathematically ill-

H. Kameoka (✉) · H. Sawada · T. Higuchi
NTT Communication Science Laboratories, NTT Corporation,
3-1 Morinosato Wakamiya, Atsugi, Kanagawa 243-0198, Japan
e-mail: kameoka.hirokazu@lab.ntt.co.jp

H. Sawada
e-mail: sawada.hiroshi@lab.ntt.co.jp

T. Higuchi
e-mail: higuchi.takuya@lab.ntt.co.jp

S. Makino (ed.), *Audio Source Separation*, Signals and Communication Technology, https://doi.org/10.1007/978-3-319-73031-8_5

posed, it is generally necessary to make certain assumptions about the source signals and/or the mixing process and formulate an optimization problem using a criterion designed according to these assumptions. For example, one well-known approach involves independent component analysis (ICA) [1], which makes separation possible by estimating a separation matrix (the inverse of the mixing matrix) such that the separated signals become statistically independent of each other. ICA is known to work well under certain conditions when the microphones outnumber the sources, the positions of all the sources are fixed and there is no reverberation. However, when these conditions are not fully met (such as when the mixing process is underdetermined or time-variant), the independence assumption is too weak to achieve good separation. To handle more general cases, we must consider further assumptions or constraints in addition to independence.

The frequency domain BSS approach is particularly notable in that while it requires us to solve an additional problem called the [1]permutation alignment problem, it allows a fast implementation compared with the time domain approach. It also provides the flexibility of allowing us to utilize various models for the time-frequency representations of the source signals and/or the array responses. For example, independent vector analysis (IVA) [2, 3] allows us to efficiently solve frequency-wise source separation and permutation alignment in a joint manner by assuming that the magnitudes of the frequency components originating from the same source tend to vary coherently over time. Other frequency domain BSS approaches can be categorized according to the way in which the source signals and/or the mixing process are modeled.

For power spectrogram modeling, multichannel extensions of non-negative matrix factorization (NMF) have attracted a lot of attention in recent years [4–12]. NMF was originally applied with notable success to monaural source separation tasks [13, 14]. The idea is to approximate the power (or magnitude) spectrogram of a mixture signal, interpreted as a non-negative matrix, as the product of two non-negative matrices. This amounts to assuming that the power spectrum of a mixture signal observed at each time frame can be approximated by the linear sum of a limited number of basis spectra scaled by time-varying amplitudes. Multichannel NMF (MNMF) is an extension of this approach to a multichannel case in order to allow for the use of spatial information as an additional clue for separation. It can also be viewed as an extension of frequency domain BSS that allows the use of spectral templates as a clue for both frequency-wise source separation and permutation alignment. While MNMF assumes each basis spectrum to be static, many source signals in the real world are non-stationary and the spectral densities vary over time. To characterize this nonstationary nature of source signals reasonably, MNMF can be further extended by describing the transition of the spectral densities and the total power of each source using a hidden Markov model (HMM) [15–18].

---

[1]The permutation alignment problem refers to a problem of grouping together the separated components of different frequency bins that originate from the same source to construct a separated signal.

For mixing process modeling, several models have been proposed for the spatial covariance matrix of a source. One popular way of modeling spatial covariances involves introducing a plane wave assumption. Under a plane wave assumption, the spatial covariance matrix of each source can be described using the direction of arrival (DOA). By using a discrete set of pre-defined spatial covariance matrices each corresponding to an angle in radians, the spatial covariance matrix of each fixed source can be modeled as either a sum or a mixture of the DOA-related covariance matrices [8, 16, 18–21]. To handle a time-varying spatial covariance and thus allow each source to move, the DOA mixture model can be further extended by describing the transition of the DOAs using an HMM [22].

This chapter provides a general formulation of the frequency domain BSS that allows us to incorporate the models for the source power spectrogram and the source spatial covariance matrix. Through this formulation, we reveal the relationship between the state-of-the-art BSS approaches. We further show that combining these models allows us to solve source separation, DOA estimation, dereverberation, and voice activity detection in a unified manner.

5.2 Problem Formulation

5.2.1 Mixing Systems

The typical mixing systems used within the frequency domain BSS framework include determined or underdetermined, time-invariant or time-variant, and instantaneous, convolutive or sparse mixtures.

We consider a situation where J source signals are captured by I microphones. The time domain mixture signal $\tilde{x}_i(t)$ observed at the i-th microphone is given as a convolutive mixture of $\tilde{s}_1(t), \ldots, \tilde{s}_J(t)$

$$\tilde{x}_i(t) = \sum_{j=1}^{J} \sum_{t'=0}^{T'-1} \tilde{a}_{i,j}(t')\tilde{s}_j(t - t') + \tilde{u}(t), \tag{5.1}$$

where $\tilde{a}_{i,j}(t')$ denotes the acoustic impulse response between microphone i and source j, and $\tilde{u}(t)$ denotes background noise. Since this model involves convolution, source separation algorithms based on this model tend to be computationally demanding. Here, let $x_i(f, n)$, $s_j(f, n)$ and $u_i(f, n)$ be the short-time Fourier transform (STFT) coefficients of $\tilde{x}_i(t)$, $\tilde{s}_j(t)$ and $\tilde{u}(t)$ where $f = 0, \ldots, F - 1$ and $n = 0, \ldots, N - 1$ are the frequency and frame indices, respectively. When the length T' of the acoustic impulse response from a source to a microphone is sufficiently shorter than the frame length of the STFT, $\mathbf{x}(f, n) = [x_1(f, n), \ldots, x_I(f, n)]^{\mathsf{T}}$ can be approximated fairly well by an instantaneous mixture in the frequency domain

$$x_i(f,n) = \sum_{j=1}^{J} a_{i,j}(f)s_j(f,n) + u_i(f,n), \tag{5.2}$$

where $a_{i,j}(f)$ is the Fourier transform of $\tilde{a}_{i,j}(t)$, namely the transfer function between microphone i and source j. This approximation is called a *narrowband approximation*. (5.2) can be arranged in the following expression

$$\mathbf{x}(f,n) = \sum_{j=1}^{J} \mathbf{a}_j(f)s_j(f,n) + \mathbf{u}(f,n) \tag{5.3a}$$

$$= \mathbf{A}(f)\mathbf{s}(f,n) + \mathbf{u}(f,n), \tag{5.3b}$$

by putting $\mathbf{x}(f,n) = [x_1(f,n),\ldots,x_I(f,n)]^\mathsf{T}$, $\mathbf{a}_j(f) = [a_{1,j}(f),\ldots,a_{I,j}(f)]^\mathsf{T}$, $\mathbf{A}(f) = [\mathbf{a}_1(f),\ldots,\mathbf{a}_J(f)]$, $\mathbf{s}(f,n) = [s_1(f,n),\ldots,s_J(f,n)]^\mathsf{T}$, and $\mathbf{u}(f,n) = [u_1(f,n),\ldots,u_I(f,n)]^\mathsf{T}$. The product of $\mathbf{a}_j(f)$ and $s_j(n,f)$

$$\mathbf{c}_j(f,n) = \mathbf{a}_j(f)s_j(f,n), \tag{5.4}$$

is called the *spatial image* of source j. The BSS framework based on (5.3) is called frequency domain BSS, which allows for fast implementations compared with methods based on the time domain convolutive mixture model given in (5.1). According to several studies such as [23], a reverberation longer than the frame length of the STFT can be modeled fairly well as a convolution for each frequency-band of the STFT representation. A frequency-wise convolutive mixture model

$$\mathbf{x}(f,n) = \sum_{j=1}^{J}\sum_{m=0}^{M-1} \mathbf{a}_j(f,m)s_j(f,n-m) + \mathbf{u}(f,n) \tag{5.5a}$$

$$= \sum_{m=0}^{M-1} \mathbf{A}(f,m)\mathbf{s}(f,n-m) + \mathbf{u}(f,n), \tag{5.5b}$$

can thus be a reasonable option especially under highly reverberant conditions. In a particular case where the mixing systems (5.3) and (5.5) are exactly invertible (determined) and $\mathbf{u}(n,f) = \mathbf{0}$, we can also use the following expressions:

$$\mathbf{W}^\mathsf{H}(f)\mathbf{x}(f,n) = \mathbf{s}(f,n), \tag{5.6}$$

$$\sum_{m=0}^{M-1} \mathbf{W}^\mathsf{H}(f,m)\mathbf{x}(f,n-m) = \mathbf{s}(f,n). \tag{5.7}$$

In this chapter, when we refer to instantaneous/convolutive mixtures, we mean frequency-wise instantaneous/convolutive mixtures given by (5.3) or (5.6), and (5.5) or (5.7).

A sparse mixture refers to a mixing model where only one source is assumed to be present at each time-frequency point. This assumption is approximately true particularly for mixtures where the spectrograms of all the sources are sparse. By using $z(f, n)$ to denote the unknown index of the predominant source at (f, n), the sparse mixture model can be expressed as

$$\mathbf{x}(f, n) = \mathbf{a}_{z(f,n)}(f) s_{z(f,n)}(f, n) + \mathbf{u}(f, n). \tag{5.8}$$

As shown above, the types of mixing systems can be characterized in terms of the relationship between $\mathbf{x}(f, n)$ and $\mathbf{s}(f, n)$, which can be summarized as follows:

1. an instantaneous mixing system given by (5.3),
2. a convolutive mixing system given by (5.5),
3. an instantaneous demixing system given by (5.6),
4. a convolutive demixing system given by (5.7), and
5. a sparse mixing system given by (5.8).

Time-variant versions of these systems are obtained by simply replacing $\mathbf{W}^{\mathsf{H}}(f)$, $\mathbf{A}(f)$, $\mathbf{W}^{\mathsf{H}}(f, m)$, $\mathbf{A}(f, m)$ and $\mathbf{a}_j(f)$ with $\mathbf{W}_n^{\mathsf{H}}(f)$, $\mathbf{A}_n(f)$, $\mathbf{W}_n^{\mathsf{H}}(f, m)$, $\mathbf{A}_n(f, m)$ and $\mathbf{a}_{j,n}(f)$, respectively.

5.2.2 *Likelihood Function*

Let us now assume that $s_j(f, n)$ independently follows a zero-mean complex Gaussian distribution with variance $v_j(f, n) = \mathbb{E}[|s_j(f, n)|^2]$

$$s_j(f, n) \sim \mathcal{N}_{\mathbb{C}}(s_j(f, n)|0, v_j(f, n)). \tag{5.9}$$

Hence, the spatial image $\mathbf{c}_j(f, n)$ follows

$$\mathbf{c}_j(f, n) \sim \mathcal{N}_{\mathbb{C}}(\mathbf{c}_j(f, n)|\mathbf{0}, v_j(f, n)\mathbf{R}_j(f)), \tag{5.10}$$

where $\mathbf{R}_j(f) = \mathbf{a}_j(f)\mathbf{a}_j^{\mathsf{H}}(f)$ expresses the spatial covariance of source j. As shown above, the rank of the spatial covariance $\mathbf{R}_j(f)$ becomes exactly 1 when the narrowband approximation holds. As we shall see in the following, we can also assume $\mathbf{R}_j(f)$ to be a full-rank matrix, which is shown to be effective particularly in reverberant conditions [24]. We call (5.9) or (5.10) the *local Gaussian model (LGM)*. The LGM assumes that frequency components with different frequency bins are independent. In the following, we elaborate on how this assumption can be justified.

Let $\tilde{s}(0), \ldots, \tilde{s}(T-1)$ be the time-domain samples of a source signal in a particular time frame. We assume that $\tilde{\mathbf{s}} = [\tilde{s}(0), \ldots, \tilde{s}(T-1)]^{\mathsf{T}} \in \mathbb{R}^T$ has been drawn from a zero-mean Gaussian random process with an autocorrelation matrix $\boldsymbol{\Sigma}_{\tilde{\mathbf{s}}}$:

$$\tilde{\mathbf{s}} \sim \mathscr{N}(\tilde{\mathbf{s}}|\mathbf{0}, \boldsymbol{\Sigma}_{\tilde{\mathbf{s}}}). \tag{5.11}$$

The Fourier transform of the sequence $\tilde{\mathbf{s}}$, given by $\mathbf{s} = \mathbf{F}\tilde{\mathbf{s}} \in \mathbb{C}^T$, follows a zero-mean multivariate complex Gaussian distribution with covariance matrix $\mathbf{F}\boldsymbol{\Sigma}_{\tilde{\mathbf{s}}}\mathbf{F}^{\mathsf{H}}$, where $\mathbf{F} \in \mathbb{C}^{T\times T}$ denotes a discrete Fourier transform matrix. Here, if we assume stationarity and circularity, the covariance matrix $\boldsymbol{\Sigma}_{\tilde{\mathbf{s}}}$ belongs to the class of nonnegative definite symmetric Toeplitz circulant matrices. Note that the stationarity assumption corresponds to assuming that the autocorrelation $[\boldsymbol{\Sigma}_{\tilde{\mathbf{s}}}]_{t,t+\tau} = \mathbb{E}[\tilde{s}(t)\tilde{s}(t+\tau)]$ of $\tilde{s}(t)$ depends only on the time difference $|\tau|$. The circularity assumption further requires that the autocorrelation is given by $[\boldsymbol{\Sigma}_{\tilde{\mathbf{s}}}]_{t,t+\tau} = \mathbb{E}[\tilde{s}(t \bmod T)\tilde{s}((t+\tau) \bmod T)]$. This implies that an infinitely repeated version of the finite segment $\tilde{s}(0), \dots, \tilde{s}(T-1)$ is assumed to be stationary. $\boldsymbol{\Sigma}_{\tilde{\mathbf{s}}}$ is then shown to be exactly diagonalized by $\mathbf{F}$ so that we obtain $\mathbf{F}\boldsymbol{\Sigma}_{\tilde{\mathbf{s}}}\mathbf{F}^{\mathsf{H}} = \mathrm{Diag}(v(0), \dots, v(T-1))$ where $v(0), \dots, v(T-1)$ are the eigenvalues of $\boldsymbol{\Sigma}_{\tilde{\mathbf{s}}}$, corresponding to the power spectral densities (PSDs) of $\tilde{s}(t)$. This indicates that $s(f)$ $(f = 0, \dots, F-1)$ independently follows a zero-mean complex Gaussian distribution with variance $v(f)$:

$$s(f) \sim \mathscr{N}_{\mathbb{C}}(s(f)|0, v(f)). \tag{5.12}$$

By adding the frame index n and the source index j to $s(f)$ and $v(f)$, and by assuming that the frequency components within different time frames are independent, we obtain (5.9).

We further assume that the background noise $\mathbf{u}(f, n)$ follows a zero-mean complex Gaussian distribution with covariance $\boldsymbol{\Sigma}_{\mathbf{u}}(f, n) = \mathbb{E}[\mathbf{u}(f, n)\mathbf{u}^{\mathsf{H}}(f, n)]$

$$\mathbf{u}(f, n) \sim \mathscr{N}_{\mathbb{C}}(\mathbf{u}(f, n)|\mathbf{0}, \boldsymbol{\Sigma}_{\mathbf{u}}(f, n)). \tag{5.13}$$

Then, $\mathbf{x}(f, n)$ follows

$$\mathbf{x}(f, n) \sim \mathscr{N}_{\mathbb{C}}\left(\mathbf{x}(f, n)|\boldsymbol{\mu}_{\mathbf{x}}(f, n), \boldsymbol{\Sigma}_{\mathbf{x}}(f, n)\right), \tag{5.14}$$

where $\boldsymbol{\mu}_{\mathbf{x}}(f, n)$ and $\boldsymbol{\Sigma}_{\mathbf{x}}(f, n)$ are expressed differently according to the choice of mixing system (5.3), (5.5), (5.6), (5.7), and (5.8), respectively:

$$\boldsymbol{\mu}_{\mathbf{x}}(f, n) = \begin{cases} \mathbf{0} & \text{(5.15a)} \\ \mathbf{0} & \text{(5.15b)} \\ \mathbf{0} & \text{(5.15c)} \\ -(\mathbf{W}^{\mathsf{H}}(f, 0))^{-1} \displaystyle\sum_{m=1}^{M-1} \mathbf{W}^{\mathsf{H}}(f, m)\mathbf{x}(f, n-m), & \text{(5.15d)} \\ \mathbf{0} & \text{(5.15e)} \end{cases}$$

$$\boldsymbol{\Sigma}_{\mathbf{x}}(f,n)=\begin{cases}\sum_j v_j(f,n)\mathbf{R}_j(f)+\boldsymbol{\Sigma}_{\mathbf{u}}(f,n) & \text{(5.16a)}\\ \sum_j\sum_m v_j(f,n-m)\mathbf{R}_j(f,m)+\boldsymbol{\Sigma}_{\mathbf{u}}(f,n) & \text{(5.16b)}\\ (\mathbf{W}^{\mathsf{H}}(f))^{-1}\boldsymbol{\Sigma}_{\mathbf{s}}(f,n)\mathbf{W}^{-1}(f) & \text{(5.16c)}\\ (\mathbf{W}^{\mathsf{H}}(f,0))^{-1}\boldsymbol{\Sigma}_{\mathbf{s}}(f,n)\mathbf{W}^{-1}(f,0) & \text{(5.16d)}\\ v_{z(f,n)}(f,n)\mathbf{R}_{z(f,n)}(f)+\boldsymbol{\Sigma}_{\mathbf{u}}(f,n). & \text{(5.16e)}\end{cases}$$

$\boldsymbol{\Sigma}_{\mathbf{s}}(f,n)$ is a diagonal matrix whose diagonal entries are $v_1(f,n),\ldots,v_J(f,n)$

$$\boldsymbol{\Sigma}_{\mathbf{s}}(f,n)=\mathrm{Diag}(v_1(f,n),\ldots,v_J(f,n)). \tag{5.17}$$

Given the observation $X=[\mathbf{x}(f,n)]_{f,n}$, BSS problems can be formulated by using (5.14) as the log-likelihood function

$$\begin{aligned}\log p(X|\boldsymbol{\theta})=\sum_{f,n}\big\{&-\log\det\boldsymbol{\Sigma}_{\mathbf{x}}(f,n)\\&-(\mathbf{x}(f,n)-\boldsymbol{\mu}_{\mathbf{x}}(f,n))^{\mathsf{H}}\boldsymbol{\Sigma}_{\mathbf{x}}^{-1}(f,n)(\mathbf{x}(f,n)-\boldsymbol{\mu}_{\mathbf{x}}(f,n))\big\},\end{aligned} \tag{5.18}$$

where $\boldsymbol{\theta}$ denotes the entire set of model parameters. Note that for a convolutive demixing system with $\boldsymbol{\mu}_{\mathbf{x}}(f,n)$ and $\boldsymbol{\Sigma}_{\mathbf{x}}(f,n)$ given as (5.15d) and (5.16d), (5.14) means a conditional distribution $p(\mathbf{x}(f,n)|\mathbf{x}(f,n-1),\ldots,\mathbf{x}(f,n-M+1))$ so that (5.18) represents

$$\begin{aligned}\log p(X|\boldsymbol{\theta})&=\sum_f\log p(\mathbf{x}(f,0),\ldots,\mathbf{x}(f,N-1))\\&=\sum_f\sum_n\log p(\mathbf{x}(f,n)|\mathbf{x}(f,n-1),\ldots,\mathbf{x}(f,n-M+1)).\end{aligned} \tag{5.19}$$

Since all the variables are indexed by frequency f in the log-likelihood function described above, the optimization problem consists of an independent set of frequency-wise source separation problems, each of which is ill-posed. In the following sections, we introduce assumptions and constraints that can be incorporated into the present framework in order to obtain a reasonable solution to the current optimization problem.

5.3 Spectral and Spatial Models

5.3.1 Spectral Models

With the LGM (5.9), the power and phase of $s_j(f,n)$ follow an exponential distribution with mean $v_j(f,n)$ and a uniform distribution on the interval $[0,2\pi)$, respectively. If there is a certain assumption, constraint or structure that we want to incorporate into the power spectrogram of each source, we can employ a parametric model or a generative model to represent $v_j(f,n)$ instead of individually treating $v_j(f,n)$ as a free parameter, or introduce a properly designed prior distribution over $v_j(f,n)$.

If we can assume that the spectra of a real-world sound source can be described using a limited number of templates, one way to express the power spectrogram $v_j(f,n)$ would be

$$v_j(f,n) = b_{j,k_j(n)}(f), \tag{5.20}$$

where $b_{j,1}(f),\ldots,b_{j,K_j}(f)$ denote the spectral templates assigned to source j and $k_j(n)$ denotes the index of a spectral template selected at frame n. If we assume $k_j(n)$ to be a latent variable generated according to a categorical distribution with probabilities $\pi_{j,1},\ldots,\pi_{j,K_j}$ such that $\sum_k \pi_{j,k} = 1$:

$$k_j(n) \sim \pi_{j,k_j(n)}, \tag{5.21}$$

the generative process of the spatial image $\mathbf{c}_j(f,n)$ of source j is described as a Gaussian mixture model (GMM) [23]. Note that the spectral templates can be either pre-trained using training samples or estimated from the mixture signal in a data-driven manner. While the above model uses each template to represent a different power spectrum, it would be more reasonable to let each template represent all the power spectra that are equal up to a scale factor and treat the scale factor as an additional parameter. Here, we use $b_{j,k}(f)$ as the k-th "normalized" spectral template and describe $v_j(f,n)$ as

$$v_j(f,n) = b_{j,k_j(n)}(f)h_j(n), \tag{5.22}$$

where $h_j(n)$ denotes the time-varying amplitude. Furthermore, since the probability of a particular template being selected may depend on the templates selected in the previous frames, we can describe the generative process of $k_j(n)$ using a Markov chain:

$$k_j(n)|k_j(n-1) \sim \pi_{j,k_j(n-1),k_j(n)}, \tag{5.23}$$

These two extensions lead to the hidden Markov model (HMM) proposed in [15–18, 25] where (5.22) can be seen as the output sequence, $k_j(n)$ as the hidden state, and $\pi_{j,k,k'}$ as the state transition probability from state k to state k' (see Fig. 5.1). By properly designing the state transition network, we can flexibly assign probabilities to state durations (the durations of the self-transitions). In addition, by incorporating states associated with speech absence or silence into the state transition network, assuming the state-dependent generative process of the scale factor $h_j(n)$ to be

$$h_j(n)|k_j(n) \sim \mathscr{G}(h_j(n)|\gamma_{k_j(n)}, \beta_{k_j(n)}), \tag{5.24}$$

where $\mathscr{G}(\cdot|\gamma, \beta)$ denotes a gamma distribution with shape parameter $\gamma > 0$ and scale parameter $\beta > 0$

$$\mathscr{G}(h|\gamma, \beta) = \frac{h^{\gamma-1}e^{-h/\beta}}{\Gamma(\gamma)\beta^{\gamma}}, \tag{5.25}$$

and setting the hyperparameters γ_k and β_k so that $h_j(n)$ tends to be near zero at the states associated with speech absence, this model makes it possible to estimate voice activity segments along with solving the separation problem [15–18].

While the above models assume that only one of the spectral templates is activated at a time, another way to model the power spectrogram $v_j(f, n)$ is to express it as the linear sum of the spectral templates $b_{j,1}(f), \ldots, b_{j,K_j}(f)$ scaled by time-varying amplitudes $h_{j,1}(n), \ldots, h_{j,K_j}(n)$:

$$v_j(f, n) = \sum_{k=1}^{K_j} b_{j,k}(f)h_{j,k}(n). \tag{5.26}$$

(5.26) can be interpreted as expressing the matrix $\mathbf{V}_j = [v_j(f, n)]_{f,n}$ as a product of two matrices $\mathbf{B}_j = [b_{j,k}(f)]_{f,k}$ and $\mathbf{H}_j = [h_{j,k}(n)]_{k,n}$ (see Fig. 5.1). Multichannel source separation methods using this model or its variants are called "multichannel non-negative matrix factorization (MNMF)" [4–10]. With this model, the entire set of spectral templates are partitioned into subsets associated with the individual sources. It is also possible to allow all the spectral templates to be shared by every source and let the contribution of the k-th spectral template to source j be determined in a data-driven manner [6–10]. To do so, we drop the index j from $b_{j,k}(f)$ and $h_{j,k}(n)$, and instead introduce a continuous indicator variable $\phi_{j,k} \in [0, 1]$ that sums to unity $\sum_j \phi_{j,k} = 1$. $\phi_{j,k}$ can be interpreted as the expectation of a binary indicator variable that describes the index of the source to which the k-th template is assigned. The power spectrogram $v_j(f, n)$ of source j can thus alternatively be modeled as

$$v_j(f, n) = \sum_{k=1}^{K} \phi_{j,k} b_k(f)h_k(n). \tag{5.27}$$

Another reasonable assumption we can make about source power spectrograms is spectral continuity. This amounts to an assumption that the magnitudes of the STFT coefficients in all the frequency bands originating from the same source tend to vary coherently over time. The most naïve way would be to assume a flat spectrum with a time-varying scale [26]

$$v_j(f, n) = h_j(n). \tag{5.28}$$

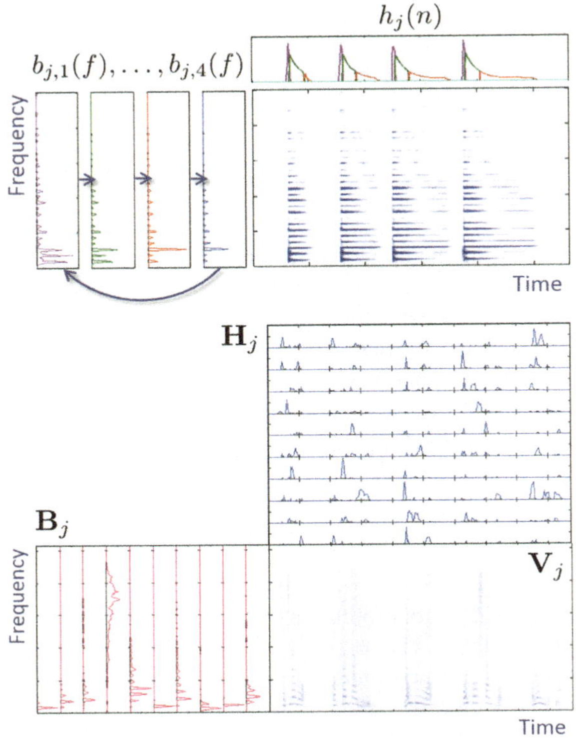

Fig. 5.1 Illustration of spectral models (5.22) and (5.26)

This is actually a particular case of the NMF model (5.26) where $K_j = 1$ and $b_{j,1}(f) = 1$, which means each source has only one flat-shaped template. Under this constraint, assuming (5.9) amounts to assuming that the norm $\|[s_j(0,n), \ldots, s_j(F-1,n)]^\mathsf{T}\|_2 = \sqrt{\sum_f |s_j(f,n)|^2}$ follows a Gaussian distribution with a time-varying variance $h_j(n)$. This is analogous to the assumption employed in independent vector analysis (IVA) [2, 3] where the norm $\|[s_j(0,n), \ldots, s_j(F-1,n)]^\mathsf{T}\|_2$ is assumed to follow a super-Gaussian distribution, which is shown to be effective in eliminating the inherent permutation indeterminacy of the frequency-domain ICA. Other representations ensuring spectral continuity include the autoregressive (AR) model (also known as the all-pole model) [27, 28]

$$v_j(f,n) = \frac{h_j(n)}{|g(e^{\mathrm{j}2\pi f/F}; \boldsymbol{\alpha}_j(n))|^2}, \tag{5.29}$$

$$g(z; \boldsymbol{\alpha}_j(n)) = 1 - \sum_{q=1}^{Q} \alpha_{j,q}(n) z^{-q}, \tag{5.30}$$

where $h_j(n)$ and $\boldsymbol{\alpha}_j(n) = (\alpha_{j,1}(n), \ldots, \alpha_{j,Q}(n))$ denote the power of the excitation signal and the AR parameter set of source j at time n, and Q is the number of poles. Employing this expression is justified by the fact that the power spectrum of speech can be approximated fairly well by an excitation-filter representation using an all-pole model as the vocal-tract filter.

A combination of the AR model and the NMF model has also been proposed [5, 29]. With this model, the power spectrum of a source is expressed as the linear sum of all possible pairs of excitation and filter templates scaled by time-varying amplitudes

$$v_j(f,n) = \sum_k \sum_{k'=1}^{K'_j} \frac{b_{j,k}(f) h_{j,k,k'}(n)}{|g(e^{j2\pi f/F}; \boldsymbol{\alpha}_{j,k'})|^2}, \tag{5.31}$$

$$g(z; \boldsymbol{\alpha}_{j,k'}) = 1 - \sum_{q=1}^{Q} \alpha_{j,k',q} z^{-q}, \tag{5.32}$$

where $b_{j,k}(f)$, $1/|g_{k'}(e^{j2\pi f/F}; \boldsymbol{\alpha}_{j,k'})|^2$ and $h_{j,k,k'}(n)$ denote the k-th excitation spectral template, the k'-th all-pole vocal-tract spectral template and the time-varying amplitude of the $\{k, k'\}$-th excitation-filter pair of source j, respectively. We can easily confirm that when $K'_j = 1$ and $Q = 0$, this model reduces to the NMF model (5.26). Another way of modeling $v_j(f,n)$ using an excitation-filter representation is to express $v_j(f,n)$ as the product of an excitation spectrogram $v_j^{\mathrm{ex}}(f,n)$ and a filter spectrogram $v_j^{\mathrm{ft}}(f,n)$

$$v_j(f,n) = v_j^{\mathrm{ex}}(f,n) v_j^{\mathrm{ft}}(f,n), \tag{5.33}$$

where $v_j^{\mathrm{ex}}(f,n)$ and $v_j^{\mathrm{ft}}(f,n)$ are expressed using the NMF models [11, 25]

$$v_j^{\mathrm{ex}}(f,n) = \sum_k b_{j,k}^{\mathrm{ex}}(f) h_{j,k}^{\mathrm{ex}}(n), \tag{5.34}$$

$$v_j^{\mathrm{ft}}(f,n) = \sum_k b_{j,k}^{\mathrm{ft}}(f) h_{j,k}^{\mathrm{ft}}(n). \tag{5.35}$$

Note that these spectral templates can also be either pre-trained using training samples or estimated from the mixture signal in an unsupervised manner.

A general flexible framework with various combinations of these spectral models is presented in [25].

5.3.2 Spatial Models

As with the source power spectrum, there are several ways to model the spatial covariance $\mathbf{R}_j(f)$ depending on assumptions we make about the properties of wave propagation.

One widely used way of modeling the spatial covariance $\mathbf{R}_j(f)$ is to constrain it to be a rank-1 matrix

$$\mathbf{R}_j(f) = \mathbf{a}_j(f)\mathbf{a}_j^{\mathsf{H}}(f). \tag{5.36}$$

As shown in (5.10), this constraint amounts to using the time-invariant instantaneous mixing system based on the narrowband approximation. We can also assume $\mathbf{R}_j(f)$ to be an unconstrained full-rank matrix, which is shown to be effective particularly under reverberant conditions [7, 24]. This can be explained by replacing $\mathbf{a}_j(f)$ with $\mathbf{a}_j(f) + \boldsymbol{\varepsilon}_j(f)$ where $\boldsymbol{\varepsilon}_j(f)$ is a random vector with mean 0 and a full-rank covariance matrix corresponding to the approximation error related to the narrowband approximation.

Another reasonable way involves describing $\mathbf{R}_j(f)$ as a function of the DOA of source j. If each source is assumed to be located far from the microphones so that the signal can be treated approximately as a plane wave, the interchannel time difference between the microphones depends only on the DOA of the source. Since the time delay between two microphones corresponds to the phase difference of the frequency response of the microphone array, the complex array response can be expressed explicitly by using the DOA. For example, with $I = 2$ microphones, the complex array response for a source at direction θ such that $0 \leq \theta < 2\pi$ is defined as a function of f depending on θ

$$\mathbf{d}(f;\theta) = \begin{bmatrix} 1 \\ e^{\mathrm{j}\omega_f B\cos\theta/C} \end{bmatrix}, \tag{5.37}$$

where ω_f is the angular frequency of the f-th frequency bin, j is the imaginary unit, B [m] is the distance between the two microphones, and C [m/s] is the speed of sound. If the DOA θ_j of source j is known, the array frequency response $\mathbf{a}_j(f)$ should be equal to $\mathbf{d}(f;\theta_j)$. Since we normally have no information about the DOA, we would like to estimate $\mathbf{a}_j(f)$ or $\mathbf{R}_j(f)$ in a data-driven manner under this constraint. By using a discrete set of pre-defined array responses $\mathbf{d}(f;\vartheta_l)$ each corresponding to an angle ϑ_l in radians, the array response $\mathbf{a}_j(f)$ of each fixed source can be modeled as a Gaussian mixture of the DOA-related array responses [20]

$$\mathbf{a}_j(f) \sim \mathcal{N}_{\mathbb{C}}(\mathbf{a}_j(f)|\mathbf{d}(f;\vartheta_{l_j}), \boldsymbol{\Sigma}_{\mathbf{a}}) \tag{5.38a}$$

$$l_j \sim \pi_{l_j} \tag{5.38b}$$

or the spatial covariance matrix $\mathbf{R}_j(f)$ can be modeled as a Wishart mixture of the DOA-related covariance matrices [18, 21]:

$$\mathbf{R}_j(f) \sim \mathscr{W}_{\mathbb{C}}(\mathbf{R}_j(f)|\nu, \boldsymbol{\Sigma}_{\mathbf{d}}(f; \vartheta_{l_j}) + \varepsilon\mathbf{I}) \tag{5.39a}$$

$$l_j \sim \pi_{l_j}, \tag{5.39b}$$

where $\boldsymbol{\Sigma}_{\mathbf{d}}(f; \vartheta_{l_j}) = \mathbf{d}(f; \vartheta_{l_j})\mathbf{d}^{\mathsf{H}}(f; \vartheta_{l_j})$ and $l_j \in \{1, \cdots, L\}$ denotes the index of the predefined DOA assigned to source j, which is assumed to have been drawn from a categorical distribution $\boldsymbol{\pi} = (\pi_1, \ldots, \pi_L)$ and $\mathscr{W}_{\mathbb{C}}$ denotes the complex Wishart distribution

$$\mathscr{W}_{\mathbb{C}}(\mathbf{R}|\nu, \boldsymbol{\Psi}) \propto |\mathbf{R}|^{\nu-I} e^{-\mathrm{tr}(\mathbf{R}\boldsymbol{\Psi}^{-1})}. \tag{5.40}$$

Here, the term $\varepsilon\mathbf{I}$ is added to ensure that $\boldsymbol{\Sigma}_{\mathbf{d}}(f; \vartheta_{l_j}) + \varepsilon\mathbf{I}$ is invertible. The spatial covariance matrix $\mathbf{R}_j(f)$ can also be modeled as a weighted sum of the DOA-related covariance matrices [8, 17]

$$\mathbf{R}_j(f) = \sum_l q_{j,l}\boldsymbol{\Sigma}_{\mathbf{d}}(f; \vartheta_l), \tag{5.41}$$

where $q_{j,l} \geq 0$ denotes the contribution of the l-th predefined DOA to source j. We call (5.38) and (5.39) the "DOA mixture model" and call (5.41) the "DOA kernel model". With the DOA mixture model, l_j is treated as a latent variable to be marginalized out, whereas with the DOA kernel model, $q_{j,l}$ is treated as a parameter to be estimated subject to non-negativity.

To handle a time-varying spatial covariance and thus allow each source to move, the DOA mixture model can be further extended by describing the transition of the DOAs using an HMM [22] (Fig. 5.2):

$$\mathbf{a}_{j,n}(f) = \mathscr{N}_{\mathbb{C}}(\mathbf{a}_{j,n}(f)|\mathbf{d}(f; \vartheta_{l_j(n)}), \boldsymbol{\Sigma}_{\mathbf{a}}), \tag{5.42}$$

$$l_j(n)|l_j(n-1) \sim \pi_{j,l_j(n-1),l_j(n)}. \tag{5.43}$$

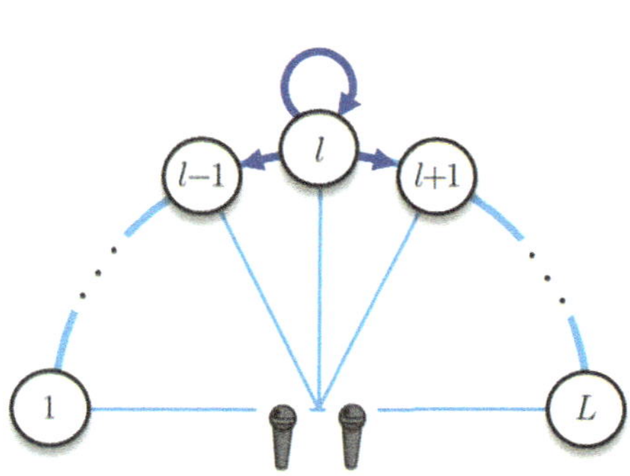

Fig. 5.2 Illustration of DOA-HMM [22]

Other approaches include placing an inverse-Wishart chain prior over the sequence $\mathbf{R}_{j,0}(f), \ldots, \mathbf{R}_{j,N-1}(f)$ [30] or a Gaussian chain prior over the sequence $\mathbf{a}_{j,0}(f), \ldots, \mathbf{a}_{j,N-1}(f)$ [12] in order to ensure that $\mathbf{R}_{j,n}(f)$ and $\mathbf{a}_{j,n}(f)$ vary smoothly over time.

5.4 Parameter Estimation and Signal Separation

5.4.1 Parameter Estimation

Once the likelihood function is defined according to the choice of mixing system, source spectral model and spatial model, there are several ways to estimate $\boldsymbol{\theta}$ given the STFT coefficients of observed signals $X = [\mathbf{x}(f, n)]_{f,n}$. The parameter estimation problems can be primarily divided into maximum likelihood (or maximum a posteriori) estimation and Bayesian inference problems. The aim of the former problem is to find the estimate of $\boldsymbol{\theta}$ that maximizes the likelihood function (or the posterior distribution) of $\boldsymbol{\theta}$ whereas the aim of the latter is to infer the posterior distribution of $\boldsymbol{\theta}$, given an observation $X = [\mathbf{x}(f, n)]_{f,n}$. In this section, we briefly introduce the general principles of the majorization-minimization (MM) algorithm (also known as the auxiliary function-based approach) [31, 32] as a representative example of the approaches for the former type and the variational inference algorithm as a representative example of the approaches for the latter type. Detailed derivations of some examples of parameter estimation algorithms will be presented in Sect. 5.6.

5.4.1.1 Majorization-Minimization Algorithm

An MM algorithm refers to an iterative algorithm that searches for a stationary point of a cost function by iteratively minimizing an auxiliary function called a "majorizer" that is guaranteed to never become below the objective function. When constructing an MM algorithm for a certain minimization problem, the main issue is to design the majorizer. If a majorizer is properly designed, the algorithm is guaranteed to converge to a stationary point of the cost function. It should be noted that this concept has been adopted in many existing algorithms. For example, the expectation-maximization (EM) algorithm [33] is a special case of the MM algorithm. It is also well known for its use in an algorithm for NMF [34–36]. In general, if we can build a tight majorizer/minorizer that is easy to optimize, we can expect to obtain a fast-converging algorithm.

Suppose $\mathscr{C}(\boldsymbol{\theta})$ is a cost function that we want to minimize with respect to $\boldsymbol{\theta}$. A [2]majorizer $\mathscr{D}(\boldsymbol{\theta}, \boldsymbol{\alpha})$ is defined as a function satisfying

[2]If we want to maximize $\mathscr{C}(\boldsymbol{\theta})$, we will use a minorizer instead, which is defined as $\mathscr{C}(\boldsymbol{\theta}) = \max_{\boldsymbol{\alpha}} \mathscr{D}(\boldsymbol{\theta}, \boldsymbol{\alpha})$.

$$\mathscr{C}(\boldsymbol{\theta}) = \min_{\boldsymbol{\alpha}} \mathscr{D}(\boldsymbol{\theta}, \boldsymbol{\alpha}), \tag{5.44}$$

where $\boldsymbol{\alpha}$ is an auxiliary variable. $\mathscr{C}(\boldsymbol{\theta})$ is then shown to be non-increasing under the updates,

$$\boldsymbol{\theta} \leftarrow \operatorname*{argmin}_{\boldsymbol{\theta}} \mathscr{D}(\boldsymbol{\theta}, \boldsymbol{\alpha}), \tag{5.45}$$

$$\boldsymbol{\alpha} \leftarrow \operatorname*{argmin}_{\boldsymbol{\alpha}} \mathscr{D}(\boldsymbol{\theta}, \boldsymbol{\alpha}). \tag{5.46}$$

This can be proved as follows. Let us denote the iteration number by ℓ, set $\boldsymbol{\theta}$ at an arbitrary value $\boldsymbol{\theta}^{(\ell)}$ and define $\boldsymbol{\alpha}^{(\ell+1)} = \operatorname{argmin}_{\boldsymbol{\alpha}} \mathscr{D}(\boldsymbol{\theta}^{(\ell)}, \boldsymbol{\alpha})$ and $\boldsymbol{\theta}^{(\ell+1)} = \operatorname{argmin}_{\boldsymbol{\theta}} \mathscr{D}(\boldsymbol{\theta}, \boldsymbol{\alpha}^{(\ell+1)})$. First, it is obvious that $\mathscr{C}(\boldsymbol{\theta}^{(\ell)}) = \mathscr{D}(\boldsymbol{\theta}^{(\ell)}, \boldsymbol{\alpha}^{(\ell+1)})$. Next, we can confirm that $\mathscr{D}(\boldsymbol{\theta}^{(\ell)}, \boldsymbol{\alpha}^{(\ell+1)}) \geq \mathscr{D}(\boldsymbol{\theta}^{(\ell+1)}, \boldsymbol{\alpha}^{(\ell+1)})$ since $\boldsymbol{\theta}^{(\ell+1)}$ is the minimizer of $\mathscr{D}(\boldsymbol{\theta}, \boldsymbol{\alpha}^{(\ell+1)})$ with respect to $\boldsymbol{\theta}$. By definition, it is obvious that $\mathscr{D}(\boldsymbol{\theta}^{(\ell+1)}, \boldsymbol{\alpha}^{(\ell+1)}) \geq \mathscr{C}(\boldsymbol{\theta}^{(\ell+1)})$ and so we can finally show that $\mathscr{C}(\boldsymbol{\theta}^{(\ell)}) \geq \mathscr{C}(\boldsymbol{\theta}^{(\ell+1)})$. A sketch of this proof can be found in Fig. 5.3.

Here, we briefly show that the EM algorithm is a special case of the MM algorithm. Let X be an observed data set, $p(X|\boldsymbol{\theta})$ be a likelihood function that we want to maximize with respect to a parameter set $\boldsymbol{\theta}$, and Z be a set of hidden or latent variables. Note that the latent variables can be either discrete or continuous. While here we consider the discrete case, the following also applies to the continuous case by simply replacing the summation over all members of Z with an integral. First, we can show that

$$\log p(X|\boldsymbol{\theta}) = \log \sum_{Z} p(X, Z|\boldsymbol{\theta}) = \log \sum_{Z} \lambda(Z) \frac{p(X, Z|\boldsymbol{\theta})}{\lambda(Z)} \tag{5.47}$$

$$\geq \sum_{Z} \lambda(Z) \log \frac{p(X, Z|\boldsymbol{\theta})}{\lambda(Z)}, \tag{5.48}$$

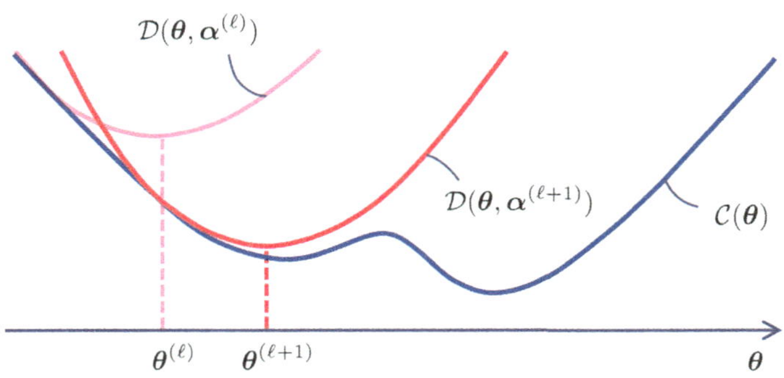

Fig. 5.3 Illustration of the majorization-minimization algorithm

where $\lambda(Z)$ is an arbitrary non-negative weight function that is subject to the normalization constraint

$$\sum_Z \lambda(Z) = 1. \tag{5.49}$$

(5.48) follows from Jensen's inequality by using the fact that the logarithmic function is a concave function. We can use the right-hand side of this inequality as the minorizer of the log-likelihood $\log p(X|\boldsymbol{\theta})$. Thus, we can show that $\log p(X|\boldsymbol{\theta})$ is non-decreasing under the updates

$$\lambda \leftarrow \underset{\lambda}{\operatorname{argmax}} \sum_Z \lambda(Z) \log \frac{p(X, Z|\boldsymbol{\theta})}{\lambda(Z)} = p(Z|X, \boldsymbol{\theta}) \tag{5.50}$$

$$\boldsymbol{\theta} \leftarrow \underset{\boldsymbol{\theta}}{\operatorname{argmax}} \sum_Z \lambda(Z) \log p(X, Z|\boldsymbol{\theta}). \tag{5.51}$$

(5.50) can be confirmed using the fact that the equality of (5.48) holds when $p(X, Z|\boldsymbol{\theta})/\lambda(Z)$ becomes equal for any Z

$$\frac{p(X, Z|\boldsymbol{\theta})}{\lambda(Z)} = \xi(X, \boldsymbol{\theta}). \tag{5.52}$$

Thus, we obtain

$$\lambda(Z) = \frac{p(X, Z|\boldsymbol{\theta})}{\xi(X, \boldsymbol{\theta})} \tag{5.53}$$

$$\Rightarrow \quad \sum_Z \lambda(Z) = \frac{1}{\xi(X, \boldsymbol{\theta})} \sum_Z p(X, Z|\boldsymbol{\theta}) = 1 \tag{5.54}$$

$$\Rightarrow \quad \xi(X, \boldsymbol{\theta}) = \sum_Z p(X, Z|\boldsymbol{\theta}) = p(X|\boldsymbol{\theta}) \tag{5.55}$$

$$\Rightarrow \quad \lambda(Z) = \frac{p(X, Z|\boldsymbol{\theta})}{p(X|\boldsymbol{\theta})} = p(Z|X, \boldsymbol{\theta}). \tag{5.56}$$

Dempster et al. called $Q(\boldsymbol{\theta}, \boldsymbol{\theta}') = \sum_Z p(Z|X, \boldsymbol{\theta}') \log p(X, Z|\boldsymbol{\theta})$ the "Q function" where $\boldsymbol{\theta}'$ denotes the estimate of $\boldsymbol{\theta}$ at the previous iteration. The EM algorithm consists of computing $p(Z|X, \boldsymbol{\theta}')$ and maximizing $Q(\boldsymbol{\theta}, \boldsymbol{\theta}')$. We can confirm that (5.50) and (5.51), respectively, correspond to these steps.

In Sect. 5.6, we show detailed derivations of MM-based BSS algorithms [7, 18].

5.4.1.2 Variational Inference Algorithm

The aim of the variational inference algorithm is to approximate the true posterior distributions of all the random variables involved in the generative model.

Let $\boldsymbol{\theta} = [\theta_m]_m$ be the entire set of the variables of interest and X be an observed data set. Our goal is to compute the posterior

$$p(\boldsymbol{\theta}|X) = \frac{p(\boldsymbol{\theta}, X)}{p(X)}. \tag{5.57}$$

The joint distribution $p(\boldsymbol{\theta}, X)$ can usually be written explicitly according to the assumed generative model. However, to obtain the exact posterior $p(\boldsymbol{\theta}|X)$, we must compute $p(X)$, which typically involves many intractable integrals. Instead of obtaining the exact posterior, the variational Bayesian approach approximates this posterior variationally by solving an optimization problem:

$$\hat{q}(\boldsymbol{\theta}) = \underset{q}{\operatorname{argmin}}\, \mathrm{KL}[q(\boldsymbol{\theta}) \| p(\boldsymbol{\theta}|X)], \tag{5.58}$$

subject to

$$\int q(\boldsymbol{\theta}) \mathrm{d}\boldsymbol{\theta} = 1, \tag{5.59}$$

where $\mathrm{KL}[\cdot \| \cdot]$ denotes the Kullback-Leibler (KL) divergence between its two arguments, i.e.,

$$\mathrm{KL}[q(\boldsymbol{\theta}) \| p(\boldsymbol{\theta}|X)] = \int q(\boldsymbol{\theta}) \log \frac{q(\boldsymbol{\theta})}{p(\boldsymbol{\theta}|X)} \mathrm{d}\boldsymbol{\theta}. \tag{5.60}$$

By restricting the class of the approximate distributions to those that factorize into independent factors:

$$q(\boldsymbol{\theta}) = \prod_m q(\theta_m), \quad \int q(\theta_m) \mathrm{d}\theta_m = 1, \tag{5.61}$$

we can use a simple coordinate ascent algorithm to find a local optimum of (5.58). It can be shown using the calculus of variations that the "optimal" distribution for each of the factors can be expressed as:

$$\hat{q}(\theta_m) \propto \exp \mathbb{E}_{\boldsymbol{\theta} \backslash \theta_m}[\log p(\boldsymbol{\theta}, X)], \tag{5.62}$$

where θ_m indicates one of the factors and $\mathbb{E}_{\boldsymbol{\theta} \backslash \theta_m}[\log p(\boldsymbol{\theta}, X)]$ is the expectation of the joint probability of the data and latent variables, taken over all variables except θ_m.

5.4.2 Signal Separation

Once the parameter set $\boldsymbol{\theta}$ is estimated, we can obtain the estimates of the source signals in different ways according to the assumed mixing systems. With an instantaneous mixing system, a typical choice would be the minimum mean square error (MMSE) estimator of $\mathbf{s}(f, n)$ [4]:

$$\hat{\mathbf{s}}(f, n) = \mathbb{E}[\mathbf{s}(f, n)|\mathbf{x}(f, n)] = \mathbf{G}(f, n)\mathbf{x}(f, n), \tag{5.63}$$

where

$$\mathbf{G}(f, n) = \Sigma_{\mathbf{s}}(f, n)\mathbf{A}_n^{\mathsf{H}}(f)(\mathbf{A}_n(f)\Sigma_{\mathbf{s}}(f, n)\mathbf{A}_n^{\mathsf{H}}(f) + \boldsymbol{\Sigma}_{\mathbf{u}}(f, n))^{-1} \tag{5.64}$$

is the well-known multichannel Wiener filter. (5.64) can be derived by using the fact that with jointly Gaussian random variables

$$\begin{bmatrix}\mathbf{x}\\ \mathbf{s}\end{bmatrix} \sim \mathcal{N}\left(\begin{bmatrix}\mathbf{x}\\ \mathbf{s}\end{bmatrix} \middle| \begin{bmatrix}\boldsymbol{\mu}_{\mathbf{x}}\\ \boldsymbol{\mu}_{\mathbf{s}}\end{bmatrix}, \begin{bmatrix}\boldsymbol{\Sigma}_{\mathbf{xx}} & \boldsymbol{\Sigma}_{\mathbf{xs}}\\ \boldsymbol{\Sigma}_{\mathbf{sx}} & \boldsymbol{\Sigma}_{\mathbf{ss}}\end{bmatrix}\right), \tag{5.65}$$

the conditional expectation $\mathbb{E}[\mathbf{s}|\mathbf{x}]$ is given as [37]

$$\mathbb{E}[\mathbf{s}|\mathbf{x}] = \boldsymbol{\mu}_{\mathbf{s}} + \boldsymbol{\Sigma}_{\mathbf{sx}}\boldsymbol{\Sigma}_{\mathbf{xx}}^{-1}(\mathbf{x} - \boldsymbol{\mu}_{\mathbf{x}}), \tag{5.66}$$

and that

$$\mathbb{E}[\mathbf{x}(f, n)\mathbf{x}^{\mathsf{H}}(f, n)] = \mathbf{A}_n(f)\Sigma_{\mathbf{s}}(f, n)\mathbf{A}_n^{\mathsf{H}}(f) + \boldsymbol{\Sigma}_{\mathbf{u}}(f, n), \tag{5.67}$$

$$\mathbb{E}[\mathbf{s}(f, n)\mathbf{x}^{\mathsf{H}}(f, n)] = \Sigma_{\mathbf{s}}(f, n)\mathbf{A}_n^{\mathsf{H}}(f). \tag{5.68}$$

When using the full-rank spatial covariance model, it may be convenient to use the MMSE estimator of the spatial image $\hat{\mathbf{c}}_j(f, n, 0) = \mathbf{a}_{j,n}(f, 0)s_j(f, n)$ [16, 24]:

$$\begin{aligned}\hat{\mathbf{c}}_j(f, n, 0) &= \mathbb{E}[\mathbf{c}_j(f, n, 0)|\mathbf{x}(f, n)]\\ &= v_j(f, n)\mathbf{R}_{j,n}(f, 0)\left(\sum_{j'=1}^{J}\sum_{m=0}^{M-1} v_{j'}(f, n-m)\mathbf{R}_{j',n}(f, m) + \boldsymbol{\Sigma}_{\mathbf{u}}(f, n)\right)^{-1}\mathbf{x}(f, n).\end{aligned} \tag{5.69}$$

This estimator can be derived in the same way using the fact that $\mathbf{x}(f, n)$, $\mathbf{c}_j(f, n, m)$, and $\mathbf{u}(f, n)$ are jointly Gaussian and

$$\mathbf{x}(f, n) = \sum_{j=1}^{J}\sum_{m=0}^{M-1} \mathbf{c}_j(f, n, m) + \mathbf{u}(f, n). \tag{5.70}$$

With an instantaneous and convolutive demixing systems, we can directly use (5.6) and (5.7) [5, 9, 10, 27, 28] once we obtain the demixing filter $\mathbf{W}^{\mathsf{H}}(f)$ or $\mathbf{W}^{\mathsf{H}}(f,n)$.

With a sparse mixing system, we can use the posterior source presence probability $\gamma_j(f,n)$ [20, 38]

$$\hat{s}_j(f,n) = \gamma_j(f,n)\frac{\mathbf{a}_j^{\mathsf{H}}(f,n)\boldsymbol{\Sigma}_{\mathbf{u}}^{-1}(f,n)\mathbf{x}(f,n)}{\mathbf{a}_j^{\mathsf{H}}(f,n)\boldsymbol{\Sigma}_{\mathbf{u}}^{-1}(f,n)\mathbf{a}_j(f,n)}. \tag{5.71}$$

5.5 Categorization of State-of-the-art Approaches

Many state-of-the-art approaches can be categorized according to the choice of mixing system, source spectral model, and spatial model. Tables 5.1 and 5.2 show the

Table 5.1 Different approaches categorized according to $\boldsymbol{\mu}_{\mathbf{x}}(f,n)$ and $\boldsymbol{\Sigma}_{\mathbf{x}}(f,n)$ in (5.14)

Method	$\boldsymbol{\mu}_{\mathbf{x}}(f,n)$	$\boldsymbol{\Sigma}_{\mathbf{x}}(f,n)$
Attias (2003)	$\mathbf{0}$	$\sum_j \sum_m v_j(f,n-m)\mathbf{R}_j(f,m)$
Izumi et al. (2007)	$\mathbf{a}_{z(f,n)}(f)\mu(f,n)$	$\boldsymbol{\Sigma}_{\mathbf{u}}$
Ozerov & Févotte (2010)	$\mathbf{0}$	$\sum_j v_j(f,n)\mathbf{R}_j(f) + \boldsymbol{\Sigma}_{\mathbf{u}}$
Duong et al. (2010)	$\mathbf{0}$	$\sum_j v_j(f,n)\mathbf{R}_j(f)$
Kameoka et al. (2010)	$\mathbf{W}^{-1}(f,0)\sum_{m=1}^{M-1}\mathbf{W}^{\mathsf{H}}(f,m)\mathbf{x}(f,n-m)$	$(\mathbf{W}^{\mathsf{H}}(f,0))^{-1}\boldsymbol{\Sigma}_{\mathbf{s}}\mathbf{W}^{-1}(f,0)$
Yoshioka et al. (2011)	$\mathbf{W}^{-1}(f,0)\sum_{m=1}^{M-1}\mathbf{W}^{\mathsf{H}}(f,m)\mathbf{x}(f,n-m)$	$(\mathbf{W}^{\mathsf{H}}(f,0))^{-1}\boldsymbol{\Sigma}_{\mathbf{s}}\mathbf{W}^{-1}(f,0)$
Ozerov et al. (2011)	$\mathbf{0}$	$\sum_j v_j(f,n)\mathbf{R}_j(f) + \boldsymbol{\Sigma}_{\mathbf{u}}$
Ono et al. (2012)	$\mathbf{0}$	$(\mathbf{W}^{\mathsf{H}}(f))^{-1}\boldsymbol{\Sigma}_{\mathbf{s}}\mathbf{W}^{-1}(f)$
Kameoka et al. (2012)	$\mathbf{a}_{z(f,n)}(f)\mu(f,n)$	$\boldsymbol{\Sigma}_{\mathbf{u}}$
Sawada et al. (2013)	$\mathbf{0}$	$\sum_j v_j(f,n)\mathbf{R}_j(f)$
Higuchi et al. (2014a)	$\mathbf{a}_{z(f,n)}(f)\mu(f,n)$	$\boldsymbol{\Sigma}_{\mathbf{u}}$
Higuchi et al. (2014b)	$\mathbf{0}$	$\sum_j \sum_m v_j(f,n)\mathbf{R}_j(f)$
Higuchi et al. (2014c)	$\mathbf{0}$	$\sum_j \sum_m v_j(f,n-m)\mathbf{R}_j(f,m)$
Otsuka et al. (2014)	$\mathbf{0}$	$v_{z(f,n)}(f,n)\mathbf{R}_{z(f,n)}(f)$
Higuchi & Kameoka (2015)	$\mathbf{0}$	$\sum_j \sum_m v_j(f,n-m)\mathbf{R}_j(f,m)$
Kitamura et al. (2015)	$\mathbf{0}$	$(\mathbf{W}^{\mathsf{H}}(f))^{-1}\boldsymbol{\Sigma}_{\mathbf{s}}\mathbf{W}^{-1}(f)$
Adiloğlu & Vincent (2016)	$\mathbf{0}$	$\sum_j v_j(f,n)\mathbf{R}_j(f)$
Kounades-Bastian et al. (2016)	$\mathbf{0}$	$\sum_j v_j(f,n)\mathbf{R}_j(f,n)$

Table 5.2 Different approaches categorized according to the constraints on $v_j(f,n)$ and $\mathbf{R}_{j,n}(f,m)$

Method	$v_j(f,n)$	$\mathbf{R}_{j,n}(f,m)$
Attias (2003)	$b_{j,k_j(n)}(f)$	$\mathbf{a}_j(f,m)\mathbf{a}_j^{\mathsf{H}}(f,m)$
Izumi et al. (2007)	none	$\mathbf{a}_j(f) = \mathbf{d}(f;\vartheta_{l_j})$
Ozerov & Févotte (2010)	$\sum_k b_{j,k}(f)h_{j,k}(n)$	$\mathbf{a}_j(f)\mathbf{a}_j^{\mathsf{H}}(f)$
Duong et al. (2010)	$v_j(f,n)$	$\mathbf{R}_j(f)$
Kameoka et al. (2010)	$\sum_{k,r} \frac{b_{j,k}(f)h_{j,k,r}(n)}{\lvert 1-\sum_q \alpha_{j,r,q}e^{-\mathrm{j}2\pi fq/F}\rvert^2}$	$\mathbf{a}_j(f,m)\mathbf{a}_j^{\mathsf{H}}(f,m)$
Duong et al. (2011)	$v_j(f,n)$	$\mathbf{R}_{j,n}(f) \sim \mathscr{IW}_{\mathbb{C}}(\nu, \mathbf{R}_{j,n-1}(f))$
Yoshioka et al. (2011)	$\frac{\eta_j(n)}{\lvert 1-\sum_q \alpha_q(n)e^{-\mathrm{j}2\pi fq/F}\rvert^2}$	$\mathbf{a}_j(f,m)\mathbf{a}_j^{\mathsf{H}}(f,m)$
Ozerov et al. (2011)	$\sum_k \phi_{j,k}b_k(f)h_k(n)$	$\mathbf{R}_j(f)$
Ono et al. (2012)	$h_j(n)$	$\mathbf{a}_j(f)\mathbf{a}_j^{\mathsf{H}}(f)$
Kameoka et al. (2012)	none	$\mathbf{a}_j(f) \sim \mathscr{N}_{\mathbb{C}}(\mathbf{d}(f;\vartheta_{l_j}), \boldsymbol{\Sigma}_{\mathbf{a}})$
Sawada et al. (2013)	$\sum_k \phi_{j,k}b_k(f)h_k(n)$	$\mathbf{R}_j(f)$
Higuchi et al. (2014a)	none	$\mathbf{a}_{j,n}(f) \sim \mathscr{N}_{\mathbb{C}}(\mathbf{d}(f;\vartheta_{l_j(n)}), \boldsymbol{\Sigma}_{\mathbf{a}})$
Nikunen & Virtanen (2014)	$\sum_k \phi_{j,k}b_k(f)h_k(n)$	$\sum_l q_{j,l}\boldsymbol{\Sigma}_{\mathbf{d}}(f;\vartheta_l)$
Higuchi et al. (2014b)	$b_{j,k_j(n)}(f)h_j(n)$	$\mathbf{R}_j(f)$
Higuchi et al. (2014c)	$b_{j,k_j(n)}(f)h_j(n)$	$\mathbf{R}_j(f,0) \sim \mathscr{W}_{\mathbb{C}}(\nu, \sum_l q_{j,l}\boldsymbol{\Sigma}_{\mathbf{d}}(f;\vartheta_l) + \varepsilon\mathbf{I})$
Otsuka et al. (2014)	$\lVert\mathbf{x}(f,n)\rVert_2^2$ (fixed)	$\mathbf{R}_j(f) \sim \mathscr{W}_{\mathbb{C}}(\nu, \boldsymbol{\Sigma}_{\mathbf{d}}(f;\vartheta_{l_j}) + \varepsilon\mathbf{I})$
Higuchi & Kameoka (2015)	$b_{j,k_j(n)}(f)h_j(n)$	$\mathbf{R}_j(f,0) \sim \mathscr{W}_{\mathbb{C}}(\nu, \boldsymbol{\Sigma}_{\mathbf{d}}(f;\vartheta_{l_j}) + \varepsilon\mathbf{I})$
Kitamura et al. (2015)	$\sum_k \phi_{j,k}b_k(f)h_k(n)$	$\mathbf{a}_j(f)\mathbf{a}_j^{\mathsf{H}}(f)$
Adiloğlu & Vincent (2016)	$v_j^{\mathrm{ex}}(f,n)v_j^{\mathrm{ft}}(f,n)$	$\mathbf{a}_j(f)\mathbf{a}_j^{\mathsf{H}}(f)$
Kounades-Bastian et al. (2016)	$\sum_k b_{j,k}(f)h_{j,k}(n)$	$\mathbf{a}_{j,n}(f) \sim \mathscr{N}_{\mathbb{C}}(\mathbf{a}_{j,n-1}(f), \boldsymbol{\Sigma}_{\mathbf{a}})$

relationships between the state-of-the-art methods according to the definitions of $\boldsymbol{\mu}_{\mathbf{x}}(f,n)$ and $\boldsymbol{\Sigma}_{\mathbf{x}}(f,n)$ in (5.14) and the constraints on $v_j(f,n)$ and $\mathbf{R}_{j,n}(f,m)$. Here, $\mathbf{R}_{j,n}(f,m)$ represents the covariance matrix of $\mathbf{a}_{j,n}(f,m) + \boldsymbol{\varepsilon}_{j,n}(f,m)$ where $\boldsymbol{\varepsilon}_{j,n}(f,m)$ denotes the approximation error related to the narrowband approximation, which we assume to be a random vector with mean 0 and a full-rank covariance matrix. If we assume that the narrowband approximation holds such that $\boldsymbol{\varepsilon}_{j,n}(f,m) = \mathbf{0}$, $\mathbf{R}_{j,n}(f,m)$ becomes equal to $\mathbf{a}_{j,n}(f,m)\mathbf{a}_{j,n}(f,m)^{\mathsf{H}}$. If we assume the mixing system to be time-invariant, the index n in $\mathbf{R}_{j,n}(f,m)$ can be dropped, i.e., $\mathbf{R}_{j,n}(f,m) = \mathbf{R}_j(f,m)$. If we assume an instantaneous mixing model, the index m in $\mathbf{R}_{j,n}(f,m)$ can be dropped, i.e., $\mathbf{R}_{j,n}(f,m) = \mathbf{R}_{j,n}(f)$. Table 5.3 categorizes the state-of-the-art methods according to the type of inference algorithm where EM, CD, MM, VB and GS stand for the EM algorithm, the coordinate descent algorithm, the MM algorithm, the variational inference algorithm, and the Gibbs sampling algorithm, respectively.

Table 5.3 Different approaches categorized according to the types of inference algorithms. EM, CD, MM, VB and GS stand for the EM algorithm, the coordinate descent algorithm, the MM algorithm, the variational inference algorithm, and the Gibbs sampling algorithm, respectively

Method	Algorithms
Attias (2003)	EM
Izumi et al. (2007)	EM
Ozerov & Févotte (2010)	EM
Duong et al. (2010)	EM
Kameoka et al. (2010)	EM
Duong et al. (2011)	EM
Yoshioka et al. (2011)	CD
Ozerov et al. (2011)	EM
Ono et al. (2012)	MM
Kameoka et al. (2012)	VB
Sawada et al. (2013)	MM
Higuchi et al. (2014a)	VB
Nikunen & Virtanen (2014)	MM
Higuchi et al. (2014b)	MM
Higuchi et al. (2014c)	MM
Otsuka et al. (2014)	GS
Higuchi & Kameoka (2015)	MM
Kitamura et al. (2015)	MM
Adiloğlu & Vincent (2016)	VB
Kounades-Bastian et al. (2016)	VB

5.6 Derivations of MNMF and MFHMM Algorithms

5.6.1 *MNMF Algorithm*

Here, we give a detailed description of the MM-based algorithm for MNMF, which we presented in [7]. This algorithm is an extension of the MM-based algorithm originally developed for solving general model-fitting problems using the Itakura-Saito (IS) divergence [39]. First we show the basic idea behind obtaining the MM-based algorithm for the single-channel NMF with the IS divergence and then show how it can be extended to a multichannel case.

The cost function for the single-channel NMF with the IS divergence is

$$\mathscr{C}_{\mathrm{NMF}}(\boldsymbol{\theta}) = \sum_{n,f} \left(\frac{|x(f,n)|^2}{v(f,n)} + \log v(f,n) \right), \tag{5.72}$$

where $x(f, n)$ is the observed STFT coefficient, $v(f, n) = \sum_k b_k(f)h_k(n)$ and $\boldsymbol{\theta}$ is a set consisting of $\mathbf{B} = [b_k(f)]_{k,f}$ and $\mathbf{H} = [h_k(n)]_{k,n}$ [14]. Although it is difficult to obtain an analytical expression of the global optimum solution, an auxiliary function of $\mathscr{C}_{\mathrm{NMF}}(\boldsymbol{\theta})$ can be obtained by following the idea in [39] as follows. First, by using the fact that a reciprocal function $f(x) = 1/x$ is convex for $x > 0$, we can use Jensen's inequality to obtain

$$\frac{|x(f,n)|^2}{v(f,n)} \leq \sum_k \rho_k(f,n)\frac{|x(f,n)|^2}{b_k(f)h_k(n)/\rho_k(f,n)} = \sum_k \rho_k^2(f,n)\frac{|x(f,n)|^2}{b_k(f)h_k(n)}, \tag{5.73}$$

where $0 \leq \rho_k(f, n) \leq 1$ is an arbitrary weight that must satisfy $\sum_k \rho_k(f, n) = 1$. It can be shown that the equality of this inequality holds when

$$\rho_k(f,n) = \frac{b_k(f)h_k(n)}{\sum_{k'} b_{k'}(f)h_{k'}(n)}. \tag{5.74}$$

Next, since the logarithmic function $f(x) = \log x$ is concave for $x > 0$, the tangent line to $f(x)$ is guaranteed never to lie below $f(x)$. Thus, we have

$$\log v(f,n) \leq \frac{v(f,n) - \kappa(f,n)}{\kappa(f,n)} + \log \kappa(f,n), \tag{5.75}$$

for any $\kappa(f, n) > 0$. The equality of this inequality holds when

$$\kappa(f,n) = v(f,n). \tag{5.76}$$

By combining these inequalities, we have

$$\mathscr{C}_{\mathrm{NMF}}(\boldsymbol{\theta}) \leq \sum_{f,n}\left(\sum_k \rho_k^2(f,n)\frac{|x(f,n)|^2}{b_k(f)h_k(n)} + \frac{v(f,n) - \kappa(f,n)}{\kappa(f,n)} + \log \kappa(f,n)\right). \tag{5.77}$$

Hence, we can use the right-hand side of this inequality as a majorizer for $\mathscr{C}_{\mathrm{NMF}}(\theta)$ where $\rho = [\rho_k(f, n)]_{k,f,n}$ and $\kappa = [\kappa(f, n)]_{f,n}$ are auxiliary variables. Here, (5.74) and (5.76) correspond to the update rules for the auxiliary variables. What is particularly notable about this majorizer is that while $\mathscr{C}_{\mathrm{NMF}}(\boldsymbol{\theta})$ involves the nonlinear interaction of $b_1(f)h_1(n), \ldots, b_K(f)h_K(n)$, it is given in a separable form expressed as the linear sum of the $1/b_k(f)h_k(n)$ and $b_k(f)h_k(n)$ terms, which is relatively easy to optimize with respect to $b_k(f)$ and $h_k(n)$. By differentiating this majorizer with respect to $b_k(f)$ and $h_k(n)$, and setting the results at zero, we obtain the following update rules for $b_k(f)$ and $h_k(n)$:

$$b_k(f) = \sqrt{\frac{\sum_n \rho_k^2(f,n)|x(f,n)|^2/h_k(n)}{\sum_n h_k(n)/\kappa(f,n)}}, \tag{5.78}$$

$$h_k(n) = \sqrt{\frac{\sum_f \rho_k^2(f,n)|x(f,n)|^2/b_k(f)}{\sum_f b_k(f)/\kappa(f,n)}}. \tag{5.79}$$

Now, let us turn to the objective function for MNMF. By substituting $\boldsymbol{\mu}_{\mathbf{x}}(f,n) = \mathbf{0}$, $\boldsymbol{\Sigma}_{\mathbf{x}}(f,n) = \sum_j v_j(f,n)\mathbf{R}_j(f)$, and $v_j(f,n) = \sum_k \phi_{j,k} b_k(f) h_k(n)$ into the log-likelihood (5.18), reversing the sign and neglecting the constant terms, we obtain the objective function to be minimized as

$$\mathscr{C}_{\mathrm{MNMF}}(\boldsymbol{\theta}) = \sum_{f,n} \left\{ \mathrm{tr}\left(\widehat{\boldsymbol{\Sigma}}_{\mathbf{x}}(f,n)\boldsymbol{\Sigma}_{\mathbf{x}}^{-1}(f,n)\right) + \log\det \boldsymbol{\Sigma}_{\mathbf{x}}(f,n) \right\}, \tag{5.80}$$

where

$$\widehat{\boldsymbol{\Sigma}}_{\mathbf{x}}(f,n) = \mathbf{x}(f,n)\mathbf{x}^{\mathsf{H}}(f,n), \tag{5.81}$$

$$\boldsymbol{\Sigma}_{\mathbf{x}}(f,n) = \sum_j \sum_k \phi_{j,k} b_k(f) h_k(n) \mathbf{R}_j(f). \tag{5.82}$$

We can confirm that when the number of channels and sources is $I = 1$ and $J = 1$, respectively, and $\phi_{j,k} = 1$, this objective function reduces to the objective (5.72). We can obtain an auxiliary function given in a separable form in the same way as the single channel case. By analogy with (5.73), we have

$$\mathrm{tr}\left(\widehat{\boldsymbol{\Sigma}}_{\mathbf{x}}(f,n)\boldsymbol{\Sigma}_{\mathbf{x}}^{-1}(f,n)\right) \le \sum_{j,k} \frac{\mathrm{tr}(\widehat{\boldsymbol{\Sigma}}_{\mathbf{x}}(f,n)\mathbf{P}_{j,k}(f,n)\mathbf{R}_j^{-1}(f)\mathbf{P}_{j,k}(f,n))}{\phi_{j,k} b_k(f) h_k(n)}, \tag{5.83}$$

for the first term with an arbitrary complex matrix $\mathbf{P}_{j,k}(f,n) \in \mathbb{C}^{I \times I}$ such that $\sum_{j,k} \mathbf{P}_{j,k}(f,n) = \mathbf{I}$ where $\mathbf{I}$ is an identity matrix, and

$$\log\det(\boldsymbol{\Sigma}_{\mathbf{x}}(f,n)) \le \mathrm{tr}(\mathbf{K}^{-1}(f,n)\boldsymbol{\Sigma}_{\mathbf{x}}(f,n)) + \log\det \mathbf{K}(f,n) - I, \tag{5.84}$$

for the second term with a positive definite matrix $\mathbf{K}(f,n)$ [7, 22]. We can show that the equalities of (5.83) and (5.84) hold when

$$\mathbf{P}_{j,k}(f,n) = \phi_{j,k} b_k(f) h_k(n) \mathbf{R}_j(f) \boldsymbol{\Sigma}_{\mathbf{x}}^{-1}(f,n), \tag{5.85}$$

$$\mathbf{K}(f,n) = \boldsymbol{\Sigma}_{\mathbf{x}}(f,n). \tag{5.86}$$

By combining these inequalities, we have

$$\mathscr{C}_{\mathrm{MNMF}}(\boldsymbol{\theta}) \le \sum_{f,n} \left\{ \sum_{j,k} \frac{\mathrm{tr}(\widehat{\boldsymbol{\Sigma}}_{\mathbf{x}}(f,n)\mathbf{P}_{j,k}(f,n)\mathbf{R}_j^{-1}(f)\mathbf{P}_{j,k}(f,n))}{\phi_{j,k} b_k(f) h_k(n)} \right.$$
$$\left. + \mathrm{tr}(\mathbf{K}^{-1}(f,n)\boldsymbol{\Sigma}_{\mathbf{x}}(f,n)) + \log\det \mathbf{K}(f,n) - I \right\}. \qquad (5.87)$$

Hence, we can use the right-hand side of this inequality as a majorizer for $\mathscr{C}_{\mathrm{MNMF}}(\boldsymbol{\theta})$ where $P = [\mathbf{P}_{j,k}(f,n)]_{j,k,f,n}$ and $K = [\mathbf{K}(f,n)]_{f,n}$ are auxiliary variables. Here, (5.85) and (5.86) correspond to the update rules for the auxiliary variables. As in the single channel case, this majorizer is given in a separable form, which is relatively easy to optimize with respect to $\boldsymbol{\Phi} = [\phi_{j,k}]_{j,k}$, $\mathbf{B} = [b_k(f)]_{k,f}$, $\mathbf{H} = [h_k(n)]_{k,n}$ and $R = [\mathbf{R}_j(f)]_{j,f}$. By differentiating this majorizer with respect to $b_k(f)$ and $h_k(n)$, and setting the results at zero, we obtain the following update rules for $b_k(f)$ and $h_k(n)$:

$$b_k(f) = \sqrt{\frac{\sum_{n,j} \frac{1}{\phi_{j,k} h_k(n)} \mathrm{tr}(\widehat{\boldsymbol{\Sigma}}_{\mathbf{x}}(f,n)\mathbf{P}_{j,k}(f,n)\mathbf{R}_j^{-1}(f)\mathbf{P}_{j,k}(f,n))}{\sum_{n,j} \phi_{j,k} h_k(n) \mathrm{tr}(\mathbf{K}^{-1}(f,n)\boldsymbol{\Sigma}_{\mathbf{x}}(f,n))}}, \qquad (5.88)$$

$$h_k(n) = \sqrt{\frac{\sum_{f,j} \frac{1}{\phi_{j,k} b_k(f)} \mathrm{tr}(\widehat{\boldsymbol{\Sigma}}_{\mathbf{x}}(f,n)\mathbf{P}_{j,k}(f,n)\mathbf{R}_j^{-1}(f)\mathbf{P}_{j,k}(f,n))}{\sum_{f,j} \phi_{j,k} b_k(f) \mathrm{tr}(\mathbf{K}^{-1}(f,n)\boldsymbol{\Sigma}_{\mathbf{x}}(f,n))}}. \qquad (5.89)$$

As regards $\phi_{j,k}$, although it is necessary to take account of the unit sum constraint, here we update $\phi_{j,k}$ at

$$\phi_{j,k} = \sqrt{\frac{\sum_{n,f} \frac{1}{b_k(f) h_k(n)} \mathrm{tr}(\widehat{\boldsymbol{\Sigma}}_{\mathbf{x}}(f,n)\mathbf{P}_{j,k}(f,n)\mathbf{R}_j^{-1}(f)\mathbf{P}_{j,k}(f,n))}{\sum_{n,f} b_k(f) h_k(n) \mathrm{tr}(\mathbf{K}^{-1}(f,n)\boldsymbol{\Sigma}_{\mathbf{x}}(f,n))}}, \qquad (5.90)$$

which minimizes the majorizer, project it onto the constraint space $\phi_{j,k} \leftarrow \phi_{j,k} / \sum_{j'} \phi_{j',k}$, and rescale $b_k(f)$ and $h_k(n)$. As regards $\mathbf{R}_j(f)$, the optimal update is given as the solution to an algebraic Riccati equation

$$\mathbf{R}_j(f)\boldsymbol{\Psi}_j(f)\mathbf{R}_j(f) = \boldsymbol{\Omega}_j(f), \qquad (5.91)$$

where the coefficient matrices are given as

$$\boldsymbol{\Psi}_j(f) = \sum_{k,n} \phi_{j,k} b_k(f) h_k(n) \mathbf{K}^{-1}(f,n), \tag{5.92}$$

$$\boldsymbol{\Omega}_j(f) = \sum_{k,n} \frac{\mathbf{P}_{j,k}(f,n)\widehat{\boldsymbol{\Sigma}}_{\mathbf{x}}(f,n)\mathbf{P}_{j,k}(f,n)}{\phi_{j,k} b_k(f) h_k(n)}. \tag{5.93}$$

Since there is a scale indeterminacy between $\mathbf{R}_j(f)$ and $\phi_{j,k} b_k(f) h_k(n)$, a convenient way to eliminate the indeterminacy would be to update $\mathbf{R}_j(f)$ using the above equation and then perform unit-trace normalization: $\mathbf{R}_j(f) \leftarrow \mathbf{R}_j(f)/\mathrm{tr}(\mathbf{R}_j(f))$.

Comparisons of the convergences of the EM algorithm and the MM algorithm for MNMF can be found in [7].

5.6.2 MFHMM Algorithm

In [18], we proposed a method that makes it possible to simultaneously perform source separation, DOA estimation, dereverberation and voice activity detection (VAD) by using a composite model of the convolutive mixing model, the HMM-based spectral model and the DOA mixture model. Since this model can be viewed as an extension of the factorial hidden Markov model (FHMM) [37] for modeling multichannel signals, we call it the multichannel FHMM (MFHMM). We describe the generative process of $v_j(f,n)$ as (5.22), (5.23), and (5.24), and the generative process of $\mathbf{R}_j(f,0)$ as (5.39). By substituting $\boldsymbol{\mu}_{\mathbf{x}}(f,n) = \mathbf{0}$, $\boldsymbol{\Sigma}_{\mathbf{x}}(f,n) = \sum_j \sum_m v_j(f,n-m)\mathbf{R}_j(f,m)$, and $v_j(f,n) = b_{j,k_j(n)}(f) h_j(n)$ into the log-likelihood (5.18), adding the log-prior terms

$$\log p(H) = \sum_j \log \sum_{K_j} p(H_j|K_j) p(K_j), \tag{5.94}$$

$$\log p(R) = \sum_j \log \sum_{z_j} p(R_j|z_j) p(z_j), \tag{5.95}$$

where $H_j = [h_j(n)]_n$, $K_j = [k_j(n)]_n$, and $R_j = [\mathbf{R}_j(f,0)]_f$, reversing the sign and neglecting the constant terms, we obtain the objective function to be minimized as

$$\begin{aligned}\mathscr{C}_{\mathrm{MFHMM}}(\boldsymbol{\theta}) = &\sum_f \sum_n \left\{ \mathrm{tr}\left(\widehat{\boldsymbol{\Sigma}}_{\mathbf{x}}(f,n)\boldsymbol{\Sigma}_{\mathbf{x}}^{-1}(f,n)\right) + \log\det \boldsymbol{\Sigma}_{\mathbf{x}}(f,n) \right\} \\ &- \sum_j \log \sum_{K_j} p(H_j|K_j) p(K_j) - \sum_j \log \sum_{z_j} p(R_j|z_j) p(z_j). \end{aligned} \tag{5.96}$$

Here, $p(H_j|K_j)$, $p(K_j)$, $p(R_j|z_j)$ and $p(z_j)$ are given by

$$p(H_j|K_j) = \prod_{n=0}^{N-1} p(h_j(n)|k_j(n)) \tag{5.97}$$

$$p(h_j(n)|k_j(n)) = \mathscr{G}(h_j(n)|\gamma_{j,k_j(n)}, \beta_{j,k_j(n)}), \tag{5.98}$$

$$p(K_j) = p(k_j(0)) \prod_{n=1}^{N-1} p(k_j(n)|k_j(n-1)) \tag{5.99}$$

$$p(k_j(n)|k_j(n-1)) = \pi_{j,k_j(n-1),k_j(n)}, \tag{5.100}$$

$$p(R_j|z_j) = \prod_f p(\mathbf{R}_j(f,0)|z_j) \tag{5.101}$$

$$p(\mathbf{R}_j(f,0)|z_j) = \mathscr{W}_{\mathbb{C}}(\mathbf{R}_j(f,0)|\nu, \mathbf{\Sigma}_{\mathbf{d}}(f;\vartheta_{z_j}) + \varepsilon\mathbf{I}), \tag{5.102}$$

$$p(z_j) = \psi_{z_j}. \tag{5.103}$$

Since the logarithmic function is concave, we can use Jensen's inequality to obtain majorizers for the log-prior terms

$$-\log \sum_{K_j} p(H_j|K_j)p(K_j) \leq -\sum_{K_j} \lambda_h(K_j) \log \frac{p(H_j|K_j)p(K_j)}{\lambda_h(K_j)}, \tag{5.104}$$

$$-\log \sum_{z_j} p(R_j|z_j)p(z_j) \leq -\sum_{z_j} \lambda_R(z_j) \log \frac{p(R_j|z_j)p(z_j)}{\lambda_R(z_j)}, \tag{5.105}$$

where $\lambda_h(K_j)$ and $\lambda_R(z_j)$ are non-negative weights that must satisfy

$$\sum_{K_j} \lambda_h(K_j) = 1, \quad \sum_{z_j} \lambda_R(z_j) = 1. \tag{5.106}$$

A majorizer for the first term of $\mathscr{C}_{\mathrm{MFHMM}}(\boldsymbol{\theta})$ can be obtained in the same way as the previous section using (5.83) and (5.84). By combining these majorizers, we obtain a majorizer for the objective $\mathscr{C}_{\mathrm{MFHMM}}(\boldsymbol{\theta})$, which allows us to obtain closed-form update equations for the model parameters [18].

5.6.3 Demixing Filter Estimation Algorithm

For the instantaneous demixing system (5.6), one popular way of estimating the demixing filters $\mathbf{W}^{\mathsf{H}}(f)$ involves the natural gradient method [40]. For the convolutive case (5.7), the log-likelihood function can be written equivalently as

$$\log p(X|\boldsymbol{\theta}) = \sum_{f,n} \Bigg\{ 2 \log \det \mathbf{W}^{\mathsf{H}}(f,0) - \sum_j \log v_j(f,n) \\ - \mathbf{y}(f,n)^{\mathsf{H}} \mathbf{W}(f,0) \boldsymbol{\Sigma}_{\mathbf{s}}^{-1}(f,n) \mathbf{W}^{\mathsf{H}}(f,0) \mathbf{y}(f,n) \Bigg\}, \quad (5.107)$$

where

$$\mathbf{y}(f,n) = \mathbf{x}(f,n) - \sum_{m=1}^{M-1} \mathbf{D}^{\mathsf{H}}(f,m)\mathbf{x}(f,n-m), \quad (5.108)$$

$$\mathbf{D}^{\mathsf{H}}(f,m) = -(\mathbf{W}^{\mathsf{H}}(f,0))^{-1}\mathbf{W}^{\mathsf{H}}(f,m). \quad (5.109)$$

Once $\mathbf{y}(f,n)$ and $\mathbf{W}^{\mathsf{H}}(f,0)$ are obtained, $\mathbf{s}(f,n)$ can be obtained by

$$\mathbf{s}(f,n) = \mathbf{W}^{\mathsf{H}}(f,0)\mathbf{y}(f,n). \quad (5.110)$$

Here, (5.108) can be seen as the dereverberation process of the observed mixture signal $\mathbf{x}(f,n)$ described as a multichannel autoregressive (AR) system with regression matrices $D = [\mathbf{D}^{\mathsf{H}}(f,m)]_{f,m}$ whereas (5.110) can be seen as the instantaneous demixing process of the dereverberated mixture signal $\mathbf{y}(f,n)$. When $\mathbf{W}^{\mathsf{H}}(f,0)$ is fixed, it can be shown that the log-likelihood function of D becomes equal up to a sign to the objective function of a vector version of the linear prediction (multichannel linear prediction), which can be maximized with respect to D by solving a Yule-Walker equation. On the other hand, when D is fixed, the log-likelihood function (5.107) can be locally maximized with respect to $\mathbf{W}^{\mathsf{H}}(f,0)$ using the natural gradient method. Thus, we can find the estimates of $\mathbf{W}^{\mathsf{H}}(f,0)$ and D by sequentially optimizing one at a time while keeping the other fixed [5, 28].

5.7 Conclusion

This chapter introduced a general formulation of the frequency domain BSS that allows the incorporation of a composite model combining a mixing process model, a source spectral model, and a spatial model. We showed that combining these models allows us to design various BSS methods with different properties and characteristics including multichannel extensions of NMF variants. Through this formulation, we revealed the relationship between the state-of-the-art BSS approaches. We also showed the derivations of the MNMF and MFHMM algorithms.

References

1. A. Hyvärinen, J. Karhunen, E. Oja, *Independent Component Analysis* (Wiley, New York, 2001)
2. A. Hiroe, Solution of permutation problem in frequency domain ICA using multivariate probability density functions, in *Proceedings International Conference on Independent Component Analysis and Blind Source Separation (ICA)* (2006), pp. 601–608
3. T. Kim, T. Eltoft, T.-W. Lee, Independent vector analysis: An extension of ICA to multivariate components, in *Proceedings of International Conference on Independent Component Analysis and Blind Source Separation (ICA)* (2006), pp. 165–172
4. A. Ozerov, C. Févotte, Multichannel nonnegative matrix factorization in convolutive mixtures for audio source separation. IEEE Trans. Audio Speech Lang. Process. **18**(3), 550–563 (2010). Mar
5. H. Kameoka, T. Yoshioka, M. Hamamura, J. Le Roux, K. Kashino, Statistical model of speech signals based on composite autoregressive system with application to blind source separation, in *Proceedings of International Conference on Latent Variable Analysis and Signal Separation (LVA/ICA)* (2010), pp. 245–253
6. A. Ozerov, C. Févotte, R. Blouet, J.-L. Durrieu, Multichannel nonnegative tensor factorization with structured constraints for user-guided audio source separation, in *Proceedings of IEEE International Conference on Acoustics, Speech and Signal Processing (ICASSP)*, Prague, May 2011, pp. 257–260
7. H. Sawada, H. Kameoka, S. Araki, N. Ueda, Multichannel extensions of non-negative matrix factorization with complex-valued data. IEEE Trans. Audio Speech Lang. Process. **21**(5), 971–982 (2013). May
8. J. Nikunen, T. Virtanen, Direction of arrival based spatial covariance model for blind sound source separation. IEEE/ACM Trans. Audio Speech Lang. Process. **22**(3), 727–739 (2014). Mar
9. D. Kitamura, N. Ono, H. Sawada, H. Kameoka, H. Saruwatari, Efficient multichannel nonnegative matrix factorization exploiting rank-1 spatial model, in *Proceedings of IEEE International Conference on Acoustics, Speech and Signal Processing (ICASSP)*, April 2015, pp. 276–280
10. D. Kitamura, N. Ono, H. Sawada, H. Kameoka, H. Saruwatari, Determined blind source separation unifying independent vector analysis and nonnegative matrix factorization. IEEE/ACM Trans. Audio Speech Lang. Process. **24**(9), 1626–1641 (2016)
11. K. Adiloğlu, E. Vincent, Variational Bayesian inference for source separation and robust feature extraction. IEEE/ACM Trans. Audio Speech Lang. Process. **24**, 1746–1758 (2016)
12. D. Kounades-Bastian, L. Girin, X. Alameda-Pineda, S. Gannot, R. Horaud, A variational EM algorithm for the separation of time-varying convolutive audio mixtures. IEEE/ACM Trans. Audio Speech Lang. Process. **24**(8), 1408–1423 (2016)
13. P. Smaragdis, J.C. Brown, Non-negative matrix factorization for polyphonic music transcription, in *Proceedings of IEEE Workshop on Applications of Signal Processing to Audio and Acoustics (WASPAA)* (2003), pp. 177–180
14. C. Févotte, N. Bertin, J.-L. Durrieu, Nonnegative matrix factorization with the Itakura-Saito divergence. With application to music analysis. Neural Comput. **21**(3), 793–830 (2009). Mar
15. T. Higuchi, H. Takeda, T. Nakamura, H. Kameoka, A unified approach for underdetermined blind signal separation and source activity detection by multichannel factorial hidden Markov models, in *Proceedings of Annual Conference of the International Speech Communication Association (Interspeech)* (2014), pp. 850–854
16. T. Higuchi, H. Kameoka, Joint audio source separation and dereverberation based on multichannel factorial hidden Markov model, in *Proceedings of IEEE International Workshop on Machine Learning for Signal Processing (MLSP)* (2014)
17. T. Higuchi, H. Kameoka, Unified approach for underdetermined BSS, VAD, dereverberation and DOA estimation with multichannel factorial HMM, in *Proceedings of IEEE Global Conference on Signal and Information Processing (GlobalSIP)* (2014)

18. T. Higuchi, H. Kameoka, Unified approach for audio source separation with multichannel factorial HMM and DOA mixture model, in *Proceedings of European Signal Processing Conference (EUSIPCO)*, August 2015
19. H. Kameoka, M. Sato, T. Ono, N. Ono, S. Sagayama, Blind separation of infinitely many sparse sources, in *Proceedings of International Workshop on Acoustic Signal Enhancement (IWAENC)* (2012)
20. H. Kameoka, M. Sato, T. Ono, N. Ono, S. Sagayama, Bayesian nonparametric approach to blind separation of infinitely many sparse sources. IEICE Trans. Fundamentals Electronics **E96-A**(10), 1928–1937 (2013)
21. T. Otsuka, K. Ishiguro, H. Sawada, H.G. Okuno, Bayesian nonparametrics for microphone array processing. IEEE/ACM Trans. Audio Speech Lang. Process. **22**(2), 493–504 (2014)
22. T. Higuchi, N. Takamune, T. Nakamura, H. Kameoka, Underdetermined blind separation and tracking of moving sources based on DOA-HMM, in *Proceedings of IEEE International Conference on Acoustics, Speech and Signal Processing (ICASSP)* (2014), pp. 3215–3219
23. H. Attias, New EM algorithms for source separation and deconvolution with a microphone array, in *Proceedings of IEEE International Conference on Acoustics, Speech and Signal Processing (ICASSP)*, vol. V (2003), pp. 297–300
24. N.Q.K. Duong, E. Vincent, R. Gribonval, Under-determined reverberant audio source separation using a full-rank spatial covariance model. IEEE Trans. Audio Speech Lang. Process. **18**(7), 1830–1840 (2010)
25. A. Ozerov, E. Vincent, F. Bimbot, A general flexible framework for the handling of prior information in audio source separation. IEEE Trans. Audio Speech Lang. Process. **20**(4), 1118–1133 (2012)
26. T. Ono, N. Ono, S. Sagayama, User-guided independent vector analysis with source activity tuning, in *Proceedings of IEEE International Conference on Acoustics, Speech and Signal Processing (ICASSP)* (2012), pp. 2417–2420
27. S. Dégerine, A. Zaïdi, Separation of an instantaneous mixture of gaussian autoregressive sources by the exact maximum likelihood approach. IEEE Trans. Sig. Process. **52**(6), 1499–1512 (2004)
28. T. Yoshioka, T. Nakatani, M. Miyoshi, H.G. Okuno, Blind separation and dereverberation of speech mixtures by joint optimization. IEEE Trans. Audio Speech Lang. Process. **19**(1), 69–84 (2011). Mar.
29. H. Kameoka, K. Kashino, Composite autoregressive system for sparse source-filter representation of speech, in *Proceedings of IEEE International Symposium on Circuits and Systems (ISCAS)* (2009), pp. 2477–2480
30. N.Q.K. Duong, H. Tachibana, E. Vincent, N. Ono, R. Gribonval, S. Sagayama, Multichannel harmonic and percussive component separation by joint modeling of spatial and spectral continuity, in *Proceedings of IEEE International Conference on Acoustics, Speech and Signal Processing (ICASSP)* (2011), pp. 205–208
31. J.D. Leeuw, W.J. Heiser, Convergence of correction matrix algorithms for multidimensional scaling, in *Geometric representations of relational data*, ed. by J.C. Lingoes, E.E. Roskam, I. Borg (Mathesis Press, Ann Arbor, MI, 1977)
32. D.R. Hunter, K. Lange, A tutorial on MM algorithms. Am. Statistician **58**(1), 30–37 (2004). Feb.
33. A.P. Dempster, N.M. Laird, D.B. Rubin, Maximum likelihood from incomplete data via the EM algorithm. J. R. Statistical Soc. Series B **39**, 1–38 (1977)
34. D.D. Lee, H.S. Seung, Algorithms for non-negative matrix factorization, in *Advances in Neural Information Processing Systems*, vol. 13 (2001)
35. M. Nakano, H. Kameoka, J. Le Roux, Y. Kitano, N. Ono, S. Sagayama, Convergence-guaranteed multiplicative algorithms for non-negative matrix factorization with beta-divergence, in *Proceedings of IEEE International Workshop on Machine Learning for Signal Processing* (2010), pp. 283–288
36. C. Févotte, J. Idier, Algorithms for nonnegative matrix factorization with the β-divergence. Neural Comput. **23**(9), 2421–2456 (2011)
37. C. Bishop, *Pattern Recognit. Mach. Learn.* (Springer-Verlag, New York, 2006)

38. Y. Izumi, N. Ono, S. Sagayama, Sparseness-based 2ch BSS using the EM algorithm in reverberant environment, in *Proceedings of IEEE Workshop on Applications of Signal Processing to Audio and Acoustics (WASPAA)* (2007), pp. 147–150
39. H. Kameoka, M. Goto, S. Sagayama, Selective amplifier of periodic and non-periodic components in concurrent audio signals with spectral control envelopes, in *IPSJ SIG Technical Reports*, vol. 2006-MUS-66-13 (2006), pp. 77–84, in Japanese
40. S. Amari, A. Cichocki, H.H. Yang, A new learning algorithm for blind signal separation, in *Advances in Neural Information Processing Systems* (MIT Press, 1996), pp. 757–763

Chapter 6
Determined Blind Source Separation with Independent Low-Rank Matrix Analysis

Daichi Kitamura, Nobutaka Ono, Hiroshi Sawada, Hirokazu Kameoka and Hiroshi Saruwatari

Abstract In this chapter, we address the determined blind source separation problem and introduce a new effective method of unifying independent vector analysis (IVA) and nonnegative matrix factorization (NMF). IVA is a state-of-the-art technique that utilizes the statistical independence between source vectors. However, since the source model in IVA is based on a spherically symmetric multivariate distribution, IVA cannot utilize the characteristics of specific spectral structures such as various sounds appearing in music signals. To solve this problem, we introduce NMF as the source model in IVA to capture the spectral structures. Since this approach is a natural extension of the source model from a vector to a low-rank matrix represented by NMF, the new method is called independent low-rank matrix analysis (ILRMA). We also reveal the relationship between IVA, ILRMA, and multichannel NMF (MNMF), namely, IVA and ILRMA are identical to a special case of MNMF, which employs a rank-1 spatial model. Experimental results show the efficacy of ILRMA compared with IVA and MNMF in terms of separation accuracy and convergence speed.

D. Kitamura (✉) · H. Saruwatari
The University of Tokyo, 7-3-1 Hongo, Bunkyo,
Tokyo 113-8656, Japan
e-mail: d-kitamura@ieee.org

N. Ono
Tokyo Metropolitan University, 6-6 Asahigaoka, Hino, Tokyo 191-0065, Japan

H. Sawada
NTT Communication Science Laboratories, 2-4 Hikaridai, Seika, Soraku,
Kyoto 619-0237, Japan

H. Kameoka
NTT Communication Science Laboratories, 3-1 Morinosato Wakamiya,
Atsugi, Kanagawa 243-0198, Japan

S. Makino (ed.), *Audio Source Separation*, Signals and Communication Technology, https://doi.org/10.1007/978-3-319-73031-8_6

6.1 Introduction

Blind source separation (BSS) is a technique for separating specific sources from a recorded sound without any information about the recording environment, mixing system, or source locations. In a determined or overdetermined situation (number of microphones $\geq$ number of sources), independent component analysis (ICA) [1–5] is the method most commonly used to solve the BSS problem, and many ICA-based techniques have been proposed [6–10]. On the other hand, for an underdetermined situation (number of microphones $<$ number of sources) including monaural recording, nonnegative matrix factorization (NMF) [11–13] with both blind and informed source separation techniques has received much attention [14–18]. BSS is generally used to solve speech separation problems, but recently the use of BSS for music signals has also become an active research area [19–22].

ICA-based BSS assumes independence between the sources to estimate a demixing matrix. In frequency domain ICA (FDICA), the permutation ambiguity of ICA in each frequency bin must be aligned so that a separated signal in the time domain contains frequency components of the same source signal. This problem is called the permutation problem, for which many solvers have been proposed (e.g., [8, 9, 23–26]). Independent vector analysis (IVA) [27–29] is a popular method simultaneously solving the separation and permutation problems. IVA assumes source vector variables and their generative model with a spherically symmetric multivariate distribution to ensure higher-order correlations between frequency bins in each source. This generative source model does not include any specific information on the spectral structures of sources, meaning that it can be generally used for various types of sound. However, some sources have specific spectral structures such as the harmonic structure of instrumental sounds or music tones. Therefore, the introduction of a better source model has the potential to improve the source separation performance. In this chapter, we only focus on the BSS problem in the determined situation, and introduce a new effective method of unifying IVA and NMF [30, 31]. The new method exploits NMF decomposition to capture the spectral structures of each source as the generative source model in IVA. Since this approach is a natural extension of the source model from a vector to a low-rank matrix represented by NMF, we call the new method *independent low-rank matrix analysis*. Intriguingly, the formulation of the new method coincides with a special case of the multichannel extension of NMF [32–36]. This fact reveals the relationship between a multichannel extension of NMF, IVA, and ILRMA.

The contents in this chapter are partially based on [30, 31] written by the authors. Note that ILRMA was called *determined rank-1 multichannel NMF* in these papers. We have renamed the method to clarify that ILRMA is a natural extension of the source model in IVA.

6.2 Generative Source Models in IVA and NMF Based on Itakura–Saito Divergence

In this section, we explain and compare the generative source models in IVA and NMF based on Itakura–Saito divergence (hereafter referred to as *Itakura–Saito NMF*).

6.2.1 Formulation

Let the numbers of sources and microphones (channels) be N and M, respectively. The source, observed, and separated signals in each time-frequency slot are described as

$$\mathbf{s}_{ij} = (s_{ij,1} \ \cdots \ s_{ij,N})^{\mathrm{T}}, \tag{6.1}$$
$$\mathbf{x}_{ij} = (x_{ij,1} \ \cdots \ x_{ij,M})^{\mathrm{T}}, \tag{6.2}$$
$$\mathbf{y}_{ij} = (y_{ij,1} \ \cdots \ y_{ij,N})^{\mathrm{T}}, \tag{6.3}$$

where $i = 1, \cdots, I$; $j = 1, \cdots, J$; $n = 1, \cdots, N$; and $m = 1, \cdots, M$ are the integral indexes of the frequency bins, time frames, sources, and channels, respectively, ^{T} denotes the vector transpose, and all the entries of these vectors are complex values. When the window length in a short-time Fourier transform (STFT) is sufficiently long compared with the impulse responses between sources and microphones, the instantaneous mixture in the frequency domain becomes valid while deriving the following expression for the mixing system:

$$\mathbf{x}_{ij} = \mathbf{A}_i \mathbf{s}_{ij}, \tag{6.4}$$

where $\mathbf{A}_i = (\mathbf{a}_{i,1} \ \cdots \ \mathbf{a}_{i,N})$ is an $M \times N$ mixing matrix and $\mathbf{a}_{i,n} = (a_{i,n1} \ \cdots \ a_{i,nM})^{\mathrm{T}}$ is the steering vector for each source. In the case of an overdetermined signal ($M > N$), a standard approach is to apply principal component analysis (PCA) in advance to reduce the dimension of $\mathbf{x}_{ij}$ so that $M = N$. If the mixing matrix $\mathbf{A}_i$ is invertible and $M = N$, we can define the demixing matrix $\mathbf{W}_i = (\mathbf{w}_{i,1} \ \cdots \ \mathbf{w}_{i,N})^{\mathrm{H}}$ as the inverse of the mixing matrix, and the separated signal can be represented as

$$\mathbf{y}_{ij} = \mathbf{W}_i \mathbf{x}_{ij}, \tag{6.5}$$

where $\mathbf{w}_{i,n} = (w_{i,n1} \ \cdots \ w_{i,nM})^{\mathrm{T}}$ is the demixing filter for each source and ^{H} denotes the Hermitian transpose. In this chapter, hereafter, we only focus on the determined situation ($N = M$) and we use indexes n and m to distinguish sources and channels, respectively.

6.2.2 IVA

IVA [27–29] is a multivariate extension of FDICA and can solve the BSS problem while avoiding the permutation problem. ICA-based methods including IVA can only be applied to the determined situation ($M = N$) with the mixing assumption (6.4) because they estimate the demixing matrix $\mathbf{W}_i$ for the separation. In IVA, we assume the multivariate source vector $\mathbf{s}_{j,n}$, observed vector $\mathbf{x}_{j,m}$, and separated vector $\mathbf{y}_{j,n}$, which consist of all the frequency bins, as

$$\mathbf{s}_{j,n} = (s_{1j,n} \cdots s_{Ij,n})^{\mathrm{T}}, \tag{6.6}$$

$$\mathbf{x}_{j,m} = (x_{1j,m} \cdots x_{Ij,m})^{\mathrm{T}}, \tag{6.7}$$

$$\mathbf{y}_{j,n} = (y_{1j,n} \cdots y_{Ij,n})^{\mathrm{T}}. \tag{6.8}$$

Figure 6.1 shows the mixing and demixing model in IVA, where $N = M = 2$. In IVA, all the source, observed, and separated signals are represented as frequency vector variables, whereas FDICA independently models each of the frequency components resulting in the permutation problem. In addition, higher-order correlations between the frequency components in each source (or separated) vector are introduced by assuming spherically symmetric multivariate source distributions $p(\mathbf{s}_{j,n}) \approx p(\mathbf{y}_{j,n}) = p(y_{1j,n}, \cdots, y_{Ij,n})$, where the spherically symmetric property means that the distribution is a function of only the norm of multivariate vector variable, i.e., $p(\mathbf{y}_{j,n}) = f(\|\mathbf{y}_{j,n}\|)$.

In the literature [27–29], a spherically symmetric multivariate Laplace distribution [37, 38] was exploited as a super-Gaussian source distribution for modeling speech sources. This distribution is shown in Fig. 6.2 and is defined as

$$p(\mathbf{s}_{j,n}) \approx p(\mathbf{y}_{j,n}) = \rho \exp\left(-\sqrt{\sum_i \left|\frac{y_{ij,n}}{r_{i,n}}\right|^2}\right), \tag{6.9}$$

where ρ is a normalization term and $r_{i,n}$ is the scale, which determines the signal scale of $y_{ij,n}$. Since the source distribution has a spherically symmetric property, higher-order correlations between the frequency components in each source are assumed, which results in avoiding the permutation problem. Hereafter, IVA based on the source distribution (6.9) is referred to as Laplace IVA.

From the generative source model $p(\mathbf{s}_{j,1}, \cdots, \mathbf{s}_{j,N}) \approx p(\mathbf{y}_{j,1}, \cdots, \mathbf{y}_{j,N})$ and the demixing system (6.5), $p(\mathbf{x}_{j,1}, \cdots, \mathbf{x}_{j,M})$ can be obtained by multiplying $p(\mathbf{y}_{j,1}, \cdots, \mathbf{y}_{j,N})$ by the Jacobian

$$\frac{\partial(\mathbf{y}_{j,1}, \cdots, \mathbf{y}_{j,N})}{\partial(\mathbf{x}_{j,1}, \cdots, \mathbf{x}_{j,M})} = \prod_i |\det \mathbf{W}_i|^2; \tag{6.10}$$

note that the Jacobian for a complex-valued variable is the square of the Jacobian for a real-valued variable [39]. Therefore, the likelihood function $\mathscr{L}(\mathbf{W})$ of the parameter

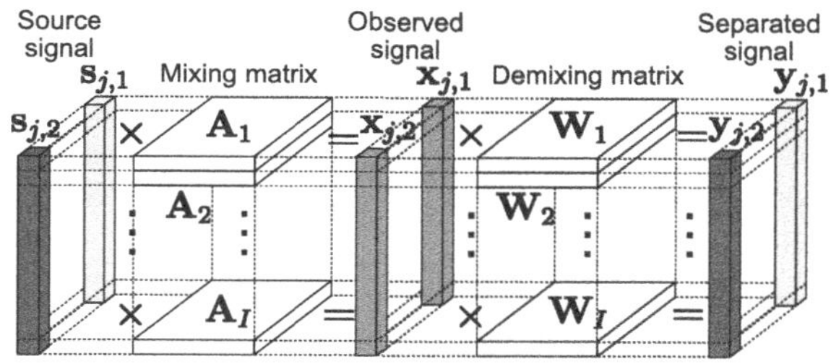

Fig. 6.1 Mixing and demixing model in IVA, where $N = M = 2$

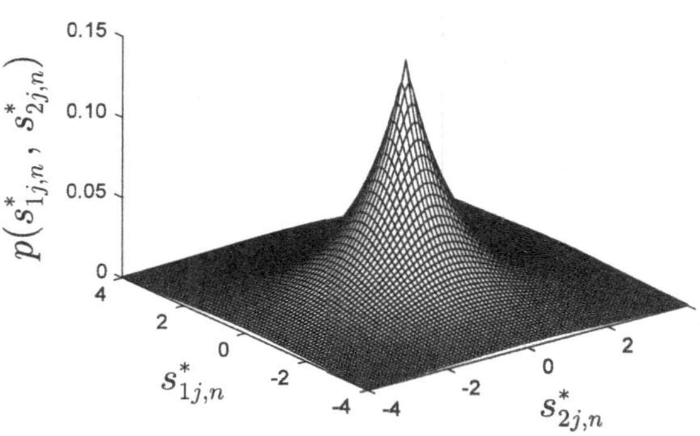

Fig. 6.2 Spherically symmetric multivariate Laplace distribution, where $s^*_{ij,n}$ can be considered as either real or imaginary part of $s_{ij,n}$ and $I = 2$. Two frequency components $s_{1j,n}$ and $s_{2j,n}$ are uncorrelated but have mutual dependences, which is called higher-order correlation

set $\mathbf{W} = \{\mathbf{W}_i | i = 1, \cdots, I\}$ is given as

$$\begin{aligned}\mathscr{L}(\mathbf{W}) &= \prod_j p(\mathbf{x}_{j,1}, \cdots, \mathbf{x}_{j,M} | \mathbf{W}) \\ &= \prod_j \left[p(\mathbf{y}_{j,1}, \cdots, \mathbf{y}_{j,N}) \cdot \prod_i |\det \mathbf{W}_i|^2 \right] \\ &= \prod_j \left\{ \left[\prod_n p(\mathbf{y}_{j,n}) \right] \cdot \prod_i |\det \mathbf{W}_i|^2 \right\}, \end{aligned} \tag{6.11}$$

where $p(\mathbf{y}_{j,1}, \cdots, \mathbf{y}_{j,N}) = \prod_n p(\mathbf{y}_{j,n})$ is obtained by assuming mutual independence between $\mathbf{y}_{j,n}$ for all the sources. The negative log-likelihood function can be calculated as

$$\begin{aligned} -\log \mathscr{L}(\mathbf{W}) &= -\sum_{i,j} \log |\det \mathbf{W}_i|^2 - \sum_{j,n} \log p(\mathbf{y}_{j,n}) \\ &= -2J \sum_i \log |\det \mathbf{W}_i| + \sum_{j,n} G(\mathbf{y}_{j,n}), \end{aligned} \tag{6.12}$$

where $G(\mathbf{y}_{j,n}) = -\log p(\mathbf{y}_{j,n})$ is called the contrast function, which depends on the source distribution $p(\mathbf{y}_{j,n})$. Note that since $y_{ij,n} = \mathbf{w}_{i,n}^{\mathrm{H}}\mathbf{x}_{ij}$, the separated signal $\mathbf{y}_{j,n}$ includes the optimization variable $\mathbf{W}_i$. The maximum log-likelihood (ML) estimation based on (6.12) is equivalent to the well-known estimation [3, 5] that maximizes the independence between all the sources with the Kullback–Leibler divergence $\mathscr{D}_{\mathrm{KL}}$ as follows:

$$\begin{aligned}
&\sum_j \mathscr{D}_{\mathrm{KL}}\left(p(\mathbf{y}_{j,1}, \cdots, \mathbf{y}_{j,N}) \| \prod_n p(\mathbf{y}_{j,n})\right) \\
&= \sum_j \int p(\mathbf{y}_{j,1}, \cdots, \mathbf{y}_{j,N}) \log \frac{p(\mathbf{y}_{j,1}, \cdots, \mathbf{y}_{j,N})}{\prod_n p(\mathbf{y}_{j,n})} d\mathbf{y}_{j,1} \cdots d\mathbf{y}_{j,N} \\
&= \text{const.} - 2J \sum_i \log |\det \mathbf{W}_i| + \sum_{j,n} G(\mathbf{y}_{j,n}).
\end{aligned} \tag{6.13}$$

On the basis of the source distribution (6.9), the contrast function $G(\mathbf{y}_{j,n})$ and the cost function in Laplace IVA can be obtained as follows:

$$G(\mathbf{y}_{j,n}) = -\log \rho + \|\mathbf{y}_{j,n}\|_2, \tag{6.14}$$

$$-\log \mathscr{L}(\mathbf{W}) = \text{const.} - 2J \sum_i \log |\det \mathbf{W}_i| + \sum_{j,n} \|\mathbf{y}_{j,n}\|_2, \tag{6.15}$$

where $\|\cdot\|_2$ denotes the ℓ_2 norm. Also, the scale is set to $r_{i,n} = 1$ for all i and n because the scales of separated signals cannot be determined by ICA or IVA, and they can be recovered by a back-projection technique [24] after the separation. For the minimization of (6.15), fast and stable update rules called iterative projection (IP) based on the auxiliary function technique have been proposed [40–42].

6.2.3 *Time-Varying Gaussian IVA*

Laplace IVA employs the spherically symmetric Laplace distribution as a super-Gaussian source distribution. The model ensures that all the frequency components in the same source have higher-order correlation. As another super-Gaussian source model with the higher-order correlation, in [43], the circularly symmetric complex Gaussian distribution with time-varying variance $r_{j,n}$ is introduced to conventional IVA instead of the stationary distribution:

$$\begin{aligned}
p(\mathbf{y}_{1,n}, \cdots, \mathbf{y}_{J,n}) &= \prod_j p(\mathbf{y}_{j,n}) \\
&= \prod_j \frac{1}{\pi r_{j,n}} \exp\left(-\frac{\|\mathbf{y}_{j,n}\|_2^2}{r_{j,n}}\right),
\end{aligned} \tag{6.16}$$

where the time-varying variance $r_{j,n}$ is shared over the frequency bins in each time frame. Similar to (6.9), the distribution (6.16) has the spherically symmetric property for the multivariate vector $\mathbf{y}_{j,n}$ because $p(\mathbf{y}_{j,n})$ only depends on the vector norm $\|\mathbf{y}_{j,n}\|_2$. Also, the distribution is assumed to be mutually independent for time frames and sources. Whereas the temporal source model $p(\mathbf{y}_{j,n})$ is based on the Gaussian distribution, the global source model $p(\mathbf{y}_{1,n}, \cdots, \mathbf{y}_{J,n})$ becomes the super-Gaussian distribution because of the time-varying variance $r_{j,n}$ [42]. This time-varying Gaussian source model has been adopted for many techniques, e.g., BSS [44, 45] and dereverberation of speech signals [46]. Hereafter, IVA based on the source distribution (6.16) is referred to as time-varying Gaussian IVA.

6.2.4 Itakura–Saito NMF

When we apply NMF to an acoustic signal, the power spectrogram obtained via STFT is considered as an observed nonnegative matrix and can be decomposed into two nonnegative matrices as

$$|\mathbf{D}|^{.2} \approx \mathbf{TV}, \tag{6.17}$$

where $\mathbf{D} \in \mathbb{C}^{I \times J}$ is a complex-valued spectrogram, and the absolute value $|\cdot|$ and the dotted exponent for matrices denote the entrywise absolute value and the entrywise exponent, respectively, $\mathbf{T} \in \mathbb{R}_{\geq 0}^{I \times L}$ is a basis matrix, which includes bases (frequently appearing spectral patterns in $|\mathbf{D}|^{.2}$) as column vectors, and $\mathbf{V} \in \mathbb{R}_{\leq 0}^{L \times J}$ is an activation matrix, which involves time-varying gains of each basis in $\mathbf{T}$ as row vectors. Also, L is the number of bases, which should be set to a much smaller value than I or J. Figure 6.3 depicts the decomposition model of NMF, where L is set to two. In this figure, the basis matrix includes two types of spectral pattern as the bases to represent the observed matrix using time-varying gains in the activation matrix. In the decomposition of NMF, the variables $\mathbf{T}$ and $\mathbf{V}$ are optimized by minimizing the cost function based on the divergence between the nonnegative observation $|\mathbf{D}|^{.2}$ and the model $\mathbf{TV}$. In particular, Itakura–Saito NMF has a special generative model, which has been given by Févotte et al. [47], as described below.

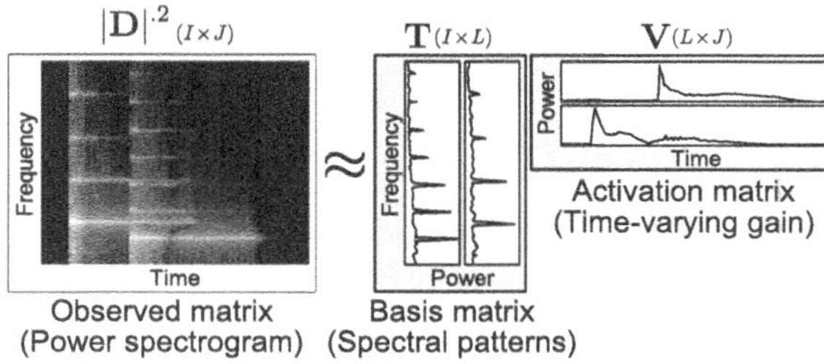

Fig. 6.3 Decomposition model of simple NMF, where $L = 2$. Basis matrix involves representative spectral patterns, and activation matrix represents time-varying gains for each basis

Fig. 6.4 Circularly symmetric complex Gaussian distribution. Probability does not depend on phase $\arg(q_{ijl})$ but only depends on amplitude $|q_{ijl}|$ or power $|q_{ijl}|^2$ because of circularly symmetric property

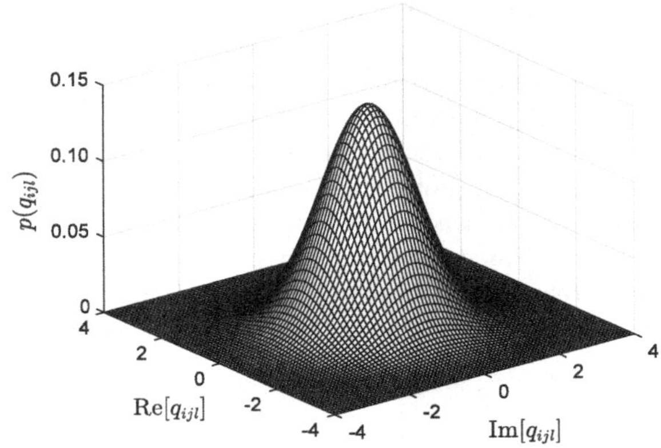

Let us assume that L complex-valued spectrograms $q_{ij1}, \cdots, q_{ijL}$ are generated from circularly symmetric (isotropic) complex Gaussian distribution [48], which are independently defined in each time-frequency slot as follows:

$$p(q_{ijl}) = \frac{1}{\pi r_{ijl}} \exp\left(-\frac{|q_{ijl}|^2}{r_{ijl}}\right), \tag{6.18}$$

where $l = 1, \cdots, L$ is the integral index of L components and r_{ijl} is the nonnegative variance of each distribution. Figure 6.4 shows the circularly symmetric complex Gaussian distribution. Since the distribution has a circularly symmetric property in the complex plane, the probability does not depend on the phase $\arg(q_{ijl})$ and only depends on the amplitude $|q_{ijl}|$ or power $|q_{ijl}|^2$. Note that the variance r_{ijl} corresponds to the expectation value of the power spectrum $|q_{ijl}|^2$, namely, $r_{ijl} = \mathrm{E}[|q_{ijl}|^2]$. When the variance r_{ijl} is large, the distribution becomes wider, and the complex-valued spectrum q_{ijl} with a large power can easily be generated, while the phase of q_{ijl} is always uniformly distributed. In addition, if we assume that the observation d_{ij}, which is the complex-valued entry of $\mathbf{D}$, is the sum of the components q_{ijl}, namely, $d_{ij} = \sum_l q_{ijl}$, the following generative model can also be assumed because of the reproductive property in complex Gaussian distributions:

$$\begin{aligned} p(\mathbf{D}) &= \prod_{i,j} p(d_{ij}) \\ &= \prod_{i,j} \frac{1}{\pi r_{ij}} \exp\left(-\frac{|d_{ij}|^2}{r_{ij}}\right), \end{aligned} \tag{6.19}$$

where $r_{ij} = \sum_l r_{ijl}$. This fact means that the additivity of power spectra $|q_{ijl}|^2$ is held only in the expectation sense. Now, the likelihood function of $\mathbf{T}$ and $\mathbf{V}$ can be obtained as follows by putting $r_{ijl} = t_{il} v_{lj}$;

$$\mathscr{L}(\mathbf{T}, \mathbf{V}) = p(\mathbf{D}|\mathbf{T}, \mathbf{V}) = \prod_{i,j} \frac{1}{\pi \sum_l t_{il} v_{lj}} \exp\left(-\frac{|d_{ij}|^2}{\sum_l t_{il} v_{lj}}\right), \tag{6.20}$$

where t_{il} and v_{lj} are the nonnegative entries of $\mathbf{T}$ and $\mathbf{V}$, respectively. The negative log-likelihood function is

$$-\log \mathscr{L}(\mathbf{T}, \mathbf{V}) = \sum_{i,j} \left(\log \pi + \log \sum_l t_{il} v_{lj} + \frac{|d_{ij}|^2}{\sum_l t_{il} v_{lj}}\right). \tag{6.21}$$

It is clear that the ML estimation based on (6.21) is equivalent to the minimization of the Itakura–Saito divergence $\mathscr{D}_{\mathrm{IS}}$ [49] between $|\mathbf{D}|^{\cdot 2}$ and $\mathbf{TV}$:

$$\begin{aligned}\mathscr{D}_{\mathrm{IS}}\left(|\mathbf{D}|^{\cdot 2} \| \mathbf{TV}\right) &= \sum_{i,j} \left(\frac{|d_{ij}|^2}{\sum_l t_{il} v_{lj}} - \log \frac{|d_{ij}|^2}{\sum_l t_{il} v_{lj}} - 1\right) \\ &= \text{const.} + \sum_{i,j} \left(\frac{|d_{ij}|^2}{\sum_l t_{il} v_{lj}} + \log \sum_l t_{il} v_{lj}\right).\end{aligned} \tag{6.22}$$

Thus, when Itakura–Saito NMF is applied to the observed power spectrogram $|\mathbf{D}|^{\cdot 2}$, it is assumed that d_{ij} follows the generative model (6.19) and the components q_{ijl} are mutually independent. The multiplicative update rules for $\mathbf{T}$ and $\mathbf{V}$ that minimize (6.21) or (6.22) are given by [50]

$$t_{il} \leftarrow t_{il} \sqrt{\frac{\sum_j |d_{ij}|^2 v_{lj} \left(\sum_{l'} t_{il'} v_{l'j}\right)^{-2}}{\sum_j v_{lj} \left(\sum_{l'} t_{il'} v_{l'j}\right)^{-1}}}, \tag{6.23}$$

$$v_{lj} \leftarrow v_{lj} \sqrt{\frac{\sum_i |d_{ij}|^2 t_{il} \left(\sum_{l'} t_{il'} v_{l'j}\right)^{-2}}{\sum_i t_{il} \left(\sum_{l'} t_{il'} v_{l'j}\right)^{-1}}}. \tag{6.24}$$

These update rules are called the multiplicative update (MU) and guarantee a monotonic decrease in cost function.

Figure 6.5a shows the source model (variance structure in a time-frequency region) assumed in time-varying Gaussian IVA. Since the variance $r_{j,n}$ is shared over the frequency bins, it can be interpreted as an uniform (flat) spectral basis. On the other hand, Itakura–Saito NMF has a more flexible source model because the variance r_{ij} is independently defined in each time-frequency slot as shown in Fig. 6.5b. It allows us to model the specific time-frequency structure with limited numbers of bases and activations.

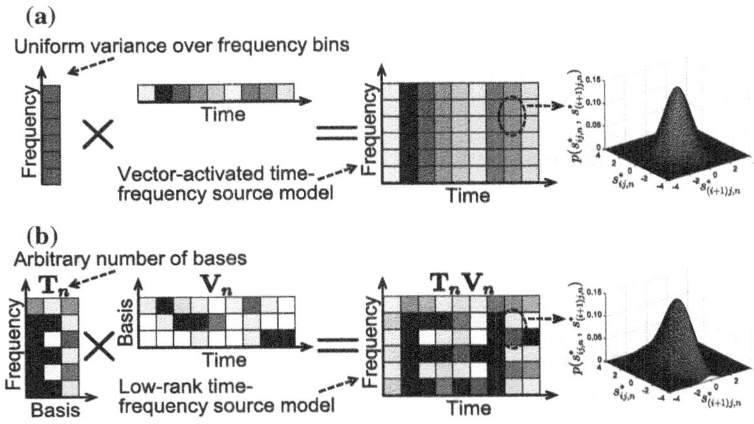

Fig. 6.5 Comparison of source models (variance structures) in **a** time-varying Gaussian IVA and **b** Itakura–Saito NMF, where grayscale in each time-frequency slot indicates scale of variance. Time-varying Gaussian IVA has uniform variance over frequency bins, and all the frequency bins have the same activations (time-varying gains), whereas Itakura–Saito NMF employs limited number of bases to capture low-rank structure, resulting in more flexible source model

6.3 Independent Low-Rank Matrix Analysis: A Unification of IVA and Itakura–Saito NMF

6.3.1 Motivation and Strategy

For speech signal separation, Laplace IVA or time-varying Gaussian IVA can achieve better performance than FDICA. However, since only the higher-order correlation defined in (6.9) or (6.16) is utilized as a spectral structure in the source model, IVA cannot treat the specific harmonic structures of each source and lacks flexibility, as shown in Fig. 6.5. For this reason, IVA is not suitable for sources that have characteristic (specific) spectral structures, such as instrumental sounds or music signals. NMF decomposition is suitable for modeling the spectrogram of music or instrumental signals because such signals typically consist of a limited number of components, for example, steady musical tones, discrete pitches, and discrete notes. This property means that the spectrogram of a music signal tends to be a low-rank matrix compared with a speech spectrogram.

In [43], the temporal power variation of sources provided by a user is exploited as the prior distribution of the time-varying gain $r_{j,n}$, which is defined as an inverse gamma distribution. In [51], a new multichannel source separation method with external model information has been proposed, which is called model-based IVA. In this approach, we consider that the time-frequency variance $r_{ij,n}$ for each source is given by another technique (e.g., single-channel spectral subtraction, voice activity detection, or time-frequency binary masking) applied in advance. The demixing matrix $\mathbf{W}_i$ is estimated on the basis of the independence between sources taking the given variance $r_{ij,n}$ into account. These approaches show that the estimation of $\mathbf{W}_i$ based on a correct and precise variance will provide better separation performance.

On the basis of these ideas, in this chapter, we introduce Itakura–Saito NMF to IVA for decomposing the sourcewise variance $r_{ij,n}$ using a limited number of NMF bases, where the demixing matrix $\mathbf{W}_i$ and the source model $p(\mathbf{y}_{j,1}, \cdots, \mathbf{y}_{j,N})$ with the NMF variables are simultaneously estimated in a fully blind manner. This approach is a natural extension of time-varying Gaussian IVA because we extend the vector source model (frequency-uniform variance) to the low-rank matrix source model (NMF decomposition) as shown in Fig. 6.5. For this reason, hereafter, we call this method *independent low-rank matrix analysis (ILRMA)*. Similarly to standard FDICA or IVA, ILRMA is applicable to the determined case ($M = N$). In the overdetermined case ($M > N$), dimensionality reduction using PCA should be applied so that $M = N$.

6.3.2 Derivation of Cost Function

In ILRMA, similarly to Itakura–Saito NMF, the circularly symmetric complex Gaussian distribution is independently assumed to be as follows in each time-frequency slot as the source model of the separated signal;

$$
\begin{aligned}
p(\mathbf{y}_{j,1}, \cdots, \mathbf{y}_{j,N}) &= \prod_n p(\mathbf{y}_{j,n}) \\
&= \prod_{n,i} \frac{1}{\pi r_{ij,n}} \exp\left(-\frac{|y_{ij,n}|^2}{r_{ij,n}}\right),
\end{aligned} \tag{6.25}
$$

where $r_{ij,n}$ is the sourcewise variance that corresponds to the expectation of the power spectrogram, namely, $r_{ij,n} = \mathrm{E}[|y_{ij,n}|^2]$. The contrast function and the negative log-likelihood function of the parameter set $\mathbf{W}$ and $\mathbf{R} = \{r_{ij,n} | i = 1, \cdots, I; j = 1, \cdots J; n = 1, \cdots N\}$ are given as

$$
\begin{aligned}
G(\mathbf{y}_{j,n}) &= \sum_i \left(\log \pi r_{ij,n} + \frac{|y_{ij,n}|^2}{r_{ij,n}}\right) \\
&= I \log \pi + \sum_i \left(\log r_{ij,n} + \frac{|y_{ij,n}|^2}{r_{ij,n}}\right),
\end{aligned} \tag{6.26}
$$

$$
\begin{aligned}
-\log \mathscr{L}(\mathbf{W}, \mathbf{R}) = \text{const.} &- 2J \sum_i \log |\det \mathbf{W}_i| \\
&+ \sum_{i,j,n} \left(\log r_{ij,n} + \frac{|y_{ij,n}|^2}{r_{ij,n}}\right).
\end{aligned} \tag{6.27}
$$

Here, we consider two types of $r_{ij,n}$ decomposition depending on the presence of a partitioning function:

$$r_{ij,n} = \sum_{l} t_{il,n} v_{lj,n}, \tag{6.28}$$

$$r_{ij,n} = \sum_{k} z_{nk} t_{ik} v_{kj}, \tag{6.29}$$

where $t_{il,n}$ and $v_{lj,n}$ are the nonnegative entries of $\mathbf{T}_n \in \mathbb{R}_{\geq 0}^{I \times L}$ and $\mathbf{V}_n \in \mathbb{R}_{\geq 0}^{I \times L}$ that are the sourcewise basis and activation matrices, and t_{ik} and v_{kj} are the nonnegative entries of $\mathbf{T}$ and $\mathbf{V}$ that include K bases and activations, respectively. Moreover, $z_{nk} \in [0, 1]$ is the entry of $\mathbf{Z} = (\mathbf{z}_1 \cdots \mathbf{z}_N)^{\mathrm{T}} \in \mathbb{R}_{[0,1]}^{N \times K}$, which is a partitioning function that clusters K bases into N sources and satisfies $\sum_n z_{nk} = 1$, and $k = 1, \cdots, K$ is the new basis index. In (6.28), a fixed number of bases, L, is utilized to decompose each separated source spectrogram $|y_{ij,n}|^2$. On the other hand, we can adaptively determine the number of bases for each separated source spectrogram by employing the partitioning function z_{nk} as (6.29). In this model, we only set the total number of bases to K. This approach is reasonable because the optimal number of bases will depend on the time-frequency structure of each source. For a source that consists of a low-rank power spectrogram, such as an instrumental signal, the number of bases should be small, whereas a speech or vocal spectrogram may require more bases for its precise representation. The cost function in ILRMA can be obtained by substituting (6.28) or (6.29) into (6.27).

In Laplace IVA, the variance $r_{i,n}$ is uniformly set to unity over the frequency bins, and is not estimated. This is because the variance only determines the signal scale of $y_{ij,n}$, and it can be restored by the back-projection technique. In time-varying Gaussian IVA, only the activation for the uniform variance is estimated based on the prior information given by users. On the other hand, the variance in ILRMA, $r_{ij,n}$, is blindly estimated by low-rank decomposition using NMF (6.28) or (6.29) to capture the time-frequency structure as shown in Fig. 6.5b. It is clear that when the number of bases is set to one for every source and all bases have a flat spectrum, the source models in time-varying Gaussian IVA and ILRMA become identical. This fact shows that ILRMA includes time-varying Gaussian IVA as a special case.

6.3.3 Update Rules

For the optimization of ICA or IVA, update rules based on the auxiliary function technique have been proposed [40–43, 51, 52], and it has been reported that these update rules are faster and more stable than those for a conventional update scheme (e.g., natural gradient method [53, 54]) and that the step size parameter can be omitted in each iteration. Regarding the estimation of $\mathbf{W}_i$, the differential of (6.27) w.r.t. $\mathbf{W}_i$ becomes equivalent to that of the auxiliary bounding function in Laplace IVA [40].

For this reason, the update rules of $\mathbf{W}_i$ based on IP can easily be derived as follows:

$$\mathbf{U}_{i,n} = \frac{1}{J}\sum_j \frac{1}{r_{ij,n}}\mathbf{x}_{ij}\mathbf{x}_{ij}^{\mathrm{H}}, \tag{6.30}$$

$$\mathbf{w}_{i,n} \leftarrow \left(\mathbf{W}_i\mathbf{U}_{i,n}\right)^{-1}\mathbf{e}_n, \tag{6.31}$$

$$\mathbf{w}_{i,n} \leftarrow \mathbf{w}_{i,n}\left(\mathbf{w}_{i,n}^{\mathrm{H}}\mathbf{U}_{i,n}\mathbf{w}_{i,n}\right)^{-\frac{1}{2}}, \tag{6.32}$$

where $\mathbf{e}_n$ denotes the $N \times 1$ unit vector with the nth element equal to unity. After the update of $\mathbf{W}_i$, the separated signal $\mathbf{y}_{ij}$ should be updated as

$$y_{ij,n} \leftarrow \mathbf{w}_{i,n}^{\mathrm{H}}\mathbf{x}_{ij}. \tag{6.33}$$

If we eliminate the partitioning function z_{nk}, which is ILRMA with (6.28), the differential of (6.27) w.r.t. $t_{il,n}$ or $v_{lj,n}$ becomes identical to the differential of the cost function in Itakura–Saito NMF (6.22). Therefore, the update rules of $t_{il,n}$ and $v_{lj,n}$ are given as

$$t_{il,n} \leftarrow t_{il,n}\sqrt{\frac{\sum_j |y_{ij,n}|^2 v_{lj,n} r_{ij,n}^{-2}}{\sum_j v_{lj,n} r_{ij,n}^{-1}}}, \tag{6.34}$$

$$v_{lj,n} \leftarrow v_{lj,n}\sqrt{\frac{\sum_i |y_{ij,n}|^2 t_{il,n} r_{ij,n}^{-2}}{\sum_i t_{il,n} r_{ij,n}^{-1}}}. \tag{6.35}$$

The estimated source model $r_{ij,n}$ should be updated by (6.28) after each update of $t_{il,n}$ and $v_{lj,n}$. Alternatively, if we employ the partitioning function z_{nk} to cluster K bases into N specific sources, which is ILRMA with (6.29), we can derive the auxiliary-function-based update rules of z_{nk}, t_{ik}, and v_{kj} by minimizing (6.27) in a similar way to in [12, 50] as

$$z_{nk} \leftarrow z_{nk}\sqrt{\frac{\sum_{i,j} |y_{ij,n}|^2 t_{ik} v_{kj} r_{ij,n}^{-2}}{\sum_{i,j} t_{ik} v_{kj} r_{ij,n}^{-1}}}, \tag{6.36}$$

$$z_{nk} \leftarrow \frac{z_{nk}}{\sum_{n'} z_{n'k}}, \tag{6.37}$$

$$t_{ik} \leftarrow t_{ik}\sqrt{\frac{\sum_{j,n} |y_{ij,n}|^2 z_{nk} v_{kj} r_{ij,n}^{-2}}{\sum_{j,n} z_{nk} v_{kj} r_{ij,n}^{-1}}}, \tag{6.38}$$

$$v_{kj} \leftarrow v_{kj}\sqrt{\frac{\sum_{i,n} |y_{ij,n}|^2 z_{nk} t_{ik} r_{ij,n}^{-2}}{\sum_{i,n} z_{nk} t_{ik} r_{ij,n}^{-1}}}, \tag{6.39}$$

where (6.37) is calculated to ensure $\sum_n z_{nk} = 1$. The estimated source model $r_{ij,n}$ should be updated by (6.29) after each update of z_{nk}, t_{ik}, and v_{kj}. The derivation of (6.36)–(6.39) is described in [31].

Thus, we can estimate all the variables that minimize (6.27) by iterating these update rules. Note that a scale ambiguity exists between $\mathbf{W}_i$ and $r_{ij,n}$ because both of them can determine the scale of the separated signal $y_{ij,n}$. Therefore, $\mathbf{W}_i$ or $r_{ij,n}$ has a risk of diverging during the optimization. To avoid this problem, the following normalization should be applied at each iteration:

$$\mathbf{w}_{i,n} \leftarrow \mathbf{w}_{i,n}\lambda_n^{-1}, \tag{6.40}$$

$$y_{ij,n} \leftarrow y_{ij,n}\lambda_n^{-1}, \tag{6.41}$$

$$r_{ij,n} \leftarrow r_{ij,n}\lambda_n^{-2}, \tag{6.42}$$

and

$$t_{il,n} \leftarrow t_{il,n}\lambda_n^{-2}, \tag{6.43}$$

should be applied for ILRMA without a partitioning function, or

$$t_{ik} \leftarrow t_{ik}\sum_n z_{nk}\lambda_n^{-2}, \tag{6.44}$$

$$z_{nk} \leftarrow \frac{z_{nk}\lambda_n^{-2}}{\sum_{n'} z_{n'k}\lambda_{n'}^{-2}}, \tag{6.45}$$

should be applied for ILRMA with a partitioning function, where λ_n is an arbitrary sourcewise normalization coefficient, such as the sourcewise average power $\lambda_n = [(IJ)^{-1}\sum_{i,j}|y_{ij,n}|^2]^{(1/2)}$. These normalizations do not change the value of the cost function (6.27). The scale of the separated signal $y_{ij,n}$ can be restored by applying the following back-projection technique [24] after the optimization:

$$\hat{\mathbf{y}}_{ij,n} = \mathbf{W}_i^{-1}\left(\mathbf{e}_n \circ \mathbf{y}_{ij}\right), \tag{6.46}$$

where $\hat{\mathbf{y}}_{ij,n} = (\hat{y}_{ij,n1} \ \cdots \ \hat{y}_{ij,nM})^{\mathrm{T}}$ is a separated source image whose scale is fitted to the observed signals at each microphone and $\circ$ denotes the Hadamard product (entrywise multiplication).

6.3.4 Summary of Algorithm

The detailed algorithm of ILRMA is summarized in Algorithms 1 and 2, where $\max(\cdot,\cdot)$ returns a matrix with the larger elements taken from two inputs in each entry, ε denotes the machine epsilon, $\mathbf{1}^{(\mathrm{size})}$ denotes matrix of ones whose size is denoted as the superscript, the quotient symbol for matrices denote the entrywise division, and $\mathsf{X} \in \mathbb{C}^{I\times J\times M}$, $\mathsf{Y} \in \mathbb{C}^{I\times J\times N}$, $\mathsf{P} \in \mathbb{R}^{I\times J\times N}$, and $\mathsf{R} \in \mathbb{R}^{I\times J\times N}$ are third-order tensors whose entries are $x_{ij,m}$, $y_{ij,n}$, $p_{ij,n}$, and $r_{ij,n}$, respectively. In addition, the third-order tensor with a subscript denotes the sliced matrix or the fiber vector in the original tensor [55]. For example, $\mathsf{X}_{i::}$, $\mathsf{X}_{:j:}$, and $\mathsf{X}_{::m}$ denote the $J \times M$, $I \times M$, and $I \times J$ sliced matrices in X, respectively. Also, $\mathsf{X}_{ij:}$, $\mathsf{X}_{i:m}$, and $\mathsf{X}_{:jm}$ denote the $M \times 1$, $J \times 1$, and $I \times 1$ fiber (column) vectors in X, respectively. To avoid division by zero, flooring with the machine epsilon is performed in the update of the NMF variables.

Algorithm 1 ILRMA without partitioning function

Initialize $\mathbf{W}_i$ with identity matrix and $\mathbf{T}_n$ and $\mathbf{V}_n$ with nonnegative random values
Calculate $\mathbf{y}_{ij} = \mathbf{W}_i\mathbf{x}_{ij}$ for all i and j
Calculate $\mathsf{P}_{::n} = |\mathsf{Y}_{::n}|^{\cdot 2}$ and $\mathsf{R}_{::n} = \mathbf{T}_n\mathbf{V}_n$ for all n, respectively
repeat
  for $n = 1$ to N **do**
    $\mathbf{T}_n \leftarrow \max\left(\mathbf{T}_n \circ \left[\frac{(\mathsf{P}_{::n}\circ\mathsf{R}_{::n}^{\cdot -2})\mathbf{V}_n^{\mathrm{T}}}{\mathsf{R}_{::n}^{\cdot -1}\mathbf{V}_n^{\mathrm{T}}}\right]^{\cdot\frac{1}{2}}, \varepsilon\right)$
    $\mathsf{R}_{::n} = \mathbf{T}_n\mathbf{V}_n$
    $\mathbf{V}_n \leftarrow \max\left(\mathbf{V}_n \circ \left[\frac{\mathbf{T}_n^{\mathrm{T}}(\mathsf{P}_{::n}\circ\mathsf{R}_{::n}^{\cdot -2})}{\mathbf{T}_n^{\mathrm{T}}\mathsf{R}_{::n}^{\cdot -1}}\right]^{\cdot\frac{1}{2}}, \varepsilon\right)$
    $\mathsf{R}_{::n} = \mathbf{T}_n\mathbf{V}_n$
    for $i = 1$ to I **do**
      $\mathbf{U}_{i,n} = \frac{1}{J}\left\{\mathsf{X}_{i::}^{\mathrm{H}}\left[\mathsf{X}_{i::}\circ\left(\mathsf{R}_{i:n}^{\cdot -1}\mathbf{1}^{(1\times M)}\right)\right]\right\}^{\mathrm{T}}$
      $\mathbf{w}_{i,n} \leftarrow (\mathbf{W}_i\mathbf{U}_{i,n})^{-1}\mathbf{e}_n$
      $\mathbf{w}_{i,n} \leftarrow \mathbf{w}_{i,n}(\mathbf{w}_{i,n}^{\mathrm{H}}\mathbf{U}_{i,n}\mathbf{w}_{i,n})^{-\frac{1}{2}}$
    end for
  end for
  Calculate $\mathbf{y}_{ij} = \mathbf{W}_i\mathbf{x}_{ij}$ for all i and j
  Calculate $\mathsf{P}_{::n} = |\mathsf{Y}_{::n}|^{\cdot 2}$ for all n
  for $n = 1$ to N **do**
    $\lambda_n = \sqrt{\frac{1}{IJ}\sum_{i,j} p_{ij,n}}$
    for $i = 1$ to I **do**
      $\mathbf{w}_{i,n} \leftarrow \mathbf{w}_{i,n}\lambda_n^{-1}$
    end for
    $\mathsf{P}_{::n} \leftarrow \mathsf{P}_{::n}\lambda_n^{-2}$
    $\mathsf{R}_{::n} \leftarrow \mathsf{R}_{::n}\lambda_n^{-2}$
    $\mathbf{T}_n \leftarrow \mathbf{T}_n\lambda_n^{-2}$
  end for
until converge
Calculate $\hat{\mathbf{y}}_{ij,n} = \mathbf{W}_i^{-1}\left(\mathbf{e}_n \circ \mathbf{y}_{ij}\right)$ for all i, j, and n

---

Algorithm 2 ILRMA with partitioning function

Initialize $\mathbf{W}_i$ with identity matrix, $\mathbf{T}$ and $\mathbf{V}$ with nonnegative random values, and $\mathbf{Z}$ with random values in range [0, 1]
$\mathbf{Z} \leftarrow \mathbf{Z} \circ \left(\mathbf{1}^{(N\times N)}\mathbf{Z}\right)^{\cdot -1}$
Calculate $\mathbf{y}_{ij} = \mathbf{W}_i \mathbf{x}_{ij}$ for all i and j
Calculate $\mathsf{P}_{::n} = |\mathsf{Y}_{::n}|^{\cdot 2}$ and $\mathsf{R}_{::n} = \left[\left(\mathbf{1}^{(I\times 1)}\mathbf{z}_n^{\mathrm{T}}\right) \circ \mathbf{T}\right]\mathbf{V}$ for all n, respectively
repeat
  for $n = 1$ to N **do**
    $\mathbf{b}_n^{(\mathbf{Z})} = \left(\dfrac{\{[\mathbf{T}^{\mathrm{T}}(\mathsf{P}_{::n}\circ\mathsf{R}_{::n}^{\cdot -2})]\circ\mathbf{V}\}\mathbf{1}^{(J\times 1)}}{\left[\left(\mathbf{T}^{\mathrm{T}}\mathsf{R}_{::n}^{\cdot -1}\right)\circ\mathbf{V}\right]\mathbf{1}^{(J\times 1)}}\right)^{\cdot\frac{1}{2}}$
  end for
  $\mathbf{Z} \leftarrow \max\left(\mathbf{Z}\circ\mathbf{B}^{(\mathbf{Z})}, \varepsilon\right)$, where $\mathbf{B}^{(\mathbf{Z})} = (\mathbf{b}_1^{(\mathbf{Z})} \cdots \mathbf{b}_N^{(\mathbf{Z})})^{\mathrm{T}}$
  $\mathbf{Z} \leftarrow \mathbf{Z} \circ \left(\mathbf{1}^{(N\times N)}\mathbf{Z}\right)^{\cdot -1}$
  Calculate $\mathsf{R}_{::n} = \left[\left(\mathbf{1}^{(I\times 1)}\mathbf{z}_n^{\mathrm{T}}\right) \circ \mathbf{T}\right]\mathbf{V}$ for all n
  for $i = 1$ to I **do**
    $\mathbf{b}_i^{(\mathbf{T})} = \left(\dfrac{\left\{\left[\mathbf{V}\left(\mathsf{P}_{i::}\circ\mathsf{R}_{i::}^{\cdot -2}\right)\right]\circ\mathbf{Z}^{\mathrm{T}}\right\}\mathbf{1}^{(N\times 1)}}{\left[\left(\mathbf{V}\mathsf{R}_{i::}^{\cdot -1}\right)\circ\mathbf{Z}^{\mathrm{T}}\right]\mathbf{1}^{(N\times 1)}}\right)^{\cdot\frac{1}{2}}$
  end for
  $\mathbf{T} \leftarrow \max\left(\mathbf{T}\circ\mathbf{B}^{(\mathbf{T})}, \varepsilon\right)$, where $\mathbf{B}^{(\mathbf{T})} = (\mathbf{b}_1^{(\mathbf{T})} \cdots \mathbf{b}_I^{(\mathbf{T})})^{\mathrm{T}}$
  Calculate $\mathsf{R}_{::n} = \left[\left(\mathbf{1}^{(I\times 1)}\mathbf{z}_n^{\mathrm{T}}\right) \circ \mathbf{T}\right]\mathbf{V}$ for all n
  for $j = 1$ to J **do**
    $\mathbf{b}_j^{(\mathbf{V})} = \left(\dfrac{\left\{\left[\mathbf{T}^{\mathrm{T}}\left(\mathsf{P}_{:j:}\circ\mathsf{R}_{:j:}^{\cdot -2}\right)\right]\circ\mathbf{Z}^{\mathrm{T}}\right\}\mathbf{1}^{(N\times 1)}}{\left[\left(\mathbf{T}^{\mathrm{T}}\mathsf{R}_{:j:}^{\cdot -1}\right)\circ\mathbf{Z}^{\mathrm{T}}\right]\mathbf{1}^{(N\times 1)}}\right)^{\cdot\frac{1}{2}}$
  end for
  $\mathbf{V} \leftarrow \max\left(\mathbf{V}\circ\mathbf{B}^{(\mathbf{V})}, \varepsilon\right)$, where $\mathbf{B}^{(\mathbf{V})} = (\mathbf{b}_1^{(\mathbf{V})} \cdots \mathbf{b}_J^{(\mathbf{V})})$
  Calculate $\mathsf{R}_{::n} = \left[\left(\mathbf{1}^{(I\times 1)}\mathbf{z}_n^{\mathrm{T}}\right) \circ \mathbf{T}\right]\mathbf{V}$ for all n
  for $n = 1$ to N **do**
    for $i = 1$ to I **do**
      $\mathsf{U}_{i,n} = \frac{1}{J}\left\{\mathsf{X}_{i::}^{\mathrm{H}}\left[\mathsf{X}_{i::}\circ\left(\mathsf{R}_{i:n}^{\cdot -1}\mathbf{1}^{(1\times M)}\right)\right]\right\}^{\mathrm{T}}$
      $\mathbf{w}_{i,n} \leftarrow (\mathbf{W}_i \mathsf{U}_{i,n})^{-1}\mathbf{e}_n$
      $\mathbf{w}_{i,n} \leftarrow \mathbf{w}_{i,n}(\mathbf{w}_{i,n}^{\mathrm{H}}\mathsf{U}_{i,n}\mathbf{w}_{i,n})^{-\frac{1}{2}}$
    end for
  end for
  Calculate $\mathbf{y}_{ij} = \mathbf{W}_i \mathbf{x}_{ij}$ for all i and j
  Calculate $\mathsf{P}_{::n} = |\mathsf{Y}_{::n}|^{\cdot 2}$ for all n
  for $n = 1$ to N **do**
    $\lambda_n = \sqrt{\frac{1}{IJ}\sum_{i,j} p_{ij,n}}$
    for $i = 1$ to I **do**
      $\mathbf{w}_{i,n} \leftarrow \mathbf{w}_{i,n}\lambda_n^{-1}$
    end for
    $\mathsf{P}_{::n} \leftarrow \mathsf{P}_{::n}\lambda_n^{-2}$
    $\mathsf{R}_{::n} \leftarrow \mathsf{R}_{::n}\lambda_n^{-2}$
  end for
  Calculate $t_{ik} \leftarrow t_{ik}\sum_n z_{nk}\lambda_n^{-2}$ for all i and k
  Calculate $z_{nk} \leftarrow z_{nk}\dfrac{\lambda_n^{-2}}{\sum_{n'} z_{n'k}\lambda_{n'}^{-2}}$ for all n and k
until converge
Calculate $\hat{\mathbf{y}}_{ij,n} = \mathbf{W}_i^{-1}\left(\mathbf{e}_n \circ \mathbf{y}_{ij}\right)$ for all i, j, and n

6.4 Relationship Between Time-Varying Gaussian IVA, ILRMA, and Multichannel NMF

In NMF-based source separation, the decomposed bases and activations must be clustered in every source to achieve source separation. One effective way of achieving this is to utilize a sample sound of the target signal [16–18]. However, such supervision cannot be utilized in BSS. To solve this problem, multichannel NMF (MNMF) has been proposed [32, 34–36, 56–58]. In particular, MNMF methods [32, 34–36] treat convolutive mixtures similarly to FDICA, IVA, and ILRMA and estimate a mixing system for the sources, which is utilized for the clustering of bases. In these MNMFs, the spatial covariance [59, 60], which is the covariance matrix of a zero-mean multivariate Gaussian distribution, has been utilized to model the mixing conditions of the recording environment. In this section, the relationship between time-varying Gaussian IVA, ILRMA, and MNMF is revealed from the viewpoint of their assumed generative models.

6.4.1 Generative Model in MNMF and Spatial Covariance

In MNMF [32, 34–36] and its related methods [59, 60], the probability distribution of multichannel STFT coefficients $\mathbf{x}_{ij}$ is modeled by a circularly symmetric multivariate complex Gaussian distribution with a time-frequency-variant covariance matrix as follows:

$$p(\mathbf{x}_{ij}) = \frac{1}{\pi^M \det \mathbf{R}_{ij}^{(\mathrm{x})}} \exp\left(-\mathbf{x}_{ij}^{\mathrm{H}} \mathbf{R}_{ij}^{(\mathrm{x})^{-1}} \mathbf{x}_{ij}\right), \tag{6.47}$$

where $\mathbf{R}_{ij}^{(\mathrm{x})}$ is called the spatial covariance [59, 60] of the observed multichannel signal $\mathbf{x}_{ij}$, namely, $\mathbf{R}_{ij}^{(\mathrm{x})} = \mathrm{E}[\mathbf{x}_{ij}\mathbf{x}_{ij}^{\mathrm{H}}]$. This spatial covariance can be decomposed into the time-invariant source covariance $\mathbf{R}_{i,n}^{(\mathrm{s})}$, the time-variant scalar variance $r_{ij,n}$, and the time-invariant noise covariance $\mathbf{R}_i^{(\mathrm{n})}$ that contributes to additional noise $\mathbf{n}_{ij}$, as

$$\mathbf{R}_{ij}^{(\mathrm{x})} = \sum_n r_{ij,n} \mathbf{R}_{i,n}^{(\mathrm{s})} + \mathbf{R}_i^{(\mathrm{n})}. \tag{6.48}$$

The spatial covariance $\mathbf{R}_{i,n}^{(\mathrm{s})}$ represents the spatial position and the spatial spread of the nth source. In particular, if the mixing system can be modeled by the mixing matrix $\mathbf{A}_i$ as (6.4) with a noiseless assumption, the spatial covariance $\mathbf{R}_{i,n}^{(\mathrm{s})}$ is equal to the rank-1 matrix

$$\mathbf{R}_{i,n}^{(\mathrm{s})} = \mathbf{a}_{i,n}\mathbf{a}_{i,n}^{\mathrm{H}}. \tag{6.49}$$

This mixing model is the called *rank-1 spatial model*, which is identical to the assumption of an instantaneous mixture in the frequency domain. In contrast, if the mixing system cannot be modeled by (6.4) owing to, for example, strong reverberation in the recording environment, the rank of $\mathbf{R}_{i,n}^{(\mathrm{s})}$ increases so that it becomes a full-rank spatial covariance [59, 60].

6.4.2 Existing MNMF Models

Existing MNMF models and their related works can be characterized in terms of two features: models of spatial covariance $\mathbf{R}_{ij}^{(\mathrm{x})}$ and source spectrograms. Table 6.1 summarizes the existing methods. The models proposed in [59, 60] have the most general representations. Several types of $\mathbf{R}_{i,n}^{(\mathrm{s})}$ have been investigated including rank-1 and full-rank matrices. MNMF in [34] (hereafter referred to as *Ozerov's MNMF*) was the first method to model a power spectrogram $r_{ij,n}$ using NMF decomposition. In this method, the sourcewise spatial covariance $\mathbf{R}_{i,n}^{(\mathrm{s})}$ is constrained by a rank-1 matrix, and an additive noise component $\mathbf{n}_{ij}$ is also assumed. The update rules of the variables based on both expectation-maximization (EM) and MU algorithms have been derived. Ozerov's MNMF was extended to a full-rank spatial model in [32]. Also, a more flexible source model with a partitioning function z_{nk} was introduced in [35]. As another optimization scheme, an MU algorithm based on an auxiliary function technique was proposed in [36] (hereafter referred to as *Sawada's MNMF*). It also employs the full-rank $\mathbf{R}_{i,n}^{(\mathrm{s})}$ and the flexible source model with z_{nk} and NMF variables. Note that all the existing MNMFs estimate the sourcewise mixing system $\mathbf{R}_{i,n}^{(\mathrm{s})}$ to achieve separation via multichannel Wiener filtering [61], whereas ILRMA estimates the demixing matrix $\mathbf{W}_i$.

6.4.3 Equivalence Between ILRMA and MNMF with Rank-1 Spatial Model

From (6.47), the likelihood function of the observed spatial covariance $\mathbf{R}^{(\mathrm{x})} = \{\mathbf{R}_{ij}^{(\mathrm{x})} | i = 1, \cdots, I; j = 1, \cdots, J\}$ is given as

$$\begin{aligned}\mathscr{L}(\mathbf{R}^{(\mathrm{x})}) &= \prod_{i,j} p(\mathbf{x}_{ij} | \mathbf{R}_{ij}^{(\mathrm{x})}) \\ &= \prod_{i,j} \frac{1}{\pi^M \det \mathbf{R}_{ij}^{(\mathrm{x})}} \exp\left(-\mathbf{x}_{ij}^{\mathrm{H}} \mathbf{R}_{ij}^{(\mathrm{x})^{-1}} \mathbf{x}_{ij}\right),\end{aligned} \tag{6.50}$$

and the negative log-likelihood function is

Table 6.1 Models of mixing system, spatial covariance, power spectrogram, and their optimization in each method

Literature	Model of $\mathbf{R}_{ij}^{(\mathbf{x})}$	Spatial covariance	Power spectrogram	Optimization
Ozerov and Févotte [34]	$\sum_{n,l} t_{il,n} v_{lj,n} \mathbf{R}_{i,n}^{(\mathbf{s})} + \mathbf{R}_i^{(\mathbf{n})}$ $(\mathbf{x}_{ij} = \mathbf{A}_i \mathbf{s}_{ij} + \mathbf{n}_{ij})$	Rank-1 matrix $\mathbf{R}_{i,n}^{(\mathbf{s})}$ and diagonal matrix $\mathbf{R}_i^{(\mathbf{n})}$	NMF w/o partitioning function	EM and MU for $\mathbf{A}_i$, $\mathbf{R}_i^{(\mathbf{n})}$, $\mathbf{T}_n$, and $\mathbf{V}_n$
Arberet et al. [32]	$\sum_{n,l} t_{il,n} v_{lj,n} \mathbf{R}_{i,n}^{(\mathbf{s})} + \mathbf{R}_i^{(\mathbf{n})}$	Full-rank matrix $\mathbf{R}_{i,n}^{(\mathbf{s})}$ and diagonal matrix $\mathbf{R}_i^{(\mathbf{n})}$	NMF w/o partitioning function	EM for $\mathbf{R}_{i,n}^{(\mathbf{s})}$, $\mathbf{R}_i^{(\mathbf{n})}$, $\mathbf{T}_n$, and $\mathbf{V}_n$
Duong et al. [60]	$\sum_n r_{ij,n} \mathbf{R}_{i,n}$	Several types of $\mathbf{R}_{i,n}^{(\mathbf{s})}$ including rank-1 and full-rank matrices	$r_{ij,n}$ (w/o NMF)	EM for $\mathbf{R}_{i,n}^{(\mathbf{s})}$
Ozerov et al. [35]	$\sum_n \mathbf{R}_{i,n}^{(\mathbf{s})} \sum_k z_{nk} t_{ik} v_{kj} + \mathbf{R}_i^{(\mathbf{n})}$ $(\mathbf{x}_{ij} = \mathbf{A}_i \mathbf{s}_{ij} + \mathbf{n}_{ij})$	Rank-1 matrix $\mathbf{R}_{i,n}^{(\mathbf{s})}$ and diagonal matrix $\mathbf{R}_i^{(\mathbf{n})}$	NMF with partitioning function	EM and MU for $\mathbf{A}_i$, $\mathbf{R}_i^{(\mathbf{n})}$, $\mathbf{Z}$, $\mathbf{T}$, and $\mathbf{V}$
Sawada et al. [36]	$\sum_n \mathbf{R}_{i,n}^{(\mathbf{s})} \sum_k z_{nk} t_{ik} v_{kj}$	Full-rank matrix $\mathbf{R}_{i,n}^{(\mathbf{s})}$	NMF with partitioning function	MU for $\mathbf{R}_{i,n}^{(\mathbf{s})}$, $\mathbf{Z}$, $\mathbf{T}$, and $\mathbf{V}$
Kitamura et al. [30, 31]	$\sum_n \mathbf{R}_{i,n}^{(\mathbf{s})} \sum_k z_{nk} t_{ik} v_{kj}$ $(\mathbf{x}_{ij} = \mathbf{A}_i \mathbf{s}_{ij})$	Rank-1 matrix $\mathbf{R}_{i,n}^{(\mathbf{s})}$	NMF with partitioning function	IP for $\mathbf{W}_i = \mathbf{A}_i^{-1}$MU for $\mathbf{Z}$, $\mathbf{T}$, and $\mathbf{V}$

$$
\begin{aligned}
-\log \mathscr{L}(\mathbf{R}^{(\mathbf{x})}) &= \sum_{i,j} \left[M \log \pi + \log \det \mathbf{R}_{ij}^{(\mathbf{x})} + \mathbf{x}_{ij}^{\mathrm{H}} \mathbf{R}_{ij}^{(\mathbf{x})^{-1}} \mathbf{x}_{ij} \right] \\
&= \mathrm{const.} + \sum_{i,j} \left[\log \det \mathbf{R}_{ij}^{(\mathbf{x})} + \mathrm{tr}\left(\mathbf{X}_{ij} \mathbf{R}_{ij}^{(\mathbf{x})^{-1}} \right) \right], \qquad (6.51)
\end{aligned}
$$

where $\mathbf{X}_{ij} = \mathbf{x}_{ij}\mathbf{x}_{ij}^{\mathrm{H}}$ is an observed instantaneous covariance matrix. Similar to Itakura–Saito NMF in Sect. 6.2.4, the ML estimation based on (6.51) is identical to the multichannel Itakura–Saito divergence $\mathscr{D}_{\mathrm{MIS}}$ [36], which is known as Stein's loss [62] in the statistics field or the log-determinant divergence [63] in the machine learning field:

$$\sum_{i,j} \mathscr{D}_{\mathrm{MIS}}(\mathbf{X}_{ij} \| \mathbf{R}_{ij}^{(\mathbf{x})}) = \sum_{i,j} \left[\mathrm{tr}\left(\mathbf{X}_{ij} \mathbf{R}_{ij}^{(\mathbf{x})^{-1}} \right) - \log\det \mathbf{X}_{ij} \mathbf{R}_{ij}^{(\mathbf{x})^{-1}} - M \right]$$
$$= \mathrm{const.} + \sum_{i,j} \left[\log\det \mathbf{R}_{ij}^{(\mathbf{x})} + \mathrm{tr}\left(\mathbf{X}_{ij} \mathbf{R}_{ij}^{(\mathbf{x})^{-1}} \right) \right]. \quad (6.52)$$

In FDICA, IVA, and ILRMA, the mixing model (6.4) with a noiseless assumption is used, which results in the rank-1 spatial model (6.49). On the basis of this assumption, the covariance matrix $\mathbf{R}_{ij}^{(\mathbf{x})}$ can be rewritten using the mixing matrix $\mathbf{A}_i$ as

$$\mathbf{R}_{ij}^{(\mathbf{x})} = \sum_n r_{ij,n} \mathbf{a}_{i,n} \mathbf{a}_{i,n}^{\mathrm{H}}$$
$$= \mathbf{A}_i \mathbf{D}_{ij} \mathbf{A}_i^{\mathrm{H}}, \quad (6.53)$$

where

$$\mathbf{D}_{ij} = \begin{pmatrix} r_{ij,1} & 0 & \cdots & 0 \\ 0 & r_{ij,2} & \ddots & \vdots \\ \vdots & \ddots & \ddots & 0 \\ 0 & \cdots & 0 & r_{ij,N} \end{pmatrix}. \quad (6.54)$$

If we substitute (6.53) into the cost function in MNMF (6.51), we obtain

$$-\log \mathscr{L}(\mathbf{R}^{(\mathbf{x})}) = \mathrm{const.} + \sum_{i,j} \left[\log\det \mathbf{A}_i \mathbf{D}_{ij} \mathbf{A}_i^{\mathrm{H}} + \mathrm{tr}\left(\mathbf{X}_{ij} \left(\mathbf{A}_i^{\mathrm{H}} \right)^{-1} \mathbf{D}_{ij}^{-1} \mathbf{A}_i^{-1} \right) \right]$$
$$= \mathrm{const.} + \sum_{i,j} \Big[\log(\det \mathbf{A}_i)(\det \mathbf{D}_{ij})(\det \mathbf{A}_i)^{\mathrm{H}}$$
$$+ \mathrm{tr}\left(\mathbf{W}_i^{-1} \mathbf{y}_{ij} \mathbf{y}_{ij}^{\mathrm{H}} \left(\mathbf{W}_i^{-1} \right)^{\mathrm{H}} \mathbf{W}_i^{\mathrm{H}} \mathbf{D}_{ij}^{-1} \mathbf{W}_i \right) \Big]$$
$$= \mathrm{const.} + \sum_{i,j} \Big[\log|\det \mathbf{A}_i|^2 + \log\det \mathbf{D}_{ij}$$
$$+ \mathrm{tr}\left(\mathbf{W}_i \mathbf{W}_i^{-1} \mathbf{y}_{ij} \mathbf{y}_{ij}^{\mathrm{H}} \mathbf{D}_{ij}^{-1} \right) \Big]$$
$$= \mathrm{const.} - 2J \sum_i \log|\det \mathbf{W}_i|$$
$$+ \sum_{i,j} \left[\log \prod_n r_{ij,n} + \mathrm{tr}\left(\mathbf{y}_{ij} \mathbf{y}_{ij}^{\mathrm{H}} \mathbf{D}_{ij}^{-1} \right) \right]$$
$$= \mathrm{const.} - 2J \sum_i \log|\det \mathbf{W}_i| + \sum_{i,j,n} \left[\log r_{ij,n} + \frac{|y_{ij,n}|^2}{r_{ij,n}} \right], \quad (6.55)$$

where we used $\mathbf{x}_{ij} = \mathbf{W}_i^{-1}\mathbf{y}_{ij}$ and $\mathbf{W}_i = \mathbf{A}_i^{-1}$ to transform the variables. Thus, it is revealed that the cost function in MNMF with the rank-1 spatial model is identical to (6.27), the cost function in ILRMA, because the same spatial and source models are assumed.

Figure 6.6 shows the relationship between IVA, ILRMA, and MNMF. MNMF with a rank-1 spatial model, which assumes an instantaneous mixture in the frequency domain, is essentially equivalent to ILRMA, which is IVA with a flexible source model using NMF decomposition. Therefore, ILRMA can be considered as an intermediate model between IVA and MNMF in terms of the model flexibility. From the IVA side, we introduced the source model using NMF with bases to capture the specific spectral patterns, and from the MNMF side, a rank-1 spatial model was introduced to transform the variable $\mathbf{A}_i$ into $\mathbf{W}_i$ and to make the optimization more efficient.

6.5 Experiments on Speech and Music Separation

In this section, we evaluate the separation performance of Laplace IVA [40], ILRMA [31] without and with a partitioning function, Ozerov's MNMF [34], and Sawada's MNMF [36] for a convolutive mixture of a speech or music signal. Note that a more substantial evaluation of ILRMA compared with other methods can be found in [31].

6.5.1 Datasets

We investigated two cases: speech signal and music signal cases. In the speech signal case, we used live recorded mixture signals obtained from an underdetermined BSS task in SiSEC2011 [19]. This dataset includes 12 mixture signals (*dev1* and *dev2* datasets) with female and male speech, where the reverberation time is 130/250 ms and the microphone spacing is 1 m/5 cm. Details of the other conditions for this dataset can be found in [19]. Note that since this dataset is for underdeter-

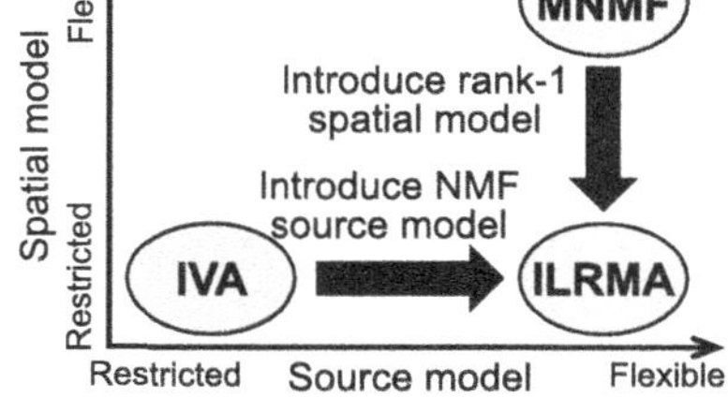

Fig. 6.6 Relationship between IVA, ILRMA, and MNMF from viewpoint of flexibility of spatial and source models

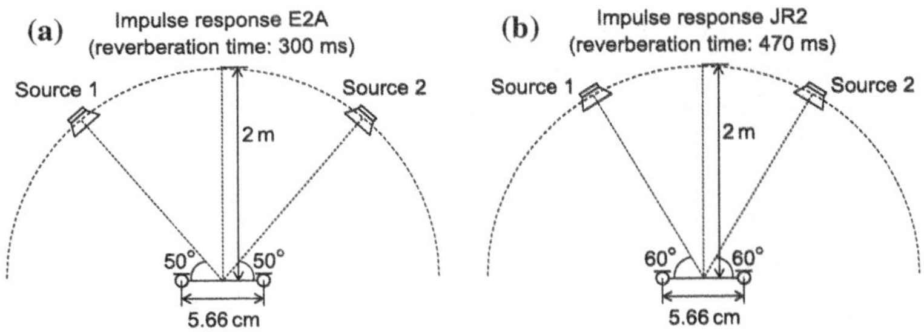

Fig. 6.7 Recording conditions of impulse responses **a** E2A and **b** JR2

Table 6.2 Music sources

ID	Song name	Source (1/2)
1	bearlin-roads	acoustic_guit_main/vocals
2	another_dreamer-the_ones_we_love	guitar/vocals
3	fort_minor-remember_the_name	violins_synth/vocals
4	ultimate_nz_tour	guitar/synth

Table 6.3 Experimental conditions

Sampling frequency	16 kHz
FFT length	256 ms in speech signal case and 512 ms in music signal case
Window shift length	128 ms in both speech and music signal cases
Initialization	$\mathbf{W}_i$: identity matrix
	NMF variables: uniform random values $[\varepsilon, 1]$
Number of iterations	200

mined BSS, three sources ($N = 3$) are provided as stereo recordings ($M = 2$). In this experiment, we used only the first and second speech sources to make the task determined ($N = M = 2$). In the music signal case, the observed signals were produced by convoluting the impulse response *E2A* or *JR2*, which was obtained from the RWCP database [64], with each source. Figure 6.7 shows the recording conditions of impulse responses E2A and JR2. As the music sources, we used professionally produced music obtained from a music separation task in SiSEC2011. The titles of the music and the instruments used are shown in Table 6.2.

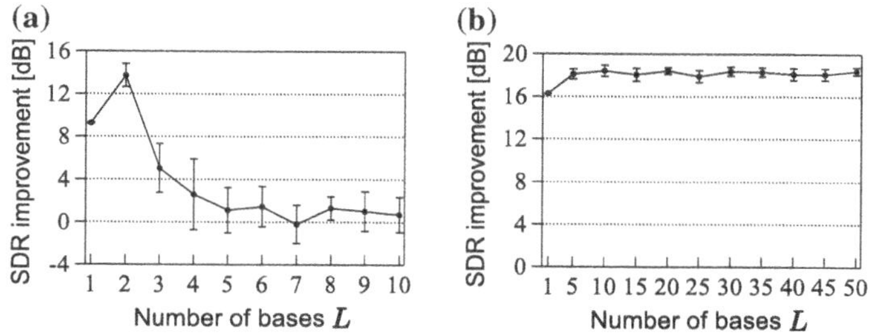

Fig. 6.8 Average SDR improvements for **a** dev1 female speech with 1 m microphone spacing and 130 ms reverberation time and **b** song ID4 with impulse response E2A

6.5.2 Experimental Analysis of Optimal Number of Bases for ILRMA

We first give an experimental analysis of the optimal number of bases for ILRMA. Since NMF decomposition is more suitable for music than speech because of the stable pitch of instruments, we expect that the optimal number of bases will be different between them. For this reason, we evaluated the separation performance of ILRMA without a partitioning function using various numbers of bases for each source, where this method models all the sources with the same fixed number of bases L. The experimental conditions used are shown in Table 6.3. As the evaluation score, we used the improvement of the signal-to-distortion ratio (SDR) proposed in [65], which indicates the total separation performance including the degree of separation and the quality of the separated sources.

Figure 6.8 shows the average SDR improvements and their deviations in 10 trials with different various pseudorandom seeds. From these results, we confirm that ILRMA cannot achieve a good separation performance for speech signals when the number of bases is large. This is due to the structural complexity of the speech spectrogram. Figure 6.9 shows cumulative singular values of each source spectrogram in the speech and music signals. The speech sources require more than 50 bases to represent the spectrogram while the music sources are saturated with 25 bases. Because of the time-varying pitch, it is difficult to capture speech spectrograms using NMF decomposition. If ILRMA fails to capture the correct spectrogram of each speech in the optimization, the demixing matrix will be trapped at a poor solution (local minimum). On the other hand, owing to the low rank of music spectrograms, ILRMA gives a better performance for music separation even if the number of bases increases.

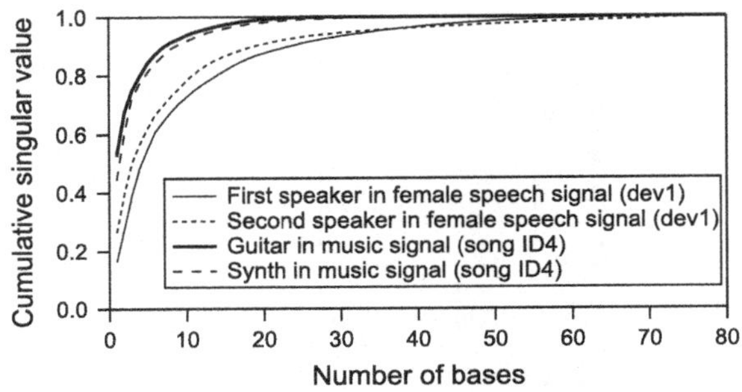

Fig. 6.9 Cumulative singular values of each source spectrogram in dev1 female speech and song ID4 music, where all sources are truncated to be the same signal length

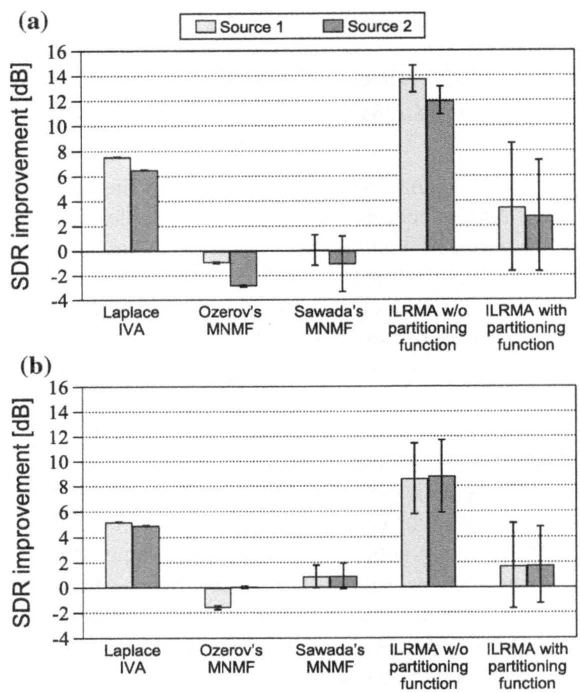

Fig. 6.10 Average SDR improvements for female speech (dev1) with 1 m microphone spacing, where reverberation time is **a** 130 ms and **b** 250 ms

6.5.3 Comparison of Separation Performance

We next compare the separation performance of each method. In Ozerov's MNMF, we used the experimental conditions described in [34]. In the other methods, the experimental conditions shown in Table 6.3 were used. On the basis of the results in Sect. 6.5.2, we set the number of bases of each source to $L = 2$ for the speech signals and $L = 30$ for the music signals in ILRMA without a partitioning function. In ILRMA with a partitioning function and Sawada's MNMF, we set the total number of bases to $K = 2 \times N$ for the speech signals and $K = 30 \times N$ for the music signals.

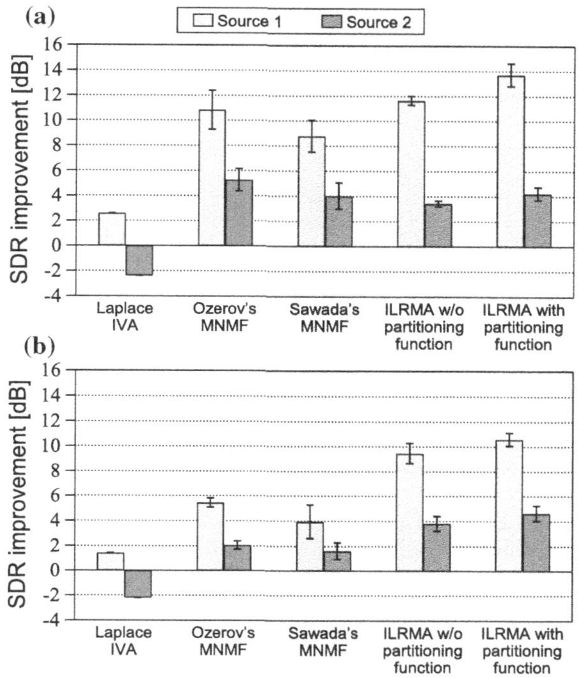

Fig. 6.11 Average SDR improvements for music signal song ID3 with impulse response **a** E2A and **b** JR2

Figures 6.10 and 6.11 respectively show typical examples of results for speech and music signals given by the average SDR improvements and their deviations in 10 trials with different pseudorandom seeds. The total average scores are shown in Tables 6.4 and 6.5. From these results, we confirm that Laplace IVA cannot achieve satisfactory separation because the source model in Laplace IVA is not flexible. Ozerov's MNMF outperforms Laplace IVA for the music signals, but the separation performance for speech signals is inferior to that of Laplace IVA. Sawada's MNMF gives better performance than Laplace IVA and Ozerov's MNMF for the music signals. ILRMA achieves a high and stable performance. For the speech signals, the partitioning function causes instability in the separation. This might be due to the sensitivity of the performance to the number of bases, as discussed in Sect. 6.5.2. In contrast, for the music signals, ILRMA with a partitioning function exhibits slightly higher performance than ILRMA without a partitioning function. This improvement is achieved by modeling the sources with the optimal number of bases using the partitioning function z_{nk}. For music signals with impulse response JR2, the SDRs of ILRMA are markedly degraded compared with those with impulse response E2A because the reverberation time is longer than impulse response E2A and close to the length of the window function in the STFT. Even if Sawada's MNMF has the potential to model such a mixing system by employing a full-rank spatial model, it is a very difficult problem to find the optimal $\mathbf{R}^{(\mathrm{s})}_{i,n}$. Figure 6.12 shows an example

Table 6.4 Averaged SDR improvements (dB) over various speech signals and sources with same recording conditions

Recording conditions (rev. time and mic. spacing)	Laplace IVA	Ozerov's MNMF	Sawada's MNMF	ILRMA w/o partitioning function	ILRMA with partitioning function
130 ms and 1 m	2.98	1.35	0.68	11.91	4.88
130 ms and 5 cm	2.86	2.13	1.13	8.97	3.48
250 ms and 1 m	2.03	0.49	0.48	7.34	2.09
250 ms and 5 cm	2.43	0.91	0.47	6.43	1.91

Table 6.5 Averaged SDR improvements (dB) over various music signals and sources with same impulse response

Impulse response	Laplace IVA	Ozerov's MNMF	Sawada's MNMF	ILRMA w/o partitioning function	ILRMA with partitioning function
E2A	5.72	5.73	10.32	12.29	12.29
JR2	1.77	2.37	6.11	6.62	7.40

of the SDR convergence and the actual computational time for each method in the case of a music signal, where the calculations were performed using MATLAB 8.3 (64-bit) with an Intel Core i7-4790 (3.60 GHz) CPU. Both Laplace IVA and ILRMA show much faster convergence than MNMFs. Sawada's MNMF requires a longer computational time because the eigenvalue decomposition of a $2M \times 2M$ matrix is required for each update iteration of $\mathbf{R}_{i,n}^{(\mathrm{s})}$.

Figure 6.13 shows the result of a subjective evaluation, where we presented 48 pairs of separated speech and 48 pairs of separated music signals in random order to 14 examinees, who selected which signal they preferred from the viewpoint of the total quality of the separated sounds. We can confirm that Laplace IVA is better than MNMF for the speech signals. In contrast, Sawada's MNMF achieves a better result for music signals owing to the suitable representation using NMF. ILRMA is the most preferable method for the high-quality separation of both speech and music signals. Similarly to FDICA and Laplace IVA, ILRMA employs the demixing matrix $\mathbf{W}_i$ for the separation, which is essentially equivalent to the spatial linear filter [66] in beamforming techniques [67, 68], and it is more difficult for such linear filtering to generate artificial noise than for time-frequency mask separation techniques including MNMF with multichannel Wiener filtering. Thus, the quality of separated sources via ILRMA from the viewpoint of human perception might be better than that via MNMF.

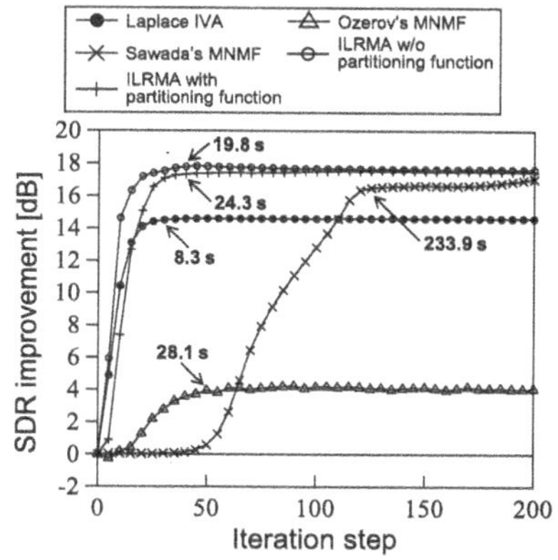

Fig. 6.12 SDR convergence and examples of actual calculation time for guitar source in song ID4 with impulse response E2A, where signal length is 18.6 s. In Laplace IVA and ILRMA, SDR instantly converges with better performance owing to fast and stable optimization of demixing matrix, whereas Sawada's MNMF requires many iterations for separation

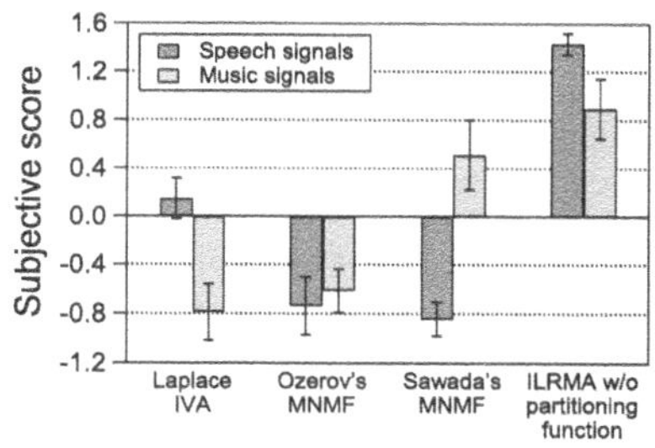

Fig. 6.13 Results of subjective scores obtained by Thurstone pairwise comparison method, where we presented 48 pairs of separated speech and 48 pairs of separated music signals in random order to 14 examinees, who selected which signal they preferred from the viewpoint of total quality of separated sound. Scores show relative tendency of selection

6.6 Conclusions

In this chapter, we introduced a new determined BSS technique that extends a source model in IVA from a vector to a low-rank matrix using the NMF representation. Also, the relationship between conventional MNMF and IVA was revealed: ILRMA is equivalent to MNMF with a rank-1 spatial model, and time-varying Gaussian IVA can be thought of as a special case of ILRMA, namely, ILRMA can be thought of as IVA with increased flexibility of the model. ILRMA can be optimized using fast update rules based on the auxiliary function technique. The experimental results show that ILRMA achieves faster convergence and better results than the conventional BSS techniques. A further extension of ILRMA can be found in [69], which relaxes the rank-1 constraint of the spatial covariance in ILRMA using extra observations for overdetermined cases such as when $M = 2N$ or $3N$.

Acknowledgements This work was partially supported by Grant-in-Aid for JSPS Fellows Grant Number 26 · 10796, and SECOM Science and Technology Foundation.

References

1. P. Comon, Independent component analysis, a new concept? Signal Process. **36**(3), 287–314 (1994)
2. A.J. Bell, T.J. Sejnowski, An information-maximization approach to blind separation and blind deconvolution. Neural Comput. **7**(6), 1129–1159 (1995)
3. J.-F. Cardoso, Infomax and maximum likelihood for blind source separation. IEEE Signal Process. Lett. **4**(4), 112–114 (1997)
4. S. Haykin (ed.), *Unsupervised Adaptive Filtering (Volume I: Blind Source Separation)* (Wiley-Interscience, 2000)
5. A. Hyvärinen, J. Karhunen, E. Oja, *Independent Component Analysis* (Wiley-Interscience, 2001)
6. P. Smaragdis, Blind separation of convolved mixtures in the frequency domain. Neurocomputing **22**(1), 21–34 (1998)
7. S. Araki, R. Mukai, S. Makino, T. Nishikawa, H. Saruwatari, The fundamental limitation of frequency domain blind source separation for convolutive mixtures of speech. IEEE Trans. Speech and Audio Process. **11**(2), 109–116 (2003)
8. H. Sawada, R. Mukai, S. Araki, S. Makino, Convolutive blind source separation for more than two sources in the frequency domain, in *Proceeding ICASSP* (2004), pp. III-885–III-888
9. H. Buchner, R. Aichner, W. Kellerman, A generalization of blind source separation algorithms for convolutive mixtures based on second order statistics. IEEE Trans. Speech and Audio Process. **13**(1), 120–134 (2005)
10. H. Saruwatari, T. Kawamura, T. Nishikawa, A. Lee, K. Shikano, Blind source separation based on a fast-convergence algorithm combining ICA and beamforming. IEEE Trans. Speech and Audio Process. **14**(2), 666–678 (2006)
11. D.D. Lee, H.S. Seung, Learning the parts of objects by non-negative matrix factorization. Nature **401**(6755), 788–791 (1999)
12. D.D. Lee, H.S. Seung, Algorithms for non-negative matrix factorization, in *Proceedings NIPS* (2000), pp. 556–562
13. A. Cichocki, R. Zdunek, A.H. Phan, S. Amari, *Nonnegative Matrix and Tensor Factorizations: Applications to Exploratory Multi-way Data Analysis and Blind Source Separation* (Wiley, 2009)
14. T. Virtanen, Monaural sound source separation by nonnegative matrix factorization with temporal continuity and sparseness criteria. IEEE Trans. Audio, Speech, and Lang. Process. **15**(3), 1066–1074 (2007)
15. A. Ozerov, C. Févotte, M. Charbit, Factorial scaled hidden Markov model for polyphonic audio representation and source separation, in *Proceedings WASPAA* (2009), pp. 121–124
16. P. Smaragdis, B. Raj, M. Shashanka, Supervised and semi-supervised separation of sounds from single-channel mixtures, in *Proceedings ICA* (2007), pp. 414–421
17. D. Kitamura, H. Saruwatari, K. Yagi, K. Shikano, Y. Takahashi, K. Kondo, Music signal separation based on supervised nonnegative matrix factorization with orthogonality and maximum-divergence penalties. IEICE Trans. Fundam. Electron. Commun. Comput. Sci. **E97-A**(5), 1113–1118 (2014)

18. D. Kitamura, H. Saruwatari, H. Kameoka, Y. Takahashi, K. Kondo, S. Nakamura, Multichannel signal separation combining directional clustering and nonnegative matrix factorization with spectrogram restoration. IEEE/ACM Trans. Audio, Speech, and Lang. Process. **23**(4), 654–669 (2015)
19. S. Araki, F. Nesta, E. Vincent, Z. Koldovský, G. Nolte, A. Ziehe, A. Benichoux, The 2011 signal separation evaluation campaign (SiSEC2011):-audio source separation, in *Proceedings LVA/ICA* (2012), pp. 414–422
20. N. Ono, Z. Koldovský, S. Miyabe, N. Ito, The 2013 signal separation evaluation campaign (SiSEC2013), in *Proceedings MLSP* (2013)
21. N. Ono, Z. Rafii, D. Kitamura, N. Ito, A. Liutkus, The 2015 signal separation evaluation campaign, in *Proceedings LVA/ICA* (2015), pp. 387–395
22. A. Liutkus, F.-R. Stöter, Z. Rafii, D. Kitamura, B. Rivet, N. Ito, N. Ono, J. Fontecave, The 2016 signal separation evaluation campaign, in *Proceedings LVA/ICA* (2017)
23. S. Kurita, H. Saruwatari, S. Kajita, K. Takeda, F. Itakura, Evaluation of blind signal separation method using directivity pattern under reverberant conditions, in *Proceedings ICASSP* (2000), pp. 3140–3143
24. N. Murata, S. Ikeda, A. Ziehe, An approach to blind source separation based on temporal structure of speech signals. Neurocomputing **41**(1–4), 1–24 (2001)
25. H. Sawada, R. Mukai, S. Araki, S. Makino, A robust and precise method for solving the permutation problem of frequency-domain blind source separation. IEEE Trans. Speech and Audio Process. **12**(5), 530–538 (2004)
26. H. Sawada, S. Araki, S. Makino, Measuring dependence of bin-wise separated signals for permutation alignment in frequency-domain BSS, in *Proceedings ISCAS* (2007), pp. 3247–3250
27. A. Hiroe, Solution of permutation problem in frequency domain ICA using multivariate probability density functions, in *Proceedings ICA* (2006), pp. 601–608
28. T. Kim, T. Eltoft, T.-W. Lee, Independent vector analysis: an extension of ICA to multivariate components, in *Proceedings ICA* (2006), pp. 165–172
29. T. Kim, H.T. Attias, S.-Y. Lee, T.-W. Lee, Blind source separation exploiting higher-order frequency dependencies. IEEE Trans. Audio, Speech, and Lang. Process. **15**(1), 70–79 (2007)
30. D. Kitamura, N. Ono, H. Sawada, H. Kameoka, H. Saruwatari, Efficient multichannel nonnegative matrix factorization exploiting rank-1 spatial model, in *Proceedings ICASSP* (2015), pp. 276–280
31. D. Kitamura, N. Ono, H. Sawada, H. Kameoka, H. Saruwatari, Determined blind source separation unifying independent vector analysis and nonnegative matrix factorization. IEEE/ACM Trans. Audio, Speech, and Lang. Process. **24**(9), 1626–1641 (2016)
32. S. Arberet, A. Ozerov, N.Q.K. Duong, E. Vincent, R. Gribonval, F. Bimbot, P. Vandergheynst, Nonnegative matrix factorization and spatial covariance model for under-determined reverberant audio source separation, in *Proceedings ISSPA* (2010), pp. 1–4
33. H. Kameoka, T. Yoshioka, M. Hamamura, J. Le Roux, K. Kashino, Statistical model of speech signals based on composite autoregressive system with application to blind source separation, in *Proceedings LVA/ICA* (2010), pp. 245–253
34. A. Ozerov, C. Févotte, Multichannel nonnegative matrix factorization in convolutive mixtures for audio source separation. IEEE Trans. Audio, Speech, and Lang. Process. **18**(3), 550–563 (2010)
35. A. Ozerov, C. Févotte, R. Blouet, J.-L. Durrieu, Multichannel nonnegative tensor factorization with structured constraints for user-guided audio source separation, in *Proceedings ICASSP* (2011), pp. 257–260
36. H. Sawada, H. Kameoka, S. Araki, N. Ueda, Multichannel extensions of non-negative matrix factorization with complex-valued data. IEEE Trans. Audio, Speech, and Lang. Process. **21**(5), 971–982 (2013)
37. T. Eltoft, T. Kim, T.-W. Lee, On the multivariate Laplace distribution. IEEE Signal Process. Lett. **13**(5), 300–303 (2006)

38. S. Kotz, T.J. Kozubowski, K. Podgórski, Symmetric multivariate Laplace distribution, in *The Laplace Distribution and Generalizations*, chap. 5 (Birkhäuser, Basel, 2001), pp. 231–238
39. T. Adali, H. Ki, J.-F. Cardoso, Complex ICA using nonlinear functions. IEEE Trans. Signal Process. **56**(9), 4536–4544 (2008)
40. N. Ono, Stable and fast update rules for independent vector analysis based on auxiliary function technique, in *Proceedings WASPAA* (2011), pp. 189–192
41. N. Ono, Fast stereo independent vector analysis and its implementation on mobile phone, in *Proceedings IWAENC* (2012)
42. N. Ono, Auxiliary-function-based independent vector analysis with power of vector-norm type weighting functions, in *Proceedings APSIPA ASC* (2012)
43. T. Ono, N. Ono, S. Sagayama, User-guided independent vector analysis with source activity tuning, in *Proceedings ICASSP* (2012), pp. 2417–2420
44. K. Hild, H.T. Attias, S. Nagarajan, An expectation-maximization method for spatio-temporal blind source separation using an AR-MOG source model. IEEE Trans. Neural Netw. **19**(3), 508–519 (2008)
45. C. Févotte, J.-F. Cardoso, Maximum likelihood approach for blind audio source separation using time-frequency Gaussian source models, in *Proceedings WASPAA* (2005), pp. 78–81
46. T. Nakatani, B.-H. Juang, T. Yoshioka, K. Kinoshita, M. Delcroix, M. Miyoshi, Speech dereverberation based on maximum-likelihood estimation with time-varying Gaussian source model. IEEE Trans. Audio, Speech, and Lang. Process. **16**(8), 1512–1527 (2008)
47. C. Févotte, N. Bertin, J.-L. Durrieu, Nonnegative matrix factorization with the Itakura-Saito divergence. With application to music analysis. Neural Comput. **21**(3), 793–830 (2009)
48. F.D. Neeser, J.L. Massey, Proper complex random processes with applications to information theory. IEEE Trans. Inf. Theory **39**(4), 1293–1302 (1993)
49. F. Itakura, S. Saito, Analysis synthesis telephony based on the maximum likelihood method, in *Proceedings ICA* (1968), pp. C-17–C-20
50. M. Nakano, H. Kameoka, J. Le Roux, Y. Kitano, N. Ono, S. Sagayama, Convergence-guaranteed multiplicative algorithms for nonnegative matrix factorization with beta-divergence, in *Proceedings MLSP* (2010), pp. 283–288
51. A.R. López, N. Ono, U. Remes, K. Palomäki, M. Kurimo, Designing multichannel source separation based on single-channel source separation, in *Proceedings ICASSP* (2015), pp. 469–473
52. N. Ono, S. Miyabe, Auxiliary-function-based independent component analysis for super-Gaussian sources, in *Proceedings LVA/ICA* (2010), pp. 165–172
53. S. Amari, A. Cichocki, H.H. Yang, A new learning algorithm for blind signal separation, in *Proceedings NIPS* (1996), pp. 757–763
54. A. Cichocki, S. Amari, *Adaptive Blind Signal and Image Processing: Learning Algorithms and Applications*, vol. 1 (Wiley, 2002)
55. T.G. Kolda, B.W. Bader, Tensor decompositions and applications. SIAM Rev. **51**(3), 455–500 (2009)
56. D. FitzGerald, M. Cranitch, E. Coyle, Non-negative tensor factorisation for sound source separation, in *Proceedings ISSC* (2005), pp. 8–12
57. R.M. Parry, I.A. Essa, Estimating the spatial position of spectral components in audio, in *Proceedings ICA* (2006), pp. 666–673
58. Y. Mitsufuji, A. Roebel, Sound source separation based on non-negative tensor factorization incorporating spatial cue as prior knowledge, in *Proceedings ICASSP* (2013), pp. 71–75
59. N.Q.K. Duong, E. Vincent, R. Gribonval, Spatial covariance models for under-determined reverberant audio source separation, in *Proceedings WASPAA* (2009), pp. 129–132
60. N.Q.K. Duong, E. Vincent, R. Gribonval, Under-determined reverberant audio source separation using a full-rank spatial covariance model. IEEE Trans. Audio, Speech, and Lang. Process. **18**(7), 1830–1840 (2010)
61. K.U. Simmer, J. Bitzer, C. Marro, Post-filtering techniques, in *Microphone Arrays: Signal Processing Techniques and Applications*, ed. by M. Brandstein, D. Ward, chap. 3 (Springer, Heidelberg, 2001), pp. 39–60

62. W. James, C. Stein, Estimation with quadratic loss, in *Proceedings Berkeley Symposium on Mathematical Statistics and Probability*, vol. 1 (1961), pp. 361–379
63. B. Kulis, M. Sustik, I. Dhillon, Learning low-rank kernel matrices, in *Proceedings ICML* (2006), pp. 505–512
64. S. Nakamura, K. Hiyane, F. Asano, T. Nishiura, T. Yamada, Acoustical sound database in real environments for sound scene understanding and hands-free speech recognition, in *Proceedings LREC* (2000), pp. 965–968
65. E. Vincent, R. Gribonval, C. Févotte, Performance measurement in blind audio source separation. IEEE Trans. Audio, Speech, and Lang. Process. **14**(4), 1462–1469 (2006)
66. S. Araki, S. Makino, Y. Hinamoto, R. Mukai, T. Nishikawa, H. Saruwatari, Equivalence between frequency-domain blind source separation and frequency-domain adaptive beamforming for convolutive mixtures. EURASIP J. Adv. Signal Process. **2003**(11), 1–10 (2003)
67. J.-F. Cardoso, A. Souloumiac, Blind beamforming for non-Gaussian signals. IEE Proc. F - Radar and Signal Process. **140**(6), 362–370 (1993)
68. D.B. Ward, R.A. Kennedy, R.C. Williamson, Constant directivity beamforming, in *Microphone Arrays: Signal Processing Techniques and Applications*, ed. by M. Brandstein, D. Ward, chap. 1 (Springer, Heidelberg, 2001), pp. 3–17
69. D. Kitamura, N. Ono, H. Sawada, H. Kameoka, H. Saruwatari, Relaxation of rank-1 spatial constraint in overdetermined blind source separation, in *Proceedings EUSIPCO* (2015), pp. 1271–1275

Chapter 7
Deep Neural Network Based Multichannel Audio Source Separation

Aditya Arie Nugraha, Antoine Liutkus and Emmanuel Vincent

Abstract This chapter presents a multichannel audio source separation framework where deep neural networks (DNNs) are used to model the source spectra and combined with the classical multichannel Gaussian model to exploit the spatial information. The parameters are estimated in an iterative expectation-maximization (EM) fashion and used to derive a multichannel Wiener filter. Different design choices and their impact on the performance are discussed. They include the cost functions for DNN training, the number of parameter updates, the use of multiple DNNs, and the use of weighted parameter updates. Finally, we present its application to a speech enhancement task and a music separation task. The experimental results show the benefit of the multichannel DNN-based approach over a single-channel DNN-based approach and the multichannel nonnegative matrix factorization based iterative EM framework.

7.1 Introduction

Audio source separation aims to recover the underlying constitutive source signals of an observed mixture signal [1–5]. Related research can be divided into speech separation and music separation. Speech separation aims to recover the speech signal from a mixture containing multiple background noise sources with possibly interfering speech. This is important for speech enhancement (including in hearing aids) and noise-robust automatic speech recognition (ASR). Music separation aims to recover singing voice and musical instruments from a mixture. This has various applications, including music editing (remixing, upmixing, etc.), music information retrieval, and dialogue extraction (e.g. from a mixture containing music accompaniment).

A. A. Nugraha (✉) · E. Vincent
Inria Nancy, Grand Est, 54600 Villers-lès-Nancy, France
e-mail: aditya.nugraha@inria.fr

A. Liutkus
Inria Sophia Antipolis, Mèditerranèe, 34392 Montpellier, France
e-mail: antoine.liutkus@inria.fr

S. Makino (ed.), *Audio Source Separation*, Signals and Communication Technology, https://doi.org/10.1007/978-3-319-73031-8_7

Acquiring (recording) audio using a single microphone and performing single-channel separation on the acquired signal is practical, especially for speech in real-world environments. However, along with the technology development, it is getting easier and cheaper to acquire audio using multiple microphones. The acquired multichannel signal captures additional information useful for separation. As a simple analogy, humans are able to predict the direction from which a sound comes based on the difference of amplitude and phase between the signals captured by the left and right ears. These inter-channel differences are related to the position of the sound sources relative to the microphones and can be exploited in multichannel separation to provide better results than single-channel separation. Multichannel separation is also preferable for music since most professionally-produced recordings available nowadays are in stereo (two-channel) format. If we consider the application on film audio tracks, we will deal with either six- or eight-channel surround sound formats.

Recent studies have shown that deep neural networks (DNNs) are able to model complex functions and perform well on various tasks, notably ASR [6, 7]. DNNs also have been applied to single-channel speech enhancement and shown to provide a significant increase in ASR performance compared to earlier approaches based on beamforming or nonnegative matrix factorization (NMF) [8]. The DNNs typically operate on magnitude or log-magnitude spectra in the Mel domain or the short time Fourier transform (STFT) domain. Various other features have been studied [9]. The DNNs can be used either to predict the source spectrograms [10–15] whose ratio yields a time-frequency mask or directly to predict a time-frequency mask [16–22]. The estimated source signal is then obtained as the product of the input mixture signal and the estimated time-frequency mask. Various DNN architectures and training criteria have been investigated and compared [19, 21, 23]. Although the authors in [13] considered both speech and music separation, most studies focused either on speech separation [10, 12, 14, 16–22] or on music separation [11, 15]. The approaches above considered single-channel source separation, where the input signal is either one of the channels of the original multichannel mixture signal or the result of delay-and-sum (DS) beamforming [19]. As a result, they do not fully exploit the benefits of multichannel data as achieved by multichannel filtering [1, 4].

There also exist a few approaches exploiting multichannel data. These approaches can be divided into three categories: (1) the ones deriving DNN input features from multichannel data for estimating a single-channel mask [14, 18]; (2) the ones estimating multichannel filters directly from time-domain signals using DNNs [24]; and (3) the ones estimating mask or spectra using DNNs which then are used to derive multichannel filters [25, 26].

In this chapter, we discuss the DNN-based multichannel source separation framework proposed in [26] which belongs to the third category. In this framework, DNNs are used to model the source spectra and combined with the classical multichannel Gaussian model to exploit the spatial information. These spectral and spatial parameters are then re-estimated in an iterative expectation-maximization (EM) fashion and used to derive a multichannel Wiener filter. This framework is built upon the classical iterative EM framework in [27], which was also used up to some variants in [28–33]. This chapter summarizes and reuses the materials from our works in [26,

34, 35]. This chapter presents a brief study of the impact of different design choices on the performance of the DNN-based framework, including the cost function used for the DNN training, the number of spatial parameter updates, the number of EM iterations, the use of weighted parameter updates, and the use of multiple DNNs. Finally, this chapter also presents the application of the DNN-based framework to a speech enhancement task and a music separation task.

The rest of this chapter is organized as follows. Section 7.2 formulates the problem of multichannel audio source separation and describes the classical iterative EM framework, which is the basis for the DNN-based iterative framework described in Sect. 7.3. Section 7.4 presents the application of the framework to a speech enhancement task and a music separation task. We also present the impact of different design choices and the comparison to other separation techniques in these sections. Finally, Sect. 7.5 concludes the chapter.

7.2 Background

In this section, the problem of multichannel audio source separation is formulated. Following this formulation, the classical iterative EM framework is then described.

7.2.1 Problem Formulation

Let I denote the number of channels, J the number of sources, $\mathbf{x}(t) \in \mathbb{R}^{I\times 1}$ the observed I-channel mixture signal, and $\mathbf{c}_j(t) \in \mathbb{R}^{I\times 1}$ the I-channel spatial image of source j. This source spatial image $\mathbf{c}_j(t)$ is a version of the signal of source j which presents in the observed mixture $\mathbf{x}(t)$. Both $\mathbf{x}(t)$ and $\mathbf{c}_j(t)$ are in the time domain and related by

$$\mathbf{x}(t) = \sum_{j=1}^{J} \mathbf{c}_j(t). \tag{7.1}$$

Figure 7.1 shows an example of studio music production where a song consists of vocal, guitars, and drums. As defined by (7.1), this stereo song is composed by the stereo spatial images of vocals, guitars, and drums. These stereo spatial images are not necessarily the original recordings. In most cases, sound engineers modify and mix the original sources. Since they can do downmixing (combining several channels to get fewer channels) or upmixing (splitting few channels to get more channels), the number of channels of the spatial images and the mixture may be chosen freely. In contrast, for the cases of speech recordings in real-world environments or live music recordings, the number of channels of the spatial images and the mixture corresponds to the number of microphones. Each channel of a source spatial image is the signal captured by each microphone after traveling from the source.

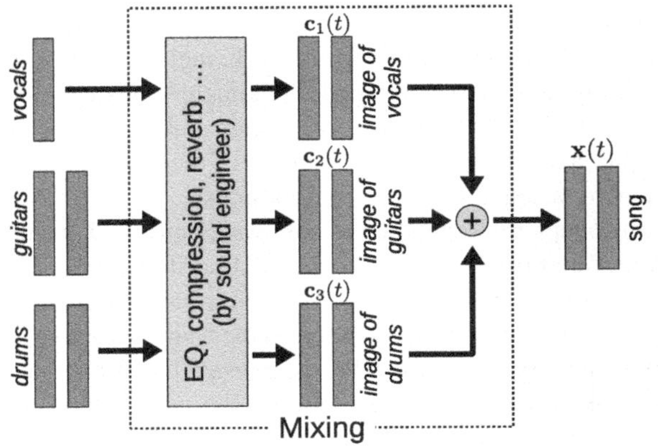

Fig. 7.1 Illustration of a studio music production

Multichannel audio source separation aims to recover the multichannel source spatial images $\mathbf{c}_j(t)$ from the observed multichannel mixture signal $\mathbf{x}(t)$.

7.2.2 Multichannel Gaussian Model

Let $\mathbf{x}(f,n) \in \mathbb{C}^{I \times 1}$ and $\mathbf{c}_j(f,n) \in \mathbb{C}^{I \times 1}$ denote the STFT coefficients [36] of $\mathbf{x}(t)$ and $\mathbf{c}_j(t)$, respectively, for frequency bin f and time frame n. Also, let F be the number of frequency bins and N the number of time frames.

We assume that $\mathbf{c}_j(f,n)$ are independent for different j, f, or n and follow a multivariate complex-valued zero-mean isotropic Gaussian distribution [27, 37]

$$\mathbf{c}_j(f,n) \sim \mathcal{N}_c\left(\mathbf{0}, v_j(f,n)\mathbf{R}_j(f)\right), \tag{7.2}$$

where $v_j(f,n) \in \mathbb{R}_+$ denotes the power spectral density (PSD) of source j for frequency bin f and time frame n, and $\mathbf{R}_j(f) \in \mathbb{C}^{I \times I}$ is the spatial covariance matrix of source j for frequency bin f. This $I \times I$ matrix represents spatial information by encoding the spatial position and the spatial width of the corresponding source [27]. Since the mixture $\mathbf{x}(f,n)$ is the sum of $\mathbf{c}_j(f,n)$, it is consequently distributed as

$$\mathbf{x}(f,n) \sim \mathcal{N}_c\left(\mathbf{0}, \sum_{j=1}^{J} v_j(f,n)\mathbf{R}_j(f)\right). \tag{7.3}$$

Given the PSDs $v_j(f,n)$ and the spatial covariance matrices $\mathbf{R}_j(f)$ of all sources, the spatial source images can be estimated in the minimum mean squared error sense using multichannel Wiener filtering [27]

$$\widehat{\mathbf{c}}_j(f,n) = \mathbf{W}_j(f,n)\mathbf{x}(f,n), \tag{7.4}$$

where the Wiener filter $\mathbf{W}_j(f,n)$ is given by

$$\mathbf{W}_j(f,n) = v_j(f,n)\mathbf{R}_j(f)\left(\sum_{j'=1}^{J} v_{j'}(f,n)\mathbf{R}_{j'}(f)\right)^{-1}. \tag{7.5}$$

Finally, the time-domain source estimates $\widehat{\mathbf{c}}_j(t)$ are recovered from $\widehat{\mathbf{c}}_j(f,n)$ by inverse STFT.

Following this formulation, source separation becomes the problem of estimating the PSDs $v_j(f,n)$ and the spatial covariance matrices $\mathbf{R}_j(f)$ of all sources.

7.2.3 General Iterative EM Framework

The general iterative EM framework for estimating the PSDs $v_j(f,n)$ and the spatial covariance matrices $\mathbf{R}_j(f)$ of all sources is summarized in Algorithm 1. For the following discussions, the term 'spectral parameters' refers to the PSDs and they are used interchangeably. Further, they are also used interchangeably with 'spectrograms', which are the graphical loosely represent PSD estimates. Likewise, the term 'spatial parameters' refers to the spatial covariance matrices.

In the beginning, the estimated PSDs $v_j(f,n)$ are initialized in the *spectrogram initialization* step. This can be done, for instance, by computing the PSD of the mixture and then dividing it with the number of sources, which implies each source contributes equally to the mixture. Initialization is also done for the estimated spatial covariance matrices $\mathbf{R}_j(f)$, for instance by assigning $I \times I$ identity matrices.

The following iterations can be divided into E-step and M-step. In the E-step, given the estimated parameters $v_j(f,n)$ and $\mathbf{R}_j(f)$ of each source, the source image estimates $\widehat{\mathbf{c}}_j(f,n)$ are obtained by multichannel Wiener filtering (7.4) and the posterior second-order raw moments of the spatial source images $\widehat{\mathbf{R}}_{\mathbf{c}_j}(f,n)$ are computed as

$$\widehat{\mathbf{R}}_{\mathbf{c}_j}(f,n) = \widehat{\mathbf{c}}_j(f,n)\widehat{\mathbf{c}}_j^H(f,n) + \left(\mathbf{I} - \mathbf{W}_j(f,n)\right) v_j(f,n)\mathbf{R}_j(f), \tag{7.6}$$

where $\mathbf{I}$ denotes the $I \times I$ identity matrix and $\cdot^H$ is the Hermitian transposition.

In the M-step, the spatial covariance matrices $\mathbf{R}_j(f)$ are updated as

$$\mathbf{R}_j(f) = \frac{1}{N}\sum_{n=1}^{N} \frac{1}{v_j(f,n)}\widehat{\mathbf{R}}_{\mathbf{c}_j}(f,n). \tag{7.7}$$

The source PSDs $v_j(f,n)$ are first estimated without constraints as

Algorithm 1 General iterative EM framework [27]

Inputs:
STFT of mixture $\mathbf{x}(f, n)$
Number of channels I and number of sources J
Number of EM iterations L
Spectral models $\mathrm{M}_0, \mathrm{M}_1, \ldots, \mathrm{M}_J$

for each source j of J **do**
    Initialize the spectrogram: $v_j(f, n) \leftarrow$ *spectrogram initialization*
    Initialize the spatial covariance matrix: $\mathbf{R}_j(f) \leftarrow I \times I$ identity matrix
end for
for each EM iteration l of L **do**
    Compute the mixture covariance matrix:
    $\mathbf{R}_\mathbf{x}(f, n) \leftarrow \sum_{j=1}^{J} v_j(f, n)\, \mathbf{R}_j(f)$
    for each source j of J **do**
        Compute the Wiener filter gain:
        $\mathbf{W}_j(f, n) \leftarrow$ Eq. (7.5) given $v_j(f, n), \mathbf{R}_j(f), \mathbf{R}_\mathbf{x}(f, n)$
        Compute the spatial image:
        $\widehat{\mathbf{c}}_j(f, n) \leftarrow$ Eq. (7.4) given $\mathbf{x}(f, n), \mathbf{W}_j(f, n)$
        Compute the posterior second-order raw moment of the spatial image:
        $\widehat{\mathbf{R}}_{\mathbf{c}_j}(f, n) \leftarrow$ Eq. (7.6) given $v_j(f, n), \mathbf{R}_j(f), \mathbf{W}_j(f, n), \widehat{\mathbf{c}}_j(f, n)$
        Update the spatial covariance matrix:
        $\mathbf{R}_j(f) \leftarrow$ Eq. (7.7) given $v_j(f, n), \widehat{\mathbf{R}}_{\mathbf{c}_j}(f, n)$
        Compute the unconstrained spectrogram:
        $z_j(f, n) \leftarrow$ Eq. (7.8) given $\mathbf{R}_j(f), \widehat{\mathbf{R}}_{\mathbf{c}_j}(f, n)$
        Update the spectrogram:
        $v_j(f, n) \leftarrow$ *spectrogram fitting* given $z_j(f, n), \mathrm{M}_j$
    end for
end for
for each source j of J **do**
    Compute the final spatial image:
    $\widehat{\mathbf{c}}_j(f, n) \leftarrow$ Eq. (7.4) given all $v_j(f, n)$, all $\mathbf{R}_j(f), \mathbf{x}(f, n)$
end for

Outputs:
All spatial source images $[\widehat{\mathbf{c}}_1(f, n), \ldots, \widehat{\mathbf{c}}_J(f, n)]$

$$z_j(f, n) = \frac{1}{I}\mathrm{tr}\left(\mathbf{R}_j^{-1}(f)\widehat{\mathbf{R}}_{\mathbf{c}_j}(f, n)\right), \tag{7.8}$$

where tr(·) denotes the trace of a matrix. Then, they are updated according to a given spectral model by fitting $v_j(f, n)$ from $z_j(f, n)$ in the *spectrogram fitting* step. The spectrogram initialization and the spectrogram fitting steps depend on how the spectral parameters are modeled. Spectral models used in this context (denoted by $\mathrm{M}_0, \mathrm{M}_1, \ldots, \mathrm{M}_J$ in Algorithm 1) may include NMF [29], which is a linear model with nonnegativity constraints, kernel additive model (KAM) [32, 33], which relies on the local regularity of the sources, and continuity models [28]. In this chapter, we present the use of DNNs for this purpose.

7.3 DNN-Based Multichannel Source Separation

In this section, the DNN-based multichannel source separation framework is presented. The cost functions for DNN training are also discussed. Finally, the weighted spatial parameter updates is introduced.

7.3.1 Algorithm

As indicated in the end of previous section, in this DNN-based framework, DNNs are used to model the source specta $v_j(f, n)$. This framework works with a single DNN or multiple DNNs. In the single DNN case, the DNN is used for spectrogram initialization (without any following spectrogram fitting). In the multiple DNNs case, one DNN is used for spectrogram initialization and one or more DNNs are used for spectrogram fitting. We can train different DNNs for spectrogram fitting at different iterations. Thus, the maximum number of DNNs for spectrogram fitting is equal to the number of iterations L. Let DNN_0 and DNN_l be the DNNs used for spectrogram initialization and spectrogram fitting, respectively. DNN_0 estimates the source spectra from the observed mixture and DNN_l aims to improve the source spectra estimated at iteration l. DNN_0 and DNN_l estimate the spectra of all sources simultaneously. This is similar to the DNNs used in the context of single-channel source separation in [10, 12, 13]. Besides, DNN_l is similar to the DNNs used in the context of single-channel speech enhancement in [38, 39] since they estimate clean spectra from the corresponding noisier spectra.

In this chapter, we consider features in the magnitude STFT domain as the inputs and outputs of DNNs. The inputs of DNN_0 and DNN_l are denoted by $\sqrt{z_x(f, n)}$ and $\sqrt{z_j(f, n)}$, respectively. The outputs of both types of DNNs are denoted by $\sqrt{v_j(f, n)}$ and the training targets are denoted by $\sqrt{\tilde{v}_j(f, n)}$. DNN_0 takes the magnitude spectrum $\sqrt{z_x(f, n)}$ and yields the initial magnitude spectra $\sqrt{v_j(f, n)}$ for all sources simultaneously. Then, DNN_l takes the estimated magnitude spectra $\sqrt{z_j(f, n)}$ of all sources and yields the improved magnitude spectra $\sqrt{v_j(f, n)}$ for all sources simultaneously.

Instead of alternately doing spatial and spectral parameter updates as in classical EM iterations (see Algorithm 1), the spatial parameters are updated several times before the spectral parameters are updated. This is motivated by the use of DNN for spectrogram initialization. The initial spectrograms should be close to the targets already, while the initial spatial covariance matrices are far.

The DNN-based iterative framework is described in Algorithm 2.

7.3.2 Cost Functions

In this chapter, we present the use of the following cost functions for the DNN training. These cost functions measure the difference between the target $\sqrt{\tilde{v}_j(f,n)}$ and the estimate $\sqrt{v_j(f,n)}$.

Algorithm 2 DNN-based iterative framework

Inputs:

STFT of mixture $\mathbf{x}(f,n)$
Number of channels I and number of sources J
Number of spatial updates K and number of EM iterations L
DNN spectral models $\text{DNN}_0, \text{DNN}_1, \ldots, \text{DNN}_L$

Do pre-processing on the observed mixture:
  $\sqrt{z_x(f,n)} \leftarrow \text{preprocess}(\mathbf{x}(f,n))$
Initialize all source spectrograms simultaneously:
  $[v_1(f,n), \ldots, v_J(f,n)] \leftarrow \text{DNN}_0\left(\sqrt{z_x(f,n)}\right)^2$
for each source j of J **do**
  Initialize the spatial covariance matrix: $\mathbf{R}_j(f) \leftarrow I \times I$ identity matrix
end for
for each EM iteration l of L **do**
  for each spatial update k of K **do**
    Compute the mixture covariance matrix:
      $\mathbf{R}_\mathbf{x}(f,n) \leftarrow \sum_{j=1}^{J} v_j(f,n)\,\mathbf{R}_j(f)$
    for each source j of J **do**
      Compute the Wiener filter gain:
        $\mathbf{W}_j(f,n) \leftarrow$ Eq. (7.5) given $v_j(f,n)$, $\mathbf{R}_j(f)$, $\mathbf{R}_\mathbf{x}(f,n)$
      Compute the spatial image:
        $\widehat{\mathbf{c}}_j(f,n) \leftarrow$ Eq. (7.4) given $\mathbf{x}(f,n)$, $\mathbf{W}_j(f,n)$
      Compute the posterior second-order raw moment of the spatial image:
        $\widehat{\mathbf{R}}_{\mathbf{c}_j}(f,n) \leftarrow$ Eq. (7.6) given $v_j(f,n)$, $\mathbf{R}_j(f)$, $\mathbf{W}_j(f,n)$, $\widehat{\mathbf{c}}_j(f,n)$
      Update the spatial covariance matrix:
        $\mathbf{R}_j(f) \leftarrow$ Eq. (7.7) given $v_j(f,n)$, $\widehat{\mathbf{R}}_{\mathbf{c}_j}(f,n)$
    end for
  end for
  for each source j of J **do**
    Compute the unconstrained source spectrogram:
      $z_j(f,n) \leftarrow$ Eq. (7.8) given $\mathbf{R}_j(f)$, $\widehat{\mathbf{R}}_{\mathbf{c}_j}(f,n)$
  end for
  Update all source spectrograms simultaneously:
    $[v_1(f,n), \ldots, v_J(f,n)] \leftarrow \text{DNN}_l\left(\left[\sqrt{z_1(f,n)}, \ldots, \sqrt{z_J(f,n)}\right]\right)^2$
end for
for each source j of J **do**
  Compute the final spatial image:
    $\widehat{\mathbf{c}}_j(f,n) \leftarrow$ Eq. (7.4) given all $v_j(f,n)$, all $\mathbf{R}_j(f)$, $\mathbf{x}(f,n)$
end for

Outputs:

All spatial source images $[\widehat{\mathbf{c}}_1(f,n), \ldots, \widehat{\mathbf{c}}_J(f,n)]$

---

1. The *Itakura-Saito (IS)* divergence [40] is expressed as

$$\mathscr{D}_{\text{IS}} = \frac{1}{JFN} \sum_{j,f,n} \left(\frac{\widetilde{v}_j(f,n)}{v_j(f,n)} - \log \frac{\widetilde{v}_j(f,n)}{v_j(f,n)} - 1 \right). \tag{7.9}$$

   Since this metric is known to yield signals with good perceptual quality, it becomes a popular metric in the audio processing community, including for NMF-based audio source separation [40–42]. From the theoretical point of view of the framework presented in this chapter, this metric is attractive because it results in maximum likelihood (ML) estimation of the spectra [40] and the whole Algorithm 2 then achieves ML estimation.
2. The (generalized) *Kullback-Leibler (KL)* divergence [43] is expressed as

$$\mathscr{D}_{\text{KL}} = \frac{1}{JFN} \sum_{j,f,n} \left(\sqrt{\widetilde{v}_j(f,n)} \log \frac{\sqrt{\widetilde{v}_j(f,n)}}{\sqrt{v_j(f,n)}} - \sqrt{\widetilde{v}_j(f,n)} + \sqrt{v_j(f,n)} \right). \tag{7.10}$$

   This metric is also a popular choice for NMF-based audio source separation [40] and has been shown to be effective for DNN training [11].
3. The *Cauchy cost function* is expressed as

$$\mathscr{D}_{\text{Cau}} = \frac{1}{JFN} \sum_{j,f,n} \left(\frac{3}{2} \log \left(\widetilde{v}_j(f,n) + v_j(f,n) \right) - \log \sqrt{v_j(f,n)} \right). \tag{7.11}$$

   This metric has been proposed recently for NMF-based audio source separation and advocated as performing better than the IS divergence in some cases [44].
4. The *phase-sensitive (PS) cost function* is defined as

$$\mathscr{D}_{\text{PS}} = \frac{1}{2JFN} \sum_{j,f,n} |m_j(f,n)\widetilde{x}(f,n) - \widetilde{c}_j(f,n)|^2, \tag{7.12}$$

   where $m_j(f,n) = v_j(f,n)/\sum_{j'} v_{j'}(f,n)$ is the single-channel Wiener filter [8, 23], while $\widetilde{x}(f,n)$ and $\widetilde{c}_j(f,n)$ are the single-channel versions of the multichannel mixture $\mathbf{x}(f,n)$ and the multichannel ground truth source spatial images $\mathbf{c}_j(f,n)$, respectively. These single-channel signals can be obtained, for instance, by DS beamforming [45, 46]. This metric minimizes the error in the complex-valued STFT domain, as opposed to the error in the magnitude STFT domain as the other cost functions considered here.
5. The *mean squared error (MSE)* [40] is expressed as

$$\mathscr{D}_{\text{MSE}} = \frac{1}{2JFN} \sum_{j,f,n} \left(\sqrt{\widetilde{v}_j(f,n)} - \sqrt{v_j(f,n)} \right)^2. \tag{7.13}$$

This metric is the most widely used cost function for various optimization processes, including DNN training for regression tasks.

7.3.3 *Weighted Spatial Parameter Updates*

We also consider a general form of spatial parameter update as

$$\mathbf{R}_j(f) = \left(\sum_{n=1}^{N} \omega_j(f,n)\right)^{-1} \sum_{n=1}^{N} \frac{\omega_j(f,n)}{v_j(f,n)} \widehat{\mathbf{R}}_{\mathbf{c}_j}(f,n), \tag{7.14}$$

where $\omega_j(f,n)$ denotes the weight of source j for frequency bin f and frame n. When $\omega_j(f,n)$ is set to be equal for all time-frequency (TF) bins (f,n), e.g. $\omega_j(f,n) = 1$, it reduces to the exact EM formulation (7.7). When $\omega_j(f,n) = v_j(f,n)$, as used in [33, 34], it reduces to

$$\mathbf{R}_j(f) = \left(\sum_{n=1}^{N} v_j(f,n)\right)^{-1} \sum_{n=1}^{N} \widehat{\mathbf{R}}_{\mathbf{c}_j}(f,n). \tag{7.15}$$

Experience shows that these weights are able to handle bad estimates $v_j(f,n)$. This weighting trick mitigates the importance of problematic underestimated TF bins. When $v_j(f,n)$ for a specific TF bin (f,n) is very low, its value of $v_j(f,n)^{-1}\widehat{\mathbf{R}}_{\mathbf{c}_j}(f,n)$ will be very big and cause a detrimental effect to the following computations (e.g. $\mathbf{R}_j(f)$ becomes ill-conditioned) and, ultimately, the performance. This weighting trick also increases the importance of high energy TF bins, whose value of $v_j(f,n)^{-1}\widehat{\mathbf{R}}_{\mathbf{c}_j}(f,n)$ is closer to the true $\mathbf{R}_j(f)$ on average.

7.4 Experimental Evaluation

In this section, we present the application of the DNN-based iterative framework for speech enhancement in the context of the CHiME-3 Challenge [47] and music separation in the context of the SiSEC-2016 Campaign [48]. Section 7.4.1 describes the general system design which applies to both experiment categories. Then, Sects. 7.4.2 and 7.4.3 present the specific framework settings and the experimental results for speech and music separation, respectively.

7.4.1 General System Design

7.4.1.1 Framework

The framework can be divided into three main successive steps as follows.

Pre-processing This step is required to prepare the real-valued input of DNN_0 $\sqrt{z_x(f, n)}$ by deriving it from the complex-valued STFT coefficients of the multichannel mixture signal $\mathbf{x}(f, n)$.

Initialization In this step, the initial source PSDs are estimated simultaneously by DNN_0 given the input prepared above. Besides, the source spatial covariance matrices are initialized as $I \times I$ identity matrix.

Multichannel filtering The source PSDs and spatial covariance matrices are then re-estimated and updated using the iterative framework (Algorithm 2), in which DNN_l is employed for spectrogram fitting at iteration l. In order to avoid numerical instabilities due to the use of single precision, the PSDs $v_j(f, n)$ are floored to 10^{-5} in the parameter update iterations.

7.4.1.2 DNN spectral models

Four design aspects are discussed below: the architecture, the inputs and outputs, the training criterion, and the training algorithm.

Architecture
The DNNs follow a fully-connected feedforward network architecture. The number of hidden layers and the number of units in each input or hidden layer may vary. The number of units in the output layer equals the dimension of spectra multiplied by the number of sources. The activation functions of the hidden and output layers are rectified linear units (ReLUs) [49]. Other network architectures, e.g. recurrent neural network and convolutional neural network, may be used instead of the one used here. The performance comparison with different architectures is beyond the scope of this chapter.

Inputs and outputs
In order to provide temporal context, the input frames are concatenated into *supervectors* consisting of a center frame, left context frames, and right context frames. In choosing the context frames, we use every second frame relative to the center frame in order to reduce the redundancies caused by the windowing of STFT. Although this causes some information loss, this enables the supervectors to represent a longer context [15, 50]. In addition, we do not use the magnitude spectra of the context frames directly, but the difference of magnitude between the context frames and the center frame. These differences act as complementary features similar to delta features [26].

The dimension of the supervectors is reduced by principal component analysis (PCA) to the dimension of the DNN input. Dimensionality reduction by PCA

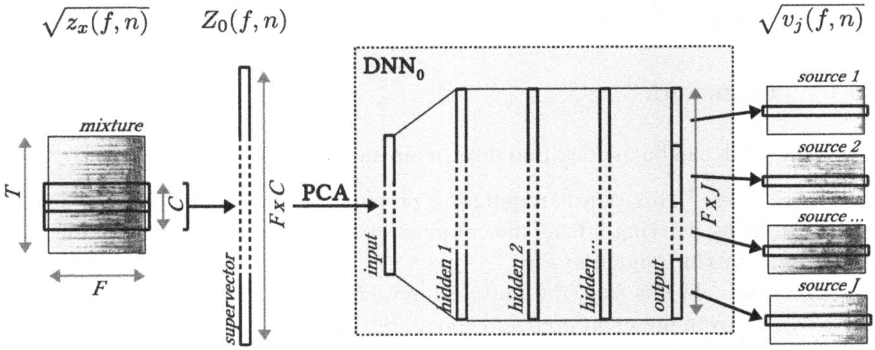

Fig. 7.2 Illustration of the inputs and outputs of the DNN for spectrogram initialization. Inputs: magnitude spectrum of the mixture (left). Outputs: magnitude spectra of the sources (right)

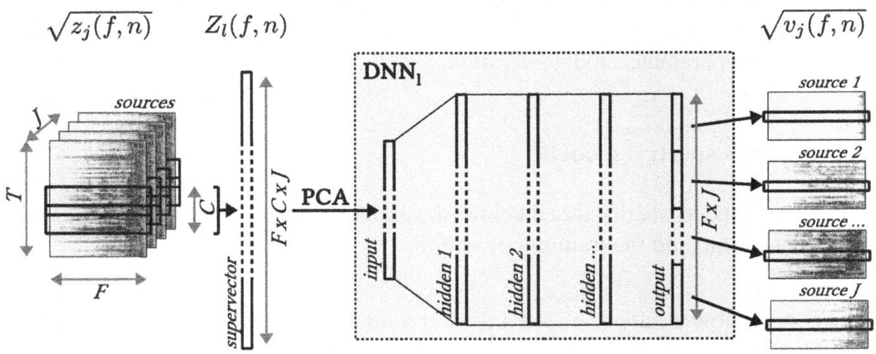

Fig. 7.3 Illustration of the inputs and outputs of the DNNs for spectrogram fitting. Inputs: stack of magnitude spectra of all sources (left). Outputs: magnitude spectra of the sources (right)

significantly minimizes the computational cost of DNN training with a negligible effect on the performance of DNN provided enough components are kept [51]. Standardization (zero mean, unit variance) is done element-wise before and after PCA over the training data. The standardization factors and the PCA transformation matrix are then kept for pre-processing of any input. Thus, strictly speaking, the inputs of DNNs are not the supervectors of magnitude spectra $Z_0(f, n)$ and $Z_l(f, n)$ (see Figs. 7.2 and 7.3), but their transformation into reduced dimension vectors.

Figures 7.2 and 7.3 illustrate the inputs and outputs of the DNNs for spectrogram initialization and spectrogram fitting, respectively. F denotes the dimension of the spectra, $C = 2c + 1$ the context length, and J the number of sources. In this chapter, we considered $c = 2$, so the supervectors for the input of the DNNs were composed by 5 time frames (2 left context, 1 center, and 2 right context frames).

Training criterion
Beside using a cost function from Sect. 7.3.2, an ℓ_2 weight regularization term is used to prevent overfitting [52]. It can be expressed as

$$\mathscr{D}_{\ell_2} = \frac{\lambda}{2} \sum_q w_q^2 \tag{7.16}$$

where w_q are the DNN weights and the regularization parameter is fixed to $\lambda = 10^{-5}$. No regularization is applied to the biases.

In order to avoid numerical instabilities, our implementation used regularized formulations of IS, KL, and Cauchy costs by adding the regularization parameter $\delta_{cf} = 10^{-3}$ in the logarithm computation. It should be noted that the use of regularization in this case is a common practice to avoid instabilities [42, 53]. In addition, geometric analysis on the PS cost function by considering that $m_j(f,n) \in \mathbb{R}_+^{F \times N}$ leads to a simplified formula as

$$\mathscr{D}_{\text{PS}} = \frac{1}{2JFN} \sum_{j,f,n} \left(m_j(f,n) \left|\widetilde{x}(f,n)\right| - \left|\widetilde{c}_j(f,n)\right| \cos\left(\angle\widetilde{x}(f,n) - \angle\widetilde{c}_j(f,n)\right)\right)^2, \tag{7.17}$$

where $\angle\cdot$ denotes the angle of complex-valued STFT spectra. See [26] for further implementation details.

Training algorithm
Following [54], the weights are initialized randomly from a zero-mean Gaussian distribution with standard deviation of $\sqrt{2/n_l}$, where n_l is the number of inputs to the neuron and, in this case, equals to the size of the previous layer. The biases are initialized to zero.

The DNNs are trained by greedy layer-wise supervised training [55] where the hidden layers are added incrementally. In the beginning, a network with one hidden layer is trained after random weight initialization. The output layer of this trained network is then substituted by new hidden and output layers to form a new network, while the parameters of the existing hidden layer are kept. Thus, we can view this as a pre-training method for the training of a new deeper network. After random initialization for the parameters of new layers, the new network is entirely trained. This procedure is done iteratively until the target number of hidden layers is reached.

Training is done by backpropagation with minibatches size of 100 and the ADADELTA parameter update algorithm [56]. Compared to standard stochastic gradient descent (SGD), ADADELTA employs adaptive dimension-wise learning rates and does not require manual setting of the learning rate. The hyperparameters of ADADELTA are set to $\rho = 0.95$ and $\epsilon = 10^{-6}$ following [56]. The validation error is computed every epoch and the training is stopped after 10 consecutive epochs failed to obtain better validation error. The latest model which yields the best validation error is kept. Besides, the maximum number of training epochs is set to 100.

7.4.2 Application: Speech Enhancement

7.4.2.1 Task and dataset

We consider the problem of speech enhancement in the context of the CHiME-3 Challenge. This speech separation and recognition challenge considers the use of ASR in real-world noisy environments for a multi-microphone tablet device. The challenge provides real and simulated 6-channel microphone array data in 4 varied noise settings (bus, cafe, pedestrian area, and street junction) divided into training, development, and test sets. The training set consists of 1,600 real and 7,138 simulated utterances (`tr05_real` and `tr05_simu`), the development set consists of 1,640 real and 1,640 simulated utterances (`dt05_real` and `dt05_simu`), while the test set consists of 1,320 real and 1,320 simulated utterances (`et05_real` and `et05_simu`). The utterances are taken from the 5k vocabulary subset of the Wall Street Journal corpus [57]. All data are sampled at 16 kHz. For further details, please refer to [47]. In short, we deal with the separation of two sources ($J = 2$), namely speech and noise, from a 6-channel mixture ($I = 6$).

We used the source separation performance metrics defined in the BSS Eval toolbox 3.0[1] [58] in most of the experiments presented in this section. The metrics include signal-to-distortion ratio (SDR), source-image-to-spatial-distortion ratio (ISR), signal-to-interference ratio (SIR), and signal-to-artifacts ratio (SAR). In addition, at the end of this section, we use the best speech enhancement system as the front-end, combine it with the best back-end in [34], and evaluate the ASR performance in terms of word error rate (WER).

The multichannel ground truth speech and noise signals for the real data were extracted using the baseline simulation tool provided by the challenge organizers [47]. These signals are not perfect because they are extracted based on an estimation of the impulse responses between the close-talking microphone and the microphones on the tablet device. Since the resulting source separation performance metrics for the real data are unreliable, we evaluate the separation performance on the simulated data, which has reliable ground truth signals, for studying the impact of the different design choices. Nevertheless, since the ground truth transcriptions for ASR are reliable, we evaluate the ASR performance on real data.

7.4.2.2 Algorithm settings

The DNN-based multichannel speech enhancement framework is depicted in Fig. 7.4. Following [19], the input of DNN_0 $\sqrt{z_x(f,n)} = |\widetilde{x}(f,n)|$ was the magnitude of single-channel signals obtained from the multichannel noisy signals $\mathbf{x}(f,n)$ by DS beamforming [45, 46]. In doing so, the time-varying time difference of arrivals (TDOAs) between the speaker's mouth and each of the microphones are first measured using the provided baseline speaker localization tool [47], which relies on a

[1] http://bass-db.gforge.inria.fr/bss_eval/.

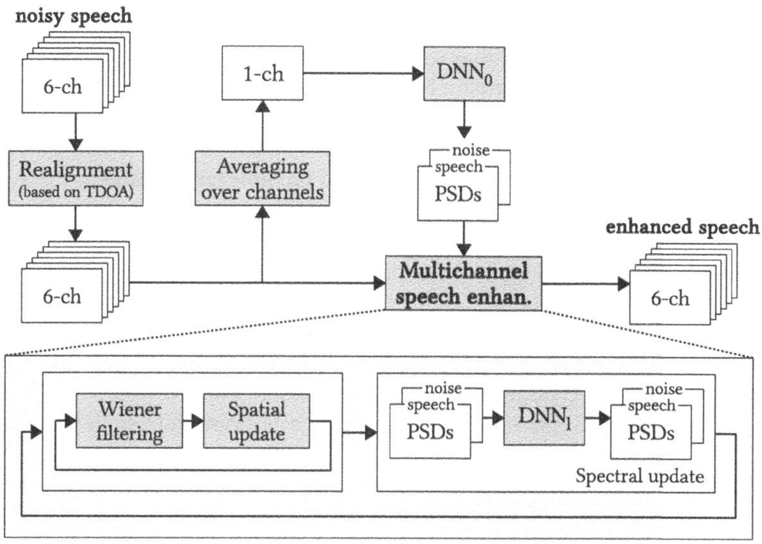

Fig. 7.4 DNN-based multichannel speech enhancement framework

nonlinear variant of steered response power using the phase transform (SRP-PHAT) [59, 60]. All channels are then aligned with each other by shifting the phase of STFT of the input noisy signal $\mathbf{x}(f,n)$ in all time-frequency bins (f,n) by the opposite of the measured delay. This preprocessing is required to satisfy the model in (7.2) which assumes that the sources do not move over time. Finally, a single-channel signal is obtained by averaging the realigned channels together. On the output side, the estimated speech spatial image are averaged over channels to obtain a single-channel signal for the speech recognition evaluation, which empirically provided better ASR performance than the use of one of the channels. Likewise, the training target $\sqrt{\widetilde{v}_j(f,n)} = |\widetilde{c}_j(f,n)|$ was the magnitude of DS beamforming outputs applied on the multichannel ground truth speech and noise signals $\mathbf{c}_j(f,n)$. Recall that the ground truth signals for the real data are unreliable, thus the training target for the real data is not as clean as it should be.

The STFT coefficients were computed using a Hamming window of length 1024 and hopsize 512 resulting in $F = 513$ frequency bins. DNN_0 and DNN_l have a similar architecture. They have an input layer, three hidden layers, and an output layer. Both types of DNNs have hidden and output layer size of $F \times J = 1026$. DNN_0 has an input layer size of $F = 513$ and DNN_l of $F \times J = 1026$.

The DNNs for *the source separation experiment* were trained on both the real and simulated training sets (`tr05_real` and `tr05_simu`) with the real and simulated development sets (`dt05_real` and `dt05_simu`) as validation data. Conversely, we trained the DNNs for *the speech recognition experiment* on the real training set only (`tr05_real`) and validated them on the real development set only (dt05_real). The same DNNs were also used for the performance comparison to the NMF-based iterative EM framework discussed in the end of this subsection. See [34] for the perfomance comparison between these two different training settings.

7.4.2.3 Impact of cost functions and spatial parameter updates

Figure 7.5 shows the performance comparison for different cost functions and also different numbers of spatial updates. In this case, the spectral parameters $v_j(f, n)$ are initialized by DNN_0 and kept fixed during the iterative procedure. In other words, the iteration only updates the spatial parameters. The performance metrics were computed on the resulting 6-channel estimated speech signals. The x-axis of each chart corresponds to the number of spatial updates k. Thus, $k = 0$ is equivalent to single-channel source separation since separation is done independently for each channel.

In general, the performance increases along spatial updates. We observe that different cost functions behave differently along these updates. The performance of the PS cost is the best according to most metrics for the first few updates, but then it saturates quickly. On the contrary, the other cost functions are still getting better. Interestingly, after many updates (in this case, after $k = 20$), we can observe that some cost functions are better than the others for some metrics. Thus, the cost function selection should depend on the task (e.g. fewer artifacts are preferable to low interference) and the computational constraints (e.g. only few updates can be done). For general purposes, KL is the most reasonable choice because it improved all of the metrics well. Although IS is theoretically-motivated, there are better alternatives.

7.4.2.4 Impact of spectral parameter updates

Figure 7.6 shows the performance comparison for different numbers of iterations after fixing the number of spatial updates $K = 20$. We trained two additional DNNs for spectrogram fitting, i.e. DNN_1 and DNN_2 for $l = 1$ and $l = 2$, respectively. This figure shows that generally the iterative spectral and spatial updates improve the performance, although we can observe that Cauchy and IS tend to saturate more quickly than other costs. Overall, the performance saturates after few iterations. Finally, the multichannel approach outperformed the single-channel DNN-based approach even when using DNN_0 only. Additional experiments where one DNN (namely DNN_1) was used for spectrogram fitting of multiple iterations are presented in [26].

7.4.2.5 Comparison to NMF-based iterative EM framework

In this subsection, we compare the performance of DNN-based framework described in this chapter to that of NMF-based framework [29]. We used the implementation of the latter framework found in the Flexible Audio Source Separation Toolbox

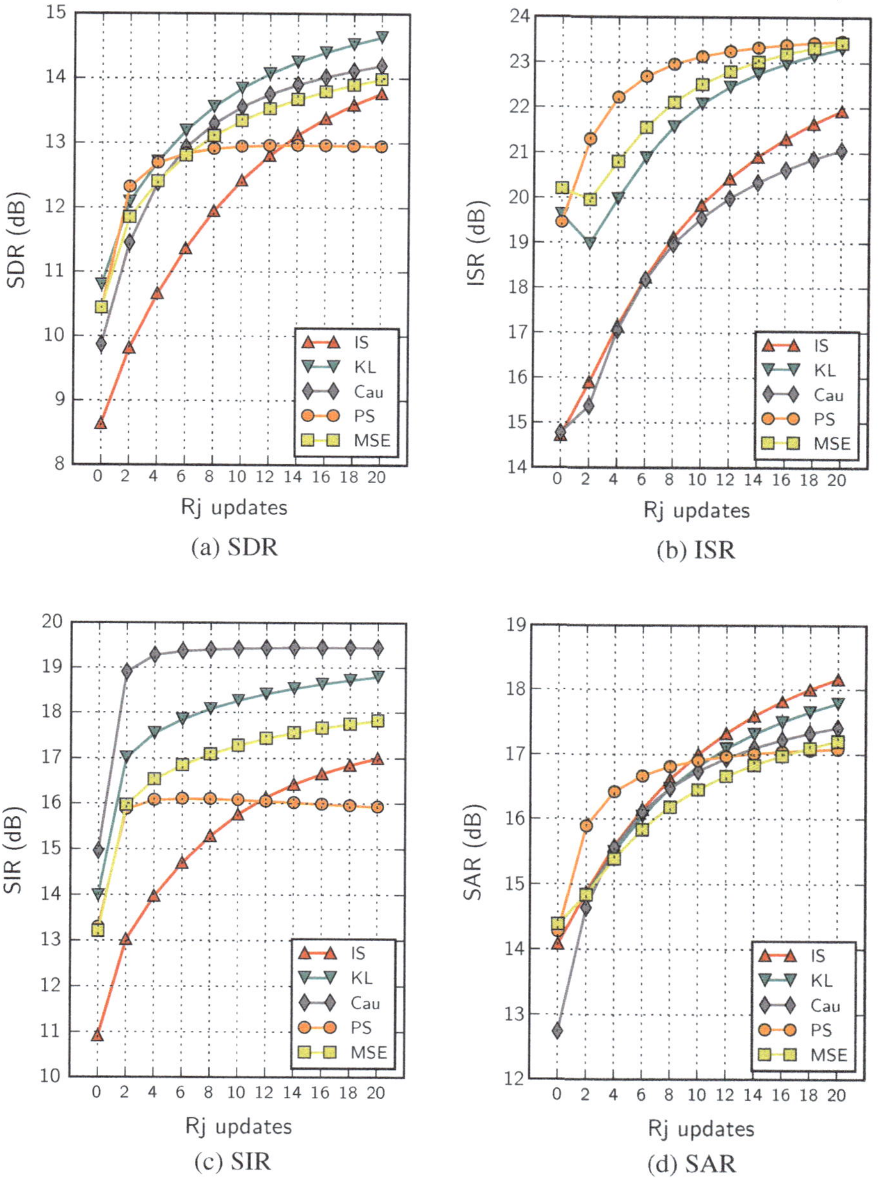

Fig. 7.5 Performance comparison for various numbers of spatial updates with the DNNs trained using different cost functions. The PSDs $v_j(f, n)$ are estimated by DNN_0 and the spatial covariance matrices $\mathbf{R}_j(f)$ are updated in the iterative procedure. The evaluation was done on the simulated test set (`et05_simu`). The figures show the mean value. Higher is better

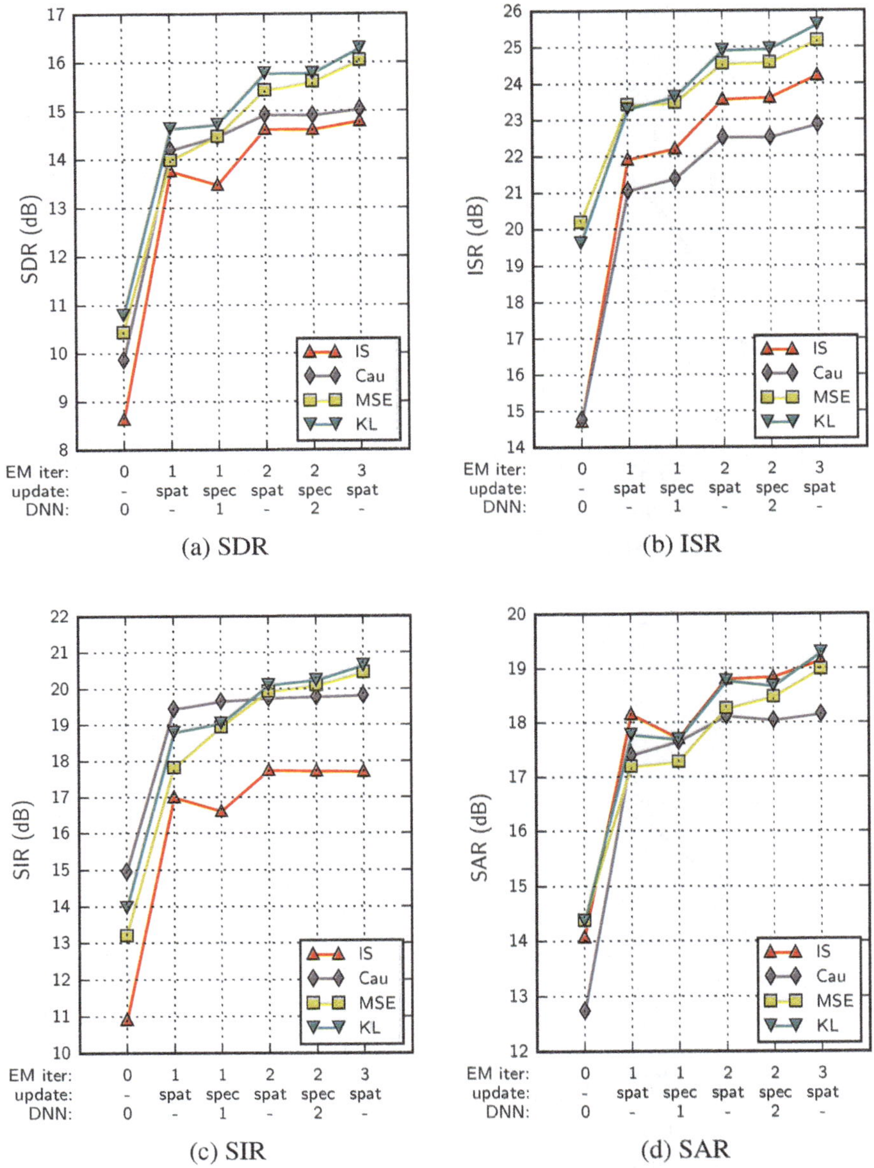

Fig. 7.6 Performance comparison for each update of the EM iterations with the DNNs trained using different cost functions. Different DNNs are used for each EM iteration. The spatial covariance matrices $\mathbf{R}_j(f)$ are updated with $K = 20$. The evaluation was done on the simulated test set (`et05_simu`). The figures show the mean value. Higher is better

Table 7.1 Performance comparison in terms of source separation metrics (in dB). The evaluation was done on the simulated test set (`et05_simu`). The table shows the mean value. Higher is better

Enhancement method	SDR	ISR	SIR	SAR
NMF-based iterative EM [29]	7.72	10.77	13.29	12.29
Proposed: KL (3 DNNs)	13.25	24.25	15.58	18.23

Table 7.2 Average WERs (%) using the DNN+sMBR back-end trained with enhanced multi-condition data followed by 5-gram KN smoothing and RNN-LM rescoring. The evaluation was done on the real sets. Boldface numbers show the best performance for each dataset. Lower is better

Enhancement method	EM iter.	Update type	Dev	Test
Observed	–	–	9.65	19.28
DS beamforming	–	–	6.35	13.70
NMF-based [29]	50	–	6.10	13.41
DNN-based: KL (3 DNNs)	0	–	6.64	15.18
	1	spatial	5.37	11.46
		spectral	5.19	11.46
	2	spatial	**4.87**	10.79
		spectral	4.99	11.12
	3	spatial	4.88	**10.14**

(FASST)[2] and followed the settings used in [61]. The speech spectral and spatial models for this framework were trained on the real training set (`tr05_real`). Besides, the noise spectral and spatial models were initialized for each mixture using 5 s of background noise context based on the available annotation. This setting is favourable to the NMF-based framework. However, because of this, the comparison is not completely fair since the DNN-based framework does not exploit this context information. As described earlier, the DNNs used in this evaluation were also trained on the real training set only. The separation results from this evaluation were then used for the following speech recognition evaluation.

Table 7.1 compares the performance of the NMF-based framework after 50 EM iterations and the performance of the DNN-based framework after the spatial update of the EM iteration $l = 3$. The DNN-based framework is clearly better than the NMF-based iterative EM framework for all metrics. This confirms that DNNs are able to model spectral parameters much better than NMF does.

Table 7.2 shows the speech recognition evaluation results in terms of WER. This evaluation followed the Kaldi setup distributed by the CHiME-3 challenge organizers[3] [47, 62]. It includes the use of (a) feature-space maximum likelihood regression features [63]; (b) acoustic models based on Gaussian Mixture Model and DNN trained with the cross entropy criterion followed by state-level minimum Bayes risk (sMBR)

[2]http://bass-db.gforge.inria.fr/fasst.

[3]https://github.com/kaldi-asr/kaldi/tree/master/egs/chime3.

criterion [64]; and (c) language models with 5-gram Kneser-Ney (KN) smoothing [65] and rescoring using recurrent neural network-based language model (RNN-LM) [66]. The acoustic models are trained on enhanced multi-condition real and simulated data. See [62] for the further details of the methods used in the evaluation. It should be noted that we did not do any further optimization on the speech recognition back-end.

The evaluation results include the baseline performance (observed), DS beamforming, and NMF-based iterative EM framework [29]. The baseline performance was measured using only channel 5 of the observed 6-channel mixture. This channel is considered as the most useful channel because the corresponding microphone faces the user and is located at the bottom-center of the tablet device. DS beamforming was performed on the 6-channel mixture. For both the NMF-based and DNN-based frameworks, we compute the average over channels of the separation results from the earlier source separation evaluation.

For the DNN-based single-channel enhancement (see EM iteration $l = 0$), the WER on the real test set decreases by 21% relative w.r.t. the observed WER. This single-channel enhancement takes the output of DS beamforming on the 6-channel mixture. However, this single-channel enhancement did not provide better performance compared to the DS beamforming alone. It indicates that proper exploitation of multichannel information is crucial. The DNN-based multichannel enhancement then decreases the WER on the real test set by 33% relative w.r.t. the corresponding single-channel enhancement, 26% relative w.r.t. the DS beamforming alone, and 24% relative w.r.t. the NMF-based iterative EM framework [29].

7.4.3 *Application: Music Separation*

7.4.3.1 Task and dataset

We also consider the problem of music separation in the context of the professionally-produced music recordings task (labeled as 'MUS') of SiSEC 2016. In this task, we want to separate music recordings into their constitutive sources, namely *vocals*, *bass*, *drums*, and *other*. The dataset, called Demixing Secrets Dataset (DSD100),[4] comprises 100 full-track songs of diverse music genres by various artists with their corresponding sources. All mixtures and sources are stereo signals sampled at 44.1 kHz. The dataset is divided evenly into development and evaluation sets. In short, we deal with the separation of four sources ($J = 4$) from a stereo mixture ($I = 2$).

[4] See MUS 2016 task on http://sisec.inria.fr.

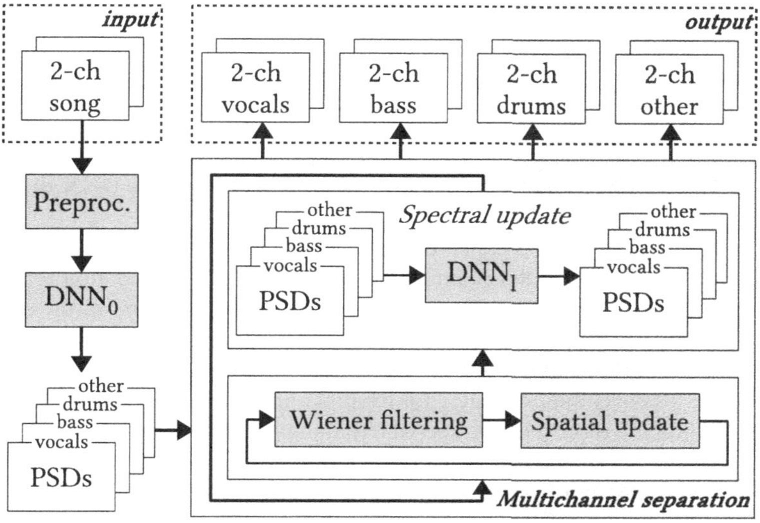

Fig. 7.7 DNN-based multichannel music separation framework

7.4.3.2 Algorithm settings

The DNN-based music separation framework is depicted in Fig. 7.7. In this evaluation, we used one DNN for spectrogram initialization and another DNN for spectrogram fitting, namely DNN_0 and DNN_1, respectively.

The STFT coefficients were computed using a Hamming window of length 2048 and hopsize 1024 resulting in $F = 1025$ frequency bins. DNN_0 has an input layer size of 2050 with three hidden layers, while DNN_1 has an input layer size of 4100 with two hidden layers. These settings are chosen based on preliminary experiments and computational considerations. The hidden layers and output layers of both DNNs have a size of $F \times J = 4100$. Dropout [67] with a rate 0.5 is implemented for all hidden layers.

The input of DNN_0 $\sqrt{z_x(f,n)}$ was derived from the multichannel mixture signal $\mathbf{x}(f,n)$ as

$$\sqrt{z_x(f,n)} = \sqrt{\frac{1}{I}\|\mathbf{x}(f,n)\|^2}. \tag{7.18}$$

The training target $\sqrt{\widetilde{v}_j(f,n)}$ was derived from the true source spatial image $\mathbf{c}_j(f,n)$ as

$$\sqrt{\widetilde{v}_j(f,n)} = \sqrt{\frac{1}{I}\text{tr}\left(\widetilde{\mathbf{R}}_j(f)^{-1}\mathbf{c}_j(f,n)\mathbf{c}_j^H(f,n)\right)}, \tag{7.19}$$

where $\widetilde{\mathbf{R}}_j(f)$ is an estimate of the true spatial covariance matrix computed as

$$\widetilde{\mathbf{R}}_j(f) = \frac{I}{N} \sum_{n=1}^{N} \frac{\mathbf{c}_j(f,n)\mathbf{c}_j^H(f,n)}{\|\mathbf{c}_j(f,n)\|^2}. \tag{7.20}$$

Compared to that was done in the speech enhancement task (Sect. 7.4.2), this provides better targets for the sources which are not mixed to the center (corresponding to $\widetilde{\mathbf{R}}_j(f) = \mathbf{I}$), e.g. *drums* and *other*, and consequently allows the DNN to provide better estimates.

We randomly divided the supervectors ($Z_0(f,n)$ and $Z_l(f,n)$ in Figs. 7.2 and 7.3) of each song from the development set into training and validation sets with a ratio of 8 to 2. By doing so, these two sets contains *different* parts of the same set of songs. However, the fact that these parts come from the same set of songs makes the DNN training prone to overfitting. Another data splitting scheme is by dividing the available 50 development songs, for example, into 40 songs for training and 10 songs for validation (note that, we keep the training-validation ratio of 8 to 2). Using this scheme, we observed that the early stopping mechanism of the DNN training was triggered too early resulting a DNN with poor performance. The cost function used for DNN training is MSE with an ℓ_2 regularization term.

In addition, after every spatial parameter update in the multichannel iteration procedure, the parameter is normalized and regularized with $\delta_{\mathbf{R}} = 10^{-5}$ as

$$\mathbf{R}_j(f) = \frac{I}{\text{tr}(\mathbf{R}_j(f))}\mathbf{R}_j(f) + \delta_{\mathbf{R}}\mathbf{I}. \tag{7.21}$$

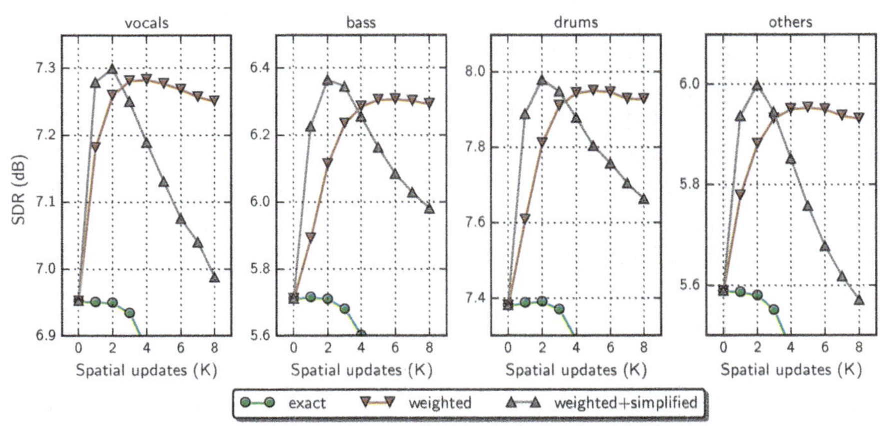

Fig. 7.8 Performance comparison on the *development* set for various numbers of spatial updates with different parameter updates. The PSDs $v_j(f,n)$ are initialized by DNN_0 and the spatial covariance matrices $\mathbf{R}_j(f)$ are updated in the iterative procedure. Higher is better

7.4.3.3 Impact of weighted spatial parameter updates

Figure 7.8 shows the impact of different spatial parameter update strategies, namely 'exact', 'weighted', and 'weighted+simplified', on the performance in terms of SDR. The performance is computed on all songs on 30 s excerpts, taken every 15 s. The differences between these strategies are listed below.

- 'exact': $\widehat{\mathbf{R}}_{\mathbf{c}_j}(f,n) \leftarrow$ (7.6), $\mathbf{R}_j(f) \leftarrow$ (7.7) (as in [27, 29] and the speech enhancement experiment discussed in Sect. 7.4.2)
- 'weighted': $\widehat{\mathbf{R}}_{\mathbf{c}_j}(f,n) \leftarrow$ (7.6), $\mathbf{R}_j(f) \leftarrow$ (7.15)
- 'weighted+simplified': $\widehat{\mathbf{R}}_{\mathbf{c}_j}(f,n) = \widehat{\mathbf{c}}_j(f,n)\widehat{\mathbf{c}}_j(f,n)^H$; $\mathbf{R}_j(f) \leftarrow$ (7.15) (as in [33, 34])

The results show that the 'exact' strategy fails to improve the performance. It should be noted that in the oracle setting, in which $v_j(f,n)$ is computed from the true source image, this 'exact' strategy works well. Hence, it does not work in this case probably because of $v_j(f,n)$ is badly estimated by the DNN. Following this reasoning, the 'weighted' and 'weighted+simplified' strategies show that the weighted spatial parameter updates handle bad estimation of $v_j(f,n)$ effectively. We observe that 'weighted' is more robust to the setting of K than 'weighted+simplified'. This also shows that the inclusion of prior information in the computation of $\widehat{\mathbf{R}}_{\mathbf{c}_j}(f,n)$ allows the system to be more stable.

It is also worth mentioning that the 'exact' strategy works for our speech enhancement task (Sect. 7.4.2). This might be because, in that task, we dealt with fewer sources, fewer frequency bins, and more training data which lead to better DNNs providing better estimation of $v_j(f,n)$. In addition, when the 'weighted' strategy is used in that speech enhancement task followed by the speech recognition evaluation, we observed an improvement of WER up to 2% absolute.

7.4.3.4 Comparison to various music separation techniques

Figure 7.9 shows the performance comparison between four systems based on the framework described in this chapter (`NUG{1,2,3,4}`) and other music separation techniques. `NUG1` and `NUG3` correspond to 'weighted+simplified' after spatial updates of EM iterations 1 and 2 with $K = 2$, respectively. Similarly, `NUG2` and `NUG4` correspond to 'weighted' with $K = 4$. To be clear, `NUG3` and `NUG4` used additional DNNs for spectrogram fitting. These systems are compared to the other techniques listed below. See [68] for the implementation details of these techniques.

- Matrix factorization systems include `OZE` [29], `DUR` [69], and `HUA`[70]
- `KAM{1,2}` are variants of KAM [32]
- `RAF{1,2,3}` are variants of REPET [71–73]
- `UHL{1,2}` are variants of the DNN-based method in [15]

Among these other techniques, `UHL{1,2}` are also DNN-based ones. We can observe that DNN-based techniques performed better than the classical techniques. At a glance, the performance of NUG{1,2,3,4} is comparable to that of UHL{1,2}. However, we could provide lower spatial error (NUG{3,4}), interference (NUG3), or artifact NUG{1,2}. The most important difference between these systems is that UHL{1,2} do single-channel filtering, instead of multichannel filtering. This shows that we can also benefit from multichannel filtering in a music separation task.

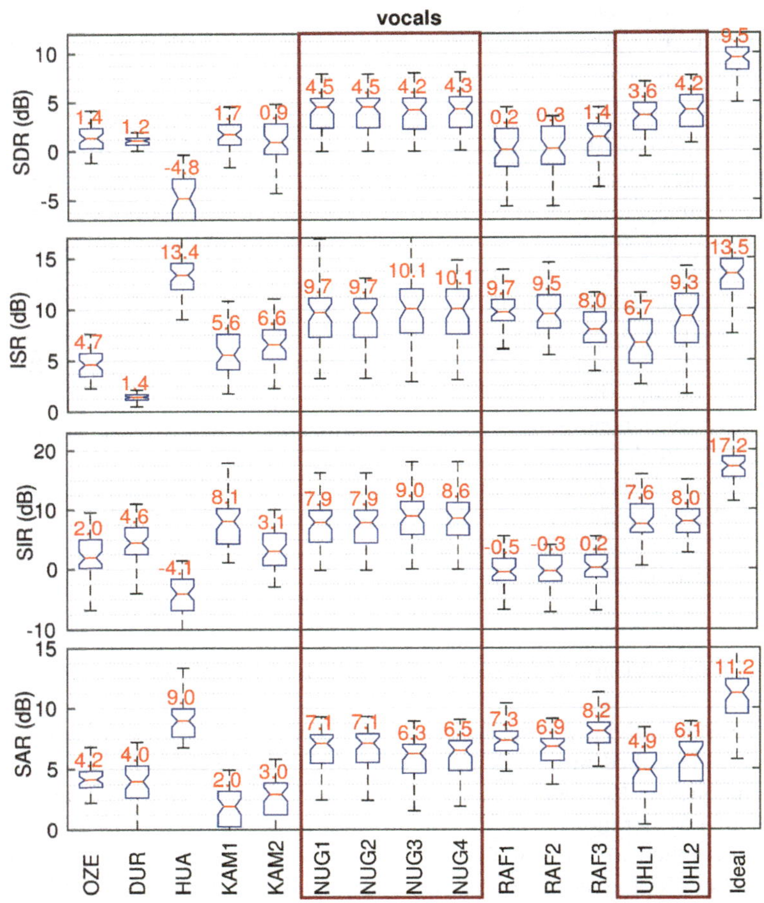

Fig. 7.9 Performance comparison on the *vocals* of *evaluation* set. The numbers shown above boxplots indicate the median values. Higher is better. The systems shown inside the red boxes are based on DNNs. NUG{1,2,3,4} are multichannel separation systems based on the framework described in this chapter, while UHL{1,2} are single-channel separation systems as in [15]

7.5 Closing Remarks

This chapter focused on the use of DNNs for multichannel audio source separation. The separation framework combines DNNs to model the source spectra and the classical multichannel Gaussian model to exploit the spatial information. Experimental results demonstrated the effectiveness of this framework for both speech enhancement and music separation tasks. Beside assessing the source separation performance, the speech recognition performance was also evaluated for the first task. Several design choices and their comparative importance are presented. Finally, the results show the benefit of the presented DNN-based multichannel approach over a single-channel DNN-based approach and multichannel NMF.

Acknowledgements The authors would like to thank the developers of Theano [74] and Kaldi [75]. Experiments presented in this article were carried out using the Grid'5000 testbed, supported by a scientific interest group hosted by Inria and including CNRS, RENATER and several Universities as well as other organizations (see https://www.grid5000.fr).

References

1. S. Makino, H. Sawada, T.-W. Lee (eds.), *Blind Speech Separation* (Springer, Dordrecht, The Netherlands, 2007)
2. M. Wölfel, J. McDonough, *Distant Speech Recognition* (Wiley, Chichester, West Sussex, UK, 2009)
3. T. Virtanen, R. Singh, B. Raj (eds.), *Techniques for Noise Robustness in Automatic Speech Recognition* (Wiley, Chicester, West Sussex, UK, 2012)
4. G.R. Naik, W. Wang (eds.), *Blind Source Separation: Advances in Theory, Algorithms and Applications* (Springer, Berlin, Germany, 2014)
5. E. Vincent, N. Bertin, R. Gribonval, F. Bimbot, From blind to guided audio source separation: How models and side information can improve the separation of sound. IEEE Signal Process. Mag. **31**(3), 107–115 (2014)
6. L. Deng, D. Yu, *Deep Learning: Methods and Applications*, Found. Trends Signal Process. vol. 7 (Now Publishers Inc., Hanover, MA, USA, 2014), pp. 3–4
7. G. Hinton, L. Deng, D. Yu, G. Dahl, A. Mohamed, N. Jaitly, A. Senior, V. Vanhoucke, P. Nguyen, T. Sainath, B. Kingsbury, Deep neural networks for acoustic modeling in speech recognition: the shared views of four research groups. IEEE Signal Process. Mag. **29**(6), 82–97 (2012)
8. F. Weninger, H. Erdogan, S. Watanabe, E. Vincent, J. Le Roux, J.R. Hershey, B. Schuller, Speech enhancement with LSTM recurrent neural networks and its application to noise-robust ASR, in *Proceeding of the Latent Variable Analysis Signal Separation (LVA/ICA), Liberec, Czech Republic, International Conference* (2015)
9. J. Chen, Y. Wang, D. Wang, A feature study for classification-based speech separation at low signal-to-noise ratios IEEE/ACM. Trans. Audio Speech Lang. Process. **22**(12), 1993–2002 (2014)
10. Y. Tu, J. Du, Y. Xu, L. Dai, C.-H. Lee, Speech separation based on improved deep neural networks with dual outputs of speech features for both target and interfering speakers in *Proceeding of the International Symposium Chinese Spoken Language Processing (ISCSLP)*, Singapore (2014), pp. 250–254
11. P.-S. Huang, M. Kim, M. Hasegawa-Johnson, P. Smaragdis, Singing-voice separation from monaural recordings using deep recurrent neural networks, in *Proceeding of the International Society for Music Information Retrieval (ISMIR)*, Taipei, Taiwan (2014), pp. 477–482

12. P.-S. Huang, M. Kim, M. Hasegawa-Johnson, P. Smaragdis, Deep learning for monaural speech separation, in *Proceeding of IEEE International Conference on Acoustics, Speech and Signal Processing (ICASSP)*, Florence, Italy (2014), pp. 1562–1566
13. P.-S. Huang, M. Kim, M. Hasegawa-Johnson, P. Smaragdis, Joint optimization of masks and deep recurrent neural networks for monaural source separation. IEEE Trans. Audio Speech Lang. Process. **23**(12), 2136–2147 (2015)
14. S. Araki, T. Hayashi, M. Delcroix, M. Fujimoto, K. Takeda, T. Nakatani, Exploring multi-channel features for denoising-autoencoder-based speech enhancement, in *Proceeding of IEEE International Conference on Acoustics, Speech and Signal Processing (ICASSP)*, Brisbane, Australia (2015), pp. 2135–2139
15. S. Uhlich, F. Giron, Y. Mitsufuji, Deep neural network based instrument extraction from music, in *Proceeding of IEEE International Conference on Acoustics, Speech and Signal Processing (ICASSP)*, Brisbane, Australia (2015), pp. 2135–2139
16. Y. Wang, D. Wang, Towards scaling up classification-based speech separation. IEEE Trans. Audio Speech Lang. Process. **21**(7), 1381–1390 (2013)
17. A. Narayanan, D. Wang, Ideal ratio mask estimation using deep neural networks for robust speech recognition, in *Proceeding of IEEE International Conference on Acoustics, Speech and Signal Processing (ICASSP)*, Vancouver, Canada (2013), pp. 7092–7096
18. Y. Jiang, D. Wang, R. Liu, Z. Feng, Binaural classification for reverberant speech segregation using deep neural networks. IEEE/ACM Trans. Audio Speech Lang. Process. **22**(12), 2112–2121 (2014)
19. F. Weninger, J. Le Roux, J.R. Hershey, B. Schuller, Discriminatively trained recurrent neural networks for single-channel speech separation, in *Proceeding of IEEE Global Conference Signal Information Processing (GlobalSIP)*, Atlanta, GA, USA (2014), pp. 577–581
20. A. Narayanan, D. Wang, Improving robustness of deep neural network acoustic models via speech separation and joint adaptive training. IEEE/ACM Trans. Audio Speech Lang. Process. **23**(1), 92–101 (2015)
21. Y. Wang, D. Wang, A deep neural network for time-domain signal reconstruction, in *Proceeding of IEEE International Conference on Acoustics, Speech and Signal Processing (ICASSP)*, Brisbane, Australia (2015), pp. 4390–4394
22. D.S. Williamson, Y. Wang, D. Wang, Complex ratio masking for monaural speech separation. IEEE/ACM Trans. Audio Speech Lang. Process. **24**(3), 483–492 (2016)
23. H. Erdogan, J.R. Hershey, S. Watanabe, J. Le Roux, Phase-sensitive and recognition-boosted speech separation using deep recurrent neural networks, in *Proceeding of IEEE International Conference on Acoustics, Speech and Signal Processing (ICASSP)*, Brisbane, Australia (2015), pp. 708–712
24. X. Xiao, S. Watanabe, H. Erdogan, L. Lu, J. Hershey, M. Seltzer, G. Chen, Y. Zhang, M. Mandel, D. Yu, Deep beamforming networks for multi-channel speech recognition, in *Proceeding of IEEE International Conference on Acoustics, Speech and Signal Processing (ICASSP)* (2016), pp. 5745–5749
25. J. Heymann, L. Drude, R. Haeb-Umbach, Neural network based spectral mask estimation for acoustic beamforming, in *Proceeding of IEEE International Conference on Acoustics, Speech and Signal Processing (ICASSP)* (2016), pp. 196–200
26. A.A. Nugraha, A. Liutkus, E. Vincent, Multichannel audio source separation with deep neural networks. IEEE/ACM Trans. Audio Speech Lang. Process. **24**(9), 1652–1664 (2016)
27. N.Q.K. Duong, E. Vincent, R. Gribonval, Under-determined reverberant audio source separation using a full-rank spatial covariance model. IEEE Trans. Audio Speech Lang. Process. **18**(7), 1830–1840 (2010)
28. N. Duong, H. Tachibana, E. Vincent, N. Ono, R. Gribonval, S. Sagayama, Multichannel harmonic and percussive component separation by joint modeling of spatial and spectral continuity, in *Proceeding of IEEE International Conference on Acoustics, Speech and Signal Processing (ICASSP)*, Prague, Czech Republic (2011), pp. 205–208
29. A. Ozerov, E. Vincent, F. Bimbot, A general flexible framework for the handling of prior information in audio source separation. IEEE Trans. Audio Speech Lang. Process. **20**(4), 1118–1133 (2012)

30. T. Gerber, M. Dutasta, L. Girin, C. Févotte, Professionally-produced music separation guided by covers, in *Proceeding of International Society of Music Information Retrieval (ISMIR)*, Porto, Portugal (2012), pp. 85–90
31. M. Togami, Y. Kawaguchi, Simultaneous optimization of acoustic echo reduction, speech dereverberation, and noise reduction against mutual interference. IEEE/ACM Trans. Audio Speech Lang. Process. **22**(11), 1612–1623 (2014)
32. A. Liutkus, D. Fitzgerald, Z. Rafii, B. Pardo, L. Daudet, Kernel additive models for source separation. IEEE Trans. Signal Process. **62**(16), 4298–4310 (2014)
33. A. Liutkus, D. Fitzgerald, Z. Rafii, Scalable audio separation with light kernel additive modelling, in *Proceeding of IEEE International Conference on Acoustics, Speech and Signal Processing (ICASSP)*, Brisbane, Australia (2015), pp. 76–80
34. S. Sivasankaran, A.A. Nugraha, E. Vincent, J.A. Morales-Cordovilla, S. Dalmia, I. Illina, A. Liutkus, Robust ASR using neural network based speech enhancement and feature simulation, in *Proceeding of IEEE Automatic Speech Recognition and Understanding Workshop (ASRU)*, Scottsdale, AZ, USA (2015), pp. 482–489
35. A.A. Nugraha, A. Liutkus, E. Vincent, Multichannel music separation with deep neural networks, in *Proceeding of European Signal Processing Conference (EUSIPCO)*, Budapest, Hungary (2016), pp. 1748–1752
36. J.O. Smith, *Spectral Audio Signal Processing*. (W3K Publishing, 2011)
37. E. Vincent, M.G. Jafari, S.A. Abdallah, M.D. Plumbley, M.E. Davies, Probabilistic modeling paradigms for audio source separation, in *Machine Audition: Principles, Algorithms and Systems*, ed. by W. Wang (IGI Global, Hershey, PA, USA, 2011), pp. 162–185 (ch. 7)
38. D. Liu, P. Smaragdis, M. Kim, Experiments on deep learning for speech denoising, in *Proceeding of ISCA INTERSPEECH*, Singapore (2014), pp. 2685–2688
39. Y. Xu, J. Du, L.-R. Dai, C.-H. Lee, An experimental study on speech enhancement based on deep neural networks. IEEE Signal Process. Lett. **21**(1), 65–68 (2014)
40. C. Févotte, N. Bertin, J.-L. Durrieu, Nonnegative matrix factorization with the Itakura-Saito divergence: with application to music analysis. Neural Comput. **21**(3), 793–830 (2009)
41. N. Bertin, C. Févotte, R. Badeau, A tempering approach for Itakura-Saito non-negative matrix factorization. with application to music transcription, in *Proceeding of IEEE International Conference on Acoustics, Speech and Signal Processing (ICASSP)*, Taipei, Taiwan (2009), pp. 1545–1548
42. A. Lefèvre, F. Bach, C.Févotte, Online algorithms for nonnegative matrix factorization with the Itakura-Saito divergence, in *Proceeding of IEEE Workshop on Application of Signal Processing to Audio and Acoustics (WASPAA)*, New Paltz, NY, USA (2011), pp. 313–316
43. C. Févotte, A. Ozerov, Notes on nonnegative tensor factorization of the spectrogram for audio source separation: statistical insights and towards self-clustering of the spatial cues, in *Proceeding of International Symposium on Computer Music Modeling and Retrieval*, Málaga, Spain (2010), pp. 102–115
44. A. Liutkus, D. Fitzgerald, R. Badeau, Cauchy nonnegative matrix factorization, in *Proceeding of IEEE Workshop on Application of Signal Processing to Audio and Acoustics (WASPAA)*, New Paltz, NY, USA (2015), pp. 1–5
45. J. McDonough K. Kumatani, Microphone arrays, in *Techniques for Noise Robustness in Automatic Speech Recognition*, ed. by T. Virtanen, R. Singh, B. Raj (Wiley, Chicester, West Sussex, UK, 2012) (ch. 6)
46. K. Kumatani, J. McDonough, B. Raj, Microphone array processing for distant speech recognition: from close-talking microphones to far-field sensors. IEEE Signal Process. Mag. **29**(6), 127–140 (2012)
47. J. Barker, R. Marxer, E. Vincent, S. Watanabe, The third 'CHiME' speech separation and recognition challenge: dataset, task and baselines, in *Proceeding of IEEE Automation Speech Recognition and Understanding Workshop (ASRU)*, Scottsdale, AZ, USA (2015), pp. 504–511
48. A. Liutkus, F.-R. Stöter, Z. Rafii, D. Kitamura, B. Rivet, N. Ito, N. Ono, J. Fontecave, The 2016 signal separation evaluation campaign, in *Proceeding of International Conference Latent Variable Analysis Signal Separation (LVA/ICA)*, Grenoble, France, (2017), pp. 323–332

49. X. Glorot, A. Bordes, Y. Bengio, Deep sparse rectifier networks, in *Proceeding of International Conference Artificial Intelligence and Statistics (AISTATS)*, Fort Lauderdale, FL, USA, vol. 15 (2011), pp. 315–323
50. A.A. Nugraha, K. Yamamoto, S. Nakagawa, Single-channel dereverberation by feature mapping using cascade neural networks for robust distant speaker identification and speech recognition. EURASIP. J. Audio Speech Music Process. **2014**(13) (2014)
51. X. Jaureguiberry, E. Vincent, G. Richard, Fusion methods for speech enhancement and audio source separation. IEEE/ACM Trans. Audio Speech Lang. Process. **24**(7), 1266–1279 (2016)
52. Y. Bengio, Practical recommendations for gradient-based training of deep architectures, in *Neural Networks: Tricks of the Trade*, ed. by G. Montavon, G. Orr, K.-R. Müller. Lecture Notes in Computer Science, vol. 7700 (Springer, Berlin, Germany, 2012), pp. 437–478 (ch. 19)
53. P. Sprechmann, A.M. Bronstein, G. Sapiro, Supervised non-negative matrix factorization for audio source separation, in *Excursions in Harmonic Analysis*, ed. by R. Balan, M. Begu, J.J. Benedetto, W. Czaja, K.A. Okoudjou. vol. 4 (Springer, Switzerland, 2015), pp. 407–420
54. K. He, X. Zhang, S. Ren, J. Sun, Delving deep into rectifiers: surpassing human-level performance on imagenet classification (2015), arXiv e-prints http://arXiv.org/abs/1502.01852
55. Y. Bengio, P. Lamblin, D. Popovici, H. Larochelle, Greedy layer-wise training of deep networks, in *Proceeding of Conference Neural Information Processing Systems (NIPS)*, Vancouver, Canada (2006), pp. 153–160
56. M.D. Zeiler, ADADELTA: an adaptive learning rate method (2012), arXiv e-prints http://arXiv.org/abs/1212.5701
57. J. Garofalo, D. Graff, D. Paul, D. Pallett, CSR-I (WSJ0) Complete. (Linguistic Data Consortium, Philadelphia, 2007)
58. E. Vincent, R. Gribonval, C.Févotte, Performance measurement in blind audio source separation. IEEE Trans. Audio Speech Lang. Process. **14**(4), 1462–1469 (2006)
59. B. Loesch, B. Yang, Adaptive segmentation and separation of determined convolutive mixtures under dynamic conditions, in *Proceeding of International Conference Latent Variable Analysis Signal Separation (LVA/ICA)*, Saint-Malo, France (2010), pp. 41–48
60. C. Blandin, A. Ozerov, E. Vincent, Multi-source TDOA estimation in reverberant audio using angular spectra and clustering. Signal Process. **92**(8), 1950–1960 (2012)
61. Y. Salaün, E. Vincent, N. Bertin, N. Souviraà-Labastie, X. Jaureguiberry, D.T. Tran, F. Bimbot, The Flexible Audio Source Separation Toolbox Version 2.0, In: IEEE International of Conference Acoustics Speech Signal Process. (ICASSP), Florence, Italy (2014), Show & Tell, https://hal.inria.fr/hal-00957412
62. T. Hori, Z. Chen, H. Erdogan, J.R. Hershey, J. Le Roux, V. Mitra, S. Watanabe, The MERL/SRI system for the 3rd chime challenge using beamforming, robust feature extraction, and advanced speech recognition, in *Proceeding of IEEE Automatic Speech Recognition and Understanding Workshop (ASRU)*, Scottsdale, AZ, USA (2015), pp. 475–481
63. M.J.F. Gales, Maximum likelihood linear transformations for hmm-based speech recognition. Comput. Speech Lang. **12**(2), 75–98 (1998)
64. K. Veselý, A. Ghoshal, L. Burget, D. Povey, *Proceeding of Sequence-discriminative training of deep neural networks*, Lyon, France, ISCA INTERSPEECH (2013), pp. 2345–2349
65. R. Kneser, H. Ney, Improved backing-off for M-gram language modeling, in *Proceeding of IEEE International Conference on Acoustics Speech Signal Processing (ICASSP)*, Detroit, MI, USA, vol. 1 (1995), pp. 181–184
66. T. Mikolov, M. Karafiát, L. Burget, J. Černocký, S. Khudanpur, *Prooceeding of Recurrent Neural Network Based Language Model*, Chiba, Japan, ISCA INTERSPEECH (2010), pp. 1045–1048
67. N. Srivastava, G. Hinton, A. Krizhevsky, I. Sutskever, R. Salakhutdinov, Dropout: a simple way to prevent neural networks from overfitting. J. Mach. Learn. Res. **15**, 1929–1958 (2014)
68. N. Ono, D. Kitamura, Z. Rafii, N. Ito, A. Liutkus, The 2015 signal separation evaluation campaign (SiSEC2015), in *Proceeding of International Conference Latent Variable Analysis Signal Separation (LVA/ICA)*, Liberec, Czech Republic (2015)

69. J.-L. Durrieu, B. David, G. Richard, A musically motivated mid-level representation for pitch estimation and musical audio source separation. IEEE J. Sel. Top. Signal Process. **5**(6), 1180–1191 (2011)
70. P.-S. Huang, S.D. Chen, P. Smaragdis, M. Hasegawa-Johnson, Singing-voice separation from monaural recordings using robust principal component analysis, in *Proceeding of IEEE International Conference on Acoustics, Speech Signal Processing (ICASSP)*, Kyoto, Japan (2012), pp. 57–60
71. Z. Rafii, B. Pardo, Repeating pattern extraction technique (REPET): A simple method for music/voice separation. IEEE Trans. Audio, Speech, Lang. Process. **21**(1), 73–84 (2013)
72. A. Liutkus, Z. Rafii, R. Badeau, B. Pardo, G. Richard, Adaptive filtering for music/voice separation exploiting the repeating musical structure, in *Proceeding of IEEE International Conference on Acoustics, Speech Signal Processing (ICASSP)*, Kyoto, Japan (2012), pp. 53–56
73. Z. Rafii, B. Pardo, Music/voice separation using the similarity matrix, in *Proceeding of International Society of Music Information Retrieval (ISMIR)*, Porto, Portugal (2012), pp. 583–588
74. Theano Development Team, Theano: A Python framework for fast computation of mathematical expressions (2016), arXiv e-prints http://arXiv.org/abs/1605.02688
75. D. Povey, A. Ghoshal, G. Boulianne, L. Burget, O. Glembek, N. Goel, M. Hannemann, P. Motlicek, Y. Qian, P. Schwarz, J. Silovsky, G. Stemmer, K. Vesely, The Kaldi speech recognition toolkit in *Proceeding of IEEE Automatic Speech Recognition and Understanding Workshop (ASRU)*, Hawaii, USA (2011)

Chapter 8
Efficient Source Separation Using Bitwise Neural Networks

Minje Kim and Paris Smaragdis

Abstract Efficiency is one of the key issues in single-channel source separation systems due to the fact that they are often employed for real-time processing. More computationally demanding approaches tend to produce better results, but often not fast enough to be deployed in practical systems. For example, as opposed to the iterative separation algorithms using source-specific dictionaries, a Deep Neural Network (DNN) performs separation via an iteration-free feedforward process. However, even the feedforward process can be very complex depending on the size of the network. In this chapter, we introduce Bitwise Neural Networks (BNN) as an extremely compact form of neural networks, whose feedforward pass uses only efficient bitwise operations (e.g. XNOR instead of multiplication) on binary weight matrices and quantized input signals. As a result, we show that BNNs can perform denoising with a negnigible loss of quality as compared to a corresponding network with the same structure, while reducing the network complexity significantly.

8.1 Introduction

A real-world monophonic audio signal, i.e. an observation made by a single microphone, often contains more than one source, which makes it difficult for a computer system to understand it. In contract to visual objects, which when aligned occlude each other, audio sources observed by the same microphone superimpose,

M. Kim (✉)
Department of Intelligent Systems Engineering, School of Informatics, Computing, and Engineering, Indiana University, Bloomington, IN, USA
e-mail: minje@indiana.edu

P. Smaragdis
Department of Computer Science and Department of Electrical and Computer Engineering, University of Illinois at Urbana-Champaign, Champaign, IL, USA
e-mail: paris@illinois.edu

P. Smaragdis
Adobe Research, Adobe Systems Inc., San Francisco, CA, USA

S. Makino (ed.), *Audio Source Separation*, Signals and Communication Technology, https://doi.org/10.1007/978-3-319-73031-8_8

necessitating the ability to perform *selective attention*. This process has long been the main challenge of source separation systems.

It was discovered that various Latent Variable Analysis (LVA) techniques are effective to resolve this issue, when it comes to analyzing a polyphonic audio source and learning a set of templates [1]. Using a matrix representation of audio, e.g. magnitude spectrograms after applying Short-Time Fourier Transform (STFT), the goal is to seek a low-rank approximation whose basis vectors form a dictionary, while their encodings hold the amount of the contribution of all the dictionary items. Nonnegative Matrix Factorization (NMF) [2, 3] and its probabilistic correspondence, Probabilistic Latent Semantic Indexing (or Probabilistic Latent Semantic Analysis) [4, 5], are common choices for this task.

A source separation scheme has also evolved from this low-rank approximation by learning source-specific dictionaries in advance during the training time, and then decompose the test mixture by using the learned dictionaries as templates [6]. Now the test-time decomposition is different from the usual matrix factorization procedure as one of the factor matrices, the dictionaries, is fixed and free from any further updates, while we focus only on the encoding matrix. Yet, we need to estimate at least some of the parameters to approximate a mixture during runtime, so a few EM-like iterations are inevitable.

Since the main issue of this chapter is efficiency of source separation algorithms during runtime, we consider these iterative updates as a disadvantage, although they are common in the dictionary-based systems. The complexity usually depends on the size of the dictionary and the number of iterations required until convergence. However, when it comes to the *semi-supervised* case the efficiency issue becomes more complicated, because it assumes that only a part of the dictionaries is available from the known source, while the other parts and all their encoding values have to be estimated during the test time from the mixture. Therefore, in the semi-supervised separation due to the lack of stopping criteria overfitting can take place. Likewise, efficiency of semi-supervised approaches to single-channel source separation needs improvement. Although there has been an effort to speed up the EM procedure by replacing the floating-point operations for nearest neighbor searches with binary operations using hashing techniques [7, 8], it still results in an iterative process, something that we would rather avoid for a real-time implementation.

Neural networks, on the other hand, can serve as an alternative solution, most of whose variants are free from iterations during runtime. The basic idea of a neural network-based source separation is to build a network that learns the mapping between the mixture signal and its source components as the input and output, respectively. Once the mapping is learned, the network can predict the source components from the unseen mixture signals (if the mixture is somewhat similar to the training data). Moreover, one can construct a Deep Neural Network (DNN) for this job, which tends to benefit from a large amount of training data when modeling complicated mapping functions. A DNN is indeed one of the very successful approaches to monaural source separation [9–13], which still retains the desirable iteration-free runtime process.

However, we would like to raise another efficiency issue that can be caused by introducing a very large and deep neural network into the procedure. Although it is clear that the complexity of a DNN does not depend on the number of iterations, the sheer number of parameters involved in the feedforward process can become an issue. For example, for a network with three hidden layers, each of which has 1,000 units, the number of floating-point operations is easily over a few millions, which can sometimes exceed the amount of computation for an NMF-based method with all its necessary iterations. This high computational cost during runtime can seriously limit the wide spread use of a DNN-based system especially for the implementations with limited resources. One can optimize this network by using fixed-point representations, but the quantization process is not straightforward, still necessitating a fair amount of computation during runtime.

In this chapter, we introduce Bitwise Neural Networks (BNN) to achieve goals in both efficiency and accuracy [14, 15]. As one of the recently proposed neural network architectures that use an extreme quantization schemes [16–18], BNNs are a highly condensed form of neural networks, where we represent all the parameters and variables with bipolar binaries (i.e. $+1$ and -1), so that the operations on them are defined in a bitwise manner as well. Our goal is to train such a bitwise network whose performance catches up its corresponding real-valued network even with the same network structure. By doing so, we can replace all the floating-point operations with cheaper bitwise ones, e.g. an XNOR gate instead of multiplication, which will be very efficient both in software and hardware implementations. We reformulate the DNN-based separation scheme to accommodate this BNN architecture, and show that the BNN-based methods are performing well enough considering their sensibly cheaper feedforward process.

8.2 A Basic Neural Network for Source Separation

In this section, we briefly review a neural network-based source separation system which will serve as a baseline system for the following BNNs. Although there are many other choices in terms of its structure, for example using Recurrent Neural Networks (RNN) [10], Long Short-Term Memory (LSTM) [19], and deep unfolding networks [12], for simplicity we will focus on the basic fully-connected structure.

There are also many possible choices when it comes to the input and output representations. Raw magnitude Fourier coefficients can be a straightforward feature [9, 10], but there are more speech-specific features, such as cochleagrams [20] and Mel-Frequency Cepstrum Coefficients (MFCC) [21, 22] as well for speech denoising applications. As for the output representation, once again the network can be formulated as a Denoising AutoEncoder (DAE) that tries to predict clean magnitude spectra from their noisy versions [9, 23]. Or, a DNN can also predict the masking values (e.g. 1 for speech and 0 for noise) for all the time-frequency bins so that they can be used to mask out the uninterested sources. Ideal Binary Masks (IBM) [11, 20] and Ideal Ratio Masks (IRM) [21, 22] are common choice along with phase

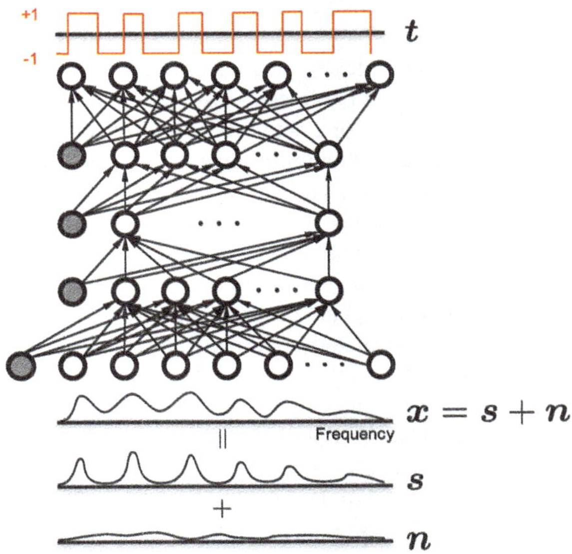

Fig. 8.1 The baseline DNN separation system

information as well [24]. All the various choices have their pros and cons, but the variety of feature representations calls for another consideration when building a BNN for source separation as to how we can come up with a scheme to effectively convert all those real-valued features into bitstrings. We will revisit this problem in Sect. 8.3.

Figure 8.1 illustrates the DNN-based separation procedure. Among all the choices, we use raw Fourier magnitudes and masks as our input and output representations without loss of generality. Let x_d denote d-th element of the Fourier spectrum at a given time frame, $\mathbf{x} \in \mathbb{C}^D$. We also define the ground-truth masks by t_d.[1] If we assume two sources, e.g. speech s_d and noise n_d, we get the speech component by masking the input:

$$s_d \approx t_d x_d = \frac{|s_d|}{|s_d + n_d|} x_d \tag{8.1}$$

We also define a weight matrix per layer, $\mathbf{W}^{(l)} \in \mathbb{R}^{K^{(l+1)} \times K^{(l)}}$, and the bias terms, $\mathbf{b}^{(l)} \in \mathbb{R}^{K^{(l+1)}}$, where $K^{(l)}$ is the number of input units for l-th layer. With these, the forwardpropagation procedure for a tanh network with L hidden layers is defined recursively as follows:

[1] Since we would like to target on real-valued masks defined between 0 and 1, we use this approximated version of masks t_d rather than the more proper complex-valued one.

$$z_d^{(0)} = |x_d|, \quad d \in \{1, \ldots, D\}, \tag{8.2}$$

$$a_j^{(l)} = \sum_{i \in \{1,\ldots,K^{(l)}\}} W_{ji}^{(l)} \cdot z_i^{(l)} + b_j^{(l)}, \tag{8.3}$$

$$z_j^{(l+1)} = \tanh\left(a_j^{(l)}\right), \tag{8.4}$$

$$a_j^{(L)} = \sum_{i \in \{1,\ldots,K^{(L)}\}} W_{ji}^{(L)} \cdot z_i^{(L)} + b_j^{(L)}, \tag{8.5}$$

$$z_j^{(L+1)} = \tanh\left(a_j^{(L)}\right). \tag{8.6}$$

Note that the first layer inputs are the magnitudes of the input mixture spectra (therefore, $K^{(0)} = D$). Throughout this chapter we use $\tanh(\cdot)$ as our activation function for all layers.

A sum-of-squared error function for the IRM target variables are defined as follows:

$$\mathscr{E}(\mathbf{z}^{(L+1)}||\mathbf{t}') = \sum_{d \in \{1,\ldots,D\}} \frac{1}{2}\left(t'_d - z_d^{(L+1)}\right)^2. \tag{8.7}$$

Note that we rescale t_d and create a new variable $t'_d = 2(t_d - 0.5)$ to make it range between -1 to $+1$ and to match it the output of the tanh network. We will revisit this particular choice of the activation function later as it is related to the use of bipolar binaries. We can easily scale back the network outputs for masking the test signals, i.e. $0.5 \cdot z_d^{(L+1)} + 0.5$.

Alternatively, we can define an IBM target variable, which can be more conveniently converted into binary variables[2]:

$$t_d = \begin{cases} 0 & \text{if } \frac{|s_d|}{|s_d + n_d|} < 0 \\ 1 & \text{otherwise} \end{cases} \tag{8.8}$$

Based on the error function, we recursively define the SGD-based backpropagation procedure as usual:

$$\delta_d^{(L)} = \left(1 - \tanh^2\left(a_d^{(L)}\right)\right) \cdot \left(t'_d - z_d^{(L+1)}\right), \tag{8.9}$$

$$\delta_i^{(l)} = \left(1 - \tanh^2\left(a_d^{(l)}\right)\right) \cdot \sum_j \delta_j^{(l+1)} \cdot W_{ji}^{(l)}, \tag{8.10}$$

$$\nabla W_{ji}^{(l)} = \delta_j^{(l)} \cdot z_i^{(l)}, \tag{8.11}$$

$$\nabla b_j^{(l)} = \delta_j^{(l)}. \tag{8.12}$$

---

[2] We call this procedure *binarization*.

There are also various options to train this kind of networks. We found that some commonly used techniques, such as Stochastic Gradient Descent (SGD), dropout [25], and momentum, are also effective.

8.3 Binary Features for Audio Signals

Before delving into the details about BNNs, we would like to discuss the meanings of binary features for audio signals. Since BNNs will eventually replace the real-valued variables and logic with bitwise arithmetic, they can be seen as a way to approximate a Boolean mapping function whose input and output are bit strings. Hence, in order to make this argument complete, we need to ensure that the input and output of the network have their corresponding binary representations as well. Since most of the time the original signals are real-valued, we need a way to convert them into bit patterns without affecting the performance of the source separation system.

There are two important criteria we can use when choosing a binary feature extraction method: whether it is efficient and invertible. First, since the conversion from the raw input (e.g. magnitudes of Fourier spectra) to any bit string is an additional procedure, it should not take up too much resources for the system-wide efficiency. Second, once the BNN produces bit strings as its output, the conversion from the output to the original signal domain must be efficient and not as lossy as possible, too. There are such binary feature extraction methods that also defines its corresponding decoding procedure, which automatically converts the bit string back to the raw signal domain. Otherwise, we need to construct a scheme that finds the most similar bit string from the database (e.g. in terms of Hamming distance), and then take the signal that has generated the matching bit pattern as the output of the network.[3] In any ways, the inversion procedure should be able to recover the original signal with least error, and as efficiently as possible.

Note that IBMs are naturally binary variables, so they can serve as the binary target variables as they are. If the system needs to use IRM or any other real-valued target variables, we need to binarize the target variables as well, which are supposed to be converted back to the raw signal using an appropriate conversion process.

In the following sections we review some existing hashing techniques to discuss their pros and cons as our candidate binary feature extraction methods.

8.3.1 Winner-Take-All Hashing

Winner-Take-All (WTA) hashing [26] is a fast and effective hashing technique whose original usage in object recognition [27] was very successful. As a follow-up it has

[3]Note that this searching task in the binary feature space is very similar to the information retrieval process using hashing.

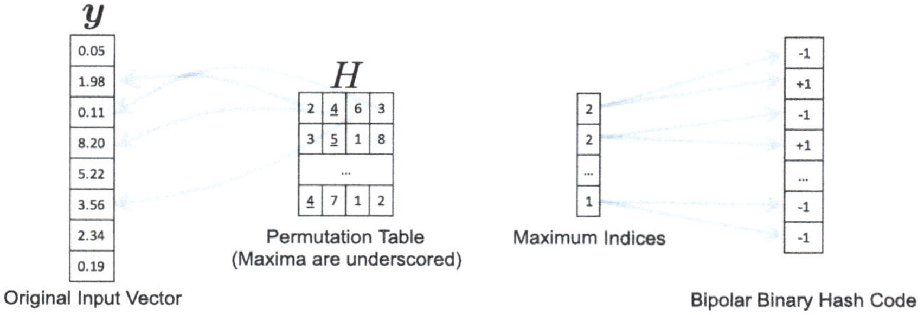

Fig. 8.2 A WTA hashing example

been also shown that WTA hashing is useful to convert audio spectra into bit patterns for the source separation applications [7, 8]. The WTA hash function consists of simple comparison-based operations and is based on random projections.

WTA hashing first defines a $M \times N$ permutation table H with random indices, each of which addresses one of the elements of the input vector. For a magnitude spectrum $\mathbf{y} \in \mathbb{R}^D$, we first choose $N << D$ elements based on the indices written in the first row of the permutation table. Then, we find the winner, the maximum of the N randomly chosen elements to encode its position. For example, when $N = 4$ suppose that the four chosen elements are $\{1.98, 8.2, 3.56, 0.11\}$. Since the second chosen element (8.2) is the maximum, we write down 01[4] on the hash code. In the next round, we see the second row of the permutation table and pick up another four values from the input to elect the maximum. We repeat this procedure for M such comparisons. Figure 8.2 shows this procedure on an input vector of 8 elements. Note that we use bipolar binaries to encode the final hash code.

For two different input vectors, if pairwise relationships between one's elements are all the same with those of the other vector, we can consider that they are very similar to each other. For example, the vector used in Fig. 8.2 has a similar shape with another vector, $[0.5, 19.7, 1.2, 81.9, 50.2, 34.4, 23.1, 1.8]^\top$, although their scales are quite different. WTA hashing approximates these pairwise relationships and encodes them as a bit string.

WTA hashing is convenient and fast because the procedure consists of indexing and comparison. Although its approximation performance is acceptable as a locality sensitive hashing family [28], it does not learn any mapping from the input to the hash codes, e.g. on the contrary to the other machine learning based ones [29]. Another disadvantage of WTA hashing for our regression purpose is the fact that its inverted conversion from the hash code back to the original input is not defined. Once again, we can prepare a dictionary of averaged spectra that share the same hash code, the conversion can be expensive due to the suboptimal, yet required Hamming distance-based search.

---

[4] $[-1, +1]$ when we use bipolar binaries.

8.3.2 Semantic Hashing

On the other hand, semantic hashing is a data-driven method [30]. It learns a deep AutoEncoder (AE), which is a DNN that learns an identity function between the training samples and themselves. The details about training this kind of network are not very different from training an ordinary networks for classification or regression except the fact that the target variables are defined by the input. However, the big difference of this AE for semantic hashing comes from the fact that we want to use the hidden unit outputs of a hidden layer as our hash codes. See Fig. 8.3 for an example.

Although real-valued, hidden unit outputs are not very far from binarization, once we choose to use sigmoid functions as the activation, which can be seen as a smoothed version of the step function (e.g. tanh versus sign). For example, if we want to use $\mathbf{z}^{(2)}$ in Fig. 8.3 as our hash codes, they can be simply rounded to have discrete values, e.g. $\text{sign}(z_d^{(2)})$. Semantic hashing is more careful in binarizing the hidden unit outputs as there is no guarantee that their distribution is extreme, i.e. most of the values are either near -1 or $+1$. To this end, semantic hashing perturb the input with Gaussian noise by turning the AE model into a denoising AE, whose job is to predict the clean input given its noisy version. Since the network has to be robust to this additive noise, its intermediate variables, i.e. hidden unit outputs, tend to have extreme values.

We would like to point out that semantic hashing can suffer from the efficiency issue, although its deep structure has a lot of potential in terms of learning the optimal embedding for the particular data set. Since it is a DNN, in order to learn a high quality hash function it may need to involve a large amount of parameters. Therefore, during the encoding procedure we are expected to run a feedforward step in this network

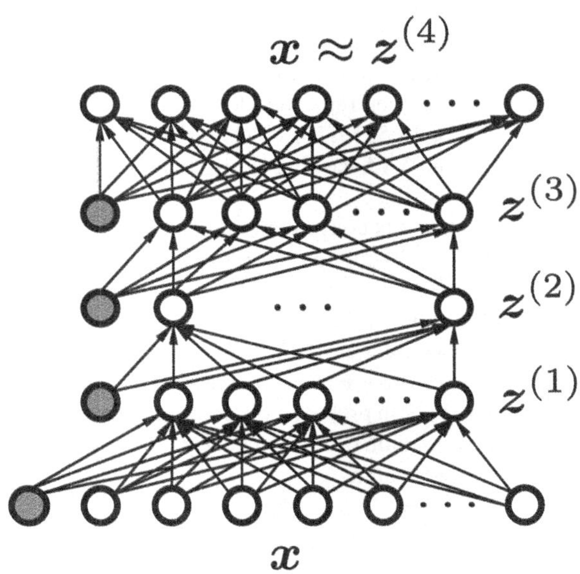

Fig. 8.3 A semantic hashing example

up to the hidden layer of interest, which is nothing but adding a few more floating-point layers underneath the BNN we would like to construct. This computationally heavy nature of semantic hashing is not a desired characteristics for our purpose, because we have to minimize the cost of this binarization phase.

8.3.3 *Quantization and Dispersion*

Quantization-and-Dispersion (QaD) is an alternative binarization technique we can use to quickly convert any real-valued signals into bit strings [15]. It first converts any real value into a fixed-point representation, i.e. an integer, by using a traditional population-based quantization scheme. Lloyd-Max algorithm can be used for this purpose [31]. Figure 8.4 shows the QaD procedure. For an N-bit encoding, first, we assign an integer to each of the 2^N ranges. Based on the sample distribution, we try to divide the sample space with regions that contain as equal amount of samples as possible (the topmost figure in Fig. 8.4). Usually, the average of the samples allocated in the same range is the corresponding representative value of the integer. We could

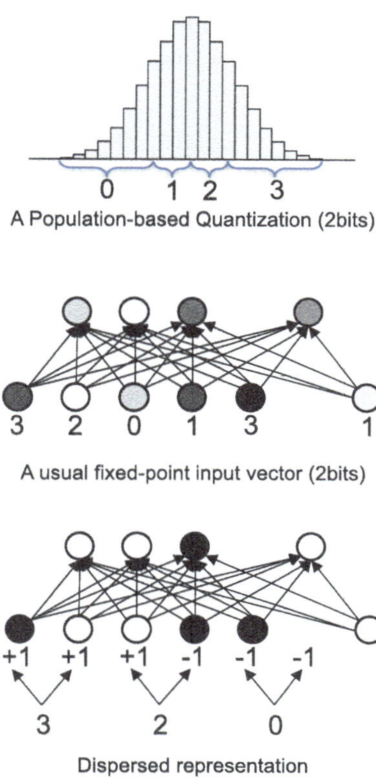

Fig. 8.4 The QaD process. Some figures are from [15]

go ahead and use these integers (or their representative values) as our input as in the middle figure, but then the input to the BNN is not binary. Instead, we represent each integer with its corresponding N-bit bit string, and then disperse them into N input units (the bottom figure). Therefore, the network needs to enlarge the size of its input layer by N times than usual to accommodate these dispersed binary values. We hope that the enlarged input layer does not increase the complexity of the BNN significantly, because it is only for the first layer and the additional weights are all binary as well.

The procedure does introduce some quantization error, but the amount of error is predictable, because it is dependent on how many bits we use to divide the sample distribution. Also, its inversion is straightforward and fast, as the Lloyd-Max algorithm produces not only the boundaries, but the range-specific averages. Furthermore, since it is a scalar quantization technique, the training part is trivial (finding the boundaries from the sample distribution for the even quantization). Therefore, we employ this technique for our universal binarization technique for BNNs.

8.4 BNN Feedforward

In this section we first introduce the feedforward operations for BNN, and then discuss about the linear separability.

8.4.1 The Feedforward Procedure

For a given C-dimensional bit string input vector, $\bar{\mathbf{x}} \in \{-1, +1\}^C$, we can define the feedforward procedure similar to (8.2)–(8.6). However, now we can define it in a more efficient way by replacing all the multiplication with an XNOR operation, $\otimes$, because XNOR between bipolar binaries are equivalent to their multiplication (see Table 8.1 for an XNOR table). Moreover, the tanh activation function is now replaced with the sign function, too.

$$\bar{z}_d^{(0)} = \bar{x}_d, \quad d \in \{1, \dots, D\}, \tag{8.13}$$

$$\bar{a}_j^{(l)} = \sum_{i \in \{1,\dots,K^{(l)}\}} \bar{W}_{ji}^{(l)} \otimes \bar{z}_i^{(l)} + \bar{b}_j^{(l)}, \tag{8.14}$$

$$\bar{z}_j^{(l+1)} = \text{sign}\big(\bar{a}_j^{(l)}\big), \tag{8.15}$$

$$\bar{a}_j^{(L)} = \sum_{i \in \{1,\dots,K^{(L)}\}} \bar{W}_{ji}^{(L)} \otimes \bar{z}_i^{(L)} + \bar{b}_j^{(L)}, \tag{8.16}$$

$$\bar{z}_j^{(L+1)} = \text{sign}\big(\bar{a}_j^{(L)}\big). \tag{8.17}$$

Table 8.1 The truth table for the XNOR operation

Input		Output
True (+1)	True (+1)	True (+1)
True (+1)	False (−1)	False (−1)
False (−1)	True (+1)	False (−1)
False (−1)	False (−1)	True (+1)

Note that now we introduce some new notations with bars on top of the symbols to distinguish them from the variables used in the DNN feedforward process (8.2)–(8.6). In practice, the sign function can be implemented by counting the number of +1's in the summation (8.14) and (8.16) checking if it exceeds a certain threshold, e.g. $(K^{(l)} + 1)/2$.

8.4.2 Linear Separability

Figure 8.5 shows some classification problems that can illustrate the performance of BNNs. First, in Fig. 8.5a we can see that the traditional XOR problem can be solved with two hyperplanes. If we are allowed to use real-valued coefficients to define the hyperplanes, there are infinitely many solutions to this problem. Interestingly, there is a BNN solution for this, where we can define two hyperplanes only with bipolar binary coefficients. In (b) we see a corresponding multilayer perceptron defined with bipolar binaries and sign functions. From this example, we see that a BNN can solve non-linear problems.

On the other hand, Fig. 8.5c shows a simple linearly separable problem that BNNs fail to solve due to the lack of flexibility to define necessary hyperplanes. Consequently, it requires two hyperplane to solve this linearly separable problem. Zero weights in BNNs can get around this problem (d), which add additional rotation options and eventually solve the problem with a single hyperplane.

Eventually, however, BNN needs more hyperplanes than a real-valued network. Figure 8.5e depicts another linearly separable case (e.g. with the red hyperplane), but there is no BNN that solves this problem with a single hyperplane even if we allow zero weights.

8.4.3 Efficiency

As a result, we might need to expect a larger network topology for a BNN to perform as good as a corresponding DNN. This fact involves more parameters to train. On top of that, the sheer number of XNOR and bit counting operations increase as

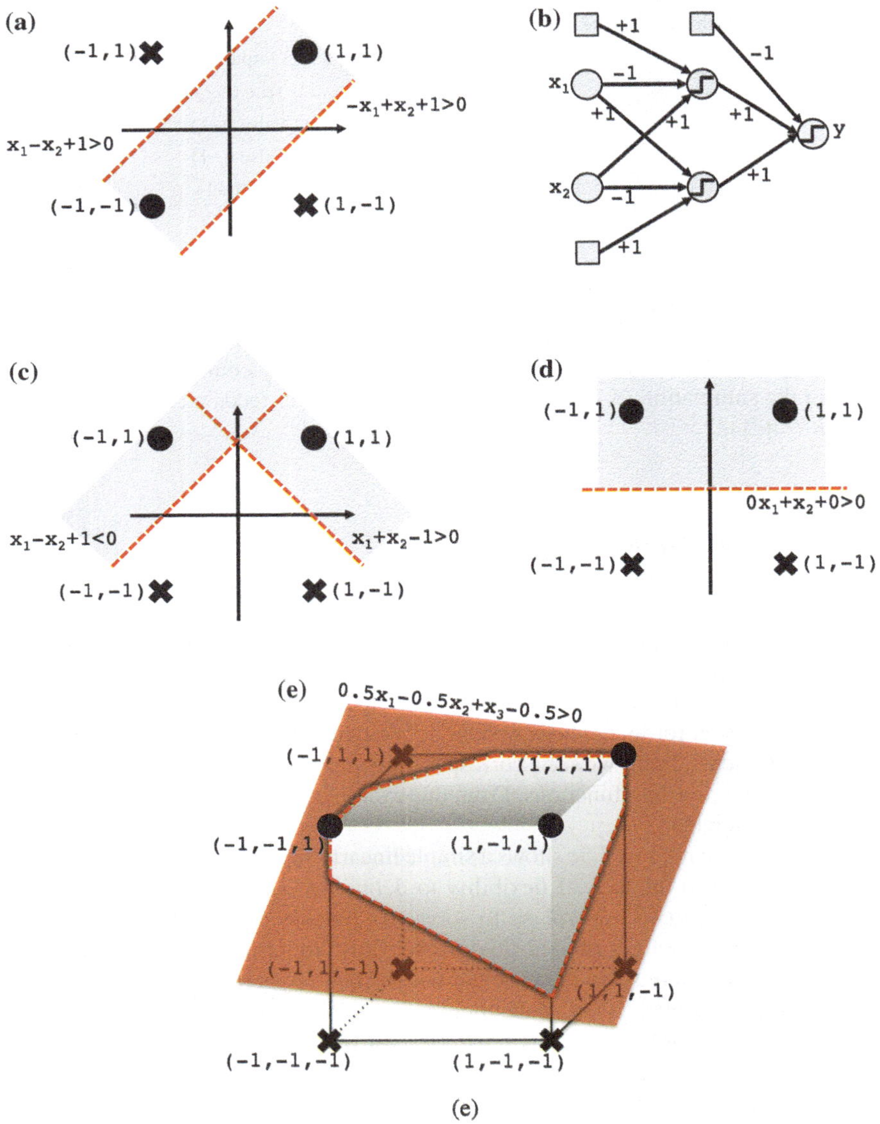

Fig. 8.5 Figures are from [15]. **a** The XOR problem and binary hyperplanes that solve it. **b** A corresponding BNN. **c** A linearly separable problem for which BNN needs a least two hyperplanes. **d** A BNN with zero weights that can define hyperplanes more flexibly and solve the problem. **e** A linearly separable problem which BNN cannot solve with a single hyperplane even with zero weights

well. However, since those operations are much cheaper than their corresponding floating-point or N-bit fixed-point operations, the increased network size does not always mean a more complex feedforward process.

Overall, in the big O notation we see the same order of complexity for both DNN and BNN once their structures are the same. However, if we compare the floating-point or fixed-point multiplications with XNOR for example, we expect that the latter spends much less amount of power. As for the spatial complexity, in theory BNN spends only one bit (or up to two bits depending on how we represent zero weights) per parameter, while an optimized fixed-point DNN calls for N bits to encode a the parameter.

8.5 BNN Training

Training BNN consists of two passes of backpropagation. The main BNN training part is the second round where we use noisy feedforward and the sparsity concept. Since, the second round (final) results rely greatly on the quality of initialization, we first train a regular network in the first round, and then use the results to initialize the actual parameters for BNN training in the second round.

8.5.1 *The First Round: Weight Compressed DNN*

In the first round, we train an ordinary DNN with tanh activation functions as discussed in Sect. 8.2. The only additional step in this part is *weight compression*, with which we can ensure that the weights and biases are bound between -1 and $+1$. The weight compression is done by wrapping the parameters with tanh, and the use the wrapped versions during feedforward. Therefore, feedforward works as follows:

$$\tilde{z}_q^{(0)} = \bar{x}_q, \quad q \in \{1, \ldots, Q\}, \tag{8.18}$$

$$\tilde{a}_j^{(l)} = \sum_{i \in \{1,\ldots,K^{(l)}\}} \tanh\left(\tilde{W}_{ji}^{(l)}\right) \cdot \tilde{z}_i^{(l)} + \tanh\left(\tilde{b}_j^{(l)}\right), \tag{8.19}$$

$$\tilde{z}_j^{(l+1)} = \tanh\left(\tilde{a}_j^{(l)}\right), \tag{8.20}$$

$$\tilde{a}_j^{(L)} = \sum_{i \in \{1,\ldots,K^{(L)}\}} \tanh\left(\tilde{W}_{ji}^{(L)}\right) \cdot \tilde{z}_i^{(L)} + \tanh\left(\tilde{b}_j^{(L)}\right), \tag{8.21}$$

$$\tilde{z}_j^{(L+1)} = \tanh\left(\tilde{a}_j^{(L)}\right). \tag{8.22}$$

Here for the first round, we use a different notation scheme with tilde on top of the symbols to distinguish them from the parameters of an ordinary DNN and the binary ones of BNN. Note that we cannot reuse an existing DNN due to the mandatory use of binarized input $\bar{\mathbf{x}}$, which has replaced the original real-valued vector. Because of the chain rule for this additional wrapping function, now backpropagation has some additional terms, i.e. the derivatives of tanh:

$$\tilde{\delta}_d^{(L)} = (1 - \tanh^2(\tilde{a}_d^{(L)})) \cdot (t'_d - \tilde{z}_d^{(L+1)}), \tag{8.23}$$

$$\tilde{\delta}_i^{(l)} = (1 - \tanh^2(\tilde{a}_d^{(l)})) \cdot \sum_j \tilde{\delta}_j^{(l+1)} \cdot \tanh\left(\tilde{W}_{ji}^{(l)}\right), \tag{8.24}$$

$$\nabla \tilde{W}_{ji}^{(l)} = \tilde{\delta}_j^{(l)} \cdot \tilde{z}_i^{(l)} \cdot \left(1 - \tanh^2\left(\tilde{W}_{ji}^{(l)}\right)\right), \tag{8.25}$$

$$\nabla \tilde{b}_j^{(l)} = \tilde{\delta}_j^{(l)} \cdot \left(1 - \tanh^2\left(\tilde{b}_j^{(l)}\right)\right). \tag{8.26}$$

Through the first round, we construct a real-valued neural network whose input vectors are the same bipolar binaries as in the following BNN. Also, we can make sure that the weights and biases used in the feedforward process are bound between -1 and $+1$.

8.5.2 The Second Round: Noisy Feedforward and Sparsity

Although there is always a trivial solution to this problem by simply enumerating all the mappings of the training set [32], learning a compact and reasonable approximation model to this kind of Boolean mapping functions through BNN is basically an NP-complete combinatorial optimization problem [33]. There has been some early effort, such as μ-perceptron networks, whose network structure is quite limited by allowing an input unit to be connected to only one hidden unit [34]. More recently, Expectation BackPropagation (EBP) algorithm was proposed as a probabilistic approach to the problem, which estimates the posterior probabilities of the discrete weights given the data [16].

The main part of BNN training is done by a procedure called *noisy feedforward*, which has been known in the literature for the fixed-point implementations of neural networks [35, 36]. It is done by using the discrete version of the real-valued network parameters during the feedforward process, while backpropagation does the update on the real-valued ones. In other words, we keep two sets of parameters, real-valued and binary, and use them for backpropagation and forwardpropagation during training, respectively. This intervention during feedforward can reduce the additional error introduced from the noisy nature of the quantized parameters, because the network is aware of it and tries to fix it during training. On the contrary, a naïve quantization cannot, because it discretizes the parameters once all the training is done. The two sets of parameters have relationships as follows:

$$\bar{W}_{ji}^{(l)} = \text{sign}\left(W_{ji}^{(l)}\right) \tag{8.27}$$

$$\bar{b}_j^{(l)} = \text{sign}\left(b_j^{(l)}\right). \tag{8.28}$$

The reason why we keep these two sets is because of the use of the non-differentiable step function, (8.15), (8.17), (8.27), and (8.28). Therefore, we use an approximation, the derivative of tanh, assuming that tanh is a smooth approximation of the step function:

$$\delta_d^{(L)} = (1 - \tanh^2 (a_d^{(L)})) \cdot (t'_d - \bar{z}_d^{(L+1)}), \tag{8.29}$$

$$\delta_i^{(l)} = (1 - \tanh^2 (a_d^{(l)})) \cdot \sum_j \delta_j^{(l+1)} \cdot \bar{W}_{ji}^{(l)}, \tag{8.30}$$

$$\nabla W_{ji}^{(l)} = \delta_j^{(l)} \cdot \bar{z}_i^{(l)} \cdot \left(1 - \tanh^2 (W_{ji}^{(l)})\right), \tag{8.31}$$

$$\nabla b_j^{(l)} = \delta_j^{(l)} \cdot \left(1 - \tanh^2 (b_j^{(l)})\right). \tag{8.32}$$

Note that $W_{ji}^{(l)}$ and $b_j^{(l)}$ have been initialized with the tanh wrapped ones we trained in the first round, e.g. $W_{ji}^{(l)} = \tanh\left(\tilde{W}_{ji}^{(l)}\right)$.

Another important factor in training BNNs is enforcing the sparsity of the weights. For a desired sparsity ρ of the l-th layer weight matrix, before we apply the binarization steps (8.27) and (8.28) we find the boundary u that satisfies the following condition:

$$\rho = \frac{\sum_j J_j^{(l)} + \sum_{ji} I_{ji}^{(l)}}{\left(K^{(l)} + 1\right) \cdot K^{(l+1)}} \tag{8.33}$$

$$I_{ji}^{(l)} = \begin{cases} 1 & \text{if } -u < W_{ji}^{(l)} < u \\ 0 & \text{otherwise} \end{cases} \tag{8.34}$$

$$J_j^{(l)} = \begin{cases} 1 & \text{if } -u < b_j^{(l)} < u \\ 0 & \text{otherwise} \end{cases} \tag{8.35}$$

In other words, the weights that are within the boundaries should be turned off to be zero, which will add an additional rotation option for the hyperplanes:

$$W_{ji}^{(l)} = \begin{cases} 0 & \text{if } -u < W_{ji}^{(l)} < u \\ \text{sign}\left(W_{ji}^{(l)}\right) & \text{otherwise} \end{cases} \tag{8.36}$$

$$b_j^{(l)} = \begin{cases} 0 & \text{if } -u < b_j^{(l)} < u \\ \text{sign}\left(b_j^{(l)}\right) & \text{otherwise} \end{cases} \tag{8.37}$$

Another way to view this is to consider these additional zeros as another quantization level so that the weights are now ternary not binary. However, we would like to think of this matrix as a sparse bipolar binary matrix, because then the feedforward procedure can benefit from the sparse coding, e.g. by skipping the zero weights. Indeed, we found that for the speech enhancement experiments the weight matrices are very sparse to produce the optimal separation results.

8.6 Experimental Results

This section presents some experimental results on single-channel speech denoising tasks, where we compare the performance of the fully bitwise neural networks and their corresponding floating-point networks.[5] Because the goal of the comparison is to see if BNNs can catch up the performance of a proper DNN, we focus on a relatively small, but big enough data set to showcase the merit of BNN. We consider the first round training results as our baseline DNN networks although their input is not a real-valued spectrum, but a bit string from the Quantization-and-Dispersion procedure. It is a fair comparison in the sense that we have to do our best to produce good first round results to guarantee the performance of the final BNNs from the second round, which are using the first round results to initialize their parameters. However, a comprehensive DNN without the QaD procedure will be slightly better than our baseline.

8.6.1 The Data Set

We construct a dataset from the TIMIT corpus and the ten non-stationary noise types used in [37]. For training, we randomly chose five utterances from a speaker in the training set. We repeat this for ten female speakers. For each of the 50 randomly chosen clean speech utterances like this, we mix it up with one of the ten noise types to build a set of 500 noisy utterances in total. As for the test signals, we go through the same procedure with only five random speakers. It is made sure that the section of the noise signal used for training mixtures does not overlap with that for test mixtures. The Signal-to-Noise Ratio (SNR) is set to be 0 dB.

8.6.2 Pre-processing

All signals go through Short-Time Fourier Transform (STFT) with a Hann window of 1024 points and a 75% overlap between frames. The magnitudes of a mixture spectrum are used as the input vector after converting them using a QaD procedure with 4 bits per magnitude. Therefore, the total number of binary input variables is $513 \times 4 = 2052$. We calculate IBMs from (8.8), and then convert their range from between 0 and 1 to between -1 and $+1$. The neural networks are trained to predict a 513-dimensional IBM vector from a 2052-dimensional QaD binary vector.

---

[5]The results of this section are mostly from [15].

8.6.3 The Setup for the First Round

- *Structure*: We build two weight-compressed networks with two hidden layers, but one has 1024 hidden units per layer (1024×2) and the other has 2048 (2048×2)
- *Dropout*: Dropout parameters are set to be 0.95 for the input layer and 0.8 for the other hidden layers.
- *Mini batch*: 100 samples
- *Momentum*: 0.95
- *Learning rate*: 10^{-7}
- *Number of epoch*: 5,000

8.6.4 The Setup for the Second Round

For the second round we stick to the same setup with the first round except the following ones:

- *Sparsity* ρ: 0.98
- *Learning rate*: 10^{-6}

8.6.5 Discussion

Figure 8.6 compares the first-round results (as our DNN baseline) and the second-round results, which starts from the first-round, but then turns all the parameters into bipolar binaries or zeros (BNN). We use the popular source separation quality measurements as proposed in [38]. First, in (a) we see that the separation quality in terms of Signal-to-Interference Ratio (SIR) for both DNN and BNN are very good although DNN clearly outperforms BNN by more than 3.5 dB. However, usually a higher SIR does not always mean a better separation, because the separation algorithm tends to introduce artifacts along the way, which can be measured by Signal-to-Artifiacts Ratio (SAR). As seen in (a) the SAR difference between DNN and BNN are not very significant. Eventually, the overall separation measure, Signal-to-Distortion Ratio (SDR), says that the difference between BNN and DNN is 0.6 dB when they share the same network structure, which is acceptable considering the amount of saving BNN introduces.

In Fig. 8.6b we compare the results from larger networks where we double the number of hidden units. Clearly, we see that the SIR value of DNN does not really change, although BNN catches up. Once again, we have to be careful about the overall separation performance which can be deteriorated by a too aggressive separation, but in (b) we see that SAR values go up, too. Consequently, we see that the performance of BNN with a larger network structure is comparable to that of smaller DNN

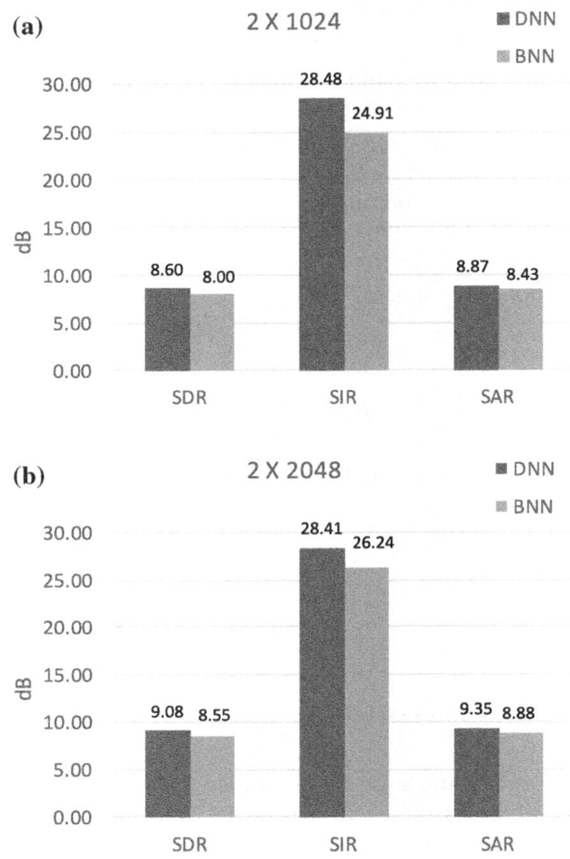

Fig. 8.6 Speech enhancement results using two different network structures (**a**) 1024 × 2 (**b**) 2048 × 2

(8.55 versus 8.60 dB in SDR). Of course the number of the weights in the larger network is roughly four times of the smaller one, but since BNNs are based on sparse bitwise computation, we believe that this does not mean that the large BNN is more complex than the smaller DNN.

8.7 Conclusion

In this chapter we presented a new efficient way to perform feedforward in DNNs by using their bitwise versions, where we replace multiplication and addition with bitwise XNOR and bit counting operations. In order to convert a real-valued network into a BNN, we define these operations on bit string input and outputs, which can be seen as binary features extracted from the original raw signals. During training, these bitwise logic along with the binary versions of the signal inject additional quantization noise so that the network is aware of the drastic quantization noise and tries to fix it through backpropagation. We shared some preliminary speech enhancement results where BNN shows a convincing performance while ensuring promising improvement in terms of efficiency. A detailed complexity analysis on real hardware implementations will follow as future work.

References

1. P. Smaragdis, J.C. Brown, Non-negative matrix factorization for polyphonic music transcription, in *Proceedings of IEEE Workshop on Applications of Signal Processing to Audio and Acoustics, New Paltz, NY* (2003), pp. 177–180
2. D.D. Lee, H.S. Seung, Learning the parts of objects by non-negative matrix factorization. Nature **401**, 788–791 (1999)
3. D.D. Lee, H.S. Seung, Algorithms for non-negative matrix factorization, in *Advances in Neural Information Processing Systems (NIPS)*, vol. 13. (2001)
4. T. Hofmann, Probablistic latent semantic analysis, in *Proceedings of the International Conference on Uncertainty in Artificial Intelligence (UAI)* (1999)
5. T. Hofmann, Probablistic latent semantic indexing, in *Proceedings of the ACM SIGIR Conference on Research and Development in Information Retrieval (SIGIR)* (1999)
6. B. Raj, P. Smaragdis, Latent variable decomposition of spectrograms for single channel speaker separation, in *Proceedings of the IEEE Workshop on Applications of Signal Processing to Audio and Acoustics* (2005), pp. 17–20
7. M. Kim, P. Smaragdis, G.J. Mysore, Efficient manifold preserving audio source separation using locality sensitive hashing, in *Proceedings of the IEEE International Conference on Acoustics, Speech, and Signal Processing (ICASSP)* (2015), pp. 479–483
8. M. Kim, P. Smaragdis, Efficient neighborhood-based topic modeling for collaborative audio enhancement on massive crowdsourced recordings, in *Proceedings of the IEEE International Conference on Acoustics, Speech, and Signal Processing (ICASSP)* (2016), pp. 41–45
9. Y. Xu, J. Du, L.-R. Dai, C.-H. Lee, An experimental study on speech enhancement based on deep neural networks. IEEE Signal Process. Lett. **21**(1), 65–68 (2014)
10. P. Huang, M. Kim, M. Hasegawa-Johnson, P. Smaragdis, Joint optimization of masks and deep recurrent neural networks for monaural source separation. IEEE/ACM Trans. Audio Speech Lang. Process. **23**(12), 2136–2147 (2015)
11. D.S. Williamson, Y. Wang, D.L. Wang, Reconstruction techniques for improving the perceptual quality of binary masked speech. J. Acoust. Soc. Am. **136**, 892–902 (2014)
12. J. LeRoux, J.R. Hershey, F. Weninger, Deep NMF for speech separation, in *Proceedings of the IEEE International Conference on Acoustics, Speech, and Signal Processing (ICASSP)* (2015), pp. 66–70
13. J.R. Hershey, Z. Chen, J. LeRoux, S. Watanabe, Deep clustering: discriminative embeddings for segmentation and separation, in *Proceedings of the IEEE International Conference on Acoustics, Speech, and Signal Processing (ICASSP)* (2016), pp. 31–35
14. M. Kim, P. Smaragdis, Bitwise neural networks, in *International Conference on Machine Learning (ICML) Workshop on Resource-Efficient Machine Learning* (2015)
15. M. Kim, Audio computing in the wild: frameworks for big data and small computers. Ph.D. dissertation, University of Illinois at Urbana-Champaign, 2016
16. D. Soudry, I. Hubara, R. Meir, Expectation backpropagation: parameter-free training of multilayer neural networks with continuous or discrete weights, in *Advances in Neural Information Processing Systems (NIPS)* (2014), pp. 963–971

17. M. Rastegari, V. Ordonez, J. Redmon, A. Farhadi, XNOR-Net: imagenet classification using binary convolutional neural networks (2016), arXiv preprint arXiv:1603.05279
18. I. Hubara, M. Courbariaux, D. Soudry, R. El-Yaniv, Y. Bengio, Binarized neural networks, in *Advances in Neural Information Processing Systems* (2016), pp. 4107–4115
19. F. Weninger, H. Erdogan, S. Watanabe, E. Vincent, J. LeRoux, J.R. Hershey, B. Schuller, Speech enhancement with LSTM recurrent neural networks and its application to noise- robust ASR, in *Proceedings of the International Conference on Latent Variable Analysis and Signal Separation (LVA/ICA)* (2015), pp. 91–99
20. Y. Wang, D.L. Wang, Towards scaling up classification-based speech separation. IEEE Trans. Audio Speech Lang. Process. **21**(7), 1381–1390 (2013)
21. A. Narayanan, D.L. Wang, Ideal ratio mask estimation using deep neural networks for robust speech recognition, in *Proceedings of the IEEE International Conference on Acoustics, Speech, and Signal Processing (ICASSP)* (2013), pp. 7092–7096
22. D.S. Williamson, Y. Wang, D.L. Wang, A two-stage approach for improving the perceptual quality of separated speech, in *Proceedings of the IEEE International Conference on Acoustics, Speech, and Signal Processing (ICASSP)* (2014), pp. 7084–7088
23. M. Kim, P. Smaragdis, Adaptive denoising autoencoders: a fine-tuning scheme to learn from test mixtures, in *Proceedings of the International Conference on Latent Variable Analysis and Signal Separation (LVA/ICA)* (2015), pp. 100–107
24. H. Erdogan, J.R. Hershey, S. Watanabe, J. Le Roux, Phase-sensitive and recognition- boosted speech separation using deep recurrent neural networks, in *Proceedings of the IEEE International Conference on Acoustics, Speech, and Signal Processing (ICASSP)* (2015), pp. 708–712
25. N. Srivastava, G. Hinton, A. Krizhevsky, I. Sutskever, R. Salakhutdinov, Dropout: a simple way to prevent neural networks from overfitting. J. Mach. Learn. Res. **15**(1), 1929–1958 (2014)
26. J. Yagnik, D. Strelow, D.A. Ross, R. Lin, The power of comparative reasoning, in *Proceedings of the International Conference on Computer Vision (ICCV)* (2011), pp. 2431–2438
27. T. Dean, M.A. Ruzon, M. Segal, J. Shlens, S. Vijayanarasimhan, J. Yagnik, Fast, accurate detection of 100,000 object classes on a single machine, in *Proceedings of the IEEE International Conference on Computer Vision and Pattern Recognition (CVPR)* (2013), pp. 1814–1821
28. P. Indyk, R. Motwani, Approximate nearest neighbor-towards removing the curse of dimensionality, in *Proceedings of the Annual ACM Symposium on Theory of Computing (STOC)* (1998), pp. 604–613
29. Y. Weiss, A. Torralba, R. Fergus, Spectral hashing, in *Advances in Neural Information Processing Systems (NIPS)* (2009), pp. 1753–1760
30. R.R. Salakhutdinov, G.E. Hinton, Semantic hashing, in *SIGIR Workshop on Information Retrieval and Applications of Graphical Models* (2007)
31. S. Lloyd, Least squares quantization in PCM. IEEE Trans. Inf. Theory **28**(2), 129–137 (1982)
32. W.S. McCulloch, W.H. Pitts, A logical calculus of the ideas immanent in nervous activity. Bull. Math. Biophys. **5**(4), 115–133 (1943)
33. L. Pitt, L.G. Valiant, Computational limitations on learning from examples. J. Assoc. Comput. Mach. **35**, 965–984 (1988)
34. M. Golea, M. March, T.R. Hancock, On learning μ-perceptron networks with binary weights, in *Advances in Neural Information Processing Systems (NIPS)* (1992), pp. 591–598
35. E. Fiesler, A. Choudry, H.J. Caulfield, Weight discretization paradigm for optical neural networks, in *The Hague, 12–16 April. International Society for Optics and Photonics* (1990), pp. 164–173
36. K. Hwang, W. Sung, Fixed-point feedforward deep neural network design using weights $+1$, 0, and -1, in *2014 IEEE Workshop on Signal Processing Systems (SiPS)* (2014)
37. Z. Duan, G.J. Mysore, P. Smaragdis, Online PLCA for real-time semi-supervised source separation, in *Proceedings of the International Conference on Latent Variable Analysis and Signal Separation (LVA/ICA)* (2012), pp. 34–41
38. E. Vincent, C. Févotte, R. Gribonval, Performance measurement in blind audio source separation. IEEE Trans. Audio Speech Lang. Process. **14**(4), 1462–1469 (2006)

Chapter 9
DNN Based Mask Estimation for Supervised Speech Separation

Jitong Chen and DeLiang Wang

Abstract This chapter introduces deep neural network (DNN) based mask estimation for supervised speech separation. Originated in computational auditory scene analysis (CASA), we treat speech separation as a mask estimation problem. Given a time-frequency (T-F) representation of noisy speech, the ideal binary mask (IBM) or ideal ratio mask (IRM) is defined to differentiate speech-dominant T-F units from noise-dominant ones. Mask estimation is then formulated as a problem of supervised learning, which learns a mapping function from acoustic features extracted from noisy speech to an ideal mask. Three main aspects of supervised learning are learning machines, training targets, and features, which are discussed in separate sections. Subsequently, we describe several representative supervised algorithms, mainly for monaural speech separation. For supervised separation, generalization to unseen conditions is a critical issue. The generalization capability of supervised speech separation is also discussed.

9.1 Speech Separation Problem

The human auditory system has the remarkable ability in separating one sound source from others. In an acoustic environment like a cocktail party, we are able to follow one speaker while filtering out others without much effort. Speech separation is called the "cocktail party problem" by Cherry in his 1953 paper [1]. The ability to separate speech from background noise is crucial for our daily communication. The speech of interest is usually corrupted by additive noises from other sound sources and reverberation from surface reflections. Although humans perform speech separation

J. Chen (✉) · D. Wang
Department of Computer Science and Engineering, Center for Cognitive and Brain Sciences, The Ohio State University, 2015 Neil Avenue, Columbus, OH 43210, USA
e-mail: chenjit@cse.ohio-state.edu

D. Wang
e-mail: dwang@cse.ohio-state.edu

S. Makino (ed.), *Audio Source Separation*, Signals and Communication Technology, https://doi.org/10.1007/978-3-319-73031-8_9

with ease, it has been very challenging to construct an automatic system to match the human auditory system in this fundamental task.

Speech separation has a wide range of applications such as robust automatic speech and speaker recognition, noise reduction for hearing aids and cochlea implants, and enhanced mobile communication. Driven by these applications, researchers have developed many techniques over the past decades. Depending on the number of sensor recordings, we categorize these techniques into multi-microphone speech separation and monaural speech separation. The dominant approach for array-based processing is beamforming [2]. Compared to array processing, monaural processing is more flexible but potentially more challenging as it operates without spatial cues. Two traditional approaches for monaural processing are speech enhancement [3] and computational auditory scene analysis (CASA) [4]. The former analyzes general statistics of speech and noise, and typically requires a noise estimate. The latter is based on auditory scene analysis principles [5] and exploits perceptual cues, such as pitch and onset.

A emerging approach is to use supervised learning to address speech separation, where the discriminative characteristics of speech and noise are learned from training data. The concept of time-frequency masking in CASA [6] has led to the original formulation of speech separation as a supervised learning problem. Supervised speech separation represents a data-driven approach and benefits from the rapid advances in machine learning.

Supervised speech separation started from the concept of the ideal binary mask. Inspired by the masking phenomenon in auditory perception and the exclusive allocation principle in auditory scene analysis [5], in 2001 Hu and Wang first suggested the ideal binary mask (IBM) as a main goal of CASA [7] (see also [8, 9]). The idea is to retain parts of a mixture where the target sound is stronger than the background noise and discard the rest. The IBM is defined from the time-frequency (T-F) representation of a mixture as follows:

$$IBM\,(t,f) = \begin{cases} 1, & \text{if } SNR(t,f) > LC \\ 0, & \text{otherwise} \end{cases} \tag{9.1}$$

where t denotes time and f denotes frequency. The IBM assigns the value 1 to a T-F unit if the local SNR within the unit exceeds a local criterion (LC), and 0 otherwise. Ideal binary masking as a way of speech separation is illustrated in Fig. 9.1. Ideal binary masking has been shown to dramatically improve speech intelligibility for normal-hearing (NH) listeners and hearing-impaired (HI) listeners [10–13].

With the IBM as the computational goal, speech separation can be naturally formulated as a binary classification problem. This formulation of the speech separation problem as supervised learning has proven to be consequential, as it now opens the separation problem to treatment by a variety of pattern classification and function approximation (or regression) algorithms. The first study in supervised speech separation was conducated in the binaural domain by Roman et al. [14, 15].

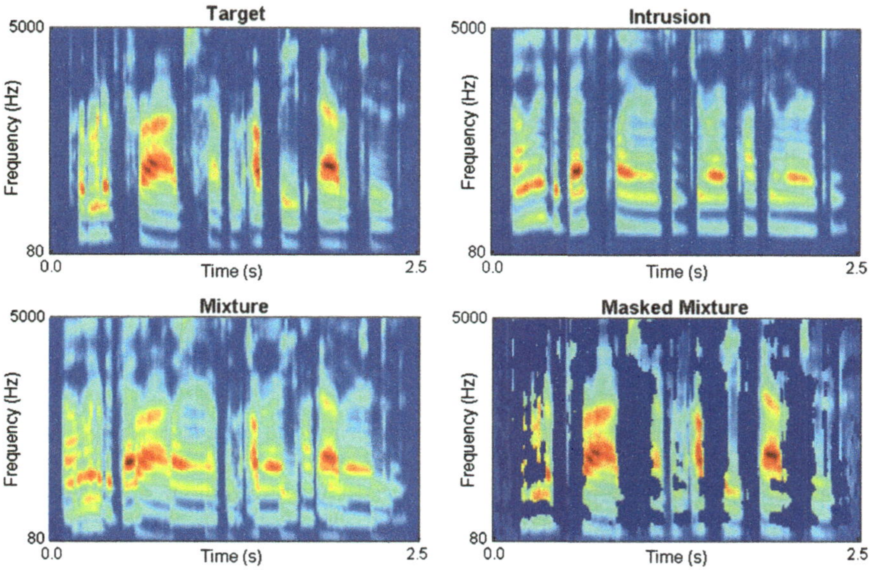

Fig. 9.1 IBM illustration. The top left plot shows a cochleagram of a target utternce, the top right shows an interfering utterance, the bottom left shows the mixture, and the bottom right shows the IBM-masked mixture

Like any supervised learning problem, supervised speech separation has three main aspects: learning machines, training targets, and features. These topics will be discussed in the next three sections. For learning machines, we will focus on DNNs (including recurrent neural networks). Section 9.5 presents representative algorithms, with a focus on monaural speech separation, and discusses the generalization issue. Section 9.6 concludes the chapter. We should note that speech separation and speech enhancement are used interchangeably in this chapter, particularly for separating speech from nonspeech interference.

9.2 Classifiers and Learning Machines

Recent advances in machine learning and neural networks have demonstrated the power of deep neural networks (DNNs) in many tasks such as image classification, automatic speech recognition and machine translation. In this section we introduce two types of deep neural networks that are effective for mask estimation in supervised speech separation. They are feedforward multilayer perceptrons (MLPs) and recurrent networks based on long short-term memory (LSTM).

9.2.1 Multilayer Perceptrons

Multiplayer perceptrons, exemplifying feedforward networks, are extended from Rosenblatt's perceptrons (or simple perceptrons). Multilayer perceptrons are powerful learning machines and can approximate any nonlinear function. A two-hidden-layer MLP is shown in Fig. 9.2. An MLP is typically trained with the backpropagation algorithm where the network weights are adjusted to minimize the prediction error, which is measured by a cost (loss) function. For example, when an MLP is used for regression, a common cost function is mean square error (MSE):

$$E(n) = \frac{1}{2}\sum_k [d_k(n) - y_k(n)]^2 \tag{9.2}$$

where $E(n)$, $d_k(n)$ and $y_k(n)$ denote the cost, desired output and predicted output at iteration n, respectively. The output layer is indexed by k. To run the backpropagation algorithm, activations are computed successively from the first hidden layer to the output layer where the prediction error is measured by (9.2). Then the error is backpropagated to adjust the weights. The backpropagation algorithm performs gradient descent over the weights to minimize the error or cost at the output layer.

The representational power of an MLP increases as the number of layers increases. However, a deep MLP is usually difficult to train from random initialization because of the gradient vanishing problem. With vanishing (small) gradients, the lower layers (near the input end) do not change their weights much during backpropagation and therefore do not perform feature learning effectively. One remedy is to perform layer-wise unsupervised pretraining with unlabeled data to properly initialize the network and then fine-tune it with labeled data. Hinton et al. [16] proposed restrictive Boltzmann machines (RBMs) to pretrain DNNs layer by layer and found that it improves subsequent supervised learning [16]. Another remedy is to use rectified linear unit (ReLU) [17] to replace the traditional sigmoid activation function. The

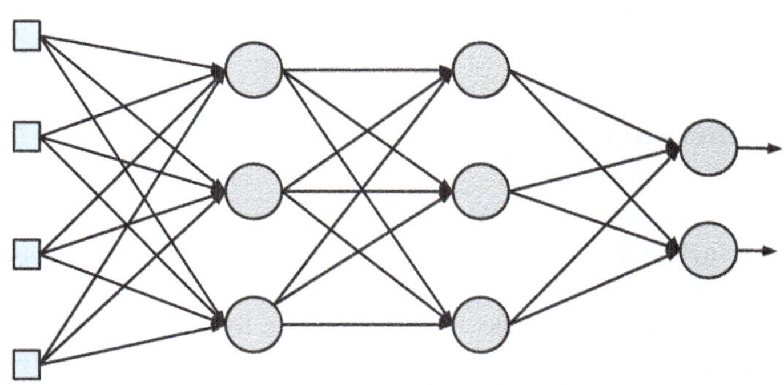

Fig. 9.2 Diagram of a multilayer perceptron

ReLU is defined as follows:

$$f(x) = \begin{cases} x, & \text{if } x > 0 \\ 0, & \text{otherwise} \end{cases} \tag{9.3}$$

Unlike the sigmoid function, which has a small gradient when the input value is large, the ReLU has the derivative of 1 for all positive values, and therefore alleviates the gradient vanishing problem. Recent practice suggests that DNNs with the ReLU can be successfully trained without unsupervised pretraining when sufficiently large training data sets are available.

9.2.2 Recurrent Neural Networks

Recurrent neural networks (RNNs) allow recurrent (feedback) connections, typically between hidden units. Unlike feedforward networks, which process each input sample independently, RNNs treat input samples as a sequence and model the changes over time. For a speech signal, a current frame is influenced by the previous ones. Therefore, RNNs may be employed to naturally learn the temporal dynamics of speech. To illustrate the information flow, we show a standard RNN unrolled over two time steps in Fig. 9.3. Clearly, the features extracted from the previous time step are used to compute the hidden activations at the current time step.

The recurrent connections are typically trained with backpropagation through time (BPTT). However, such RNN training is susceptible to the vanishing or exploding gradient problem [19]. To alleviate this problem, long short-term memory (LSTM)

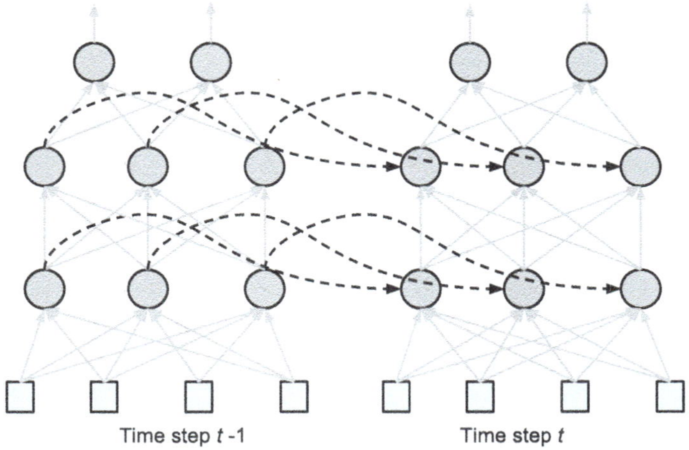

Fig. 9.3 An RNN unrolled over two time steps. The dashed lines indicate recurrent connections

introduces memory cells to store information from the past. As shown in Fig. 9.4, an LSTM block has three gates, namely input gate, forget gate and output gate. The input gate and forget gate control how the memory cell should be updated, and the output gate modulates the output of the block. The mechanism of an LSTM block can be described by the following equations:

$$i_t = \sigma(W_{ix}x_t + W_{ih}h_{t-1} + b_i) \tag{9.4}$$

$$f_t = \sigma(W_{fx}x_t + W_{fh}h_{t-1} + b_f) \tag{9.5}$$

$$o_t = \sigma(W_{ox}x_t + W_{oh}h_{t-1} + b_o) \tag{9.6}$$

$$z_t = g(W_{zx}x_t + W_{zh}h_{t-1} + b_z) \tag{9.7}$$

$$c_t = f_t \odot c_{t-1} + i_t \odot z_t \tag{9.8}$$

$$h_t = o_t \odot g(c_t) \tag{9.9}$$

$$\sigma(s) = \frac{1}{1 + e^{-s}} \tag{9.10}$$

$$g(s) = \frac{e^s - e^{-s}}{e^s + e^{-s}} \tag{9.11}$$

where i_t, f_t and o_t denote the input gate, forget gate and output gate at time t, respectively. x_t, z_t, c_t, h_t respectively represent input, block input, memory cell, and hidden activation. W's and b's represent weights and biases, respectively, and $\odot$ denotes element-wise multiplication or the gating operation.

The combination of gates and memory cells facilitates the information flow over time. When LSTM is properly trained, the input gates and forget gates maintain relevant contextual information in memory cells to improve target estimation. LSTM has been shown to be successful in learning long-term dependencies in many applications such as language modeling [20], machine translation [21] and automatic speech recognition [22]. We will revisit LSTM later in Sect. 9.5.1.

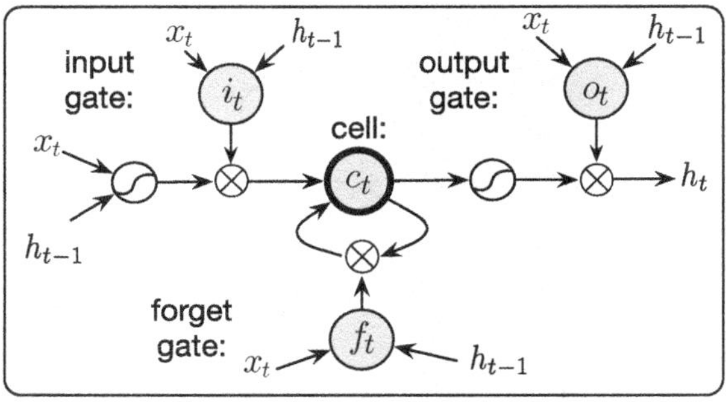

Fig. 9.4 Diagram of an LSTM block (from [18]). A circled cross denotes element-wise multiplication or gating operation, and a circled wave denotes hyperbolic tangent function

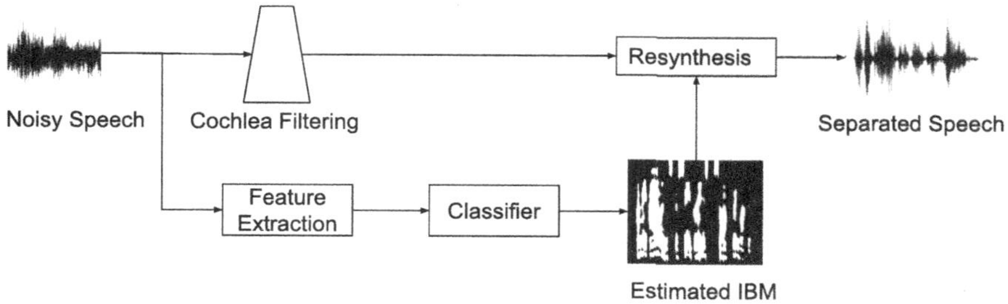

Fig. 9.5 Supervised speech separation based on classification for IBM estimation

9.3 Training Targets

In supervised speech separation, it is important to define a training target properly. Different training targets lead to different mapping functions from noisy features to separated speech. They may also lead to different levels of generalization. There are primarily two categories of training targets: masking-based targets and mapping-based targets.

Among masking-based targets, the IBM is the first and the most commonly used one. As we mentioned in Sect. 9.1, the IBM notion has led to the original formulation of speech separation as supervised learning. The IBM is typically defined on a cochleagram representation. To compute the cochleagram of a signal, we typically pass an input signal to a gammatone filterbank and compute energies in T-F units with a 20-ms window length and a 10-ms window shift [4]. To mimic human cochlea filtering, the center frequencies of a gammatone filterbank are uniformly spaced on the equivalent rectangular bandwidth (ERB) scale, leading to higher resolutions at low frequencies. The IBM can also be defined on a spectrogram or any other time-frequency representation. IBM estimation amounts to classifying T-F units of noisy speech as noise-dominant or speech-dominant. For separation, the noise-dominant T-F units are removed to suppress the background noise. An example of the IBM is shown with a 64-channel filterbank in Fig. 9.6a. A supervised speech separation system using classification-based IBM estimation is illustrated in Fig. 9.5. Acoustic features are first extracted from noisy speech (see Sect. 9.4) and sent to a classifier to produce an IBM estimate, which is then used to weight the responses from the gammatone filterbank. Finally, separated speech is resynthesized by summing processed signals across frequency channels. IBM estimation has been shown to be effective in separating target speech from background noise [23–25]. Besides the IBM, the target binary mask (TBM) has also been shown to improve speech intelligibility [26] and suggested to be a training target [27]. The TBM is a binary mask computed by comparing the target speech energy in each T-F unit with a reference speech-shaped noise. An example of the TBM is shown in Fig. 9.6b.

Instead of a binary target label on each T-F unit, a soft target label is provided by the ideal ratio mask (IRM) [28–30]:

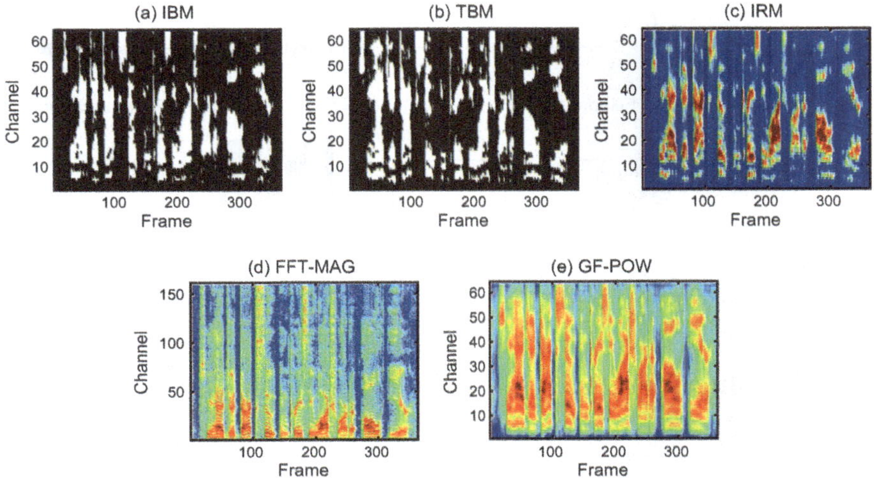

Fig. 9.6 Illustration of several training targets for a TIMIT utterance mixed with a factory noise at −5 dB SNR

$$IRM(t,f) = \sqrt{\frac{S(t,f)^2}{S(t,f)^2 + N(t,f)^2}} = \sqrt{\frac{SNR(t,f)}{SNR(t,f)+1}} \tag{9.12}$$

where $S(t,f)^2$ and $N(t,f)^2$ denote speech energy and noise energy within a T-F unit, respectively. Note that instead of a square root, other roots can be used in the IRM definition. An example of the IRM is shown in Fig. 9.6c. Different from the IBM, the IRM partially preserves a T-F unit according to the unit SNR. A recent study has shown that ratio masking leads to better speech quality than binary masking [30], and the IRM has been argued to be a better target than the IBM [30, 31].

Ideal masks can be similarly defined on a spectrogram or the short-time Fourier transform (STFT) domain. To compute the STFT, we typically use a 20-ms window length and a 10-ms window shift. With the STFT of clean speech and noisy speech, the FFT-MASK is defined as follows:

$$FFT\text{-}MASK(t,f) = \frac{S_{FFT}(t,f)}{Y_{FFT}(t,f)} \tag{9.13}$$

where, within a T-F unit, $S_{FFT}(t,f)$ and $Y_{FFT}(t,f)$ denote clean and noisy spectral magnitudes, respectively. For speech separation, the clean spectral magnitude is estimated by applying the estimated FFT-MASK, a ratio mask, to the noisy spectral magnitude. The estimated magnitude is then combined with noisy phase to derive separated speech.

In the mapping-based approach, the training targets are typically spectral representations of clean speech. A simple target is the STFT spectral magnitude (FFT-MAG) of clean speech. While the FFT-MAG is based on the spectrogram representation, another such training target called gammatone frequency power spectrum (GF-POW) is based on a cochleagram representation. Examples of the FFT-MAG and GF-POW are shown in Figs. 9.6d, e, respectively.

Wang et al. [30] have compared masking-based and mapping-based training targets in terms of speech separation performance. To benchmark the performance, they also compared with a speech enhancement algorithm [32] and a supervised non-negative matrix factorization (NMF) algorithm called ASNA-NMF [33]. In this systematic examination of training targets, training and test mixtures are created from TIMIT utterances [34] and five noises at −5 dB, 0 dB and 5 dB SNR. A DNN with three hidden layers (each layer has 1024 units) was trained and tested on different utterances and different segments of the same noises. Separated speech was evaluated using the short-time objective intelligibility (STOI) score [35] and the perceptual evaluation of speech quality (PESQ) score [36]. STOI predicts speech intelligibility by comparing envelopes of separated speech and clean speech and has a value range of [0, 1], roughly corresponding to percent correct. PESQ measures the quality of separated speech and has a value range of [−0.5, 4.5] (higher is better). We show the STOI and PESQ scores for a factory noise in Figs. 9.7 and 9.8, respectively. For the two binary masks, IBM estimation performs better in PESQ than TBM estimation. Compared to binary masking, ratio masking performs better in speech quality. Furthermore, the FFT-MASK is better than the FFT-MAG in terms of both speech intelligibility and quality. This is due to the observation that FFT-MASK estimation mainly involves a one-to-one mapping, whereas FFT-MAG estimation involves a many-to-one mapping as the FFT-MAG is insensitive to interference and mixture SNR. Many-to-one mapping appears harder to learn than one-to-one mapping.

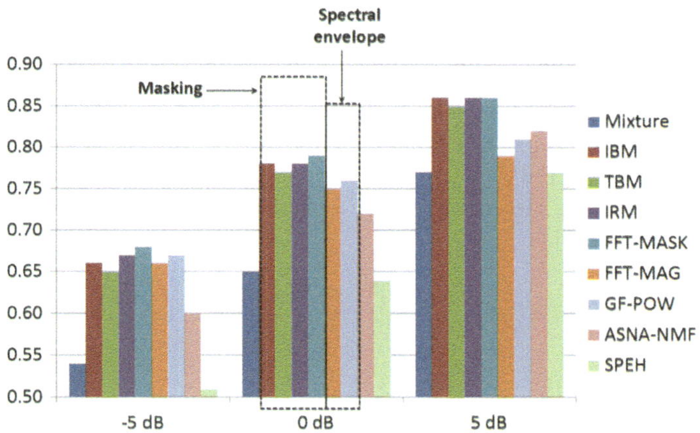

Fig. 9.7 Comparison of training targets in terms of STOI. Clean speech is mixed with a factory noise at −5 dB, 0 dB and 5 dB SNR. Results for different kinds of targets as well as a speech enhancement (SPEH) algorithm and an NMF method are highlighted for 0 dB mixtures

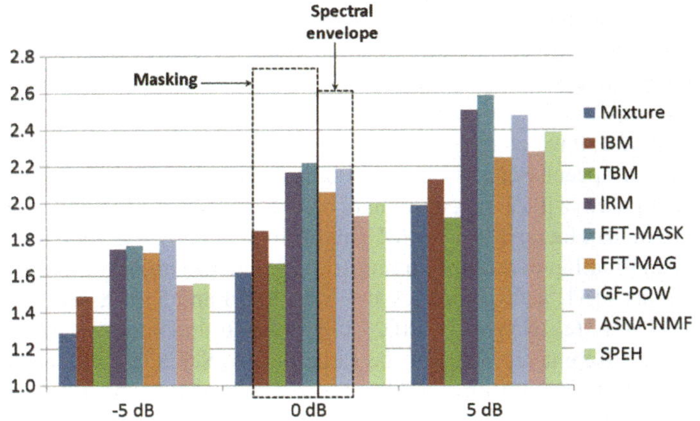

Fig. 9.8 Comparison of training targets in terms of PESQ. Clean speech is mixed with a factory noise at −5 dB, 0 dB and 5 dB SNR. See Fig. 9.7 caption for further explanations

Besides the above targets, we introduce three more recently proposed training targets (see also [37]). The first one is signal approximation which uses the following cost function [38]:

$$SA(t,f) = [RM(t,f)Y_{FFT}(t,f) - S_{FFT}(t,f)]^2 \tag{9.14}$$

where $Y_{FFT}(t,f), S_{FFT}(t,f)$ denote noisy and clean spectral magnitudes, respectively. $RM(t,f)$ can be viewed as an estimate of the IRM, which is defined on the STFT domain instead of the cochleagram domain. With signal approximation as the target, a learning machine is typically trained in two stages. In the first stage, the IRM is used as the training target to initialize the learning machine. In the second stage, the difference of masked noisy magnitude and clean magnitude is minimized. The goal is to obtain a mask estimator which leads to a good estimate of clean spectral magnitudes. Signal approximation has been shown to give some improvement over direct IRM estimation [38].

The second training target is the phase-sensitive (PS) ideal ratio mask, which is defined as follows [39]:

$$FFT\text{-}MASK_{PS}(t,f) = \frac{S_{FFT}(t,f)}{Y_{FFT}(t,f)} \cos\theta \tag{9.15}$$

where θ is the phase difference between clean speech and noisy speech within the corresponding T-F unit. Experimental results have shown that the phase sensitive target leads to a better estimate of clean speech than FFT-MASK [39].

The third training target is the complex ideal ratio mask (cIRM). The idea is to define an ideal mask that perfectly reconstructs clean speech from noisy speech:

$$S = M * Y \tag{9.16}$$

where M denotes the ideal mask to be sought for. S and Y are clean and noisy FFT signals, respectively, and $*$ represents multiplication in the complex domain. To reveal the structure in the complex data, the cIRM is defined in the Cartesian complex domain instead of the magnitude-phase domain:

$$M = \frac{Y_r S_r + Y_i S_i}{Y_r^2 + Y_i^2} + i \frac{Y_r S_i - Y_i S_r}{Y_r^2 + Y_i^2} \tag{9.17}$$

where S_r and S_i are real and imaginary components of clean speech, respectively, and Y_r and Y_i denote the real and imaginary components of noisy speech, respectively. Since M has an unbounded value, not favorable for training, the real and imaginary components are compressed to the range of $[-K, K]$ with hyperbolic tangent function:

$$cIRM_x = K \frac{1 - e^{-cM_x}}{1 + e^{-cM_x}}, x \in \{r, i\} \tag{9.18}$$

where c controls steepness and K controls the value range. cIRM estimation has been shown to improve speech quality over IRM estimation [40].

9.4 Features

Features are clearly important for supervised speech separation. Mask estimation depends on the discriminative power of acoustic features. Early studies use a few features such as interaural time differences (ITD) and interaural intensity differences (IID) [41] in binaural spearation, and pitch-based features [23] and amplitude modulation spectrogram (AMS) [24] in monaural separation. A subsequent study [42] explores more monaural features including mel-frequency cepstral coefficient (MFCC), gammatone frequency cepstral coefficient (GFCC) [43, 44], perceptual linear prediction (PLP) [45], and relative spectral transform PLP (RASTA-PLP) [46].

We have carried out a study to evaluate an extensive list of acoustic features for speech separation at low SNRs [47]. The features have been previously used for robust automatic speech recognition and classification-based speech separation. The feature list includes two mel-domain features, two linear prediction features, three gammatone-domain features, one zero-crossing feature, three autocorrelation features, two medium-term filtering features, two modulation features and a set of pitch-based features. The two mel-domain features are MFCC and delta-spectral cepstral coefficient (DSCC) [48], which is similar to MFCC except that a delta operation is applied to mel-spectrum. The two linear predication features are PLP and RASTA-PLP. The three gammatone-domain features are gammatone feature (GF), GFCC, and gammatone frequency modulation coefficient (GFMC) [49]. GF is com-

Table 9.1 Performance of a list of acoustic features in terms of classification accuracy (in %) for six noises with ARMA post-processing at −5 dB SNR. Boldface indicates best result (from [47])

	Factory	Babble	Engine	Cockpit	Vehicle	Tank	Average
MRCG	**88.0**	**79.5**	**92.2**	**92.4**	**89.9**	**90.5**	**88.8**
GF	87.6	77.4	91.9	92.1	89.9	90.2	88.2
GFCC	87.7	78.3	91.3	91.9	89.2	89.7	88.0
DSCC	86.6	77.2	90.5	90.9	88.8	88.8	87.1
MFCC	86.5	77.5	90.2	91.1	88.8	88.6	87.1
PNCC	86.6	77.2	90.1	90.9	88.6	88.3	87.0
PLP	86.9	77.4	89.5	90.9	88.7	88.2	87.0
AC-MFCC	86.7	77.0	89.3	90.5	88.7	88.1	86.7
RAS-MFCC	86.9	76.9	89.4	90.9	87.8	88.1	86.7
GFB	86.3	74.5	89.3	90.9	87.6	87.6	86.0
ZCPA	85.4	75.2	89.6	90.5	87.4	87.7	86.0
SSF	85.7	75.6	89.0	89.5	88.2	87.4	85.9
RASTA-PLP	85.9	75.9	88.2	89.7	87.9	86.8	85.7
GFMC	84.1	74.3	87.5	89.1	83.5	83.7	83.7
PITCH	85.5	69.6	84.8	88.9	79.2	82.3	81.7
AMS	82.5	74.0	84.8	87.8	75.4	79.1	80.6
PAC-MFCC	77.9	69.8	78.1	81.1	70.8	67.9	74.3

puted by passing an input signal to a gammatone filterbank and applying a decimation operation on subband signals. The zero-crossing feature, called zero-crossings with peak-amplitudes (ZCPA) [50], computes zero-crossing intervals and corresponding peak amplitudes from subband signals derived using a gammatone filterbank. The three autocorrelation features are relative autocorrelation sequence MFCC (RAS-MFCC) [51], autocorrelation sequence MFCC (AC-MFCC) [52] and phase autocorrelation MFCC (PAC-MFCC) [53], all of which apply the MFCC procedure in the autocorrelation domain. The two medium-term filtering features are power normalized cepstral coefficients (PNCC) [54] and suppression of slowly-varying components and the falling edge of the power envelope (SSF) [55]. The two modulation features are Gabor filterbank (GFB) [56] and AMS features. Pitch-based (PITCH) features calculate T-F level features based on pitch tracking and use periodicity and instantaneous frequency to discriminate speech-dominant T-F units from noise-dominant ones. In addition to existing features, we have introduced a new feature called Multi-Resolution Cochleagram (MRCG) [47], which computes four cochleagrams at different spectrotemporal resolutions to provide both local information and a broader context.

The features are post-processed with an auto-regressive moving average (ARMA) filter and then fed to a fixed MLP for IBM estimation. ARMA processing is found to improve the robustness of features in noise for automatic speech recognition [57]. The features are evaluated in terms of classification accuracy and the HIT−FA rate,

Table 9.2 Performance of a list of acoustic features in terms of HIT−FA (in %) for six noise types with ARMA post-processing at −5 dB SNR, where FA is shown in parentheses (from [47])

	Factory	Babble	Engine	Cockpit	Vehicle	Tank	Average
MRCG	**63** (7)	**49** (13)	**77** (4)	**73** (4)	**80** (10)	**77** (6)	**70** (7)
GF	61 (7)	45 (15)	75 (4)	71 (3)	80 (10)	76 (6)	68 (8)
GFCC	61 (6)	46 (14)	73 (4)	70 (3)	78 (11)	74 (6)	67 (7)
DSCC	56 (7)	42 (14)	70 (5)	66 (3)	77 (11)	73 (6)	64 (8)
MFCC	57 (7)	43 (14)	69 (5)	67 (4)	77 (11)	72 (7)	64 (8)
PNCC	56 (6)	44 (14)	69 (5)	66 (4)	77 (11)	71 (7)	64 (8)
PLP	56 (6)	41 (12)	68 (5)	66 (4)	77 (11)	71 (7)	63 (8)
AC-MFCC	56 (6)	42 (14)	67 (5)	65 (4)	77 (11)	71 (7)	63 (8)
RAS-MFCC	57 (6)	41 (14)	68 (5)	66 (4)	76 (11)	71 (7)	63 (8)
GFB	57 (7)	41 (18)	67 (5)	66 (4)	75 (12)	70 (7)	63 (9)
ZCPA	55 (8)	40 (16)	68 (5)	65 (4)	75 (13)	70 (8)	62 (9)
SSF	54 (7)	39 (15)	67 (5)	60 (4)	76 (11)	69 (7)	61 (8)
RASTA-PLP	52 (6)	38 (15)	64 (5)	61 (4)	76 (12)	67 (7)	60 (8)
GFMC	48 (7)	35 (15)	61 (6)	60 (5)	67 (17)	59 (9)	55 (10)
PITCH	46 (3)	29 (22)	50 (5)	50 (2)	59 (16)	53 (7)	48 (9)
AMS	40 (6)	27 (9)	49 (5)	52 (4)	50 (31)	45 (11)	44 (11)
PAC-MFCC	17 (5)	11 (8)	30 (9)	29 (7)	40 (48)	21 (17)	25 (16)

where HIT denotes the percent of speech-dominant T-F units in the IBM correctly classified and FA (false-alarm) refers to the percent of noise-dominant units incorrectly classified. The HIT−FA rate is found to be well correlated with speech intelligibility [24]. The training mixtures are created with 480 IEEE sentences [58] and the first halves of six nonstationary NOISEX noises [59]. The test set is created using 50 different IEEE sentences and the second halves of the six noises. The evaluation SNR is set to −5 dB. The experimental results are shown in classification accuracy and the HIT−FA in Tables 9.1 and 9.2, respectively. The MRCG feature performs the best, and gammatone-domain features (MRCG, GF and GFCC) outperform others. In addition, cepstral compaction (i.e. applying discrete cosine transform) does not help; for example, GF is better than GFCC. Furthermore, modulation-domain features are not effective. For example, GFCC performs better than GMFC, with the latter derived from the former. Finally, pitch-based features do not perform well, largely because of pitch estimation errors at the low test SNR of −5 dB.

9.5 Speech Separation Algorithms

In this section, we introduce representative DNN based algorithms for speech-nonspeech separation as well as some other speech separation or enhancement tasks.

9.5.1 Speech-Nonspeech Separation

Deep learning was introduced to the domain of speech separation/enhancement by Wang and Wang in 2013 [60]. They used DNN for subband classification to estimate the IBM. As shown in Fig. 9.9, noisy speech is first passed to a gammatone filterbank to drive subband signals, from which T-F unit-level acoustic features are extracted. These raw features are then fed to subband DNNs to learn more discriminative features. Both raw features and high-level features of the last hidden layer are used by linear support vector machines (SVMs) to efficiently estimate the subband IBM. To perform feature learning, a subband DNN is pretrained with RBM and then fine-tuned with the subband IBM as the target. With feature learning by a DNN, one can replace a kernel-based SVM, that scales badly with the training set size, with a much faster linear SVM. Therefore this system can be trained on larger training data efficiently.

To create a training set, 200 randomly chosen utterances from both male and female IEEE speakers [58] were mixed with 100 environmental noises at 0 dB SNR to produce six million training samples in each channel, with 64 channels in total. The DNN based system was tested on 20 unseen speakers mixed with 20 unseen noises at 0 dB SNR. With extensive training, the DNN based system substantially outperforms a representative speech enhancement algorithm by Hendriks et al. [32]. The SNR improvement is shown in Fig. 9.10 for 20 unseen test noises. With clean speech as the ground truth, the DNN based system gives 7.9 dB SNR gain while the speech enhancement algorithm gives 5.4 dB SNR gain. With IBM separated speech as the reference signal, the DNN based system obtains 10.5 dB SNR gain.

The above DNN based separation system was subsequently modified and then evaluated with human listeners by Healy et al. [61]. The evaluated system employs a two-stage DNN to incorporate T-F context for better IBM estimation. The first stage follows the DNN introduced in [60], both feature extraction and DNN classification

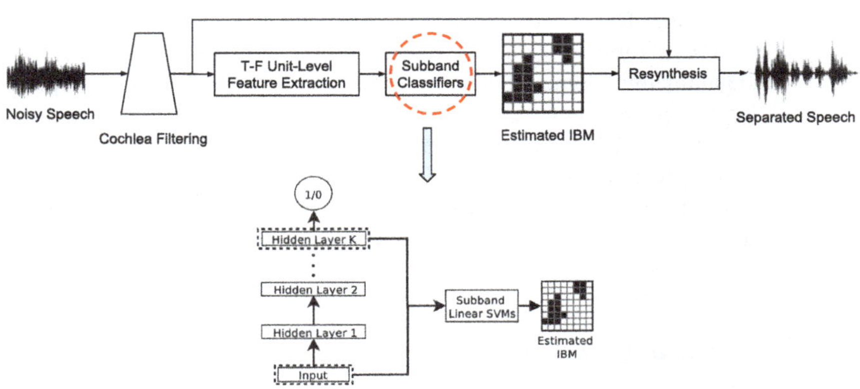

Fig. 9.9 DNN based supervised speech separation (modified from [60])

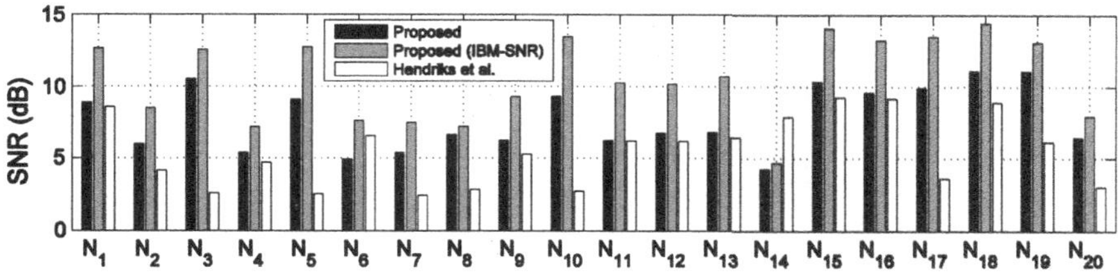

Fig. 9.10 SNR comparison between a DNN based classification algorithm and Hendriks et al.'s algorithm. "IBM-SNR" denotes the SNR with the IBM separated signal as the ground-truth (from [60])

are performed at the T-F unit level. In the second stage, a window of the output units from the first stage centered at a current T-F unit is input to another subband DNN classifier to re-estimate the IBM for that unit. The window spans five time frames and 17 (of the 64) frequency channels. Since nearby T-F units are correlated, incorporating the T-F context this way leads to improved IBM estimation. The system was trained with 100 HINT sentences [62] and tested with 160 different HINT sentences. Both training and test mixtures are created with randomly selected noise segments from a speech-shaped noise (SSN) and a multi-talker babble noise. Figure 9.11 shows the examples of clean speech, noisy speech, the IBM, the estimated mask and separated speech. The DNN estimated IBM highly resembles the IBM. The results of subject tests are shown in Fig. 9.12. Both HI and NH listeners showed substantial intelligibility improvements, with HI listeners benefiting more. It is worth stressing that HI subjects with separation outperformed NH subjects without separation. This DNN classification algorithm is the first monaural algorithm to provide demonstrated speech intelligibility improvements for HI listeners in background noise. DNN based IBM estimation has also been found to improve pitch estimation, which in turn helps speech separation [63].

For any supervised learning tasks, generalizing to unseen conditions is critical. In supervised speech separation, noise generalization and speaker generalization are two important aspects. We have conducted a recent study to address noise generalization with large scale training [64]. Our system is illustrated in Fig. 9.13, which is different from the previous systems in three major aspects. First, feature extraction is performed on a full-band signal instead of subband signals, leading to much faster processing. Second, the IRM instead of the IBM is used as the training target. Third, the DNN estimates both a current frame and neighboring frames of the IRM, leading to a smoother mask estimate. The DNN has five hidden layers with 2048 ReLUs in each layer. The input features are 64-channel gammatone filterbank energies. The training set includes 640,000 mixtures created from 560 IEEE sentences and 10,000 noises from a sound effect library (www.sound-ideas.com) at the fixed SNR of −2 dB. The total duration of the noises is about 125 hours, and the total duration of training mixtures is about 380 hours. The test set is created using 160 different IEEE sentences and two highly non-stationary noises (cafeteria and babble) at −5 dB,

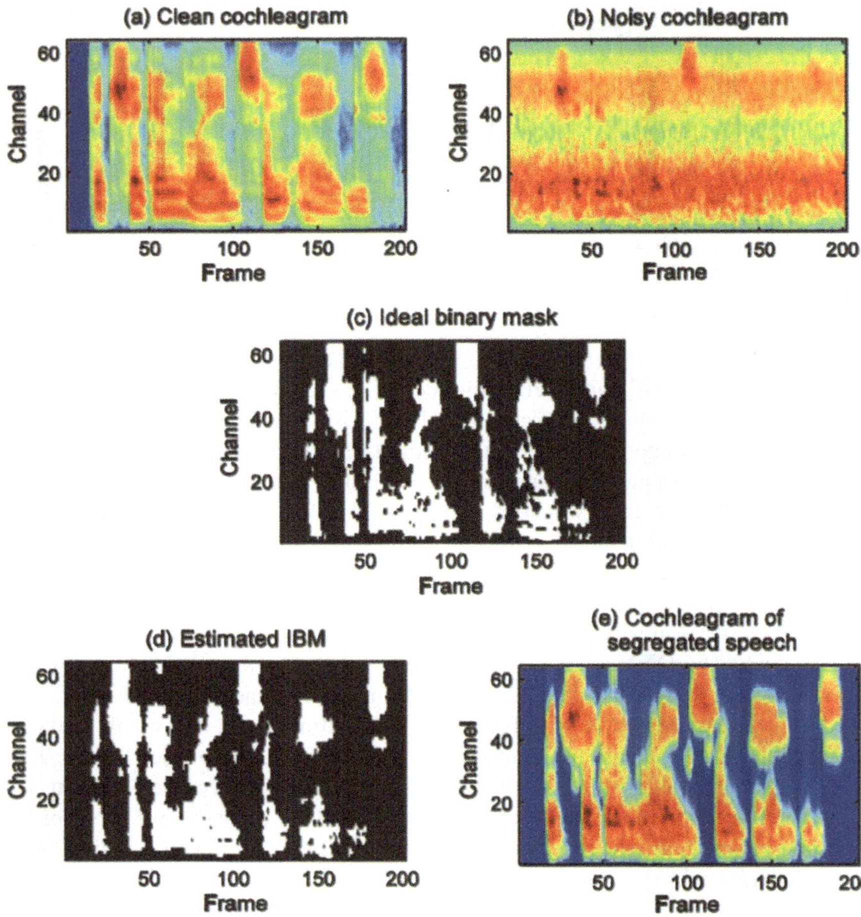

Fig. 9.11 Illustration of separating a HINT utterance from speech-shaped noise at −5 dB SNR. **a** Cochleagram of clean speech. **b** Cochleagram of noisy speech (**c**) The IBM. **d** Estimated IBM. **e** Cochleagram of separated speech after applying the estimated IBM (from [61])

−2 dB, 0 dB and 5 dB SNR. Note that neither test sentences nor test noises are used during training.

To illustrate DNN feature learning capability, we visualize the weights of the first 100 neurons in the first hidden layer. As shown in Fig. 9.14, the neurons appear to have acquired speech-specific features. For example, some neurons seem to be activated by harmonic structure (e.g., the tenth filter in the last row), while some others seem to be sensitive to formant transitions (e.g., the fifth filter in the third row). By encoding fundamental characteristics of the speech signal, the DNN learns to separate speech from unseen noises. To evaluate the effect of the number of training noises on noise generalization, we also train the same DNN with 100 environmental noises instead

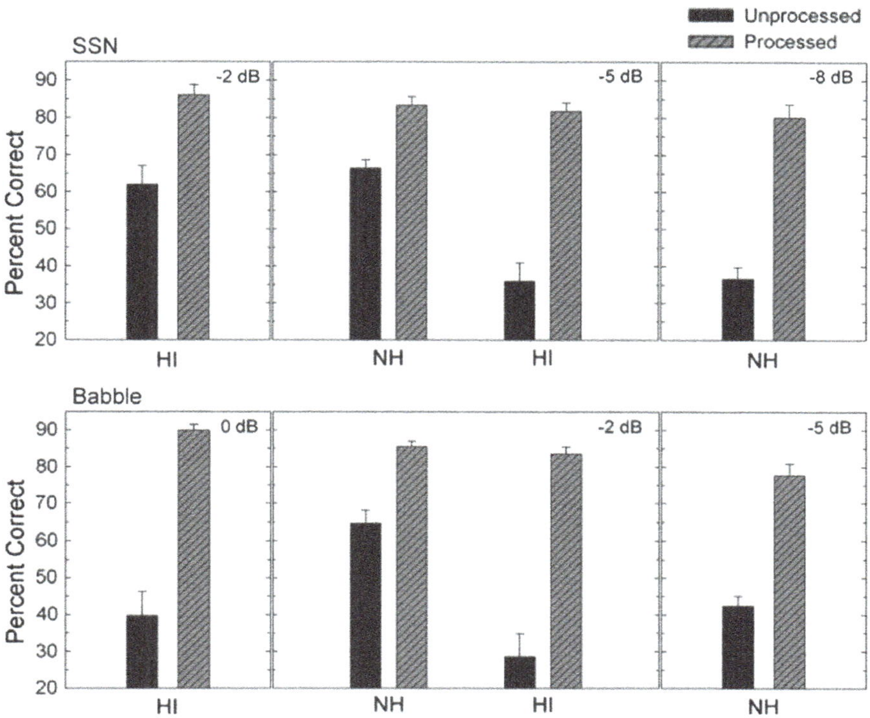

Fig. 9.12 Percent correct word recognition scores and standard errors for HINT sentences mixed with speech-shaped noise (upper panels) and multi-talker babble (lower panels), at the SNRs indicated. Intelligibility results are shown for normal-hearing and hearing-impaired listeners, both before and after algorithm processing (from [61])

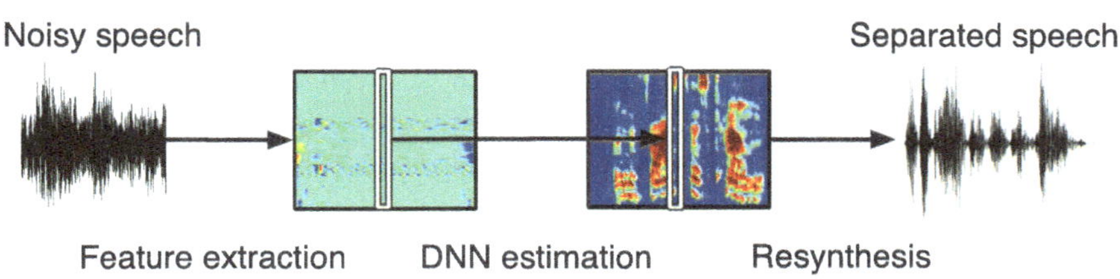

Fig. 9.13 DNN based IRM estimation with large scale training for noise generalization (from [64])

of the 10,000 noises described above, and evaluate both models with four unseen noises. As shown in Table 9.3, the 10,000-noise model substantially outperforms the 100-noise model and matches noise-dependent models, which are trained and tested on different segments of the same noise. This indicates that exposing the DNN to a large variety of noises during training is crucial for noise generalization. To evaluate SNR generalization, we additionally train a model with −5 dB mixtures, and test

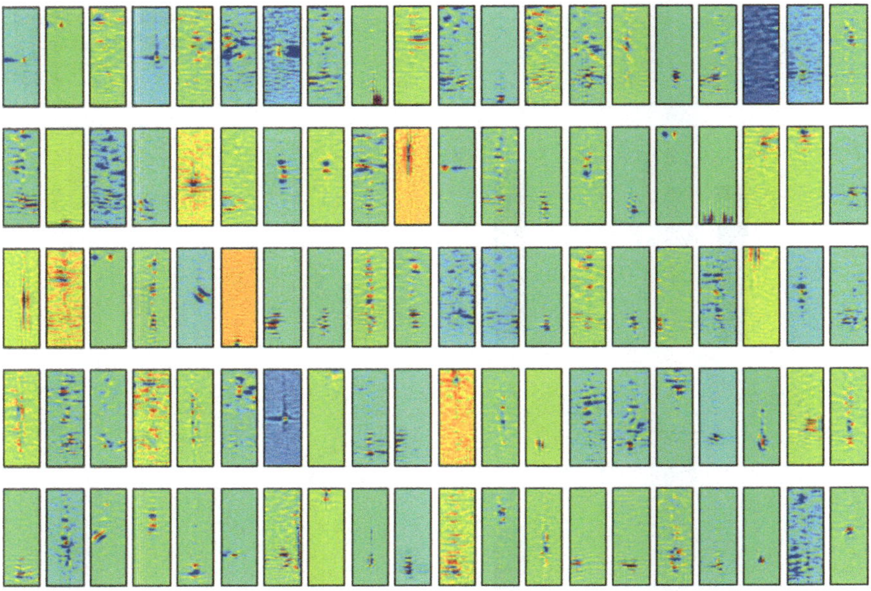

Fig. 9.14 Illustraion of 100 learned filters in the first hidden layer of a DNN trained on mixtures created using 10,000 noises. Filters or weight matrices are displayed in two dimensions: The abscissa represents time (23 frames) and the ordinate represents frequency (64 channels) (from [64])

Table 9.3 Separation performance for four unseen noises in terms of STOI at −2 dB SNR (from [64])

	Babble	Cafeteria	Factory	Babble2	Average
Unprocessed	0.612	0.596	0.611	0.611	0.608
100-noise model	0.683	0.704	0.750	0.688	0.706
10K-noise model	0.792	0.783	0.807	0.786	0.792
Noise-dependent model	0.833	0.770	0.802	0.762	0.792

both −2 and −5 dB models in both matched and unmatched SNR conditions. These test results are shown in Fig. 9.15. The DNN achieves very similar performance in matched and unmatched SNR conditions, indicating good SNR generalizability. Furthermore, we have evaluated the noise-independent model with human listeners. As shown in Fig. 9.16, both NH and HI listeners benefit from algorithm processing in all conditions, with larger benefits for HI listeners. This is the first demonstration that supervised speech separation improves speech intelligibility in completely new noises.

Besides noise generalization, speaker generalization is also an important issue. In practice, a separation system trained on a specific speaker would not work well for

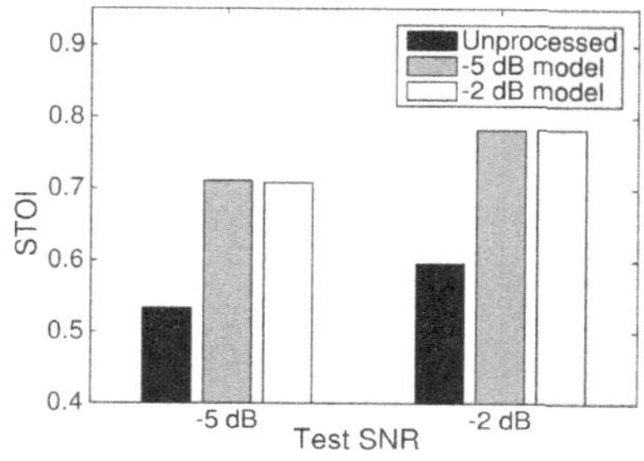

Fig. 9.15 Evaluation of SNR generalization in terms of STOI (from [64]). A model trained with −2 dB mixtures and one trained with −5 dB mixtures are tested in both matched and unmatched SNR conditions

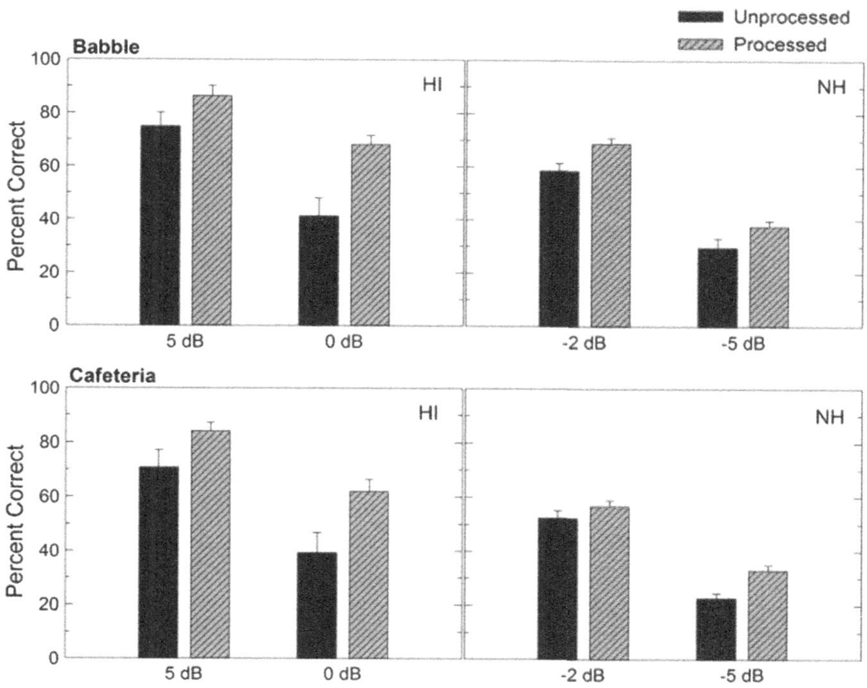

Fig. 9.16 Percent correct word recognition and standard errors for HI and NH listeners hearing unprocessed and processed noisy speech. The top and bottom show the scores for a babble noise and a cafeteria noise, respectively, at given SNRs (from [64])

an unseen speaker. A straightforward attempt for speaker generalization is to train with a large number of speakers. However, our experimental results show that feed-forward DNN appears to be incapable of modeling a large number of speakers [18]. Such a DNN typically takes a window of acoustic features for mask estimation at

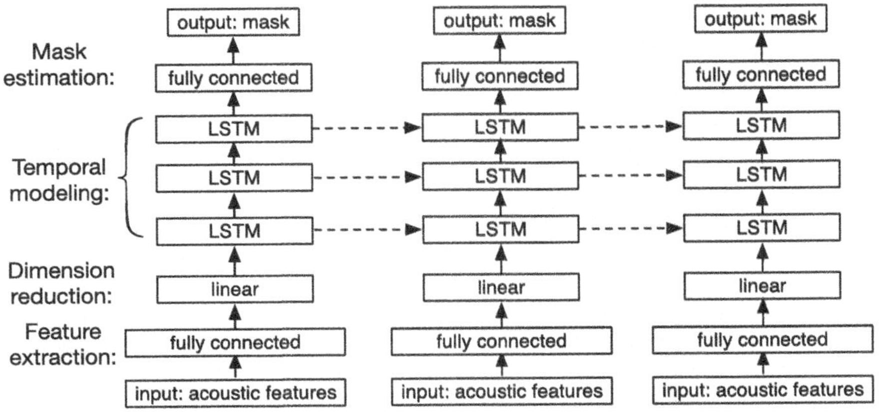

Fig. 9.17 LSTM based IRM estimation for speaker generalization (from [18])

a central frame, without using the long-term context. Unable to focus on a target speaker, a feedforward network trained on many speakers may pick up interfering speech fragments in the background noise. To better differentiate target speech from noise, a learning machine should figure out the speaker of interest from long-term observations. Therefore, RNNs, which naturally model temporal dependencies, are expected to be more suitable for speaker generalization than feedforward DNN.

We have conducted a recent study that employs an RNN with LSTM to address speaker generalization for noise-independent speech separation [18]. The separation system is illustrated in Fig. 9.17. First, the raw acoustic features are sent to a feedforward layer for feature transformation. The transformed features are then fed to three LSTM layers for temporal modeling. Finally, the IRM is estimated by a few more feedfordward layers. With LSTM layers, the network learns the characteristics of a target speaker from past observations and focuses on it during separation. To evaluate speaker generalization, we create a training set of 3,200,000 mixtures using 10,000 noises and 77 speakers from the WSJ0 corpus at random SNRs sampled from $\{-5, -4, -3, -2, -1, 0\}$ dB. Two test sets are created using unseen noises (cafeteria and babble) and 6 unseen speakers from the WSJ0 corpus at -5 dB and -2 SNR. Similarly, we create another two test sets with 6 seen speakers. To benchmark the LSTM based system, we evaluate a baseline system using a five-hidden-layer feedforward DNN. We compare the DNN and LSTM in terms of STOI improvement for both seen and unseen speakers at -5 dB SNR in Figs. 9.18 and 9.19, respectively. For seen speakers, the performance of the feedforward DNN degrades and that of LSTM improves as the number of training speakers increases. Exposed to many speakers during training, the DNN becomes more likely to mistake the background noise as target speech, while LSTM appears to focus on a target speaker by exploiting the temporal dynamics of speech. Similarly, for unseen speakers, LSTM substantially outperforms the DNN. We illustrate the estimated masks by the feedforward DNN and LSTM in Fig. 9.20. Compared to the DNN, LSTM significantly reduces false-

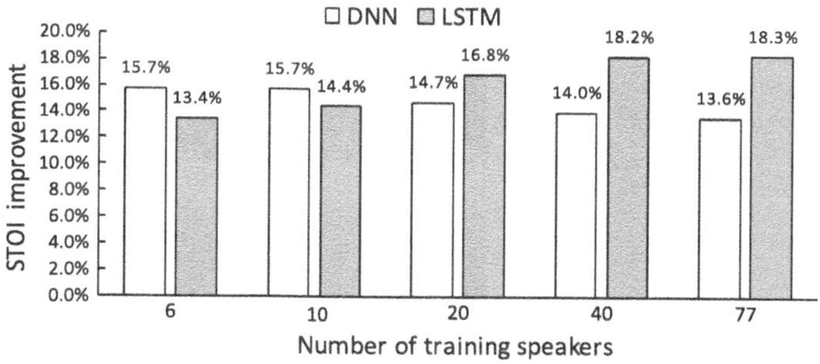

Fig. 9.18 STOI improvements of a feedforward DNN and LSTM for seen speakers at −5 dB SNR (from [18])

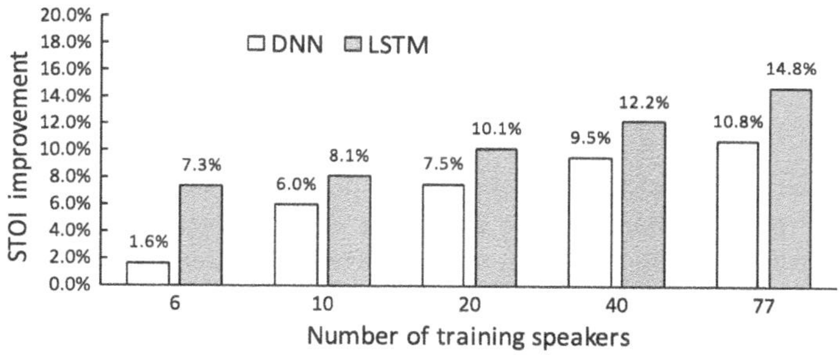

Fig. 9.19 STOI improvements of a feedforward DNN and LSTM for unseen speakers at −5 dB SNR (from [18])

alarm errors. The LSTM-based RNNs represent a promising approach for speaker- and noise-independent speech separation.

While masking-based separation algorithms lead to speech intelligibility improvement, mapping-based algorithms have been shown to improve speech quality. A recent study trains a DNN to map log power spectra of noisy speech to those of clean speech, and improves speech quality measured in PESQ [65]. However, in low SNR conditions, speech intelligibility is a more pronounced issue than speech quality. Our experimental results suggest that masking-based algorithms outperform mapping-based algorithms in terms of intelligibility, and they perform similarly in terms of speech quality, consistent with Figs. 9.7 and 9.8.

9.5.2 *Other Separation/Enhancement Tasks*

Supervised processing goes beyond monaural speech-nonspeech separation. In this section, we describe DNN-based algorithms for related speech separation or enhancement tasks.

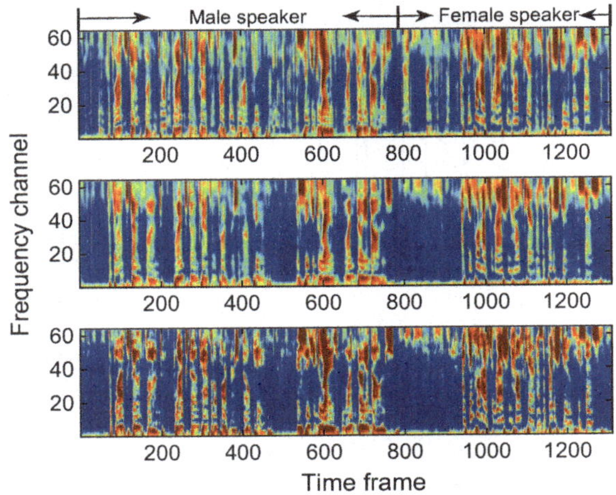

Fig. 9.20 Illustration of estimated masks by a feedforward DNN (top) and LSTM (middle) and the IRM (bottom). To create the mixture, the concatenation of a male utterance and a female utterance is mixed with babble noise at −5 dB SNR (from [18])

The first task is speech dereverberation and denoising. Reverberation is common in our daily life. For example, in a room, our ears receive both direct sound and reflections from the walls and other surfaces. Reverberation has adverse effects on speech processing such as speech communication, automatic speech recognition and speaker identification, especially when noise is also present. While traditional methods apply inverse filtering for speech dereverberation [68, 69], a recent study uses a DNN to estimate the spectral magnitudes of clean or anechoic speech from those of reverberant speech [66]. This simple spectral mapping approach has proven to be quite effective for dereverberation. This system is shown in Fig. 9.21. The DNN takes 11 frames of reverberant speech features (a current frame and 5 neighboring frame on each side) as the input, and learns to map to the current frame of anechoic speech. It is straightforward to extend the spectral mapping approach to perform dereverberation and denoising at the same time [67]. In [67], a DNN was trained on multiple T_{60}'s (0.3, 0.6 and 0.9 s) and noises (babble, factory and SSN). The test set was created using both seen noises (babble, factory and SSN) and unseen noises (white, cocktail party and playground). Experimental results have shown that the DNN based spectral mapping algorithm improves objective speech intelligibility and quality for both seen and unseen noises. An example of speech dereverberation and denoising is shown in Fig. 9.22. Both background noise and reverberation are significantly reduced following algorithm processing.

The second task is two-talker separation, where the goal is to extract two speech signals, one for each speaker, from a mixture containing both speakers. DNN based systems have been proposed to separate a target speaker from a competing speaker [70–72]. As shown in Fig. 9.23, a DNN is trained to estimate the log-power spectra of both the target speaker and interfering speaker. Experimental results have shown that this approach significantly improves objective speech intelligibility and quality. Huang et al. [73] address two-talker separation using a DNN as well as an

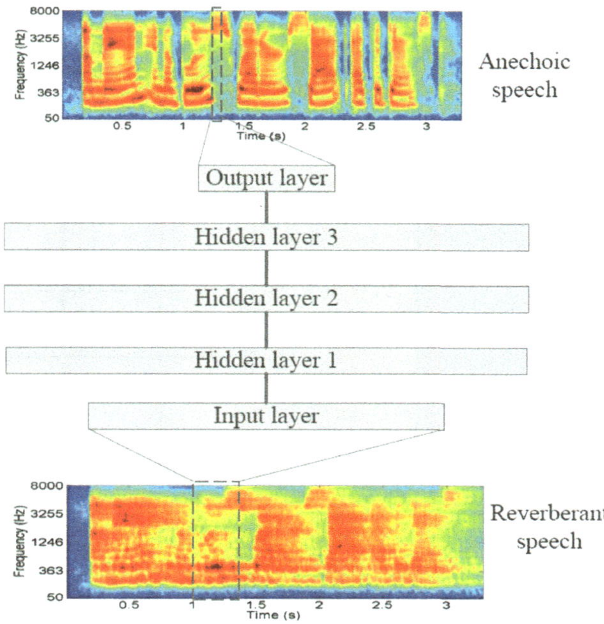

Fig. 9.21 A DNN based dereverberation system (from [66])

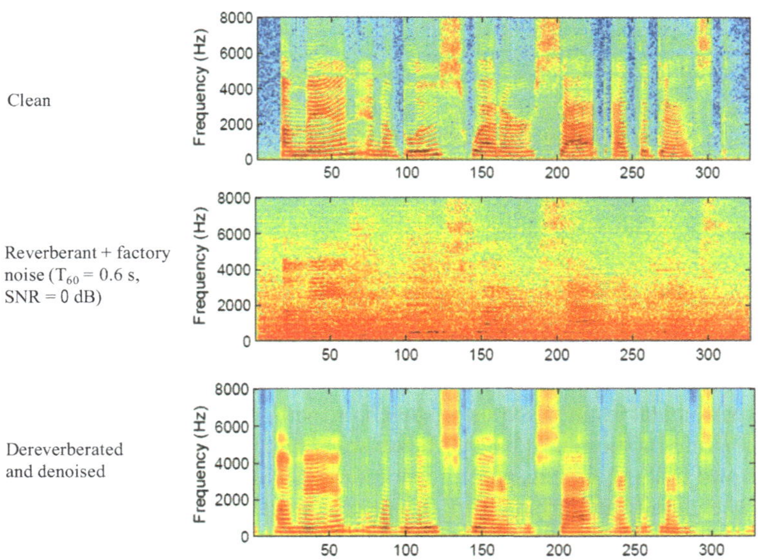

Fig. 9.22 An example of speech dereverberation and denoising using the DNN based spectral mapping method (modified from [67])

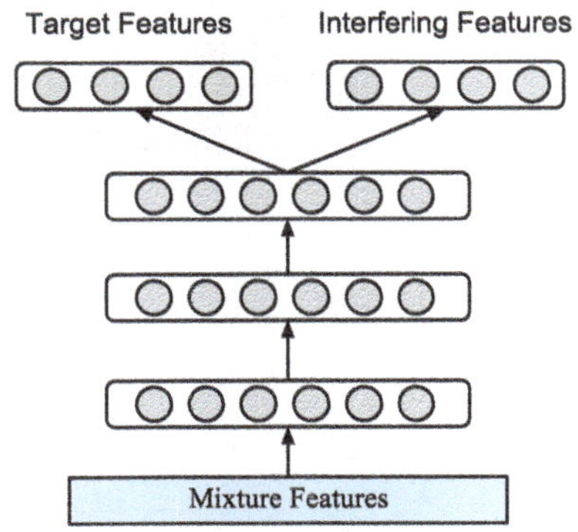

Fig. 9.23 A two-talker separation system using DNN based spectral mapping

RNN. The authors argue that directly predicting two sources $\hat{y}_{1_t}$ and $\hat{y}_{2_t}$ does not guarantee that the summation of two estimated sources equals to the mixture. Therefore, a masking layer is added to the network, which produces two final outputs shown in the following equations:

$$\tilde{y}_{1_t} = \frac{|\hat{y}_{1_t}|}{|\hat{y}_{1_t}| + |\hat{y}_{2_t}|} \odot z_t \tag{9.19}$$

$$\tilde{y}_{2_t} = \frac{|\hat{y}_{2_t}|}{|\hat{y}_{1_t}| + |\hat{y}_{2_t}|} \odot z_t \tag{9.20}$$

where $\hat{y}_{1_t}$ and $\hat{y}_{2_t}$ denote the estimated magnitude spectra for speaker 1 and speaker 2 at time t, respectively. z_t denotes the mixture magnitude spectra. This is a signal approximation training target introduced in Sect. 9.3. In addition, discriminative training is applied to maximize the difference between one speaker and the estimated version of the other speaker. During training, the following cost function is minimized:

$$\frac{1}{2}\sum_t (\|y_{1_t} - \tilde{y}_{1_t}\|^2 + \|y_{2_t} - \tilde{y}_{2_t}\|^2 - \gamma\,\|y_{1_t} - \tilde{y}_{2_t}\|^2 - \gamma\,\|y_{2_t} - \tilde{y}_{1_t}\|^2) \tag{9.21}$$

where y_{1_t} and y_{2_t} denote speaker 1 and speaker 2, respectively. γ is a tunable parameter. Experimental results have shown that both incorporation of a masking layer and discriminating training improve two-talker separation [73].

The third task is binaural speech separation, where spatial cues are used for separation. The classification framework can be readily extended to the binaural domain where features will be extracted from binaural inputs [15]. A recent study employs interaural time difference (ITD) and interaural level difference (ILD) cues and DNN for IBM estimation to perform binaural separation [74]. As shown in Fig. 9.24, the signals from two ears (or microphones) are passed to two corresponding auditory filterbanks. ITD and ILD features are extracted from T-F unit pairs and sent to a

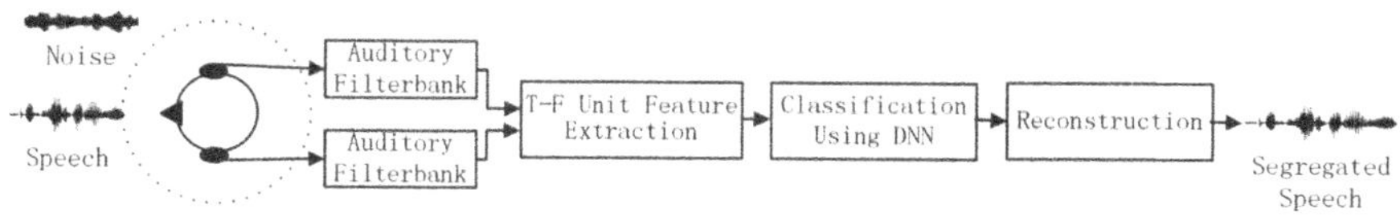

Fig. 9.24 DNN based IBM estimation for binaural speech separation (from [74])

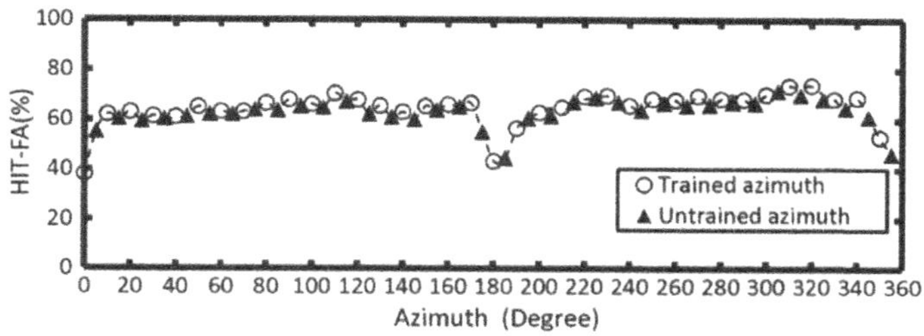

Fig. 9.25 HIT−FA rate for two-source separation at various trained and untrained azimuths in a reverberant condition (T_{60} = 0.3 s) with babble noise at 0 dB SNR (from [74])

subband DNN for IBM estimation. This is the first DNN based system for location-based speech separation. As shown in Fig. 9.25, the trained DNN generalizes well to unseen spatial configurations. It is also observed that incorporating a monaural feature improves separation performance, especially when the target and interfering sources are co-located or close to each other.

9.6 Conclusion

This chapter introduces supervised methods for speech separation or enhancement, which use DNN based mask estimation. The formulation of speech separation as a supervised learning task has enabled the use of powerful deep learning techniques and large training data. This new framework has advanced the state-of-the-art performance in speech separation by considerable margins, including the first demonstration of substantial speech intelligibility improvements in noise for hearing-impaired listeners, an achievement that has eluded traditional speech enhancement and CASA for decades.

The use of supervised learning in signal processing goes beyond speech separation and recognition, including multipitch tracking [75], voice activity detection [76] and even SNR estimation [77]. We believe that signal processing provides an important domain for supervised learning, and it in turn benefits from rapid advances in machine learning. The learning or data-driven framework will continue to push speech separation and other signal processing tasks to new performance levels in the years to come.

References

1. E.C. Cherry, Some experiments on the recognition of speech, with one and with two ears. J. Acoust. Soc. Am. **25**(5), 975–979 (1953)
2. M. Brandstein, D. Ward, *Microphone Arrays: Signal Processing Techniques and Applications* (Springer, Berlin, 2001)
3. P.C. Loizou, *Speech Enhancement: Theory and Practice* (CRC Press, Boca Raton, 2013)
4. D.L. Wang, G.J. Brown (eds.), *Computational Auditory Scene Analysis: Principles, Algorithms and Applications* (Wiley-IEEE Press, Hoboken, 2006)
5. A.S. Bregman, *Auditory Scene Analysis: The Perceptual Organization of Sound* (The MIT Press, Cambridge, 1990)
6. D.L. Wang, Time-frequency masking for speech separation and its potential for hearing aid design. Trends Amplif. **12**(4), 332–353 (2008)
7. G. Hu, D.L. Wang, Speech segregation based on pitch tracking and amplitude modulation, in *Proceedings of the WASPAA* (2001), pp. 79–82
8. G. Hu, D.L. Wang, Monaural speech segregation based on pitch tracking and amplitude modulation. IEEE Trans. Neural Netw. **15**(5), 1135–1150 (2004)
9. D.L. Wang, On ideal binary mask as the computational goal of auditory scene analysis, in *Speech Separation by Humans and Machines*, ed. by P. Divenyi (Kluwer Academic Publishers, Boston, 2005), pp. 181–197
10. D.S. Brungart, P.S. Chang, B.D. Simpson, D.L. Wang, Isolating the energetic component of speech-on-speech masking with ideal time-frequency segregation. J. Acoust. Soc. Am. **120**(6), 4007–4018 (2006)
11. N. Li, P.C. Loizou, Factors influencing intelligibility of ideal binary-masked speech: implications for noise reduction. J. Acoust. Soc. Am. **123**(3), 1673–1682 (2008)
12. M.C. Anzalone, L. Calandruccio, K.A. Doherty, L.H. Carney, Determination of the potential benefit of time-frequency gain manipulation. Ear Hear. **27**(5), 480 (2006)
13. D.L. Wang, U. Kjems, M.S. Pedersen, J.B. Boldt, T. Lunner, Speech intelligibility in background noise with ideal binary time-frequency masking. J. Acoust. Soc. Am. **125**(4), 2336–2347 (2009)
14. N. Roman, D.L. Wang, G.J. Brown, A classification-based cocktail-party processor, in *Proceedings of the NIPS-02* (2003), pp. 1425–1432
15. N. Roman, D.L. Wang, G.J. Brown, Speech segregation based on sound localization. J. Acoust. Soc. Am. **114**(4), 2236–2252 (2003)
16. G.E. Hinton, R.R. Salakhutdinov, Reducing the dimensionality of data with neural networks. Science **313**(5786), 504–507 (2006)
17. V. Nair, G.E. Hinton, Rectified linear units improve restricted boltzmann machines, in *Proceedings of the ICML* (2010), pp. 807–814
18. J. Chen, D.L. Wang, Long short-term memory for speaker generalization in supervised speech separation, in *Proceedings of the INTERSPEECH* (2016), pp. 3314–3318
19. R. Pascanu, T. Mikolov, Y. Bengio, On the difficulty of training recurrent neural networks, in *Proceedings of the ICML* (2013), pp. 1310–1318
20. M. Sundermeyer, H. Ney, R. Schlüter, From feedforward to recurrent LSTM neural networks for language modeling. IEEE/ACM Trans. Audio Speech Lang. Process. **23**(3), 517–529 (2015)
21. I. Sutskever, O. Vinyals, Q.V. Le, Sequence to sequence learning with neural networks, in *Proceedings of the NIPS* (2014), pp. 3104–3112
22. A. Graves, A. Mohamed, G. Hinton, Speech recognition with deep recurrent neural networks, in *Proceedings of the ICASSP* (2013), pp. 6645–6649
23. Z. Jin, D.L. Wang, A supervised learning approach to monaural segregation of reverberant speech. IEEE Trans. Audio Speech Lang. Process. **17**(4), 625–638 (2009)
24. G. Kim, Y. Lu, Y. Hu, P.C. Loizou, An algorithm that improves speech intelligibility in noise for normal-hearing listeners. J. Acoust. Soc. Am. **126**(3), 1486–1494 (2009)
25. K. Han, D.L. Wang, A classification based approach to speech segregation. J. Acoust. Soc. Am. **132**(5), 3475–3483 (2012)

26. U. Kjems, J.B. Boldt, M.S. Pedersen, T. Lunner, D.L. Wang, Role of mask pattern in intelligibility of ideal binary-masked noisy speech. J. Acoust. Soc. Am. **126**(3), 1415–1426 (2009)
27. S. Gonzalez, M. Brookes, Mask-based enhancement for very low quality speech, in *Proceedings of the ICASSP* (2014), pp. 7029–7033
28. S. Srinivasan, N. Roman, D.L. Wang, Binary and ratio time-frequency masks for robust speech recognition. Speech Commun. **48**(11), 1486–1501 (2006)
29. A. Narayanan, D.L. Wang, Ideal ratio mask estimation using deep neural networks for robust speech recognition, in *Proceedings of the ICASSP* (2013), pp. 7092–7096
30. Y. Wang, A. Narayanan, D.L. Wang, On training targets for supervised speech separation. IEEE/ACM Trans. Audio Speech Lang. Process. **22**(12), 1849–1858 (2014)
31. C. Hummersone, T. Stokes, T. Brookes, On the ideal ratio mask as the goal of computational auditory scene analysis, in *Blind source separation*, ed. by W.W.G.R. Naik (Springer, Berlin, 2014), pp. 349–368
32. R.C. Hendriks, R. Heusdens, J. Jensen, MMSE based noise PSD tracking with low complexity, in *Proceedings of the ICASSP* (2010), pp. 4266–4269
33. T. Virtanen, J.F. Gemmeke, B. Raj, Active-set newton algorithm for overcomplete non-negative representations of audio. IEEE/ACM Trans. Audio Speech Lang. Process. **21**(11), 2277–2289 (2013)
34. J. Garofolo, *DARPA TIMIT Acoustic-Phonetic Continuous Speech Corpus* (National Institute of Standards and Technology, Gaithersburg, 1993)
35. C.H. Taal, R.C. Hendriks, R. Heusdens, J. Jensen, An algorithm for intelligibility prediction of time-frequency weighted noisy speech. IEEE Trans. Audio Speech Lang. Process. **19**(7), 2125–2136 (2011)
36. A. Rix, J. Beerends, M. Hollier, A. Hekstra, Perceptual evaluation of speech quality (PESQ)-a new method for speech quality assessment of telephone networks and codecs, in *Proceedings of the ICASSP* (2001), pp. 749–752
37. Z. Wang, X. Wang, X. Li, Q. Fu, Y. Yan, Oracle performance investigation of the ideal masks, in *Proceedings of the IWAENC* (2016), pp. 1–5
38. F. Weninger, J.R. Hershey, J. Le Roux, B. Schuller, Discriminatively trained recurrent neural networks for single-channel speech separation, in *Proceedings of the GlobalSIP* (2014), pp. 577–581
39. H. Erdogan, J.R. Hershey, S. Watanabe, J. Le Roux, Phase-sensitive and recognition-boosted speech separation using deep recurrent neural networks, in *Proceedings of the ICASSP* (2015), pp. 708–712
40. D.S. Williamson, Y. Wang, D.L. Wang, Complex ratio masking for monaural speech separation. IEEE/ACM Trans. Audio Speech Lang. Process. **24**(3), 483–492 (2016)
41. N. Roman, D.L. Wang, Binaural tracking of multiple moving sources. IEEE/ACM Trans. Audio Speech Lang. Process. **16**(4), 728–739 (2008)
42. Y. Wang, K. Han, D.L. Wang, Exploring monaural features for classification-based speech segregation. IEEE Trans. Audio Speech Lang. Process. **21**(2), 270–279 (2013)
43. Y. Shao, D.L. Wang, Robust speaker identification using auditory features and computational auditory scene analysis, in *Proceedings of the ICASSP* (2008), pp. 1589–1592
44. X. Zhao, Y. Shao, D.L. Wang, CASA-based robust speaker identification. IEEE Trans. Audio Speech Lang. Process. **20**(5), 1608–1616 (2012)
45. H. Hermansky, Perceptual linear predictive (PLP) analysis of speech. J. Acoust. Soc. Am. **87**(4), 1738–1752 (1990)
46. H. Hermansky, N. Morgan, RASTA processing of speech. IEEE Trans. Speech Audio Process. **2**(4), 578–589 (1994)
47. J. Chen, Y. Wang, D. Wang, A feature study for classification-based speech separation at low signal-to-noise ratios. IEEE/ACM Trans. Audio Speech Lang. Process. **22**(12), 1993–2002 (2014)
48. K. Kumar, C. Kim, R.M. Stern, Delta-spectral cepstral coefficients for robust speech recognition, in *Proceedings of the ICASSP* (2011), pp. 4784–4787

49. H.K. Maganti, M. Matassoni, An auditory based modulation spectral feature for reverberant speech recognition. in *Proceedings of the INTERSPEECH* (2010), pp. 570–573
50. D.-S. Kim, S.-Y. Lee, R.M. Kil, Auditory processing of speech signals for robust speech recognition in real-world noisy environments. IEEE Trans. Speech Audio Process. **7**(1), 55–69 (1999)
51. K. Yuo, H. Wang, Robust features for noisy speech recognition based on temporal trajectory filtering of short-time autocorrelation sequences. Speech Commun. **28**(1), 13–24 (1999)
52. B.J. Shannon, K.K. Paliwal, Feature extraction from higher-lag autocorrelation coefficients for robust speech recognition. Speech Commun. **48**(11), 1458–1485 (2006)
53. S. Ikbal, H. Misra, H. Bourlard, Phase autocorrelation (PAC) derived robust speech features, in *Proceedings of the ICASSP* (2003), pp. 133–136
54. C. Kim, R. Stern, Power-normalized cepstral coefficients (PNCC) for robust speech recognition, in *Proceedings of the ICASSP* (2012), pp. 4101–4104
55. C. Kim, R.M. Stern, Nonlinear enhancement of onset for robust speech recognition. in *Proceedings of the INTERSPEECH* (2010), pp. 2058–2061
56. M.R. Schädler, B.T. Meyer, B. Kollmeier, Spectro-temporal modulation subspace-spanning filter bank features for robust automatic speech recognition. J. Acoust. Soc. Am. **131**(5), 4134–4151 (2012)
57. C. Chen, J.A. Bilmes, MVA processing of speech features. IEEE Trans. Audio Speech Lang. Process. **15**(1), 257–270 (2007)
58. IEEE, IEEE recommended practice for speech quality measurements. IEEE Trans. Audio Electroacoust. **17**(3), 225–246 (1969)
59. A. Varga, H.J. Steeneken, Assessment for automatic speech recognition: II. NOISEX-92: a database and an experiment to study the effect of additive noise on speech recognition systems. Speech Commun. **12**(3), 247–251 (1993)
60. Y. Wang, D.L. Wang, Towards scaling up classification-based speech separation. IEEE Trans. Audio Speech Lang. Process. **21**(7), 1381–1390 (2013)
61. E.W. Healy, S.E. Yoho, Y. Wang, D.L. Wang, An algorithm to improve speech recognition in noise for hearing-impaired listeners. J. Acoust. Soc. Am. **134**(4), 3029–3038 (2013)
62. M. Nilsson, S.D. Soli, J.A. Sullivan, Development of the hearing in noise test for the measurement of speech reception thresholds in quiet and in noise. J. Acoust. Soc. Am. **95**(2), 1085–1099 (1994)
63. X. Zhang, H. Zhang, S. Nie, G. Gao, W. Liu, A pairwise algorithm using the deep stacking network for speech separation and pitch estimation. IEEE/ACM Trans. Audio Speech Lang. Process. **24**(6), 1066–1078 (2016)
64. J. Chen, Y. Wang, S.E. Yoho, D.L. Wang, E.W. Healy, Large-scale training to increase speech intelligibility for hearing-impaired listeners in novel noises. J. Acoust. Soc. Am. **139**(5), 2604–2612 (2016)
65. Y. Xu, J. Du, L.-R. Dai, C.-H. Lee, An experimental study on speech enhancement based on deep neural networks. IEEE Signal Process. Lett. **21**(1), 65–68 (2014)
66. K. Han, Y. Wang, D.L. Wang, Learning spectral mapping for speech dereverberation, in *Proceedings of the ICASSP* (2014), pp. 4628–4632
67. K. Han, Y. Wang, D.L. Wang, W.S. Woods, I. Merks, T. Zhang, Learning spectral mapping for speech dereverberation and denoising. IEEE/ACM Trans. Audio Speech Lang. Process. **23**(6), 982–992 (2015)
68. C. Avendano, H. Hermansky, Study on the dereverberation of speech based on temporal envelope filtering, in *Proceedings of the ICSLP* (1996), pp. 889–892
69. M. Wu, D.L. Wang, A two-stage algorithm for one-microphone reverberant speech enhancement. IEEE Trans. Audio Speech Lang. Process. **14**(3), 774–784 (2006)
70. P.-S. Huang, M. Kim, M. Hasegawa-Johnson, P. Smaragdis, Deep learning for monaural speech separation, in *Proceedings of the ICASSP* (2014), pp. 1562–1566
71. J. Du, Y. Tu, Y. Xu, L. Dai, C.-H. Lee, Speech separation of a target speaker based on deep neural networks, in *Proceedings of the ICSP* (2014), pp. 473–477

72. Y. Tu, J. Du, Y. Xu, L. Dai, C.-H. Lee, Speech separation based on improved deep neural networks with dual outputs of speech features for both target and interfering speakers, in *Proceedings of the ISCSLP* (2014), pp. 250–254
73. P.-S. Huang, M. Kim, M. Hasegawa-Johnson, P. Smaragdis, Joint optimization of masks and deep recurrent neural networks for monaural source separation. IEEE/ACM Trans. Audio Speech Lang. Process. **23**(12), 2136–2147 (2015)
74. Y. Jiang, D.L. Wang, R. Liu, Z. Feng, Binaural classification for reverberant speech segregation using deep neural networks. IEEE/ACM Trans. Audio Speech Lang. Process. **22**(12), 2112–2121 (2014)
75. K. Han, D.L. Wang, Neural networks for supervised pitch tracking in noise, in *Proceedings of the ICASSP* (2014), pp. 1488–1492
76. X.-L. Zhang, D.L. Wang, Boosting contextual information for deep neural network based voice activity detection. IEEE/ACM Trans. Audio Speech Lang. Process. **24**(2), 252–264 (2016)
77. P. Papadopoulos, A. Tsiartas, J. Gibson, S. Narayanan, A supervised signal-to-noise ratio estimation of speech signals, in *Proceedings of the ICASSP* (2014), pp. 8237–8241

Chapter 10
Informed Spatial Filtering Based on Constrained Independent Component Analysis

Hendrik Barfuss, Klaus Reindl and Walter Kellermann

Abstract In this work, we present a linearly constrained signal extraction algorithm which is based on a Minimum Mutual Information (MMI) criterion that allows to exploit the three fundamental properties of speech and audio signals: Nonstationarity, Nonwhiteness, and Nongaussianity. Hence, the proposed method is very well suited for signal processing of nonstationary nongaussian broadband signals like speech. Furthermore, from the linearly constrained MMI approach, we derive an efficient realization in a (GSC) structure. To estimate the relative transfer functions between the microphones, which are needed for the set of linear constraints, we use an informed time-domain independent component analysis algorithm, which exploits some coarse direction-of-arrival information of the target source. As a decisive advantage, this simplifies the otherwise challenging control mechanism for simultaneous adaptation of the GSC's blocking matrix und interference and noise canceler coefficients. Finally, we establish relations between the proposed method and other well-known multichannel linear filter approaches for signal extraction based on second-order-statistics, and demonstrate the effectiveness of the proposed signal extraction method in a multispeaker scenario.

Abbreviations

ICA	Independent Component Analysis
BSS	Blind Source Separation
MWF	Multichannel Wiener Filter

H. Barfuss (✉) · K. Reindl · W. Kellermann
Chair of Multimedia Communications and Signal Processing, Friedrich-Alexander University Erlangen-Nürnberg, Cauerstrasse 7, 91058 Erlangen, Germany
e-mail: hendrik.barfuss@FAU.de

K. Reindl
e-mail: klaus.reindl@FAU.de

W. Kellermann
e-mail: walter.kellermann@FAU.de

S. Makino (ed.), *Audio Source Separation*, Signals and Communication Technology, https://doi.org/10.1007/978-3-319-73031-8_10

LCMV	Linearly Constrained Minimum Variance
DOA	Direction of Arrival
RTF	Relative Transfer Functions
SOS	Second Order Statistics
MVDR	Minimum Variance Distortionless Response
FIR	Finite Impulse Response
AIR	Acoustic Impulse Response
GSC	Generalized Sidelobe Canceler
MMI	Minimum Mutual Information
STFT	Short-Time Fourier Transform
VAD	Voice Activity Detection
SPP	Speech Presence Probability
TRINICON	TRIple-N Independent component analysis for CONvolutive mixtures
SE	Signal Extraction
NRE	Normalized RTF Estimation Error
SIR	Signal-to-Interference Ratio

10.1 Introduction

In most scenarios where signals are captured by distant microphones the desired signal components are corrupted by unwanted signal components, e.g., additional simultaneously active speakers, background noise, and often also loudspeaker signals from sound reproduction equipment. In acoustic enclosures, echoes and reverberation of all source signals add to these impairments, as illustrated in Fig. 10.1. For such scenarios, the task of extracting an ideally undistorted and interference-free version of a single desired source signal is addressed in the following.

In this chapter, we focus on the aspect of spatial filtering, i.e., the suppression of interfering point sources and background noise by a multichannel linear filter. A variety of multichannel linear filtering methods are already known from literature, see, e.g., [2, 3] and references therein. In general, the set of multichannel linear filtering methods can be divided into data-independent and data-dependent approaches [4]. In the following, we focus on data-dependent approaches, which can be further categorized into supervised and unsupervised methods, depending on whether certain reference information, e.g., a reference signal, is available.

The filter coefficients of unsupervised methods are adapted without the need of a reference signal, exploiting only the statistics of the sensor array data. Here, Independent Component Analysis (ICA) techniques [5], which are based on the underlying assumption of mutual statistically independent source signals, are employed to solve the problem of Blind Source Separation (BSS). As an advantage, no knowledge of source and microphone positions is required. When employed for acoustic signal processing, convolutive mixing and demixing systems need to be considered, see, e.g., [6–9]. Other methods like nonnegative matrix factorization or Bayesian approaches could also be used for BSS, see, e.g., [9, 10].

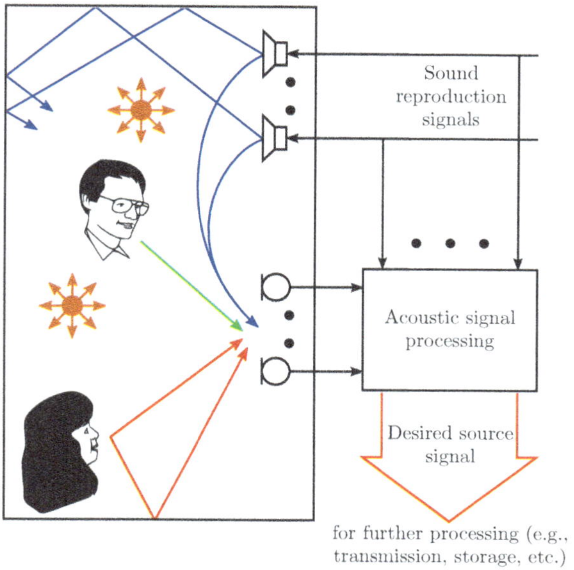

Fig. 10.1 Illustration of a typical signal extraction scenario: The desired source signal may be corrupted by interfering point sources, background noise, and sound reproduction signals (adapted from [1])

On the other hand, for supervised algorithms a reference signal must be given, defined or estimated: For the Multichannel Wiener Filter (MWF) and its variants, typically one of the microphone signals is used as a reference signal [11–13], or for Linearly Constrained Minimum Variance (LCMV)-type algorithms a reference signal is created by a data-independent beamformer which requires source position information [14, 15].

Although multichannel linear filters provide a very powerful basis for signal extraction, for application to broadband and nonstationary speech signal extraction under real acoustic conditions they exhibit specific limitations: Unsupervised methods such as BSS techniques [8, 9, 16] usually can track position changes of the sources only slowly, and are either restricted to determined situations or yield unsatisfactory performance in underdetermined scenarios. Supervised approaches such as the LCMV filter and its equivalent Generalized Sidelobe Canceler (GSC) realization are very sensitive to errors in the steering vector or linear constraints. These errors typically lead to an undesired cancellation of the desired signal components. Moreover, in time-varying environments the Blocking Matrix (BM) of the GSC needs to be adapted, which requires a sophisticated adaptation control mechanism for jointly adapting BM and Interference and Noise Canceler (INC), see, e.g., [17–19]. When using the MWF for signal extraction, accurate estimates of the desired signal and noise power spectral densities are crucial for an adequate performance. Unfortunately, adequate estimates are hard to obtain during simultaneous activity of both desired and undesired sources and especially if the sources are nonstationary, see, e.g., [20, 21].

To overcome these well-known limitations, relevant information about the specific signal extraction problem can be incorporated into the design of signal extraction algorithms, leading to improved performance. Methods exploiting prior knowledge are also referred to as *informed* spatial processing approaches [22]. The exploited knowledge could be of varying character, e.g., position information of sources in terms of Direction of Arrival (DOA) and/or Time Difference of Arrival (TDOA), coherence or diffuseness of the underlying sound field, knowledge of source activity, etc., and may be a priori known or may have to be estimated from the acquired sensor data. For example, in order to improve the performance of LCMV filtering in its equivalent GSC realization under reverberant acoustic conditions, in [23, 24], information about the acoustic environment was incorporated into the realization of the GSC, leading to the so-called Transfer Function GSC (TF-GSC). In [25, 26] prior knowledge on the activity patterns of desired speech and undesired interference signals was exploited, making it possible to estimate the required constraints for the GSC realization in the presence of nonstationary speech interference. As another example, for BSS techniques, prior knowledge on the DOA of the sources can be exploited to improve convergence speed and separation performance, see, e.g., [27, 28].

In this work, we present a signal extraction algorithm which is based in a Minimum Mutual Information (MMI) criterion that allows to exploit the three fundamental properties of speech and audio signals: Nonstationarity, Nonwhiteness, and Nongaussianity and, exploiting some coarse source position information, overcomes several limitations of both unsupervised and supervised algorithms. The MMI criterion is complemented with linear constraints analogous to LCMV filtering, which, as in [23, 29], depend on the Relative Transfer Functions (RTFs) between the microphones with respect to the desired source components. Hence, we aim at extracting the desired source signal as observed by a reference microphone. From the Linearly Constrained Minimum Mutual Information (LCMMI) approach, an efficient realization in a GSC structure is derived. To obtain a robust and reliable estimate of the required RTFs we use an informed time-domain ICA algorithm, which exploits some coarse DOA information of the target source [27, 30]. As an additional advantage, this simplifies the otherwise challenging control mechanism for simultaneously updating the BM und INC coefficients of the GSC. We establish relations between the proposed LCMMI method and other well-known Second Order Statistics (SOS)-based multichannel linear filter approaches for signal extraction, and demonstrate the effectiveness of the proposed LCMMI method in a noisy multispeaker scenario.

The remainder of this chapter is structured as follows: In Sect. 10.2, we introduce the system model underlying this chapter. Then, in Sect. 10.3, we review the well-known LCMV filter including the special case of a Minimum Variance Distortionless Response (MVDR) filter, and its efficient realization in a GSC structure, and discuss their limitations in realistic scenarios. Subsequently, we introduce the LCMMI approach in Sect. 10.4 and establish relations to well-known signal extraction algorithms presented in the previous section. Finally, we demonstrate the effectiveness of the LCMMI approach in a realistic acoustic scenario in Sect. 10.5, and give conclusions in Sect. 10.6.

10.2 Signal Model

Throughout this chapter, vectors and matrices are represented by boldface lower case and upper case letters, respectively, whereas scalars are written in italic font. Moreover, operator $(\cdot)^{\mathrm{T}}$ denotes the transposition of a vector or matrix. Additional less frequently used operators are introduced when needed.

The general linear Multiple-Input Multiple-Output (MIMO) system model considered in the following is illustrated in Fig. 10.2 analogously to [31, 32].

The acoustic mixing system is modeled by a set of generally time-varying Finite Impulse Response (FIR) filters $\mathbf{h}_{qp}[k] = [h_{qp,0}[k], h_{qp,1}[k], \ldots, h_{qp,M-1}[k]]^{\mathrm{T}}$ of length M, with time index k, modeling the Acoustic Impulse Response (AIR) between the q-th source and p-th microphone. The length-$L + D - 1$ vector $\mathbf{x}_p[k]$, where L is the length of the FIR filters of the demixing system, and D is the number of samples used to exploit nonwhiteness in Sect. 10.4, of the p-th microphone signal is given as:

$$\mathbf{x}_p[k] = \sum_{q=1}^{Q} \mathbf{H}_{qp}^{\mathrm{T}}[k]\mathbf{s}_q[k] + \mathbf{n}_p[k], \quad p \in \{1, \ldots, P\} \tag{10.1}$$

where vectors

$$\mathbf{x}_p[k] = \left[x_p[k]\; x_p[k-1]\; \ldots\; x_p[k-L-D+2]\right]^{\mathrm{T}}, \tag{10.2}$$

$$\mathbf{n}_p[k] = \left[n_p[k]\; n_p[k-1]\; \ldots\; n_p[k-L-D+2]\right]^{\mathrm{T}}, \tag{10.3}$$

are of length $L + D - 1$ and contain the p-the microphone signal and the noise components therein, respectively, and length-$M + L + D - 2$ vector

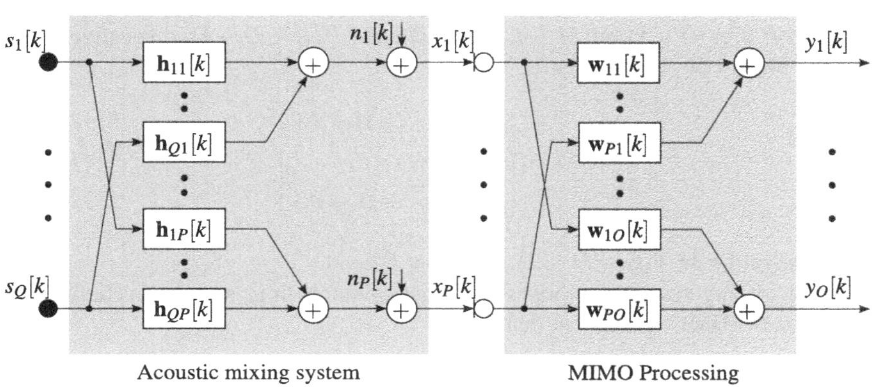

Fig. 10.2 Linear system model comprising the linear acoustic mixing system denoted by the time-varying FIR filters $\mathbf{h}_{qp}[k]$ and the MIMO processing for signal extraction, represented by the time-varying FIR filters $\mathbf{w}_{po}[k]$

$$\mathbf{s}_q[k] = \left[s_q[k]\ s_q[k-1]\ \ldots\ s_q[k-M-L-D+3]\right]^{\mathrm{T}}, \tag{10.4}$$

contains the q-th source signal. Moreover, in (10.1), Q is the number of active point sources, and the linear convolution is expressed as the multiplication of the source signal vectors $\mathbf{s}_q[k]$ with the transpose of a convolution matrix $\mathbf{H}_{qp}[k]$ of dimension $M+L+D-2 \times L+D-1$, which is defined as

$$\mathbf{H}_{qp}[k] = \begin{bmatrix} h_{qp,0}[k] & 0 & \ldots & 0 \\ h_{qp,1}[k] & h_{qp,0}[k] & \ddots & \vdots \\ \vdots & h_{qp,1}[k] & \ddots & 0 \\ h_{qp,M-1}[k] & \vdots & \ddots & h_{qp,0}[k] \\ 0 & h_{qp,M-1}[k] & \ddots & h_{qp,1[k]} \\ \vdots & \ddots & \ddots & \vdots \\ 0 & \ldots & 0 & h_{qp,M-1}[k] \end{bmatrix}. \tag{10.5}$$

Using a more compact matrix/vector notation, (10.1) can be written as

$$\mathbf{x}[k] = \mathbf{H}^{\mathrm{T}}[k]\mathbf{s}[k] + \mathbf{n}[k], \tag{10.6}$$

with vectors

$$\mathbf{x}[k] = \left[\mathbf{x}_1^{\mathrm{T}}[k],\ \mathbf{x}_2^{\mathrm{T}}[k],\ \ldots\ \mathbf{x}_P^{\mathrm{T}}[k]\right]^{\mathrm{T}}, \tag{10.7}$$

$$\mathbf{s}[k] = \left[\mathbf{s}_1^{\mathrm{T}}[k],\ \mathbf{s}_2^{\mathrm{T}}[k],\ \ldots\ \mathbf{s}_Q^{\mathrm{T}}[k]\right]^{\mathrm{T}}, \tag{10.8}$$

$$\mathbf{n}[k] = \left[\mathbf{n}_1^{\mathrm{T}}[k],\ \mathbf{n}_2^{\mathrm{T}}[k],\ \ldots\ \mathbf{n}_P^{\mathrm{T}}[k]\right]^{\mathrm{T}} \tag{10.9}$$

of length $P(L+D-1)$, $Q(M+L+D-2)$, and $P(L+D-1)$, respectively, and with the block-convolution matrix

$$\mathbf{H}[k] = \begin{bmatrix} \mathbf{H}_{11}[k] & \ldots & \mathbf{H}_{1P}[k] \\ \vdots & \ddots & \vdots \\ \mathbf{H}_{Q1}[k] & \ldots & \mathbf{H}_{QP}[k] \end{bmatrix} \tag{10.10}$$

of dimension $Q(M+L+D-2) \times P(L+D-1)$.

The demixing system comprises O output channels in general and is defined by the $PL \times O$ MIMO coefficient matrix:

$$\check{\mathbf{W}}[k] = \left[\mathbf{w}_1[k]\ \mathbf{w}_2[k]\ \ldots\ \mathbf{w}_O[k]\right] = \begin{bmatrix} \mathbf{w}_{11}[k] & \ldots & \mathbf{w}_{1O}[k], \\ \vdots & \ddots & \vdots \\ \mathbf{w}_{P1}[k] & \ldots & \mathbf{w}_{PO}[k] \end{bmatrix}, \tag{10.11}$$

where the length-L demixing filters $\mathbf{w}_{po}k$ are defined analogously to $\mathbf{h}_{qp}[k]$, i.e., $\mathbf{w}_{po}[k] = [w_{po,0}[k], w_{po,1}[k], \ldots, w_{po,L-1}[k]]^{\mathrm{T}}$. Note that we use the $\check{\cdot}$ sign to indicate the difference between a coefficient matrix and its corresponding extended block-convolution matrix used for expressing linear convolution in matrix notation, c.f. $\check{\mathbf{W}}[k]$ and $\mathbf{W}[k]$ in (10.11) and (10.18), respectively. This distinction will become important in Sect. 10.4. The o-th length-D output vector

$$\mathbf{y}_o[k] = \left[y_o[k]\ y_o[k-1]\ \ldots\ y_o[k-D+1]\right]^{\mathrm{T}} \tag{10.12}$$

containing the current and the $D-1$ previous samples can be written as

$$\mathbf{y}_o[k] = \sum_{p=1}^{P} \mathbf{W}_{po}^{\mathrm{T}}[k]\mathbf{x}_p[k], \quad o \in \{1, \ldots, O\}, \tag{10.13}$$

where the $L + D - 1 \times D$-dimensional convolution matrix $\mathbf{W}_{po}[k]$, defined analogously to $\mathbf{H}_{qp}[k]$ in (10.5), is given by

$$\mathbf{W}_{po}[k] = \begin{bmatrix} w_{po,0}[k] & 0 & \ldots & 0 \\ w_{po,1}[k] & w_{po,0}[k] & \ddots & \vdots \\ \vdots & w_{po,1}[k] & \ddots & 0 \\ w_{po,L-1}[k] & \vdots & \ddots & w_{po,0}[k] \\ 0 & w_{po,L-1}[k] & \ddots & w_{po,1}[k] \\ \vdots & \ddots & \ddots & \vdots \\ 0 & \ldots & 0 & w_{po,L-1}[k] \end{bmatrix}. \tag{10.14}$$

Using a compact notation, the length-OD signal vector

$$\mathbf{y}[k] = \left[\mathbf{y}_1^{\mathrm{T}}[k]\ \mathbf{y}_2^{\mathrm{T}}[k]\ \ldots\ \mathbf{y}_O^{\mathrm{T}}[k]\right]^{\mathrm{T}} \tag{10.15}$$

containing all length-D output signal vectors $\mathbf{y}_o[k]$ can be written as

$$\mathbf{y}[k] = \mathbf{W}^{\mathrm{T}}[k]\mathbf{x}[k], \tag{10.16}$$

where vector $\mathbf{x}[k]$ of length $P(L + D - 1)$ is defined as

$$\mathbf{x}[k] = \left[\mathbf{x}_1^{\mathrm{T}}[k]\ \mathbf{x}_2^{\mathrm{T}}[k]\ \ldots\ \mathbf{x}_P^{\mathrm{T}}[k]\right]^{\mathrm{T}}, \tag{10.17}$$

and the block-convolution matrix

$$\mathbf{W}[k] = \begin{bmatrix} \mathbf{W}_{11}[k] & \dots & \mathbf{W}_{1O}[k], \\ \vdots & \ddots & \vdots \\ \mathbf{W}_{P1}[k] & \dots & \mathbf{W}_{PO}[k] \end{bmatrix} \tag{10.18}$$

is of dimension $P(L + D - 1) \times OD$.

For the review of well-known SOS-based multichannel signal extraction algorithms, the model above is specialized to $O = 1$ and $D = 1$ in Sect. 10.3. In this case, the output signal at time index k is given as $y[k] = \mathbf{w}_1^{\mathrm{T}}[k]\mathbf{x}[k]$, where $\mathbf{x}[k]$ reduces to a length-PL vector, c.f. (10.7). An overview over the Multiple-Input Single-Output (MISO) and the general MIMO system models used in Sects. 10.3 and 10.4 is given in Fig. 10.3.

10.3 Multichannel Linear Filtering for Signal Extraction

In this section, we review the well-known SOS-based LCMV filtering concept and its efficient realization in a GSC structure. Then, we discuss their limitations in practical applications and possible countermeasures. As described above, this can be based on a simplified MISO system model (i.e., $O = 1$) with $D = 1$. For brevity, we omit the index $o = 1$ indicating the single output channel in this section.

General MISO system model, used in Sect. 10.4

$$\underbrace{\mathbf{x}[k]}_{A} = \underbrace{\mathbf{H}^{\mathrm{T}}[k]}_{A\times B}\,\underbrace{\mathbf{s}[k]}_{B} + \underbrace{\mathbf{n}[k]}_{A}, \qquad \underbrace{\mathbf{y}[k]}_{OD} = \underbrace{\mathbf{W}^{\mathrm{T}}[k]}_{OD\times A}\,\underbrace{\mathbf{x}[k]}_{A}$$

with $A = P(L + D - 1)$, $B = Q(M + L + D - 2)$

↓ $O = 1,\ D = 1$

MISO system model for SOS-based signal extraction, used in Sect. 10.3

$$\underbrace{\mathbf{x}[k]}_{A} = \underbrace{\mathbf{H}^{\mathrm{T}}[k]}_{A\times B}\,\underbrace{\mathbf{s}[k]}_{B} + \underbrace{\mathbf{n}[k]}_{A}, \qquad y[k] = \underbrace{\mathbf{w}_1^{\mathrm{T}}[k]}_{PL}\,\underbrace{\mathbf{x}[k]}_{PL}$$

with $A = PL$, $B = Q(M + L - 1)$

Fig. 10.3 Overview over the general MIMO system model introduced in Sect. 10.2 and used in Sect. 10.4, and the MISO system model used in Sect. 10.3

10.3.1 Linearly Constrained Minimum Variance Filter

The LCMV filter is a very prominent multichannel linear filtering approach to signal extraction. The FIR filters $\mathbf{w}_p[k]$ of length L are designed to minimize the total noise signal power (including background noise and interfering speakers) at the output of the LCMV filter, subject to a set of linear constraints.[1] The LCMV filter coefficients can be obtained by solving the following optimization problem [14, 33]:

$$\mathbf{w}_{\mathrm{LCMV}}[k] = \arg\min_{\mathbf{w}[k]} \mathbf{w}^{\mathrm{T}}[k]\mathbf{R}_{\tilde{\mathbf{n}}\tilde{\mathbf{n}}}[k]\mathbf{w}[k] \quad \text{s.t.} \quad \mathbf{C}\mathbf{w}[k] = \mathbf{d}. \tag{10.19}$$

The left-hand term in (10.19) represents the noise signal power (including interfering speakers and background noise) at the output of the LCMV filter, and $\mathbf{R}_{\tilde{\mathbf{n}}\tilde{\mathbf{n}}}[k]$ denotes the $PL \times PL$-dimensional correlation matrix of all interfering signal components $\tilde{\mathbf{n}}$ in the microphone signals, which are not directly suppressed at the beamformer output by one of the linear constraints. Matrix $\mathbf{R}_{\tilde{\mathbf{n}}\tilde{\mathbf{n}}}[k]$ is defined as

$$\mathbf{R}_{\tilde{\mathbf{n}}\tilde{\mathbf{n}}}[k] = \mathcal{E}\left\{\tilde{\mathbf{n}}[k]\tilde{\mathbf{n}}^{\mathrm{T}}[k]\right\}, \tag{10.20}$$

where $\mathcal{E}\{\cdot\}$ represents the expectation operator. The right-hand term in (10.19) with the $C(M+L-1) \times PL$ constraint matrix $\mathbf{C}$ and vector $\mathbf{d}$ of length $C(M+L-1)$ represents the set of C linear constraints which can be used to, e.g., preserve the desired source source components at the output and/or to suppress other interfering point sources.

The closed-form solution of (10.19) for the LCMV filter coefficients reads [14, 31, 34]:

$$\mathbf{w}_{\mathrm{LCMV}}[k] = \mathbf{R}_{\tilde{\mathbf{n}}\tilde{\mathbf{n}}}^{-1}[k]\mathbf{C}^{\mathrm{T}}(\mathbf{C}\mathbf{R}_{\tilde{\mathbf{n}}\tilde{\mathbf{n}}}^{-1}[k]\mathbf{C}^{\mathrm{T}})^{-1}\mathbf{d}. \tag{10.21}$$

Minimizing the total output signal power of the filter instead of minimizing the noise output power, i.e., using the correlation matrix $\mathbf{R}_{\mathbf{xx}}[k] = \mathcal{E}\left\{\mathbf{x}[k]\mathbf{x}^{\mathrm{T}}[k]\right\}$ instead of $\mathbf{R}_{\tilde{\mathbf{n}}\tilde{\mathbf{n}}}[k]$ in (10.19), yields to the so-called Linearly Constrained Minimum Power (LCMP) filter [14, 33]. If the linear constraints preserve the desired signal components and if desired signal and interfering signal components are orthogonal, minimizing the total output power (LCMP) is equivalent to minimizing the noise output power (LCMV) [14]. In the following we only consider the LCMV filter, but all statements for LCMV hold for LCMP as well if these assumptions hold.

The number of degrees of freedom of the LCMV filter equals PL, which implies that, if $C(M+L-1)$ equals PL, the LCMV filter coefficients are fully determined by the set of linear constraints. Then, no degrees of freedom could be used to minimize the output power.

If only a distortionless response constraint with respect to the desired source is imposed on the filter coefficients, i.e., $C=1$, the LCMV and LCMP filters specialize

---

[1] The approach of minimizing the output power in the presence of linear constraints was originally presented by Frost in [14] for use with antenna arrays, assuming free-field propagation.

to the well-known Minimum Variance Distortionless Response (MVDR) and the Minimum Power Distortionless Response (MPDR) filter, respectively [14, 33].

10.3.2 The Generalized Sidelobe Canceler

The closed-form solutions of the LCMV/LCMP filters (10.21) require the inversion of $PL \times PL$ correlation matrices, which is computationally expensive [35].

Linearly constrained iterative algorithm after Frost

To reduce computational complexity, Frost presented a linearly constrained iterative algorithm for minimizing the constrained cost function of the LCMV/LCMP filter, which does not require an inversion of the correlation matrix. It follows a first-order gradient-descent procedure. The constrained update equation of the LCMV filter can be derived to [14]

$$\mathbf{w}[k+1] = \mathbf{P}_{\mathrm{C}}^{\perp} \left(\mathbf{w}[k] - \mu \mathbf{R}_{\tilde{\mathbf{n}}\tilde{\mathbf{n}}}[k]\mathbf{w}[k]\right) + \mathbf{v}, \tag{10.22}$$

where the $PL \times PL$-dimensional matrix

$$\mathbf{P}_{\mathrm{C}}^{\perp} = \mathbf{I}_{PL \times PL} - \mathbf{C}^{\mathrm{T}} \left(\mathbf{C}\mathbf{C}^{\mathrm{T}}\right)^{-1} \mathbf{C}, \tag{10.23}$$

is the projector onto the left nullspace of $\mathbf{C}^{\mathrm{T}}$, see, e.g., [36], and vector

$$\mathbf{v} = \mathbf{C}^{\mathrm{T}} \left(\mathbf{C}\mathbf{C}^{\mathrm{T}}\right)^{-1} \mathbf{d}, \tag{10.24}$$

in (10.24) is the minimum norm vector meeting the set of linear constraints. A geometrical interpretation of Frost's constrained update (for a two-dimensional case) is given in Fig. 10.4: The unconstrained update (denoted by $\Delta\mathbf{w}[k]$ in the figure) in parentheses of (10.22) changes the filter vector $\mathbf{w}[k]$ at time instant k towards a direction minimizing the output power, which does not generally satisfy the set of linear constraints (denoted by the green line). The update filter vector is then projected onto the left nullspace (red line) of $\mathbf{C}^{\mathrm{T}}$ by multiplying the updated filter vector with $\mathbf{P}_{\mathrm{C}}^{\perp}$. Finally, the projected filter vector is complemented by $\mathbf{v}$ (blue arrow) so that the resulting vector $\mathbf{w}[k+1]$ meets the constraints again.

Generalized sidelobe canceler (GSC)

An alternative and efficient realization of Frost's constrained optimization problem, the GSC, was proposed in [15, 37, 38]. In Fig. 10.5, a block diagram of the GSC is illustrated. The key idea of the GSC is to divide the filter vector $\mathbf{w}[k+1]$ into two mutually orthogonal components [15]:

$$\mathbf{w}[k+1] = \mathbf{v} - \mathbf{B}_{\mathrm{C}}\mathbf{u}_{\mathrm{C}}^{\perp}[k+1]. \tag{10.25}$$

The first component in (10.25), i.e., the upper branch in Fig. 10.5, is the fixed filter vector $\mathbf{v}$, which can be calculated as projection of $\mathbf{w}[k+1]$ onto the column space of $\mathbf{C}^{\mathrm{T}}$:

$$\mathbf{v} = \mathbf{P}_{\mathrm{C}}\mathbf{w}[k+1] = \mathbf{C}^{\mathrm{T}}\left(\mathbf{C}\mathbf{C}^{\mathrm{T}}\right)^{-1}\mathbf{d}, \tag{10.26}$$

where

$$\mathbf{P}_{\mathrm{C}} = \mathbf{C}^{\mathrm{T}}\left(\mathbf{C}\mathbf{C}^{\mathrm{T}}\right)^{-1}\mathbf{C}. \tag{10.27}$$

The result in (10.26) is equal to (10.24). Hence, the first component of the GSC ensures that the set of linear constraints of the LCMV optimization problem is fulfilled. The second component in (10.25), which corresponds to the lower branch of the GSC structure in Fig. 10.5, consists of the $PL \times PL - C(M+L-1)$-dimensional matrix $\mathbf{B}_{\mathrm{C}}$ and a filter vector $\mathbf{u}_{\mathrm{C}}^{\perp}[k+1]$ of length $PL - C(M+L-1)$. The columns of $\mathbf{B}_{\mathrm{C}}$ form a basis of the left nullspace of $\mathbf{C}^{\mathrm{T}}$, which is orthogonal to the column space of $\mathbf{C}^{\mathrm{T}}$ [36]:

$$\mathbf{C}\mathbf{B}_{\mathrm{C}} = \mathbf{0}_{C(M+L-1)\times PL-C(M+L-1)}. \tag{10.28}$$

Therefore, if for example only a distortionless response constraint is imposed on the LCMV filter vector, $\mathbf{B}_{\mathrm{C}}$ will block the desired source components, and its output

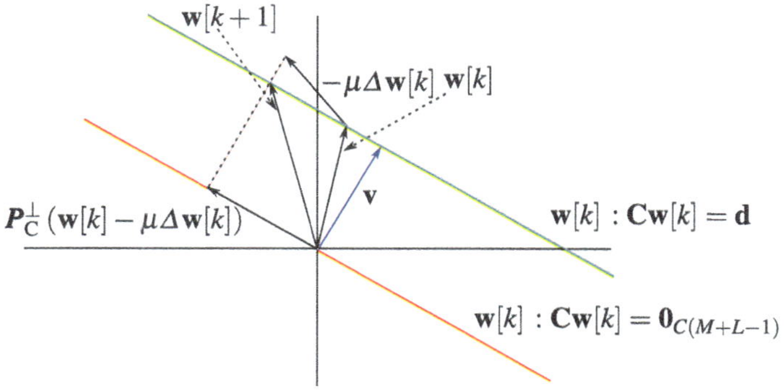

Fig. 10.4 Geometrical interpretation of Frost's linearly constrained update rule (10.22) for a two-dimensional case (after [14])

Fig. 10.5 The GSC as equivalent realization of an LCMV filter (adapted from [18]). Desired signal cancellation at the output of the GSC due to desired signal leakage into the BM is indicated by the green dashed arrows

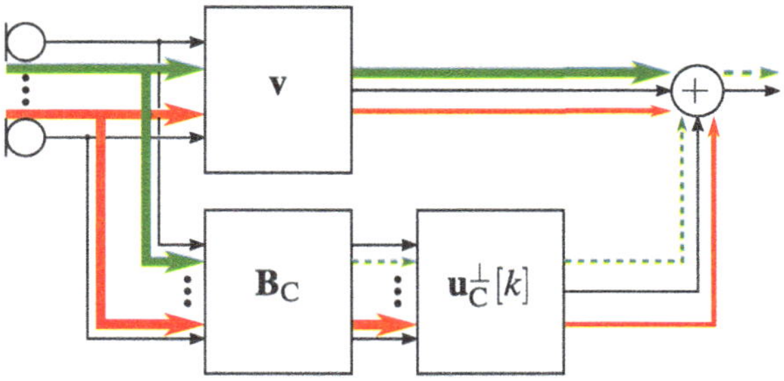

will only contain a filtered version of the interference components which are present at the microphones. Hence, $\mathbf{B}_\mathrm{C}$ is also referred to as Blocking Matrix (BM). It is assumed that $PL > C(M + L - 1)$ or equivalently, the filter length L fulfills $L > C(M - 1)/(P - C)$, such that the nullspace of $\mathbf{C}^\mathrm{T}$ is not equal to zero and the GSC structure exists. The BM can be constructed by orthonormalizing $\mathbf{P}_\mathrm{C}^\perp$ (10.23) and then choosing the $PL - C(M + L - 1)$ columns of the orthonormalized matrix [33], or by applying an orthonormalization procedure to $\mathbf{C}^\mathrm{T}$, e.g., using singular value decomposition [4]. The coefficients of the filter vector $\mathbf{u}_\mathrm{C}^\perp[k]$ are then used to minimize the overall output signal power. Since the linear constraints are ensured by the fixed filter vector in the upper branch of the GSC, an unconstrained Least Mean Square (LMS)-type update strategy can be used to update $\mathbf{u}_\mathrm{C}^\perp$:

$$\mathbf{u}_\mathrm{C}^\perp[k+1] = \mathbf{u}_\mathrm{C}^\perp[k] - \mu \mathbf{B}_\mathrm{C}^\mathrm{T}\mathbf{x}[k]y[k], \tag{10.29}$$

which leads to a less complicated and faster adaptation compared to Frost's constrained adaptive algorithm. The LCMV filter and its GSC realization are strictly equivalent, as was shown in, e.g., [39].

Application of the GSC to acoustic real-world problems and related work

Although the original GSC is a very efficient and practically relevant realization of LCMV filters, it suffers from certain limitations in acoustic real-world scenarios: There, the desired source, typically a human speaker, cannot be expected to remain at the same position. If the desired source's DOA changes, the BM has to be adapted to compensate for this change. Otherwise, the BM will not fully suppress the desired source any more, leading to leakage of desired source components into its output, which, ultimately, will lead to a cancellation of desired signal components at the output of the GSC, as illustrated in Fig. 10.5. Reverberation, which is due to multipath propagation of sound waves, is another issue: the BM should not just suppress the direct path but also reflections originating from the desired source. If this is not the case, leakage of desired signal components into the BM output and, as a consequence, cancellation of desired signal components in the GSC output will occur. In addition, the BM has to be able to adapt to a changing acoustic environment. If an adaptive BM is employed, a sophisticated adaptation control mechanism is necessary to control the adaptation of the BM and INC. More specifically, the BM should be adapted when only the desired source is active, whereas the INC should be adapted when only interfering signals are active, in order to further minimize the risk of desired signal cancellation in the GSC output [18].

Numerous methods have been proposed to mitigate the aforementioned problems of the GSC. In [40–42], the GSC with an adaptive BM was investigated for the first time. A robust GSC with spatio-temporal constraints instead of spatial constraints has then been proposed in [18, 35, 43]. To render the robust GSC more reliable to multi-speaker conditions, the adaptation of the filter coefficients of BM and INC was extended to robust statistics in [44–46], based on the Huber M-estimator [47]. In [48, 49], ICA-based techniques were exploited for adaptation of the INC, simplifying the adaptation control. To cope with the problem of reverberation, the linear constraints of the LCMV or MVDR filter should take the acoustic environment into account. Since knowledge of the AIRs cannot be assumed and their estimation is not trivial, a practically relevant approach to this is the so-called TF-GSC, where the LCMV problem is reformulated to estimate the desired signals as observed by one of the microphones [23, 29]. As a consequence, the set of linear constraints only depends on the so-called Relative Impulse Responses (RIRs) (or their Discrete-Time Fourier Transform (DTFT)-domain counterpart, the RTFs), which describe the linear dependency between microphone signals with respect to the desired source component in the reference signal. An extension of the TF-GSC to multiple linear constraints can be found in [28, 50–53], requiring the RTFs of not only the desired but also all interfering sources. A TF-GSC formulation in the signal subspace domain is given in [25, 26]. To cope with long AIRs (much longer than the block length of the Short-Time Fourier Transform (STFT)), the multiplicative model of the time-domain convolution in the STFT domain was replaced by a convolutive model in the STFT domain in [54, 55].

Although easier than estimating acoustic transfer functions (ATFs), robust and reliable estimation of RTFs is still a complicated task. Several procedures for estimating RTFs exist. SOS-based RTF estimation as multiplicative model in the STFT domain was proposed in, e.g., [23, 56–58], and explicitly exploits the nonstationarity of the speech signal for estimating the RTFs. An extension to a convolutive model in the STFT domain can be found in [54, 59, 60]. Furthermore, subspace-based RTF estimation approaches were proposed in, e.g., [25, 61, 62]. However, these approaches usually require knowledge of the activity of desired and interfering sources which is difficult to obtain in a practical multispeaker scenario. In order to avoid this dependency on Voice Activity Detection (VAD) or Speech Presence Probability (SPP) estimation (see, e.g., [63, 64]), an identification of RTFs for a determined scenario with two sources and two microphones was first addressed in [65]. In [66], an RIR estimation approach based on a constrained BSS algorithm was proposed, which does not require knowledge of the activity of the sources and can also be applied to underdetermined scenarios. This approach will be described in more detail in Sect. 10.4.7.

To summarize, the BM has to be able to cope with a reverberant acoustic environment. Incorporating information on the acoustic conditions as in [23, 29] still requires a robust and reliable estimation procedure for the RTFs needed for the linear constraints of the LCMV filter. Moreover, adaptive versions of the GSC for application to nonstationary broadband signals such as speech require an intelligent and robust adaptation control mechanism for BM and INC, in order to minimize desired signal cancellation [18, 42] and provide sufficient noise and interference suppression.

In the following, we provide an LCMMI signal extraction criterion, realized in a GSC structure, which directly accounts for multipath propagation in real acoustic environments analogously to [23, 29], and exploits fundamental properties of speech and audio signals: Nonstationariy, Nonwhiteness, Nongaussianity. Moreover, by incorporating knowledge of the DOA of the desired source, we present a robust and reliable RTF estimation procedure based on a spatially informed BSS algorithm, which can naturally cope with multiple simultaneously active sources and therefore does not rely on VAD or SPP estimation.

10.4 Linearly Constrained Minimum Mutual Information-Based Signal Extraction

In this section, we present an LCMMI-based signal extraction method, which has been published in [34, 67, 68] in more general form. It is based on the MIMO system model as introduced in Sect. 10.2. Figure 10.6 provides an overview of the derivation of the MMI-based GSC from the LCMMI cost function, as it is presented in the following. In Sect. 10.4.1 we discuss the generic LCMMI optimization criterion for signal extraction. Subsequently, we derive a constrained gradient-descent update rule for minimizing the generic cost function in Sect. 10.4.2. Via specifying the set of constraints and interpreting the constrained update rule in Sects. 10.4.3 and 10.4.4, respectively, we arrive an efficient realization of the LCMMI criterion in a GSC structure in Sect. 10.4.5. For practical use of the resulting MMI-based GSC, an efficient realization of its adaptive BM and the RTF estimates with respect to the desired source are required. This is developed in Sects. 10.4.6 and 10.4.7. Finally, Higher-Order Statistics (HOS)- and SOS-based realizations of the proposed MMI-based GSC are presented in Sect. 10.4.8. To conclude this section, links to some generic linear signal extraction methods based on second-order statistics are established in Sect. 10.4.9. Note that in the following, we assume the number of inputs being equal to the number of outputs in our system model, i.e., $O = P$.

10.4.1 Generic Optimization Criterion

The MMI-based cost function, which forms the basis of the presented signal extraction approach, is based on Shannon's mutual information [69]. It has already been used as a basis for the definition of the TRIple-N Independent component analysis for CONvolutive mixtures (TRINICON) framework, which was efficiently exploited for convolutive BSS in [8, 16] and extensively analyzed in [70]. In [71], it was described in a more general form for broadband MIMO filtering, and was originally complemented with constraints incorporating prior knowledge into the MMI-based cost function in [72]. Linear constraints were then reconsidered in [67, 68], leading to the LCMMI cost function for signal extraction in the time domain, defined as [34, 67, 68, 72]:

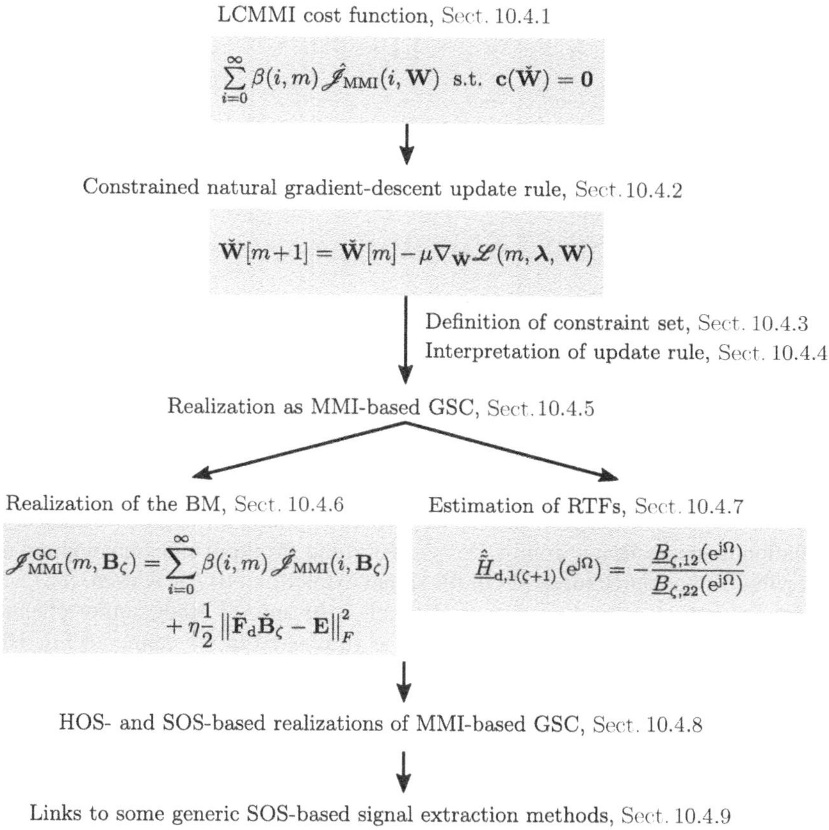

Fig. 10.6 Overview of derivation of MMI-based GSC from LCMMI cost function, as presented in Sect. 10.4

$$\mathscr{J}_{\mathrm{MMI}}(m,\mathbf{W}) = \sum_{i=0}^{\infty} \beta(i,m) \underbrace{\frac{1}{N} \sum_{k=iL}^{iL+N-1} \log \left\{ \frac{\hat{f}_{\mathbf{y},PD}(\mathbf{y}[k])}{\hat{f}_{\mathbf{y}_{\mathrm{s}},PD}(\mathbf{y}[k])} \right\}}_{\hat{\mathscr{J}}_{\mathrm{MMI}}(i,\mathbf{W})}$$
$$\text{subject to} \quad \mathbf{c}(\check{\mathbf{W}}) = \mathbf{0}_{PC(M+L-1)}, \tag{10.30}$$

where $\hat{\mathscr{J}}_{\mathrm{MMI}}(i,\mathbf{W})$ is an estimate of the Kullback-Leibler Divergence (KLD) between an estimate of the $PD \times 1$-dimensional multivariate joint Probability Density Function (PDF) $\hat{f}_{\mathbf{y},PD}(\mathbf{y}[k])$ of the output signal vector $\mathbf{y}[k]$ (10.15), and an estimate of the multivariate source model PDF $\hat{f}_{\mathbf{y}_{\mathrm{s}},PD}(\mathbf{y}[k])$.

When exploiting the cost function for BSS as in [8, 16, 70, 71], the following source model PDF is used:

$$\hat{f}^{\text{ICA}}_{\mathbf{y}_s,PD}(\mathbf{y}[k]) = \prod_{o=1}^{P} \hat{f}_{\mathbf{y}_o,D}(\mathbf{y}_o[k]). \tag{10.31}$$

In this work, we do not aim at separating all sources from each other, but at separating the desired source components from all remaining interfering signal components. To this end, we use the following source model PDF for Signal Extraction (SE) [67, 68]:

$$\hat{f}^{\text{SE}}_{\mathbf{y}_s,PD}(\mathbf{y}[k]) = \hat{f}_{\mathbf{y}_1,D}(\mathbf{y}_1[k])\hat{f}_{\mathbf{y}_{2:P},(P-1)D}(\mathbf{y}_{2:P}[k]), \tag{10.32}$$

where

$$\mathbf{y}_{2:P}[k] = \left[\mathbf{y}_2^{\text{T}}[k]\ \mathbf{y}_3^{\text{T}}[k]\ \ldots\ \mathbf{y}_P^{\text{T}}[k]\right]^{\text{T}} \tag{10.33}$$

of length$(P-1)D$ comprises the remaining $P-1$ output channels containing the undesired signal components. The differences between SOS-based versions of the cost function for the two different source model PDFs is illustrated in Fig. 10.7: The matrix on the left-hand side illustrates the general $PD \times PD$-dimensional block-correlation matrix $\mathbf{R}_{\mathbf{yy}}$ which contains the $D \times D$-dimensional correlation matrices $\mathbf{R}_{\mathbf{y}_p\mathbf{y}_q}$, $p, q \in \{1, \ldots, P\}$ of the P output channels without any processing. If the source model PDF for signal extraction (10.32) is used, the cross-correlations between the first output channel $y_1[k]$ and all other output channels $y_p[k]$, $p \in \{2, \ldots, P\}$ will be minimized, as illustrated in the center of Fig. 10.7. For the application to BSS, the source model PDF in (10.31) is used. Consequently, the cross-correlations between all output channels are minimized, leading to a block-diagonal correlation matrix at the output of the MIMO system, as illustrated in the right-hand side of Fig. 10.7.

The set of constraints $\mathbf{c}(\check{\mathbf{W}}) = \mathbf{0}_{PC(M+L-1)}$ in (10.30) is given as [66, 67, 72]:

$$\begin{aligned} &\mathbf{C}\check{\mathbf{W}} = \mathbf{D} \\ \Rightarrow\ &\mathbf{c}\left(\check{\mathbf{W}}\right) = \text{vec}\left(\mathbf{C}\check{\mathbf{W}} - \mathbf{D}\right) = \mathbf{0}_{PC(M+L-1)}, \end{aligned} \tag{10.34}$$

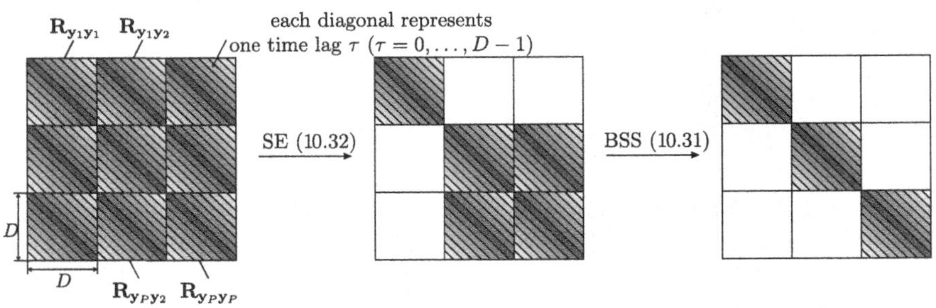

Fig. 10.7 Fundamental difference between the SOS-based versions of MMI-based signal extraction (SE) using the source model PDF in (10.32), and BSS using the source model PDF in (10.31), resulting in a specific decorrelation of the output channel signals

where $\mathbf{C}$ represents the general constraint matrix of size $C(M+L-1) \times PL$, $\mathbf{D}$ denotes the $C(M+L-1) \times P$-dimensional desired response matrix, and $\mathbf{0}_{PC(M+L-1)}$ is a vector of zeros of length $PC(M+L-1)$. The operator vec$(\cdot)$ yields the stacked columns of a matrix. Analogously to LCMV filtering, the constraints can be used to preserve the desired signal components in a specific output channel, and to explicitly suppress interfering speakers.

The generic LCMMI cost function for signal extraction (10.30) accounts for the three fundamental signal properties of speech and audio signals, i.e., Nongaussianity, Nonwhiteness and Nonstationarity, as follows [8, 16]:

(a) **Nongaussianity** can be exploited by using nongaussian PDFs in the cost function, which will be discussed in Sect. 10.4.8.
(b) **Nonwhiteness** is accounted for by considering output cross-relations over D consecutive samples for all P outputs, summarized in the length-PD output signal vector $\mathbf{y}[k]$ (10.15) (10.16). In general, the number of consecutive output samples D is chosen as $1 \leq D \leq L$. If $D = 1$, then only a single sample per output channel is considered, which will lead to an optimization criterion that disregards the nonwhiteness property of the underlying signals. When choosing $D > 1$, temporal statistical dependencies are accounted for. The statistical properties are modeled by the joint PD-variate PDF, which describes both intra-channel, i.e., temporal dependencies, and inter-channel dependencies.
(c) **Nonstationarity** of the signals is taken into account by averaging over multiple blocks of length N, each weighted by the weighting function $\beta(i, m)$ with finite support. Within the individual blocks, ergodicity is assumed and ensemble averaging typically required to estimate the KLD is replaced by time averaging over these N blocks. The weighting function $\beta(i, m)$ is normalized so that $\sum_{i=0}^{\infty} \beta(i, m) = 1$, and allows for offline, online, and block-online realizations of the signal extraction algorithm [8]. The block indices i and m refer to the blocks, which are used for estimating the multivariate PDFs.

10.4.2 Constrained Natural Gradient-Descent for Iterative Optimization Update Rule

The LCMMI cost function can be minimized using a gradient-descent approach. The demixing filter matrix $\check{\mathbf{W}}[m+1]$ at update step $m+1$ is given as

$$\check{\mathbf{W}}[m+1] = \check{\mathbf{W}}[m] - \mu \nabla_{\check{\mathbf{W}}} \mathscr{L}(m, \boldsymbol{\lambda}, \mathbf{W}), \tag{10.35}$$

where μ is the stepsize parameter, $\nabla_{\check{\mathbf{W}}} \mathscr{L}(m, \boldsymbol{\lambda}, \mathbf{W})$ is the gradient of the Lagrangian of the LCMMI cost function with respect to $\check{\mathbf{W}}$, and $\boldsymbol{\lambda}$ is a vector of Lagrange multipliers.

For an improved convergence behavior, we use a natural gradient-based update, which was originally presented in [8, 16, 73]. Applying the results therein to the problem at hand yields the natural gradient for the linearly constrained optimization problem [34, 68, 72]:

$$\nabla_{\check{\mathbf{W}}}\mathscr{L}(m,\boldsymbol{\lambda},\mathbf{W}) = \sum_{i=0}^{\infty}\beta(i,m)\mathscr{SC}\left\{\mathbf{W}\mathbf{W}^{\mathrm{T}}\nabla_{\mathbf{W}}\hat{\mathscr{J}}_{\mathrm{MMI}}(i,\mathbf{W})\right\} + \mathbf{C}^{\mathrm{T}}\boldsymbol{\Lambda}, \quad (10.36)$$

where $\mathbf{C}^{\mathrm{T}}\boldsymbol{\Lambda}$ is the partial derivative of $\mathbf{c}\left(\check{\mathbf{W}}\right)$ with respect to $\check{\mathbf{W}}$ and $\boldsymbol{\Lambda}$ represents a matrix of Lagrange multipliers of size $C(M+L-1)\times P$. The $\mathscr{S}$ylvester $\mathscr{C}$onstraint operator $\mathscr{SC}\{\cdot\}$ [8, 74] relates the $P(L+D-1)\times PD$-dimensional gradient of the Lagrangian with respect to matrix $\mathbf{W}$ to the $PL\times P$-dimensional gradient of the Lagrangian with respect to matrix $\check{\mathbf{W}}$ [34, 68]:

$$\nabla_{\check{\mathbf{W}}}\hat{\mathscr{J}}_{\mathrm{MMI}}(i,\mathbf{W}) = \mathscr{SC}\left\{\nabla_{\mathbf{W}}\hat{\mathscr{J}}_{\mathrm{MMI}}(i,\mathbf{W})\right\}. \quad (10.37)$$

A mathematical definition of $\mathscr{SC}\{\cdot\}$ can be found in, e.g., [70]. It corresponds to a sum of the diagonal elements of the $L+D-1\times D$ sub-matrices of $\nabla_{\mathbf{W}}\hat{\mathscr{J}}_{\mathrm{MMI}}(i,\mathbf{W})$, as illustrated in Fig. 10.8.

The gradient $\nabla_{\mathbf{W}}\hat{\mathscr{J}}_{\mathrm{MMI}}(i,\mathbf{W})$ of the KLD estimate with respect to $\mathbf{W}$, which is required for (10.36), can be derived to [34, 67, 68]

$$\nabla_{\mathbf{W}}\hat{\mathscr{J}}_{\mathrm{MMI}}(i,\mathbf{W}) = \frac{1}{N}\sum_{k=iL}^{iL+N-1}\mathbf{x}[k]\boldsymbol{\Phi}_{\mathrm{SE}}^{\mathrm{T}}(\mathbf{y}[k]) - \frac{\partial\log\hat{f}_{\mathbf{y},PD}(\mathbf{y}[k])}{\partial\mathbf{W}}, \quad (10.38)$$

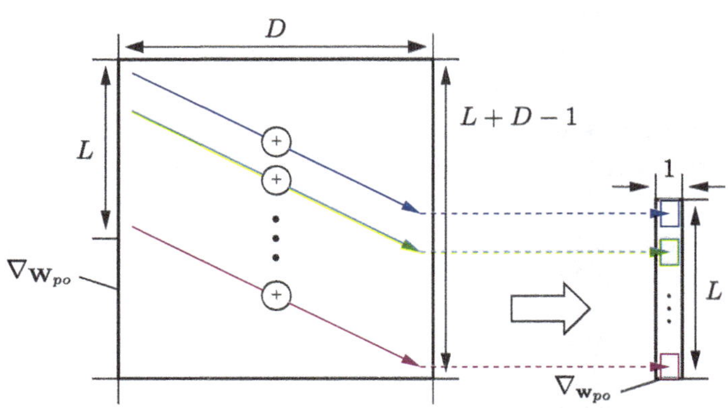

Fig. 10.8 Illustration of the $\mathscr{S}$ylvester $\mathscr{C}$onstraint operator for the gradient $\nabla_{\mathbf{W}}\hat{\mathscr{J}}_{\mathrm{MMI}}(i,\mathbf{W})$ with respect to the po-th submatrix $\mathbf{W}_{po}$ (after [68, 70, 71])

where the multivariate score function $\boldsymbol{\Phi}_{\mathrm{SE}}$, consisting of the stacked multivariate score functions of the first $(\boldsymbol{\Phi}^{\mathrm{T}}_{1,D}(\mathbf{y}_1[k]))$ and the remaining output channels $(\boldsymbol{\Phi}^{\mathrm{T}}_{2:P,(P-1)D}(\mathbf{y}_{2:P}[k]))$, is defined as [34, 67, 68, 70]

$$\begin{aligned}\boldsymbol{\Phi}_{\mathrm{SE}}(\mathbf{y}[k]) &= \left[\boldsymbol{\Phi}^{\mathrm{T}}_{1,D}(\mathbf{y}_1[k])\ \boldsymbol{\Phi}^{\mathrm{T}}_{2:P,(P-1)D}(\mathbf{y}_{2:P}[k])\right]^{\mathrm{T}} \\ &= \left[\left(-\frac{\partial \log \hat{f}_{\mathbf{y}_1,D}(\mathbf{y}_1[k])}{\partial \mathbf{y}_1[k])}\right)^{\mathrm{T}} \left(-\frac{\partial \log \hat{f}_{\mathbf{y}_{2:P},(P-1)D}(\mathbf{y}_{2:P}[k])}{\partial \mathbf{y}_{2:P}[k])}\right)^{\mathrm{T}}\right]^{\mathrm{T}}\end{aligned} \tag{10.39}$$

and corresponds to the source model PDF for signal extraction $\hat{f}^{\mathrm{SE}}_{\mathbf{y}_s,PD}(\mathbf{y}[k])$ in (10.32). As already indicated above, we will discuss the choice of the PDF and, therefore, the choice of the corresponding multivariate score function in Sect. 10.4.8.

10.4.3 Definition of the Set of Constraints

In this work, we assume knowledge of the DOA of only the desired source. We impose one constraint, i.e., $C = 1$, to each MISO subsystem $\mathbf{w}_o[k]$, $o \in \{1, \ldots, P\}$ creating the o-th output $y_o[k]$. We aim at extracting the desired source components as observed by a reference microphone (here: the first microphone) in the first output channel, while the desired signal components should be suppressed in all remaining output channels, as illustrated schematically in Fig. 10.9. This can be formulated as

$$y_{\mathrm{d},1}[k] = \mathbf{w}_1^{\mathrm{T}}[k]\mathbf{H}_{\mathrm{d}}^{\mathrm{T}}[k]\mathbf{s}_{\mathrm{d}}[k] \stackrel{!}{=} \mathbf{H}_{\mathrm{d}1}^{\mathrm{T}}[k]\mathbf{s}_{\mathrm{d}}[k] \tag{10.40}$$

$$y_{\mathrm{d},o}[k] = \mathbf{w}_o^{\mathrm{T}}[k]\mathbf{H}_{\mathrm{d}}^{\mathrm{T}}[k]\mathbf{s}_{\mathrm{d}}[k] \stackrel{!}{=} 0, \quad o \in \{2, \ldots, P\}, \tag{10.41}$$

where $y_{\mathrm{d},o}[k]$, $o \in \{1, \ldots, P\}$ represents the desired signal components in the o-th output channel, $\mathbf{w}_o[k]$ is defined in (10.11), and vector $\mathbf{s}_{\mathrm{d}}[k]$ of length $M + L - 1$ contains the desired source signal components. Moreover, matrix

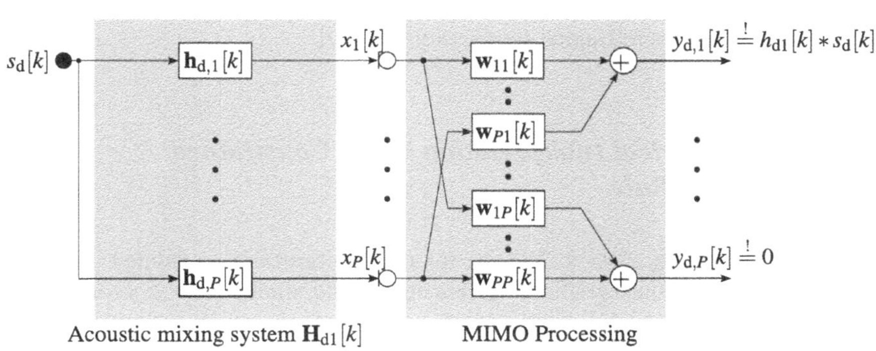

Fig. 10.9 Schematic illustration of the set of constraints defined in (10.40) and (10.41)

$$\mathbf{H}_{\mathrm{d}}[k] = \left[\mathbf{H}_{\mathrm{d}1}[k]\ \mathbf{H}_{\mathrm{d}2}[k]\ \ldots\ \mathbf{H}_{\mathrm{d}P}[k]\right] \tag{10.42}$$

of dimension $M + L - 1 \times PL$, where sub-matrices $\mathbf{H}_{\mathrm{d}p}[k]$, $p \in \{1, \ldots, P\}$ of dimension $M + L - 1 \times L$ are defined analogously to $\mathbf{H}_{qp}[k]$ in (10.5), and capture all AIRs from the desired source to the microphones.

Equation (10.40) represents a distortionless response constraint with respect to the desired signal components in the reference microphone channel, similar to the one used for MVDR filtering. In combination with the nullbeamformers with respect to the desired signal components for the remaining $P - 1$ output signals, this can be seen as the BSS solution for the desired signal components. If also DOA information on other interfering point sources is available, multiple constraints could be imposed on each output channel, as presented in [75].

Equations (10.40) and (10.41) correspond to the following choice of constraint matrix $\mathbf{C}$ and desired response matrix $\mathbf{D}$, defining the set of linear constraints (10.34) of the LCMMI optimization criterion [34, 68]:

$$\mathbf{C} = \tilde{\mathbf{H}}_{\mathrm{d}}[k] = \left[\begin{bmatrix}\mathbf{I}_{L\times L}\\ \mathbf{0}_{M-1\times L}\end{bmatrix}\ \tilde{\mathbf{H}}_{\mathrm{d},12}[k]\ \ldots\ \tilde{\mathbf{H}}_{\mathrm{d},1P}[k]\right], \tag{10.43}$$

$$\mathbf{D} = \mathbf{E} = \left[\mathbf{e}_1\ \mathbf{0}\ \ldots\ \mathbf{0}\right], \tag{10.44}$$

where sub-matrices $\tilde{\mathbf{H}}_{\mathrm{d},1p}[k]$ of dimension $M + L - 1 \times L$ are defined analogously to $\mathbf{H}_{qp}[k]$ in (10.5) and represent convolution matrices constructed from the RIRs $\tilde{\mathbf{h}}_{\mathrm{d},1p}$, which relate the desired signal components in the reference microphone channel to those contained in the p-th microphone channel. Moreover, matrices $\mathbf{I}_{L\times L}$ and $\mathbf{0}_{M-1\times L}$ in (10.43) represent an identity matrix of dimension $L \times L$ and an all-zero matrix of dimension $M - 1 \times L$, respectively, vector $\mathbf{e}_1$ is a vector of length $M + L - 1$ with a 1 as its first element and zeros elsewhere, and vectors $\mathbf{0}$ in (10.44) contain zeros and are of length $M + L - 1$. Since we construct the constraint matrix from RIRs, in analogy to [23–25, 76], the focus is on undesired signal suppression only, and no equalization of the ATF between desired source and reference microphone is performed. This has the advantage that all remaining degrees of freedom can be used for the suppression of interfering signal components, which results in optimum performance of undesired signal suppression [77–79].

10.4.4 Geometrical Interpretation of the Constrained Update Rule

From the gradient $\nabla_{\check{\mathbf{W}}}\mathcal{L}(m, \boldsymbol{\lambda}, \mathbf{W})$ (10.36) and the constraint set defined in (10.43) and (10.44), a constrained gradient-descent update rule, similar to Frost's constrained update rule (10.22)–(10.24), for the MIMO coefficient matrix $\check{\mathbf{W}}[k]$ can be derived [34, 67, 68]:

$$\check{\mathbf{W}}[m+1] = \mathbf{P}^{\perp}_{\tilde{\mathbf{H}}_d}\left(\check{\mathbf{W}}[m] - \mu \sum_{i=0}^{\infty} \beta(i,m)\mathscr{SC}\left\{\mathbf{W}\mathbf{W}^{\mathrm{T}}\nabla_{\mathbf{W}}\hat{\mathscr{J}}_{\mathrm{MMI}}(i,\mathbf{W})\right\}\right) + \mathbf{V}, \tag{10.45}$$

with

$$\mathbf{P}^{\perp}_{\tilde{\mathbf{H}}_d} = \mathbf{I}_{PL\times PL} - \tilde{\mathbf{H}}_d^{\mathrm{T}}\left(\tilde{\mathbf{H}}_d\tilde{\mathbf{H}}_d^{\mathrm{T}}\right)^{-1}\tilde{\mathbf{H}}_d, \tag{10.46}$$

$$\mathbf{V} = \tilde{\mathbf{H}}_d^{\mathrm{T}}\left(\tilde{\mathbf{H}}_d\tilde{\mathbf{H}}_d^{\mathrm{T}}\right)^{-1}\mathbf{E}. \tag{10.47}$$

Analogously to the geometrical interpretation given in Sect. 10.3.2, Fig. 10.4, after each update step m, the updated filter coefficients $\mathbf{w}_o,\ o \in \{1,\dots,P\}$ of each MISO system are projected into the left nullspace of $\tilde{\mathbf{H}}_d^{\mathrm{T}}$ by a multiplication with the projection matrix $\mathbf{P}^{\perp}_{\tilde{\mathbf{H}}_d}$. Afterwards, the projected filter coefficients are augmented with a vector $\mathbf{v}_o$ such that the linear constraints in (10.43) and (10.44) are met. Vectors $\mathbf{v}_o,\ o \in \{1,\dots,P\}$ are the columns of matrix $\mathbf{V}$ in (10.45,), i.e., $\mathbf{V} = [\mathbf{v}_1,\dots,\mathbf{v}_P]$.

As can be seen from the constraint set in (10.43) and (10.44), a distortionless response constraint with respect to the desired source is imposed on the MISO system $\mathbf{w}_1[k]$ creating the first output channel, whereas the desired signal is to be suppressed in all remaining output channels. Therefore, $\mathbf{v}_1$ is given analogously to the result in (10.24) for the LCMV filter. Hence, after augmentation with $\mathbf{v}_1$, $\mathbf{w}_1[m+1]$ will satisfy the linear constraint $\tilde{\mathbf{H}}_d\mathbf{w}_1 = \mathbf{e}_1$. On the other hand, since $\tilde{\mathbf{H}}_d\mathbf{w}_o = \mathbf{0},\ o \in \{2,\dots,P\}$ according to (10.44), the filter vectors $\mathbf{v}_o,\ o \in \{2,\dots,P\}$ are zero, and the MISO filter vectors $\mathbf{w}_o,\ o \in \{2,\dots,P\}$ will always be projected into the left nullspace of $\tilde{\mathbf{H}}_d^{\mathrm{T}}$. In this case, the constrained filter update for $\mathbf{w}_1[m]$ in (10.45) can be simplified to [34, 67, 68]

$$\mathbf{w}_1[m+1] = \mathbf{P}^{\perp}_{\tilde{\mathbf{H}}_d}\left(\mathbf{w}_1[m] - \mu\sum_{i=0}^{\infty}\beta(i,m)\mathscr{SC}\left\{\frac{1}{N}\sum_{k=iL}^{iL+N-1}\mathbf{x}[k]\boldsymbol{\Phi}^{\mathrm{T}}_{1,D}(\mathbf{y}_1[k])\right\}\right) + \mathbf{v}_1 \tag{10.48}$$

Since we are only interested in the desired signal components, it is sufficient to realize the constrained update for $\mathbf{w}_1$ (10.48) in a GSC structure.

10.4.5 Realization as Minimum Mutual Information-Based Generalized Sidelobe Canceler

In the following, the principles in [15, 38] are applied to the gradient-descent update rule for the MISO system $\mathbf{w}_1[k]$ in (10.48), analogously to Sect. 10.3.2, in order to derive an MMI-based GSC realization of the LCMMI signal extraction scheme [34, 67, 68]. The filter vector $\mathbf{w}_1[m+1]$ is divided into two mutually orthogonal components as follows:

$$\mathbf{w}_1[m+1] = \mathbf{v}_1 + \mathbf{B}_{\tilde{\mathbf{H}}_d}\mathbf{u}^{\perp}_{\tilde{\mathbf{H}}_d,\mathrm{MMI}}[m+1]. \tag{10.49}$$

The first term in (10.49), i.e., the filter vector $\mathbf{v}_1$ is obtained by the projection of $\mathbf{w}_1[m+1]$ onto the column space of $\tilde{\mathbf{H}}_{\mathrm{d}}^{\mathrm{T}}$:

$$\mathbf{v}_1 = \mathbf{P}_{\tilde{\mathrm{H}}_{\mathrm{d}}} \mathbf{w}_1[m+1] = \tilde{\mathbf{H}}_{\mathrm{d}}^{\mathrm{T}} \left(\tilde{\mathbf{H}}_{\mathrm{d}} \tilde{\mathbf{H}}_{\mathrm{d}}^{\mathrm{T}} \right)^{-1} \mathbf{e}_1 \tag{10.50}$$

with projection matrix

$$\mathbf{P}_{\tilde{\mathrm{H}}_{\mathrm{d}}} = \tilde{\mathbf{H}}_{\mathrm{d}}^{\mathrm{T}} \left(\tilde{\mathbf{H}}_{\mathrm{d}} \tilde{\mathbf{H}}_{\mathrm{d}}^{\mathrm{T}} \right)^{-1} \tilde{\mathbf{H}}_{\mathrm{d}}. \tag{10.51}$$

Analogously to (10.28), the second term in (10.49) is a linear combination of the columns of the $PL \times (P-1)L - M + 1$ BM $\mathbf{B}_{\tilde{\mathrm{H}}_{\mathrm{d}}}$. Its columns are chosen to form a basis of the left nullspace of $\tilde{\mathbf{H}}_{\mathrm{d}}^{\mathrm{T}}$, which is orthogonal to the column space of $\tilde{\mathbf{H}}_{\mathrm{d}}^{\mathrm{T}}$:

$$\tilde{\mathbf{H}}_{\mathrm{d}} \mathbf{B}_{\tilde{\mathrm{H}}_{\mathrm{d}}} = \mathbf{0}_{M+L-1 \times (P-1)L-M+1}. \tag{10.52}$$

In analogy to Sect. 10.3.2, $PL > M + L - 1$ is required or, equivalently, the filter length L has to fulfill $L > (M-1)/(P-1)$ such that the nullspace of $\tilde{\mathbf{H}}_{\mathrm{d}}^{\mathrm{T}}$ is not equal to zero and the GSC structure exists. From the set of linear constraints in (10.40) and (10.41), it can be seen that the first filter vector $\mathbf{w}_1$ of the coefficient matrix $\check{\mathbf{W}}$ provides a distortionless desired signal estimate (as received by the reference microphone), and the remaining filter vectors $\mathbf{w}_o$, $o \in \{2, \ldots, P\}$ provide $P-1$ estimates of all undesired signal components. As a consequence, the $P-1$ filter vectors $\mathbf{w}_o$, $o \in \{2, \ldots, P\}$ can be directly used as BM in the GSC structure realizing $\mathbf{w}_1$ [34]. Correspondingly, we can decompose the MIMO coefficient matrix $\check{\mathbf{W}}$ as

$$\check{\mathbf{W}} = \left[\mathbf{w}_1 | \check{\mathbf{B}} \right], \tag{10.53}$$

where the $PL \times P-1$ dimensional coefficient matrix $\check{\mathbf{B}} = [\mathbf{w}_2, \ldots, \mathbf{w}_P]$ summarizes the filter coefficients of the blocking matrix [34]. Finally, the filter vector $\mathbf{u}^{\perp}_{\tilde{\mathrm{H}}_{\mathrm{d}},\mathrm{MMI}}[m+1]$ in (10.49) of length $(P-1)L - M + 1 = (P-1)L_{\mathrm{INC}}$ is used to minimize mutual information between the signal components in the output of the BM and in the output of the GSC by the general unconstrained gradient-descent optimization procedure

$$\mathbf{u}^{\perp}_{\tilde{\mathrm{H}}_{\mathrm{d}},\mathrm{MMI}}[m+1] = \mathbf{u}^{\perp}_{\tilde{\mathrm{H}}_{\mathrm{d}},\mathrm{MMI}}[m] - \mu \sum_{i=0}^{\infty} \beta(i,m) \mathscr{S}\mathscr{C} \left\{ \frac{1}{N} \sum_{k=iL}^{iL+N-1} \mathbf{B}^{\mathrm{T}} \tilde{\mathbf{x}}[k] \boldsymbol{\Phi}^{\mathrm{T}}_{1,D}(\mathbf{y}_1[k]) \right\}, \tag{10.54}$$

where $\mathbf{B}$ is a block-convolution matrix extension of $\check{\mathbf{B}}$, and the $P-1$-channel signal $\mathbf{B}^{\mathrm{T}} \tilde{\mathbf{x}}[k]$ represents an estimate of all undesired signal components at the output of the BM. In practical applications, the filter length of the INC is approximated as $L_{\mathrm{INC}} \approx L$, as the true filter length M of the AIRs is not known [34].

10.4.6 Realization of the Blocking Matrix

For efficiently minimizing MMI, we resort to two-channel subunits well-investigated for ICA-based blind source separation and realize the BM by a set of $P-1$ parallel two-channel MIMO subsystems $\check{\mathbf{B}}_\zeta$, $\zeta \in \{1, \ldots, P-1\}$, as illustrated in Fig. 10.10. The ζ-th subsystem is applied to the reference microphone (here: the first microphone) and the $\zeta+1$-th microphone, and creates one of the $P-1$ BM output signals $y_{\hat{\mathrm{n}},\zeta}[k]$, $\zeta \in \{1, \ldots, P-1\}$ containing the noise reference. A detailed illustration of the ζ-th two-channel MIMO subsystem, which is described by the $2L \times 2$ coefficient matrix

$$\check{\mathbf{B}}_\zeta[k] = \begin{bmatrix} \mathbf{b}_{\zeta,11}[k] & \mathbf{b}_{\zeta,12}[k] \\ \mathbf{b}_{\zeta,21}[k] & \mathbf{b}_{\zeta,22}[k] \end{bmatrix}, \quad \zeta \in \{1, \ldots, P-1\} \tag{10.55}$$

is given in Fig. 10.11. The length-L FIR filters $\mathbf{b}_{\zeta,\rho 2}[k]$, $\rho \in \{1, 2\}$, which create the second output signal, suppress the desired source components as illustrated by the red arrows in Fig. 10.11, and, hence, create the noise reference signal $y_{\hat{\mathrm{n}},\zeta}[k]$. Note that only the filter vectors creating the second output signal of each subsystem are relevant for the realization of the entire BM. However, the filter vectors creating the first output channel are still required for calculating the update for the filter vectors, as presented in the following.

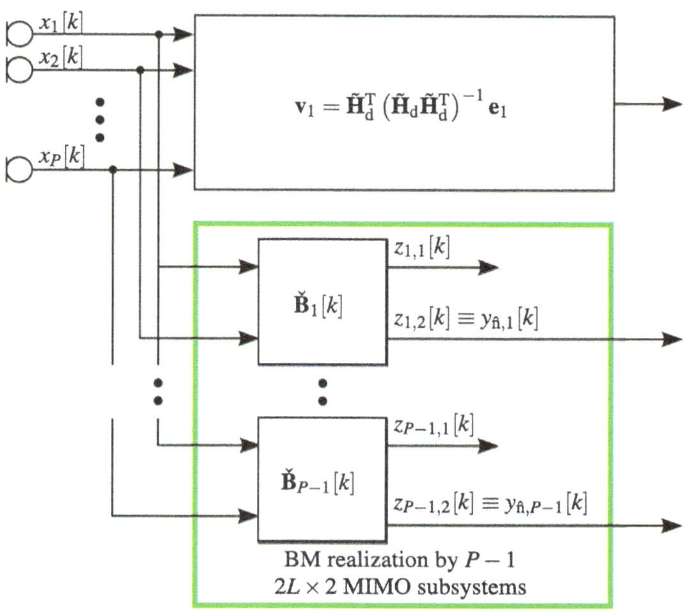

Fig. 10.10 Realization of the BM (green box) of the MMI-based GSC by $P-1$ two-channel MIMO subsystems $\check{\mathbf{B}}_\zeta$, $\zeta \in \{1, \ldots, P-1\}$ running in parallel (adapted from [34])

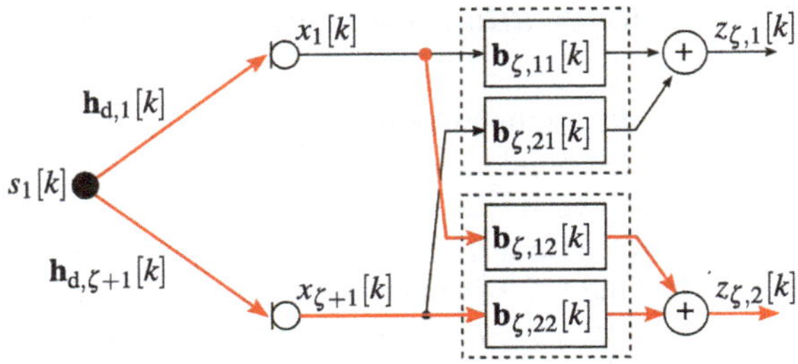

Fig. 10.11 Illustration of one of the $P-1$ two-channel MIMO subsystems $\check{\mathbf{B}}_\zeta$, used to realize the BM of the MMI-based GSC

The FIR filters $\mathbf{b}_{\zeta,\rho o}$, $\rho, o \in \{1,2\}$ are determined based on an Geometrically Constrained Minimum Mutual Information (GC-MMI) criterion [30], which exploits prior knowledge on the desired source DOA for a two-channel MMI-based BSS algorithm for estimating noise and interference components in underdetermined scenarios. The method of exploiting TDOA/DOA information for the BSS application in determined scenarios was originally proposed by [27] and further analyzed in [80–82]. The GC-MMI cost function to be minimized is given as [30]

$$\mathscr{J}_{\mathrm{MMI}}^{\mathrm{GC}}(m,\mathbf{B}_\zeta) = \sum_{i=0}^{\infty} \beta(i,m) \frac{1}{N} \sum_{k=iL}^{iL+N-1} \log\left\{ \frac{\hat{f}_{\mathbf{z}_\zeta,2D}(\mathbf{z}_\zeta[k])}{\hat{f}^{\mathrm{SE}}_{\mathbf{z}_s,2D}(\mathbf{z}_\zeta[k])} \right\} + \eta \frac{1}{2} \left\| \tilde{\mathbf{F}}_{\mathrm{d}} \check{\mathbf{B}}_\zeta - \mathbf{E} \right\|_F^2, \tag{10.56}$$

where block-convolution matrix $\mathbf{B}_\zeta$ of dimension $2(L+D-1) \times 2D$ is defined analogously to $\mathbf{W}[k]$ in (10.18), $\tilde{\mathbf{F}}_{\mathrm{d}}$ is a free-field approximation of $\tilde{\mathbf{H}}_{\mathrm{d}}$ in (10.43), $\mathbf{E}$ is defined as in (10.44), and $\|\cdot\|_F$ represents the Frobenius norm of a matrix. The scalar η denotes the weight of the penalty term, and is typically in the range $0.4 < \eta < 0.6$ [30]. The first part of (10.56) is equal to the estimate of the KLD between an estimate of the multivariate joint PDF $f_{\mathbf{z}_\zeta,2D}(\mathbf{z}_\zeta[k])$ of the length-$2D$ output signal vector $\mathbf{z}_\zeta[k]$, defined analogously to $\mathbf{y}[k]$ in (10.15):

$$\mathbf{z}_\zeta[k] = \left[\mathbf{z}^{\mathrm{T}}_{\zeta,1}[k]\ \mathbf{z}^{\mathrm{T}}_{\zeta,2}[k]\right]^{\mathrm{T}} \tag{10.57}$$

$$\mathbf{z}_{\zeta,\rho}[k] = \left[z_{\zeta,\rho}[k]\ z_{\zeta,\rho}[k-1] \ldots z_{\zeta,\rho}[k-D+1]\right]^{\mathrm{T}}, \quad \rho \in \{1,2\}, \tag{10.58}$$

and an estimate of the multivariate source model PDF $\hat{f}^{\mathrm{SE}}_{\mathbf{z}_s,2D}(\mathbf{z}_\zeta[k])$ for signal extraction (10.32). In contrast to the LCMMI criterion, the GC-MMI criterion only incorporates a soft constraint on the filter coefficients instead of the set of hard constraints of the LCMMI criterion in (10.34). The importance of the soft constraint can be adjusted by the scalar η. The soft constraint is used to suppress the direct path components of the desired source in the second output of each two-channel MIMO subsystem. Minimizing the GC-MMI cost function enforces statistical independence between the two output signals. As a consequence, also correlated echoes of the desired source signal will be identified and removed from the second output channel, which makes this approach superior to nullbeamformers which only suppress the direct path components as shown in [30, 83]. To realize the penalty term, knowledge of the direct path TDOA of the desired source at the respective microphones is sufficient, since this is required to construct matrix $\tilde{\mathbf{F}}_\mathrm{d}$. Since we only aim at suppressing the desired signal components in one output at this stage, we can achieve this goal even in underdetermined scenarios with more than two sources.

Equivalently to the LCMMI update, we use a gradient-descent strategy for updating the filter coefficients of each two-channel subsystem $\breve{\mathbf{B}}_\zeta$:

$$\breve{\mathbf{B}}_\zeta[m+1] = \breve{\mathbf{B}}_\zeta[m] - \mu \nabla_{\breve{\mathbf{B}}_\zeta} \mathscr{J}^{\mathrm{GC}}_{\mathrm{MMI}}(m, \mathbf{B}_\zeta). \tag{10.59}$$

The gradient $\nabla_{\breve{\mathbf{B}}_\zeta} \mathscr{J}^{\mathrm{GC}}_{\mathrm{MMI}}(m, \mathbf{B}_\zeta)$ of the GC-MMI criterion can be derived to [34, 66, 68]:

$$\begin{aligned}\nabla_{\breve{\mathbf{B}}_\zeta} \mathscr{J}^{\mathrm{GC}}_{\mathrm{MMI}}(m, \mathbf{B}_\zeta) = \sum_{i=0}^{\infty} \beta(i,m) \mathscr{S}\mathscr{C}\left\{\mathbf{B}_\zeta \mathbf{B}_\zeta^{\mathrm{T}} \nabla_{\mathbf{B}_\zeta} \hat{\mathscr{J}}_{\mathrm{MMI}}(i, \mathbf{B}_\zeta)\right\} + \\ \eta \tilde{\mathbf{F}}_\mathrm{d}^{\mathrm{T}} \left(\tilde{\mathbf{F}}_\mathrm{d} \breve{\mathbf{B}}_\zeta - \mathbf{E}\right),\end{aligned} \tag{10.60}$$

where $\nabla_{\mathbf{B}_\zeta} \hat{\mathscr{J}}_{\mathrm{MMI}}(i, \breve{\mathbf{B}}_\zeta)$ is defined in analogously to (10.38).

10.4.7 Estimation of the Set of Constraints

In addition to an adaptive BM which can cope with multipath propagation, estimates of the RIRs between the desired source components in the reference microphone and in all other microphones are required for realizing the MMI-based GSC. More specifically, the RIRs are required to construct the constraint matrix $\tilde{\mathbf{H}}_\mathrm{d}$ in (10.43) which is needed to realize the fixed beamformer $\mathbf{v}_1$ (10.50).

As proposed in [66], the RIR between the first and $\zeta+1$-th microphone can be estimated from the filter vectors $\mathbf{b}_{\zeta,12}[k]$ and $\mathbf{b}_{\zeta,22}[k]$ of the ζ-th two-channel subsystem $\breve{\mathbf{B}}_\zeta$ (10.55), as explained in the following. For ease of presentation, the explanation will be given here in the DTFT domain, where RTFs are estimated as DTFT transforms of RIRs. In the following, DTFT-transformed quantities are indicated with an underscore. For example, $\underline{B}_{\zeta,12}(\mathrm{e}^{\mathrm{j}\Omega})$ is the DTFT transform of $\mathbf{b}_{\zeta,12}[k]$, with normalized angular frequency $\Omega = 2\pi f/f_\mathrm{s}$, frequency f, and sampling frequency f_s. The soft constraint in (10.56) requires the desired source to be suppressed in the second output channel. Hence, the signal path from the desired signal to the second output channel of the ζ-th subsystem, highlighted by red arrows in Fig. 10.11, has to be equal to zero:

$$\underline{H}_{\mathrm{d},1}(\mathrm{e}^{\mathrm{j}\Omega}) \underline{B}_{\zeta,12}(\mathrm{e}^{\mathrm{j}\Omega}) + \underline{H}_{\mathrm{d},\zeta+1}(\mathrm{e}^{\mathrm{j}\Omega}) \underline{B}_{\zeta,22}(\mathrm{e}^{\mathrm{j}\Omega}) \stackrel{!}{=} 0, \tag{10.61}$$

Reformulating (10.61), the RTF $\underline{\tilde{H}}_{\mathrm{d},1(\zeta+1)}$ between the reference (here: first) and the $\zeta+1$-th microphone can be estimated from the DTFT-transformed FIR filters of the ζ-th two-channel MIMO subsystem $\check{\mathbf{B}}_\zeta[k]$ as follows [34, 66]:

$$\hat{\underline{\tilde{H}}}_{\mathrm{d},1(\zeta+1)}(\mathrm{e}^{\mathrm{j}\Omega}) = \frac{\underline{H}_{\mathrm{d},\zeta+1}(\mathrm{e}^{\mathrm{j}\Omega})}{\underline{H}_{\mathrm{d},1}(\mathrm{e}^{\mathrm{j}\Omega})} = -\frac{\underline{B}_{\zeta,12}(\mathrm{e}^{\mathrm{j}\Omega})}{\underline{B}_{\zeta,22}(\mathrm{e}^{\mathrm{j}\Omega})}. \tag{10.62}$$

The corresponding RIR $\hat{\tilde{\mathbf{h}}}_{\mathrm{d},1(\zeta+1)}$ can then be obtained by an inverse DTFT transform of $\hat{\underline{\tilde{H}}}_{\mathrm{d},1(\zeta+1)}$ in (10.62).

The fact that we require the RIRs between the desired source components in the first and in all other microphones with respect to the desired source to realize the fixed beamformer, and the fact that we can estimate these RIRs from the two-channel subsystems, are the reasons for using the first microphone channel as common input to the $P-1$ two-channel MIMO subsystems realizing the BM of the MMI-based GSC.

The important advantage of this method over other existing RIR/RTF estimation methods is that it does not require access to time segments in which only the desired source is active [66], i.e., it is well-suited for multi-speaker scenarios in which a reliable detection of these segments is difficult if not impossible. A detailed investigation of the GC-MMI-based RTF estimation technique can be found in [34, 66].

As the RIRs are estimated from the FIR filters of the two-channel MIMO subsystems realizing the BM, the following sequence of steps is necessary to realize the MMI-based GSC: First, all $P-1$ two-channel MIMO subsystems $\check{\mathbf{B}}_\zeta$ are updated according to (10.59) and (10.60). Second, the RTFs are estimated from the BM filter vectors according to (10.62), transformed into the time domain and the fixed beamformer $\mathbf{v}_{\mathrm{l}}$ (10.50) is realized. Finally, the filter coefficients of the INC are updated using the noise reference at the output of the BM according to (10.54) and the noise estimate is subtracted from the output of the fixed beamformer, yielding the final output signal.

10.4.8 Special Source Models

Both, the update term of the BM filter vectors (10.60) as well as the update term of the INC filter vectors (10.54) requires a multivariate score function $\boldsymbol{\Phi}_{o,D}(\mathbf{y}_o[k])$ defined in (10.39), which depends on the source model PDF. By choosing a nongaussian source model PDF, HOS can be exploited for signal extraction.

In this work, two different source model PDFs are used for SOS- and HOS-based realizations of the MMI-based GSC. Both source model PDFs are derived from the general model of a zero-mean nonwhite Spherically Invariant Random Process (SIRP) [84], which has the advantage that the estimation of the multivariate PDFs reduces to an estimation of correlation matrices only. For brevity, we only present the final results. For a detailed derivation, see, e.g., [8, 16, 34, 70].

Higher-Order Statistics (HOS)-based MMI-GSC realization

For a HOS-based realization of the MMI-based GSC, a multivariate Laplacian PDF is used as source model PDF. The corresponding score function $\boldsymbol{\Phi}_{o,D}(\mathbf{y}_o[k])$ for $\mathbf{y}_o[k]$ is derived to [8, 16, 70, 71]:

$$\boldsymbol{\Phi}_{o,D}(\mathbf{y}_o[k]) = 2 \underbrace{\frac{1}{\sqrt{2r_o[k]}} \frac{I_{D/2}\left(\sqrt{2r_o[k]}\right)}{I_{D/2-1}\left(\sqrt{2r_o[k]}\right)}}_{:=\phi_{\mathbf{y}_o}(r_o[k])} \hat{\mathbf{R}}^{-1}_{\mathbf{y}_o\mathbf{y}_o}[i]\mathbf{y}_o[k], \tag{10.63}$$

where $I_\xi(\cdot)$ denotes the ξ-th order modified Bessel function of the second kind, and $r_o[k]$ is defined as

$$r_o[k] = \mathbf{y}_o^{\mathrm{T}}[k]\hat{\mathbf{R}}^{-1}_{\mathbf{y}_o\mathbf{y}_o}[i]\mathbf{y}_o[k]. \tag{10.64}$$

Matrix $\hat{\mathbf{R}}_{\mathbf{y}_o\mathbf{y}_o}[i]$ in (10.64) represents the $D \times D$ correlation matrix estimate of the output signal vector $\mathbf{y}_o[k]$, $k = iL + j$, $j \in \{0, ..., N-1\}$ of length D, defined in (10.12). Now, let us define a $D \times D$ cross-relation matrix between the output signal vector $\mathbf{y}_p[k]$ and a nonlinearly weighted output signal vector $\mathbf{y}_o[k]$ as [8, 16]:

$$\hat{\mathbf{R}}_{\mathbf{y}_p\phi_{\mathbf{y}}(\mathbf{y}_o)}[i] = \frac{1}{N}\sum_{k=iL}^{iL+N-1} \mathbf{y}_p[k]\phi_{\mathbf{y}_o}(r_o[k])\mathbf{y}_o^{\mathrm{T}}[k] \tag{10.65}$$

with nonlinear weight $\phi_{\mathbf{y}_o}(r_o[k])$ defined in (10.63), and $r_o[k]$ defined in (10.64). Inserting $\boldsymbol{\Phi}_{o,D}(\mathbf{y}_o[k])$ (10.63) into the gradient of the two-channel BM subsystems (10.60) and applying the definition of the cross-relation matrices in (10.65) leads to the HOS-based update term for $\breve{\mathbf{B}}_\zeta$ [34, 67, 68]:

$$\begin{aligned}\nabla^{\mathrm{HOS}}_{\breve{\mathbf{B}}_\zeta} \mathscr{J}^{\mathrm{GC}}_{\mathrm{MMI}}(m, \mathbf{B}_\zeta) = \sum_{i=0}^{\infty} \beta(i,m)\mathscr{S}\mathscr{C}\left\{\mathbf{B}_\zeta \mathrm{boff}\left\{\hat{\mathbf{R}}_{\mathbf{z}_\zeta\phi_{\mathbf{z}}(\mathbf{z}_\zeta)}[i]\right\} \mathrm{bdiag}^{-1}\left\{\hat{\mathbf{R}}_{\mathbf{z}_\zeta\mathbf{z}_\zeta}[i]\right\}\right\} + \\ \eta\tilde{\mathbf{F}}_{\mathrm{d}}^{\mathrm{T}}\left(\tilde{\mathbf{F}}_{\mathrm{d}}\breve{\mathbf{B}}_\zeta - \mathbf{E}\right),\end{aligned} \tag{10.66}$$

where the operators boff$\{\cdot\}$ and bdiag$\{\cdot\}$ set the diagonal sub-matrices and the off-diagonal sub-matrices, respectively, of a block matrix to zero. The $2D \times 2D$ matrix $\hat{\mathbf{R}}_{\mathbf{z}_\zeta\phi_{\mathbf{z}}(\mathbf{z}_\zeta)}[i]$ summarizes the $D \times D$-dimensional estimated channel-wise cross-relation matrices $\hat{\mathbf{R}}_{\mathbf{z}_{\zeta,p}\phi_{\mathbf{z}}(\mathbf{z}_{\zeta,p})}[i]$ which are defined analogously to $\hat{\mathbf{R}}_{\mathbf{y}_p\phi_{\mathbf{y}}(\mathbf{y}_o)}[i]$ in (10.65). Vectors $\mathbf{z}_\zeta[k]$ and $\mathbf{z}_{\zeta,p}[k]$ are defined in (10.57) and (10.58), respectively.

Analogously, specializing the score function in the update for the INC (10.54), the HOS-based update for the INC is obtained [34, 68]:

$$\Delta\mathbf{u}^{\perp}_{\tilde{\mathrm{H}}_{\mathrm{d}},\mathrm{HOS}}[m] = \sum_{i=0}^{\infty} \beta(i,m)\mathscr{S}\mathscr{C}\left\{\hat{\mathbf{R}}_{\mathbf{y}_{\tilde{n}}\phi_{\mathbf{y}}(\mathbf{y}_1)}[i]\hat{\mathbf{R}}^{-1}_{\mathbf{y}_1\mathbf{y}_1}[i]\right\}, \tag{10.67}$$

where $\hat{\mathbf{R}}_{\mathbf{y}_{\hat{\mathrm{n}}}\phi_{\mathbf{y}}(\mathbf{y}_1)}[i]$, of dimension $(P-1)(L+D-1)\times D$ summarizes the $L+D-1\times D$ channel-wise cross-relation matrices $\hat{\mathbf{R}}_{\mathbf{y}_{\hat{\mathrm{n}},\zeta}\phi_{\mathbf{y}}(\mathbf{y}_1)}[i]$, $\zeta \in \{1,\ldots,P-1\}$ between the noise reference signal vector $\mathbf{y}_{\hat{\mathrm{n}},\zeta}[k]$, of length $L+D-1$ and the nonlinearly weighted output signal vector $\mathbf{y}_1[k]$. The required signal vectors $\mathbf{y}_{\hat{\mathrm{n}}}[k] = \mathbf{B}^{\mathrm{T}}\tilde{\mathbf{x}}[k]$ and $\mathbf{y}_{\hat{\mathrm{n}},\zeta}[k]$ are defined as:

$$\mathbf{y}_{\hat{\mathrm{n}}}^{\mathrm{T}}[k] = \left[\mathbf{y}_{\hat{\mathrm{n}},1}^{\mathrm{T}}[k]\ \mathbf{y}_{\hat{\mathrm{n}},2}^{\mathrm{T}}[k]\ \ldots\ \mathbf{y}_{\hat{\mathrm{n}},P-1}^{\mathrm{T}}[k]\right]^{\mathrm{T}} \tag{10.68}$$

$$\mathbf{y}_{\hat{\mathrm{n}},\zeta}[k] = \left[y_{\hat{\mathrm{n}},\zeta}[k]\ y_{\hat{\mathrm{n}},\zeta}[k-1]\ \ldots\ y_{\hat{\mathrm{n}},\zeta}[k-L_{\mathrm{INC}}-D+2]\right]^{\mathrm{T}},\quad \zeta\in\{1,\ldots,P-1\} \tag{10.69}$$

second order statistics (SOS)-based MMI-GSC realization

If Gaussian source models are used, SOS-based updates are obtained. In this case, the multivariate score function is given as [8, 16, 70, 71]

$$\boldsymbol{\Phi}_{o,D}(\mathbf{y}_o[k]) = \hat{\mathbf{R}}_{\mathbf{y}_o\mathbf{y}_o}^{-1}[i]\mathbf{y}_o[k], \tag{10.70}$$

and the scalar, generally nonlinear, weight simplifies to $\phi_{\mathbf{y}_o}(r_o[k]) = 1/2$. The SOS-based update term for the two-channel MIMO subsystems is given as [34, 68]

$$\begin{aligned}\nabla_{\check{\mathbf{B}}_\zeta}^{\mathrm{SOS}} \mathscr{J}_{\mathrm{MMI}}^{\mathrm{GC}}(m,\mathbf{B}_\zeta) = & \sum_{i=0}^{\infty} \beta(i,m)\mathscr{SC}\left\{\mathbf{B}_\zeta \mathrm{boff}\left\{\hat{\mathbf{R}}_{\mathbf{z}_\zeta\mathbf{z}_\zeta}[i]\right\}\mathrm{bdiag}^{-1}\left\{\hat{\mathbf{R}}_{\mathbf{z}_\zeta\mathbf{z}_\zeta}[i]\right\}\right\} + \\ & \eta\tilde{\mathbf{F}}_{\mathrm{d}}^{\mathrm{T}}\left(\tilde{\mathbf{F}}_{\mathrm{d}}\check{\mathbf{B}}_\zeta - \mathbf{E}\right),\end{aligned} \tag{10.71}$$

where the $2D \times 2D$ correlation matrices $\hat{\mathbf{R}}_{\mathbf{z}_\zeta\mathbf{z}_\zeta}[i]$ summarize the $D \times D$ cross-correlation matrices $\hat{\mathbf{R}}_{\mathbf{z}_{\zeta,\rho}\mathbf{z}_{\zeta,\rho}}[i]$ of the two output signal vectors $\mathbf{z}_{\zeta,\rho}[k]$, $\rho \in \{1,2\}$. The SOS-based update for the INC coefficients can be derived to [34, 68]:

$$\Delta\mathbf{u}_{\tilde{\mathrm{H}}_{\mathrm{d}},\mathrm{SOS}}^{\perp}[m] = \sum_{i=0}^{\infty}\beta(i,m)\mathscr{SC}\left\{\hat{\mathbf{R}}_{\mathbf{y}_{\hat{\mathrm{n}}}\mathbf{y}_1}[i]\hat{\mathbf{R}}_{\mathbf{y}_1\mathbf{y}_1}^{-1}[i]\right\}, \tag{10.72}$$

where the correlation matrix $\hat{\mathbf{R}}_{\mathbf{y}_{\hat{\mathrm{n}}}\mathbf{y}_1}[i]$ represents the channel-wise cross-relation matrices $\hat{\mathbf{R}}_{\mathbf{y}_{\hat{\mathrm{n}},\zeta}\mathbf{y}_1}[i]$, estimated from the individual BM output signal vectors $\mathbf{y}_{\hat{\mathrm{n}},\zeta}[k]$, $\zeta \in \{1,\ldots,P-1\}$ and the output signal vector $\mathbf{y}_1[k]$. Note that as opposed to the HOS-based realizations no nonlinear weighting is applied any more to the signal vectors before estimating the cross-relation matrices for calculating the SOS-based update terms.

10.4.9 Links to Some Generic Linear Signal Extraction Methods Based on Second-Order Statistics

To conclude this section, relations between the presented MMI-based GSC and the SOS-based signal extraction algorithms presented in Sect. 10.3.1 are established.

Starting point for this discussion is the SOS-based realization of the MMI-based GSC as introduced above. If we specialize to white source signals, i.e., if we approximate the source signals $s_q[k]$ and correspondingly the output signal $y_1[k]$ as normally distributed, stationary white signals, then the output signal vector reduces to $\mathbf{y}_1[k] = y_1[k]$, and the correlation matrix $\hat{\mathbf{R}}_{\mathbf{y}_1\mathbf{y}_1}[k]$ can be written as a diagonal matrix $\hat{\mathbf{R}}_{\mathbf{y}_1\mathbf{y}_1}[k] = \sigma_{y_1}^2\mathbf{I}$, where $\sigma_{y_1}^2$ denotes the power of $y_1[k]$. Moreover, due to the assumption of a stationary white output signal with $\hat{\mathbf{R}}_{\mathbf{y}_1\mathbf{y}_1}[k] = \sigma_{y_1}^2\mathbf{I}$ and $D = 1$, the $\mathcal{SC}$-operator can be omitted. Additionally, the window function $\beta(i, m)$ reduces to a rectangular window, since with the given assumptions, any pair of input signal vector and output signal sample is processed independently of any other pair. Finally, by replacing averaging over N blocks by the current estimate and assuming unit variance for $y_1[k]$, the original supervised LMS-based update rule [85, 86] for the INC coefficients results from (10.54) and (10.72):

$$\mathbf{u}_{\bar{\mathbf{H}}_\mathrm{d},\mathrm{LMS}}^{\perp}[k+1] = \mathbf{u}_{\bar{\mathbf{H}}_\mathrm{d},\mathrm{LMS}}^{\perp}[k] - \mu\check{\mathbf{y}}_{\hat{\mathrm{n}}}[k]y_1[k], \tag{10.73}$$

where $\check{\mathbf{y}}_{\hat{\mathrm{n}}}[k]$ is a truncated version of $\mathbf{y}_{\hat{\mathrm{n}}}[k]$ and $y_1[k]$ represents the error signal [86].

If we further specialize from reverberant to anechoic, i.e., free-field, acoustic conditions and assume that a delay compensation for the phase differences of the desired source is applied to the microphone signals as in [14, 15], the AIR vectors $\mathbf{h}_{1p}$, $p \in \{1, \ldots, P\}$ simplify to unit vectors. As a consequence, the quiescent vector $\mathbf{v}_1$ (10.50) results in a delay-and-sum beamformer, and the BM can be realized by a set of pairwise delay-and-subtract beamformers. Hence, the MMI-based realization of the GSC reduces to the original GSC structure proposed in [15].

Similarly, considering the constrained MMI-type update rule for the MISO system $\mathbf{w}_1[k]$ in (10.48) and approximating the source signals $s_q[k]$ (and consequently the output signal $y_1[k]$) as normally distributed stationary white signals with unit variance, and replacing the average over N blocks by the current estimate, the original linearly constrained LMS update (10.22) as proposed in [14] is obtained:

$$\mathbf{w}_1[k+1] = \mathbf{P}_{\bar{\mathbf{H}}_\mathrm{d}}^{\perp}(\mathbf{w}_1[k] - \mu\mathbf{x}[k]y_1[k]) + \mathbf{v}_1. \tag{10.74}$$

From these obtained relations, we can draw the following conclusion: Existing methods for signal extraction are based on simplistic adaptation rules, which facilitate a practical realization. However, from an MMI-based perspective, these approaches are only optimum for normally distributed and stationary white source signals, which does not reflect the characteristic properties of speech and audio signals. Accounting for these fundamental properties leads to more general and complex adaptation rules as derived above. Nevertheless, efficient realizations of the derived update rules exist [70, 87, 88]. Therefore, the LCMMI approach for signal extraction is very attractive for speech and audio signal processing applications. Future work might be the application of the proposed approach to adaptive array geometries as in, e.g., [89], as well as incorporating multiple constraints into the LCMMI cost function [75].

10.5 Experiments

In this section, we demonstrate the effectiveness of the proposed MMI-based GSC. To this end, we first describe the experimental setup in Sect. 10.5.1. Then, we demonstrate our RIR estimation method in Sect. 10.5.2, followed by an investigation of the signal enhancement performance of the proposed MMI-based GSC in Sect. 10.5.3.

10.5.1 Experimental Setup

For the evaluation, we employ a uniform linear six-element microphone array with inter-element spacing of $d = 0.042$ m and a total array aperture of 0.21 m, as illustrated in Fig. 10.12. As test scenario, we consider a two-speaker scenario, where the desired source, a male speaker, is located to the right of the microphone array ($\phi = -45°$), and the interfering source, a female speaker, is located to the left of the array at an angle of $\phi = 45°$. The source-sensor distance is 1 m for both speakers.

The microphone signals were created by convolving the clean speech signals of length 10 s and sampling rate $f_s = 16$ kHz with AIRs which were measured in an acoustic environment of dimensions 272 cm × 253 cm × 237 cm with reverberation time $T_{60} \approx 50$ ms.[2] The two source signals were of the same power.

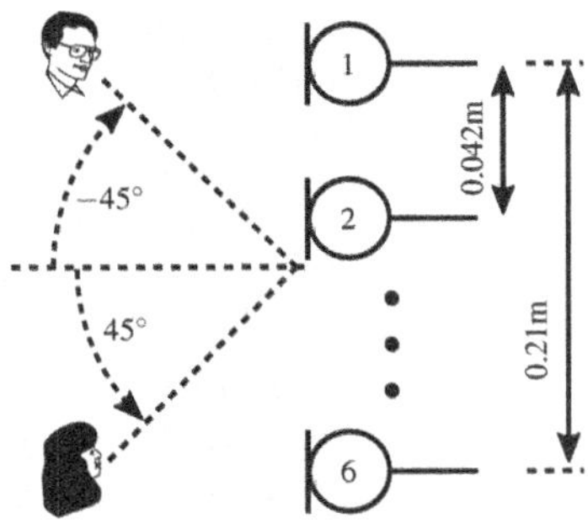

Fig. 10.12 Illustration of the employed six-channel uniform linear microphone array

[2]Note that we chose this reverberation time in order to demonstrate the advantage of the HOS-based realization over the SOS-based realization of the MMI-based GSC.

For this evaluation, we realize the algorithm as an offline or so-called batch algorithm. Then, the weighting function $\beta(i, m)$ corresponds to a rectangular window, which is described by, e.g., [8, 70]:

$$\beta(i, m) = \frac{1}{K_{\text{sig}}} \varepsilon_{0,(K_{\text{sig}}-1)}(i), \tag{10.75}$$

where $\varepsilon_{a,b}(i)$ is a rectangular window function, i.e., $\varepsilon_{a,b} = 1$ for $a \leq i \leq b$, and $\varepsilon_{a,b}(i) = 0$ elsewhere. To calculate the update for the coefficient matrix $\check{\mathbf{W}}^l$, where superscript l indexes the current offline iteration, the following steps are carried out: First, the input signal is segmented into K_{sig} blocks of length N. For each block, an individual gradient $\nabla_{\check{\mathbf{W}}} \mathcal{J}(i, \check{\mathbf{W}})$ is calculated. After this, all K_{sig} gradients are averaged, and the coefficient matrix is updated according to:

$$\check{\mathbf{W}}^{l+1} = \check{\mathbf{W}}^{l} - \mu \frac{1}{K_{\text{sig}}} \sum_{i=0}^{K_{\text{sig}}-1} \nabla_{\check{\mathbf{W}}} \mathcal{J}(i, \check{\mathbf{W}}^{l}), \tag{10.76}$$

where due to the offline processing, the update does not depend on the block-time index m any more, but on the iteration l. This procedure is repeated $l_{\max}$ times. Since in each update step, the offline algorithm uses the entire data, it will lead to the most precise estimate of the filter coefficients [8, 34]. We use here $l_{\max} = 200$ offline iterations for updating the filter coefficients of the BM and of the INC.

For both, BM and INC, a filter length of $L = 1024$ taps was used and the length of the offline blocks was set to $N = 2L = 2048$. For exploiting the nonwhiteness property, $D = L = 1024$ samples were used for adapting the BM filter coefficients. For updating the INC filter coefficients $L = 1024$ and $D = 2$ or $D = 1024$ was used. Furthermore, the stepsize parameter μ was set individually for each microphone pair, with values between $\mu = 2.5 \cdot 10^{-5}$ and $\mu = 5 \cdot 10^{-5}$, to reach convergence of each HOS-based BM two-channel subsystem. For the update of the INC coefficients, an adaptive stepsize control, with initial stepsize μ_{init} in the range between $5 \cdot 10^{-6}$ and $5 \cdot 10^{-4}$ was used for the different realizations. The main idea behind the stepsize control is to increase the stepsize if the value of the cost function $\mathcal{J}(m)$ is decreased compared to $\mathcal{J}(m-1)$ (indicating convergence of the algorithm), and to decrease $\mu(m)$ if $\mathcal{J}(m)$ exceeds $\mathcal{J}(m-1)$ by more than a specified ratio (indicating local divergence of the algorithm) [87]:

$$\mu(m+1) = \begin{cases} a \cdot \mu(m) & \text{if } \mathcal{J}(m) < \mathcal{J}(m-1) \\ b \cdot \mu(m) & \text{if } \mathcal{J}(m) \geq c \cdot \mathcal{J}(m-1) \\ \mu(m) & \text{otherwise,} \end{cases} \tag{10.77}$$

where a, b, and c were set to $a = c = 1.005$ and $b = 0.85$ in this work, respectively. Moreover, to avoid instabilities, we restricted the stepsize to a finite range $10 \cdot \mu_{\text{init}} \leq \mu \leq \mu_{\text{init}}/100$. To avoid the additional complexity of evaluating $\mathcal{J}(m)$ directly, we calculated the Frobenius norm of the output cross-correlation matrices instead. We set the weight for the penalty term of the geometric constraint to $\eta = 0.5$. As in [34, 66, 68], this weight is only used for initialization of the BM as a delay-and-subtract beamformer suppressing the direct propagation path, and is removed after the first offline iteration, i.e., the weight is set to $\eta = 0$ for all other offline iterations, since the evaluated scenario is a static scenario. If the weight for the penalty term is kept constant during the entire adaptation process, the adaptation rule needs to compromise between the MMI criterion and the penalty term in each update step, which will prevent the BM from converging to a better solution. If a dynamic scenario is evaluated, the penalty term should only be applied during runtime if the direction of the spatial null of the BM starts converging towards an interfering source [34]. For comparability, all parameter settings were used for both SOS- and HOS-based realizations of the MMI-based GSC, and are summarized in Table 10.1.

It should be noted that, for computational efficiency, we approximate the inverse of the $D \times D$ auto-correlation matrices by a narrowband approximation in the frequency domain as proposed in [88].

10.5.2 Estimation of Relative Impulse Responses

To evaluate the estimation accuracy of the required RTFs, we calculate the Normalized RTF Estimation Error (NRE) between the true RTF $\underline{\tilde{H}}_{\mathrm{d},1p}$ (calculated from the ATFs) and the estimated RTF $\hat{\underline{\tilde{H}}}_{\mathrm{d},1p}$ (estimated from the BM FIR filters as presented in Sect. 10.4.7), which is defined as

$$\mathrm{NRE}_p(\mathrm{e}^{\mathrm{j}\Omega}) = 10\log_{10}\frac{|\hat{\underline{\tilde{H}}}_{\mathrm{d},1p}(\mathrm{e}^{\mathrm{j}\Omega}) - \underline{\tilde{H}}_{\mathrm{d},1p}(\mathrm{e}^{\mathrm{j}\Omega})|^2}{|\underline{\tilde{H}}_{\mathrm{d},1p}(\mathrm{e}^{\mathrm{j}\Omega})|^2}. \tag{10.78}$$

As an example, we illustrate the RTF estimation accuracy for the microphone pair consisting of microphones 1 and 6 (c.f. Fig. 10.12) with a spacing of 0.21 m.

Table 10.1 Summary of the parameters of the setting for the evaluated offline time-domain algorithms

	BM	INC
Filter length L	1024	1024
Block length N	$N = 2L = 2048$	$N = 2L = 2048$
Number of output samples D	$D = L = 1024$	$D = L = 1024$, $D = 2$
Stepsize μ	$2.5 \cdot 10^{-5} \ldots 5 \cdot 10^{-5}$	$5 \cdot 10^{-6} \ldots 5 \cdot 10^{-4}$
Penalty term η	0.5	–
Number offline iterations $l_{\max}$	200	200

In Fig. 10.13, the blue and red curves represent the NRE calculated from the RTFs estimated using the HOS-based and the SOS-based offline realization of the GC-MMI algorithm, respectively. Moreover, the dashed black curve shows the NRE estimated from a white noise signal using the HOS-based offline realization. Looking at the performance of the HOS-based realization, the results show that the RTFs can be estimated successfully, with an NRE below −20 dB for most frequencies below 6 kHz using the HOS-based realization. For frequencies $f > 6$ kHz, a larger estimation error is visible, which is attributed to the limited support due to speech signal energy in this frequency range. When using white noise as desired signal (dashed black curve), this effect vanishes and an NRE lower or equal to −40 dB can be obtained.

Looking at the SOS-based GC-MMI realization, it can be seen that it yields an inferior RTF estimate than the HOS-based GC-MMI realization. Whereas for $f < 3.5$ kHz, the estimation error of the SOS-based GC-MMI realization is still relatively close to that of the HOS-based GC-MMI realization, which changes for higher frequencies. We would like to point out that this is due to the fact that the SOS-based realization simply did not yet converge to a good solution given the available number of iterations. This is illustrated in Fig. 10.14, which shows the desired signal cancellation DC_5^l after each of the $l_{\max}$ of offline iterations, indexed by l. The desired signal cancellation DC_ζ, $\zeta \in \{1, \ldots, P-1\}$ is calculated as

$$\mathrm{DC}_\zeta = 10\log_{10} \frac{\mathscr{E}\left\{x_{1,\mathrm{d}}^2[k]\right\}}{\mathscr{E}\left\{y_{\hat{\mathrm{n}},\zeta,\mathrm{d}}^2[k]\right\}}, \tag{10.79}$$

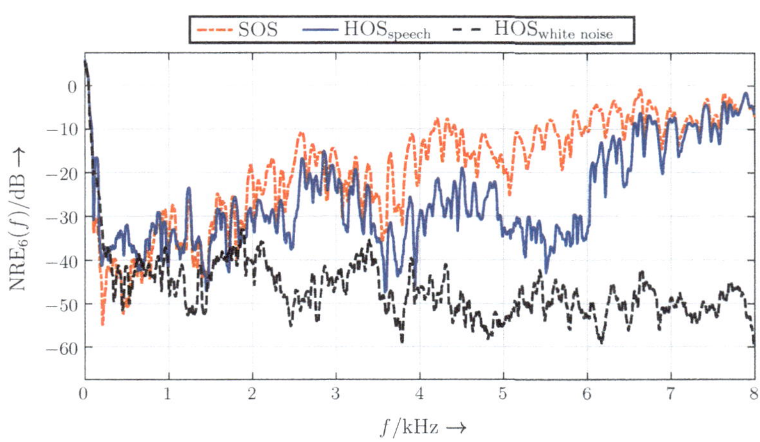

Fig. 10.13 Illustration of Normalized RTF Estimation Error (NRE) between microphones 1 and 6 with spacing $d = 0.21$ m after 200 offline iterations. The RTFs were estimated from a male speech signal using an HOS-based (blue curve) and a SOS-based (red curve) offline realization of the GC-MMI algorithm. The black curve represents the NRE calculated from an HOS-based algorithm and white noise as desired signal

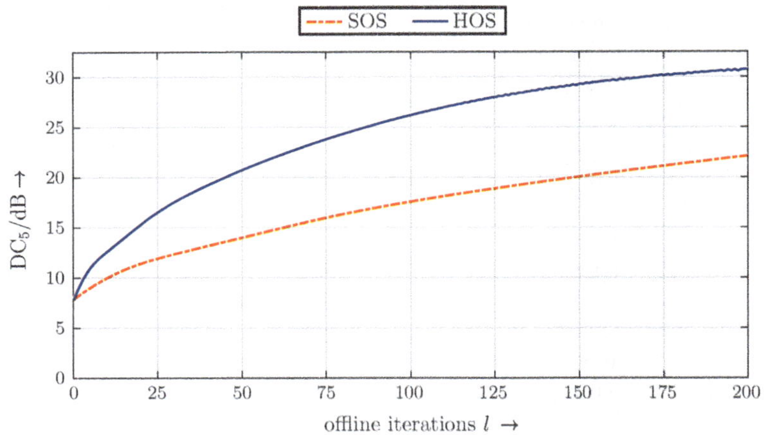

Fig. 10.14 Illustration of the desired signal cancellation DC_5 between microphones 1 and 6 with spacing $d = 0.21$ m over 200 offline iterations. The desired signal cancellation was obtained using an HOS-based (blue curve) and a SOS-based (red curve) offline realization of the GC-MMI algorithm

where $x_{1,\mathrm{d}}[k]$ denotes the desired signal components in the reference microphone channel (here: the first microphone channel), and $y_{\hat{\mathrm{n}},\zeta,\mathrm{d}}[k]$ represents the desired signal components in the ζ-th output of the BM. As the BM should provide reference signals of all undesired signal components with no desired speech components, DC_ζ should be as large as possible. Asymptotically, the quality of the SOS-based RTF estimate will be similar to that obtained with the HOS-based realization. We chose this example to explicitly show the faster convergence of the HOS-based realization.

The visual expression is also confirmed by means of the averaged NRE values $\overline{\mathrm{NRE}}_6$ which are summarized in Table 10.2. As in [66], $\overline{\mathrm{NRE}}_6$ was obtained by taking the arithmetic average of $\mathrm{NRE}_6(f)$ for frequencies $300\,\mathrm{Hz} \leq f \leq 6\,\mathrm{kHz}$.

To summarize, these results confirm the reliability and effectiveness of the presented RTF estimation approach in a realistic double-talk scenario. For a more detailed evaluation of the RTF estimation performance, we refer to [34, 66].

Table 10.2 Mean NRE values $\overline{\mathrm{NRE}}_6$ obtained with SOS- and HOS-based GC-MMI realizations

	$\overline{\mathrm{NRE}}_6$/dB
SOS-based realization, speech	−15.3
HOS-based realization, speech	−27.0
HOS-based realization, white noise	−43.3

10.5.3 Signal Enhancement

In the following, we evaluate the signal enhancement performance of the MMI-based GSC. To demonstrate the effect of exploiting the fundamental properties of speech, we evaluate the following realizations of the BM:

I SOS-based realization of the BM with $D_{\mathrm{BM}} = 1024$ combined with SOS-based realization of the INC with $D_{\mathrm{INC}} = 2$, i.e., nonwhiteness of speech signals is practically not exploited for updating the coefficients of the INC. The initial stepsize was set to $\mu_{\mathrm{init}} = 5 \cdot 10^{-4}$.

II SOS-based realization of the BM with $D_{\mathrm{BM}} = 1024$, combined with SOS-based realization of the INC with $D_{\mathrm{INC}} = 1024$, i.e., nonwhiteness of speech signals is exploited for updating the coefficients of both the BM and the INC . The initial stepsize was set to $\mu_{\mathrm{init}} = 5 \cdot 10^{-5}$.

III HOS-based realization of the MMI-GSC with $D_{\mathrm{BM}} = D_{\mathrm{INC}} = 1024$, i.e., nongaussianity and nonwhiteness are exploited for updating BM and INC. The initial stepsize was set to $\mu_{\mathrm{init}} = 5 \cdot 10^{-6}$.

Realization I only exploits nonwhiteness for calculating the update for the BM filter coefficients, whereas for the update of the INC, this is practically not the case, since $D_{\mathrm{INC}} = 2$. Note that we chose $D_{\mathrm{INC}} = 2$ instead of $D_{\mathrm{INC}} = 1$ to be able to use the same framework as for $D_{\mathrm{INC}} = L$ and, therefore, produce comparable results. In contrast to this, Realization II also exploits the nonwhiteness property for updating the INC coefficients, by taking the inter- and intra-channel correlations over $D_{\mathrm{INC}} = 1024$ time lags into account. Nonstationarity is always exploited. Thus, by comparing Realizations I and II, we can demonstrate the effect of exploiting nonwhiteness for calculating the INC update. In addition to this, Realization III also exploits nongaussianity by considering a multivariate Laplacian PDF as source model PDF, as explained in Sect. 10.4.8. Since Realization III exploits all three fundamental properties of wideband speech signals, it should yield the best signal enhancement performance.

To investigate the signal enhancement performance, we evaluate three different performance measures, which are explained in the following. At first, we calculate the desired signal cancellation DC_{ζ}, $\zeta \in \{1, \dots, P-1\}$, defined in (10.79), in each output signal of the BM. For brevity, we show the mean desired signal cancellation DC, i.e., averaged over all DC_{ζ}. Furthermore, we investigate the Signal-to-Interference Ratio (SIR) and the speech distortion (SD) obtained at the output of the GSC. The SIR is defined as

$$\mathrm{SIR} = 10\log_{10} \frac{\mathscr{E}\left\{y_{\mathrm{d}}^2[k]\right\}}{\mathscr{E}\left\{y_{\mathrm{int}}^2[k]\right\}}, \tag{10.80}$$

where $y_{\mathrm{d}}[k]$ and $y_{\mathrm{int}}[k]$ represent the desired and undesired signal components, respectively, at the output of the GSC. As for the desired signal cancellation, high SIR values are desirable. After convolution with the AIRs, an input SIR at the reference microphone of 0.4 dB was obtained. The speech distortion is defined as

Table 10.3 Desired signal cancellation (DC), speech distortion (SD), and signal-to-interference ratio (SIR) obtained at the output of the BM and at the output of the MMI-based GSC, using Realizations I, II, and III, respectively

	DC/dB	SD/dB	SIR/dB
I: SOS, $D_{BM} = L$, $D_{INC} = 2$	23.7	−24.7	19.6
II: SOS, $D_{BM} = D_{INC} = L$	23.7	−24.4	23.2
III: HOS, $D_{BM} = D_{INC} = L$	30.5	−26.2	26.4

$$\mathrm{SD} = 10\log_{10}\frac{\mathcal{E}\left\{\left(y_{\mathrm{d}}[k] - x_{1,\mathrm{d}}[k]\right)^2\right\}}{\mathcal{E}\left\{x_{1,\mathrm{d}}^2[k]\right\}}, \tag{10.81}$$

and should be as low as possible, since the desired speech signal components should be preserved at the GSC output.

In Table 10.3, the results obtained with the three different realizations of the MMI-based GSC are summarized.

As can be seen, the HOS-based realization of the BM yields a much higher average desired signal cancellation than the SOS-based realization, which demonstrates the effectiveness of exploiting the nongaussianity property of speech signals. When looking at the speech distortion and the output SIR, one can see that both SOS-based realizations (I and II) yield approximately the same speech distortion. However, due to exploitation of the nonwhiteness, the output SIR of Realization II is increased by 3.6 dB, which demonstrates the efficacy of exploiting nonwhiteness. By further exploiting nongaussianity for updating the filter coefficients of the INC, the output SIR could be further increased by another 3.2 dB, and speech distortion was decreased by 1.8 dB.

Note that the advantage of HOS-based over SOS-based methods will decrease for higher reverberation times, which is attributed to the fact that the distribution of reverberated speech comes closer to a Gaussian distribution for an increasing reverberation time, see, e.g., [48, 90].

10.6 Conclusion

In this chapter, the focus was on desired signal enhancement in the presence of other interfering point sources by a multichannel linear filter. After a brief recapitulation of the well-known LCMV filter and the GSC as its efficient realization, problems of these approaches in real-world applications and countermeasures were discussed.

Afterwards, an LCMMI signal extraction algorithm which allows to exploit the three fundamental properties of speech and audio signals: Nonstationarity, Nonwhiteness, and Nongaussianity, was presented. Analogously to previously published methods, e.g., [23, 29], the set of linear constraints depends on the RTFs between the desired signal components contained in the microphone signals. Hence, the focus of the presented algorithm is to extract the desired signal components as contained in one of the microphone channels, i.e., a reference microphone.

Subsequently, an efficient realization of the general LCMMI approach in a GSC structure was derived, where the BM is realized by multiple two-channel geometrically constrained ICA algorithms, which require only some coarse DOA information of the desired source. Furthermore, it was shown that estimates of the desired source RTFs, required for the linear constraints, i.e., to realize the fixed beamformer of the GSC, can be obtained from the filter vectors of the two-channel subsystems realizing the BM. Since these filter vectors are determined based on an ICA criterion, adaptation, and, hence, RTF estimation, can also be performed during double-talk situations without relying on activity monitoring, such as VAD or estimating SPP. Since identifying periods of target-only activity in multispeaker scenarios is a challenging task, rendering it unnecessary is a highly attractive feature of the LCMMI algorithm. As an additional advantage, this also simplifies the otherwise challenging control mechanism for simultaneously updating the BM und INC coefficients of the GSC.

Finally, we established relations between the proposed LCMMI method and other well-known SOS-based multichannel linear filter approaches for signal extraction. Establishing these relations made it obvious that, unlike the proposed LCMMI method, these approaches do not exploit nongaussianity and nonwhiteness of speech signals, and are therefore not optimum for enhancement of speech signals. We evaluated the proposed MMI-based GSC in a two-speaker scenario with respect to accuracy of RTF estimates and signal enhancement performance. The results showed that by exploiting nongaussianity and nonwhiteness in addition to nonstationarity, RTF estimation accuracy and signal enhancement performance could be increased substantially, confirming the effectiveness of the proposed LCMMI framework for signal extraction.

References

1. W. Kellermann, H. Buchner, W. Herbordt, R. Aichner, Multichannel acoustic signal processing for human/machine interfaces—Fundamental problems and recent advances, in *Proceedings of the International Conference on Acoustics (ICA)*, April 2004, pp. I–243–250
2. S. Doclo, W. Kellermann, S. Makino, S.E. Nordholm, Multichannel signal enhancement algorithms for assisted listening devices: Exploiting spatial diversity using multiple microphones. IEEE Signal Process. Mag. **32**(2), 18–30 (2015)
3. S. Gannot, E. Vincent, S. Markovich-Golan, A. Ozerov, A consolidated perspective on multimicrophone speech enhancement and source separation. IEEE/ACM Trans. Audio, Speech Lang. Process. (ASLP). **25**(4), 692–730 (2017)
4. B.D.V. Veen, K.M. Buckley, Beamforming: a versatile approach to spatial filtering. IEEE ASSP Mag. **5**(2), 4–24 (1988)
5. A. Hyvärinen, J. Karhunen, E. Oja, *Independent Component Analysis* (Wiley, 2001)
6. P. Smaragdis, Blind separation of convolved mixtures in the frequency domain. Neurocomputing **22**(1–3), 21–34 (1998)

7. L. Parra, C. Spence, Convolutive blind separation of non-stationary sources. IEEE Trans. Audio Speech Lang. Process. (ASL) **8**(3), 320–327 (2000)
8. H. Buchner, R. Aichner, W. Kellermann, Blind source separation for convolutive mixtures: a unified treatment, in *Audio Signal Processing for Next-generation Multimedia Communication Systems*, ed. by Y. Huang, J. Benesty (Kluwer Academic Publishers, 2004), pp. 255–293
9. S. Makino, T.-W. Lee, H. Sawada, *Blind Speech Separation* (Springer, 2007)
10. A. Ozerov, C. Fevotte, Multichannel nonnegative matrix factorization in convolutive mixtures for audio source separation. IEEE Trans. Audio Speech Lang. Process. (ASL) **18**(3), 550–563 (2010)
11. B. Widrow, P.E. Mantey, L.J. Griffiths, B.B. Goode, Adaptive antenna systems. Proc. IEEE **55**(12), 2143–2159 (1967)
12. A. Spriet, M. Moonen, J. Wouters, Spatially pre-processed speech distortion weighted multi-channel Wiener filtering for noise reduction. Signal Process. (SP) **84**(12), 2367–2387 (2004)
13. S. Doclo, A. Spriet, J. Wouters, M. Moonen, Frequency-domain criterion for the speech distortion weighted multichannel Wiener filter for robust noise reduction. Speech Commun. (SC) **49**(78), 636–656 (2007)
14. O.L. Frost, An algorithm for linearly constrained adaptive array processing. Proc. IEEE **60**(8), 926–935 (1972)
15. L. Griffiths, C. Jim, An alternative approach to linearly constrained adaptive beamforming. IEEE Trans. Antennas Propag. (AP) **30**(1), 27–34 (1982)
16. H. Buchner, R. Aichner, W. Kellermann, Blind source separation algorithms for convolutive mixtures exploiting nongaussianity, nonwhiteness, and nonstationarity, in *Proceedings of the International Workshop Acoustic Echo Noise Control (IWAENC)*, September 2003, pp. 275–278
17. W. Herbordt, H. Buchner, W. Kellermann, An acoustic human-machine front-end for multimedia applications. EURASIP J. Adv. Signal Process. **2003**(1), 1–11 (2003)
18. W. Herbordt, *Sound Capture for Human/Machine Interfaces—Practical Aspects of Microphone Array Signal Processing* (Springer, Heidelberg, Germany, 2005)
19. P. Oak, W. Kellermann, A calibration method for robust generalized sidelobe cancelling beamformers, in *Proceedings of the International Workshop Acoustic Echo Noise Control (IWAENC)*, September 2005, pp. 97–100
20. M. Souden, J. Chen, J. Benesty, S. Affes, An integrated solution for online multichannel noise tracking and reduction. IEEE Trans. Audio Speech Lang. Process. (ASL) **9**(7), 2159–2169 (2011)
21. R.C. Hendriks, T. Gerkmann, Noise correlation matrix estimation for multi-microphone speech enhancement. IEEE Trans. Audio Speech Lang. Process. (ASL) **20**(1), 223–233 (2012)
22. E.A.P. Habets, S. Gannot, Tutorial: Linear and parametric microphone array processing, in IEEE International Conference on Acoustics, Speech, and Signal Processing (ICASSP), Vancouver, Canada, https://www.audiolabs-erlangen.de/fau/professor/habets/activities/ICASSP-2013/
23. S. Gannot, D. Burshtein, E. Weinstein, Signal enhancement using beamforming and nonstationarity with applications to speech. IEEE Trans. Signal Process. (SP) **49**(8), 1614–1626 (2001)
24. S. Gannot, I. Cohen, Speech enhancement based on the general transfer function gsc and postfiltering. IEEE Speech Audio Process. (SAP) **12**(6), 561–571 (2004)
25. S. Markovich-Golan, S. Gannot, I. Cohen, Multichannel eigenspace beamforming in a reverberant noisy environment with multiple interfering speech signals. IEEE Trans. Audio Speech Lang. Process. (ASL) **17**(6), 1071–1086 (2009)
26. S.M. Golan, S. Gannot, I. Cohen, Subspace tracking of multiple sources and its application to speakers extraction, in *Proceedings of the IEEE International Conference on Acoustics, Speech, Signal Processing (ICASSP)*, March 2010, pp. 201–204
27. L.C. Parra, C.V. Alvino, Geometric source separation: Merging convolutive source separation with geometric beamforming. IEEE Speech Audio Process. (SAP) **10**(6), 352–362 (2002)
28. G. Reuven, S. Gannot, I. Cohen, Dual-source transfer-function generalized sidelobe canceller. IEEE Trans. Audio Speech Lang. Process. (ASL) **16**(4), 711–727 (2008)

29. S. Gannot, D. Burshtein, E. Weinstein, Analysis of the power spectral deviation of the general transfer function GSC. IEEE Trans. Signal Process. (SP) **52**(4), 1115–1120 (2004)
30. Y. Zheng, K. Reindl, W. Kellermann, BSS for improved interference estimation for blind speech signal extraction with two microphones, in *Proceedings of the International Workshop on Computational Advances in Multi-Sensor Adaptive Processing (CAMSAP)*, Aruba, Dutch Antilles, December 2009, pp. 253–256
31. J. Benesty, J. Chen, Y. Huang, J. Dmochowski, On microphone-array beamforming from a MIMO acoustic signal processing perspective. IEEE Trans. Audio Speech Lang. Process. (ASL) **15**(3), 1053–1065 (2007)
32. J. Chen, J. Benesty, Y. Huang, An acoustic MIMO framework for analyzing microphone-array beamforming, in *IEEE International Conference on Acoustics, Speech, and Signal Processing (ICASSP)*, vol. 1, April 2007, pp. I–25–I–28
33. H. Van Trees, *Detection, Estimation, and Modulation Theory, Optimum Array Processing*. Estimation, and Modulation Theory Detection (Wiley, 2004)
34. K. Reindl, *Multichannel Acoustic Signal Extraction for Reverberant Environments* (Verlag Dr. Hut, München, 2015)
35. W. Herbordt, W. Kellermann, Adaptive beamforming for audio signal acquisition, in *Adaptive Signal Processing - Applications to Real-World Problems*, ed. by Y.H.J. Benesty (Springer, Berlin, Germany, 2003), pp. 155–194
36. G. Strang, *Linear Algebra and Its Applications*, 4th edn. (Thomson, Brooks/Cole, Belmont, CA, 2006)
37. K.M. Buckley, L.J. Griffiths, An adaptive generalized sidelobe canceller with derivative constraints. IEEE Trans. Antennas Propag. (AP) **34**(3), 311–319 (1986)
38. K. Buckley, Broad-band beamforming and the generalized sidelobe canceller. IEEE Trans. Acoust. Speech Signal Process. (ASSP) **34**(5), 1322–1323 (1986)
39. B.R. Breed, J. Strauss, A short proof of the equivalence of LCMV and GSC beamforming. IEEE Signal Process. Lett. (SPL) **9**(6), 168–169 (2002)
40. O. Hoshuyama, A. Sugiyama, A robust adaptive beamformer for microphone arrays with a blocking matrix using constrained adaptive filters, in *Proceedings of the IEEE International Conference on Acoustics, Speech, and Signal Processing (ICASSP)*, May 1996, p. 925928
41. O. Hoshuyama, A. Sugiyama, A. Hirano, A robust adaptive microphone array with improved spatial selectivity and its evaluation in a real environment, in *Proceedings of the IEEE International Conference on Acoustics, Speech, and Signal Processing (ICASSP)*, April 1997, p. 367370
42. O. Hoshuyama, A. Sugiyama, A. Hirano, A robust adaptive beamformer for microphone arrays with a blocking matrix using constrained adaptive filters. IEEE Trans. Signal Process. (SP) **47**(10), 2677–2684 (1999)
43. W. Herbordt, W. Kellermann, Analysis of blocking matrices for generalized sidelobe cancellers for non-stationary broadband signals, in *Proceedings of the IEEE International Conference on Acoustics, Speech, and Signal Processing (ICASSP)*, May 2002, pp. 4187–4187
44. W. Herbordt, H. Buchner, S. Nakamura, W. Kellermann, Application of a double-talk resilient DFT-domain adaptive filter for bin-wise stepsize controls to adaptive beamforming, in *Proceedings of the International Workshop on Nonlinear Signal and Image Processing (NSIP)*, May 2005
45. W. Herbordt, H. Buchner, S. Nakamura, W. Kellermann, Outlier-robust DFT-domain adaptive filtering for bin-wise stepsize controls, and its application to a generalized sidelobe canceller, in *Proceedings of the International Workshop Acoustic Echo Noise Control (IWAENC)*, September 2005, pp. 113–116
46. W. Herbordt, H. Buchner, S. Nakamura, W. Kellermann, Multichannel bin-wise robust frequency-domain adaptive filtering and its application to adaptive beamforming. IEEE Trans. Audio Speech Lang. Process. (ASL) **15**(4), 1340–1351 (2007)
47. P.J. Huber, E.M. Ronchetti, *Robust Statistics*. Wiley Series in Probability and Statistics (Wiley, 2009)

48. K. Kumatani, T. Gehrig, U. Mayer, E. Stoimenov, J. McDonough, M. Wölfel, Adaptive beamforming with a minimum mutual information criterion. IEEE Trans. Audio Speech Lang. Process. (ASL) **15**(8), 2527–2541 (2007)
49. K. Kumatani, J. McDonough, B. Rauch, D. Klakow, P.N. Garner, W. Li, Beamforming with a maximum negentropy criterion. IEEE Trans. Audio Speech Lang. Process. (ASL) **17**(5), 994–1008 (2009)
50. G. Reuven, S. Gannot, I. Cohen, Performance analysis of dual source transfer-function generalized sidelobe canceller. Speech Commun. (SC) **49**(78), 602–622 (2007)
51. S. Markovich, S. Gannot, I. Cohen, A comparison between alternative beamforming strategies for interference cancelation in noisy and reverberant environment, in *IEEE Convention of Electrical and Electronics Engineers in Israel (IEEEI)*, December 2008, pp. 203–207
52. S. Markovich-Golan, S. Gannot, I. Cohen, A sparse blocking matrix for multiple constraints GSC beamformer, in *Proceedings of the IEEE International Conference on Acoustics, Speech, Signal Processing (ICASSP)*, March 2012, pp. 197–200
53. S. Markovich-Golan, S. Gannot, I. Cohen, Distributed multiple constraints generalized sidelobe canceler for fully connected wireless acoustic sensor networks. IEEE Trans. Audio Speech Lang. Process. (ASL) **21**(2), 343–356 (2013)
54. R. Talmon, I. Cohen, S. Gannot, Relative transfer function identification using convolutive transfer function approximation. IEEE Trans. Audio Speech Lang. Process. (ASL) **17**(4), 546–555 (2009)
55. R. Talmon, I. Cohen, S. Gannot, Convolutive transfer function generalized sidelobe canceler. IEEE Trans. Audio Speech Lang. Process. (ASL) **17**(7), 1420–1434 (2009)
56. O. Shalvi, E. Weinstein, System identification using nonstationary signals. IEEE Trans. Signal Process. (SP) **44**(8), 2055–2063 (1996)
57. I. Cohen, Relative transfer function identification using speech signals. IEEE Speech Audio Process. (SAP) **12**(5), 451–459 (2004)
58. M. Schwab, P. Noll, T. Sikora, Noise robust relative transfer function estimation," in *European Signal Processing Conference (EUSIPCO)*, September 2006, pp. 1–5
59. R. Talmon, I. Cohen, S. Gannot, Identification of the relative transfer function between microphones in reverberant environments, in *IEEE Conventional of Electrical and Electronics Engineers in Israel (IEEEI)*, December 2008, pp. 208–212
60. R. Talmon, I. Cohen, S. Gannot, *Identification of the Relative Transfer Function between Sensors in the Short-Time Fourier Transform Domain* (Springer, Berlin, Heidelberg, 2010), pp. 33–47
61. A. Krueger, E. Warsitz, R. Haeb-Umbach, Speech enhancement with a GSC-like structure employing eigenvector-based transfer function ratios estimation. IEEE Trans. Audio Speech Lang. Process. (ASL) **19**(1), 206–219 (2011)
62. X. Li, L. Girin, R. Horaud, S. Gannot, Estimation of relative transfer function in the presence of stationary noise based on segmental power spectral density matrix subtraction, in *Proceedings of the IEEE International Conference on Acoustics, Speech and Signal Processing (ICASSP)*, April 2015, pp. 320–324
63. T. Gerkmann, C. Breithaupt, R. Martin, Improved a posteriori speech presence probability estimation based on a likelihood ratio with fixed priors. IEEE Trans. Audio Speech Lang. Process. (ASL) **16**(5), 910–919 (2008)
64. T. Gerkmann, M. Krawczyk, R. Martin, Speech presence probability estimation based on temporal cepstrum smoothing, in *Proceedings of the IEEE International Conference on Acoustics, Speech and Signal Processing (ICASSP)*, March 2010, pp. 4254–4257
65. E. Weinstein, M. Feder, A.V. Oppenheim, Multi-channel signal separation by decorrelation. IEEE Speech Audio Process. (SAP) **1**(4), 405–413 (1993)
66. K. Reindl, S. Markovich-Golan, H. Barfuss, S. Gannot, W. Kellermann, Geometrically constrained TRINICON-based relative transfer function estimation in underdetermined scenarios, in *Proceedings of the IEEE Workshop Applications Signal Processing Audio Acoustics (WASPAA)*, October 2013

67. K. Reindl, W. Kellermann, Linearly-constrained multichannel interference suppression algorithms derived from a minimum mutual information criterion, in *Proceedings of the IEEE China Summit & International Conference on Communication and Signal Processing (ChinaSIP 2013)*, July 2013
68. K. Reindl, S. Meier, H. Barfuss, W. Kellermann, Minimum mutual information-based linearly constrained broadband signal extraction. IEEE/ACM Trans. Audio Speech Lang. Process. (ASLP) **22**(6), 1096–1108 (2014)
69. C. Shannon, A mathematical theory of communication. Bell Syst. Tech. J. **27**, 623–656 (1948)
70. R. Aichner, Acoustic blind source separation in reverberant and noisy environments. Ph.D. dissertation, University Erlangen-Nürnberg, Germany, May 2007
71. H. Buchner, Broadband adaptive MIMO filtering: a unified treatment and applications to acoustic human-machine interfaces. Ph.D. dissertation, University Erlangen-Nürnberg, Germany, 2010
72. H. Buchner, A systematic approach to incorporate deterministic prior knowledge in broadband adaptive MIMO systems, in *Proceedings of the (Systems and Computers (ASILOMAR), Nov, Asilomar Conference Signals)*, 2010, pp. 461–468
73. S.I. Amari, Natural gradient works efficiently in learning. Neural Comput. **10**(2), 251–276 (1998)
74. H. Buchner, R. Aichner, W. Kellermann, A generalization of blind source separation algorithms for convolutive mixtures based on second-order statistics. IEEE Speech Audio Process. (SAP) **13**(1), 120–134 (2005)
75. S. Markovich-Golan, S. Gannot, W. Kellermann, Combined LCMV-TRINICON beamforming for separating multiple speech sources in noisy and reverberant environments. IEEE/ACM Trans. Audio Speech Lang. Process. (ASLP) **25**(2), 320–332 (2017)
76. J. Chen, J. Benesty, Y. Huang, A minimum distortion noise reduction algorithm with multiple microphones. IEEE Trans. Audio Speech Lang. Process. (ASL) **16**(3), 481–493 (2008)
77. E.A.P. Habets, J. Benesty, I. Cohen, S. Gannot, On a tradeoff between dereverberation and noise reduction using the MVDR beamformer, in *Proceedings of the IEEE International Conference on Communication and Signal Processing (ICASSP)*, April 2009, p. 37413744
78. E.A.P. Habets, J. Benesty, I. Cohen, S. Gannot, J. Dmochowski, New insights into the MVDR beamformer in room acoustics. IEEE Trans. Audio Speech Lang. Process. (ASL) **18**(1), 158170 (2010)
79. E.A.P. Habets, J. Benesty, S. Gannot, I. Cohen, The MVDR beamformer for speech enhancement, in *Speech Processing in Modern Communication-Challenges and Perspectives*, ed. by I. Cohen, J. Benesty, S. Gannot (Springer, Berlin, Germany, 2010), pp. 225–254
80. M. Knaak, S. Araki, S. Makino, Geometrically constraint ICA for convolutive mixtures of sound, in *Proceedings of the IEEE International Conference on Acoustics, Speech, Signal Processing (ICASSP)*, vol. 2, April 2003, pp. II–725–728
81. M. Knaak, S. Araki, S. Makino, Geometrically constraint ICA for robust separation of sound mixtures, in *Proceedings of the International Symposium on Independent Component Analysis Blind Separation (ICA)*, April 2003, pp. 951–956
82. M. Knaak, S. Araki, S. Makino, Geometrically constrained independent component analysis. IEEE Trans. Audio Speech Lang. Process. (ASL) **15**(2), 715–726 (2007)
83. Y. Zheng, K. Reindl, W. Kellermann, Analysis of dual-channel ICA-based blocking matrix for improved noise estimation. EURASIP J. Adv. Signal Process. **2014**(4:26), 1–24 (2014)
84. H. Brehm, W. Stammler, Description and generation of spherically invariant speech-model signals. Signal Process. (SP) **12**, 119–141 (1987)
85. H. Buchner, W. Kellermann, A fundamental relation between blind and supervised adaptive filtering illustrated for blind source separation and acoustic echo cancellation, in *Joint Workshop Hands-free Speech Communication, Microphone Arrays (HSCMA)*, May 2008, pp. 17–20
86. S. Haykin, *Adaptive Filter Theory*, 4th ed. (Prentice-Hall, 2002)
87. R. Aichner, H. Buchner, F. Yan, W. Kellermann, A real-time blind source separation scheme and its application to reverberant and noisy acoustic environments. Signal Process. (SP) **86**, 1260–1277 (2006)

88. R. Aichner, H. Buchner, W. Kellermann, Exploiting narrowband efficiency for broadband convolutive blind source separation. EURASIP J. Adv. Signal Process. **2007**, 1–9 (2006)
89. H. Barfuss, W. Kellermann, An adaptive microphone array topology for target signal extraction with humanoid robots, in *Proceedings of the International Workshop Acoustic Signal Enhancement (IWAENC)*, September 2014, pp. 16–20
90. K. Kumatani, J. McDonough, B. Raj, Microphone array processing for distant speech recognition: from close-talking microphones to far-field sensors. IEEE Signal Process. Mag. **29**(6), 127–140 (2012)

Chapter 11
Recent Advances in Multichannel Source Separation and Denoising Based on Source Sparseness

Nobutaka Ito, Shoko Araki and Tomohiro Nakatani

Abstract This chapter deals with multichannel source separation and denoising based on sparseness of source signals in the time-frequency domain. In this approach, time-frequency masks are typically estimated based on clustering of source location features, such as time and level differences between microphones. In this chapter, we describe the approach and its recent advances. Especially, we introduce a recently proposed clustering method, *observation vector clustering*, which has attracted attention for its effectiveness. We introduce algorithms for observation vector clustering based on a complex Watson mixture model (cWMM), a complex Bingham mixture model (cBMM), and a complex Gaussian mixture model (cGMM). We show through experiments the effectiveness of observation vector clustering in source separation and denoising.

11.1 Introduction

When a desired sound is recorded by distant microphones, it is mixed with other sounds, which often degrade speech quality and intelligibility as well as automatic speech recognition (ASR) performance. To resolve this problem, techniques such as source separation, denoising, and dereverberation have been studied extensively. This chapter focuses on source separation and denoising; see [1] for dereverberation.

Figure 11.1 illustrates source separation and denoising we deal with in this paper. Suppose we record $N (\geq 1)$ source signals in the presence of background noise by using $M (\geq 2)$ microphones. Our goal is to estimate each source signal from the observed signals. Note that there is not only a multichannel approach [2–5] using mul-

N. Ito (✉) · S. Araki · T. Nakatani
NTT Communication Science Laboratories, NTT Corporation,
2-4 Hikaridai, Seikacho, Kyoto, Sorakugun 619-0237, Japan
e-mail: ito.nobutaka@lab.ntt.co.jp

S. Araki
e-mail: araki.shoko@lab.ntt.co.jp

T. Nakatani
e-mail: nakatani.tomohiro@lab.ntt.co.jp

S. Makino (ed.), *Audio Source Separation*, Signals and Communication Technology, https://doi.org/10.1007/978-3-319-73031-8_11

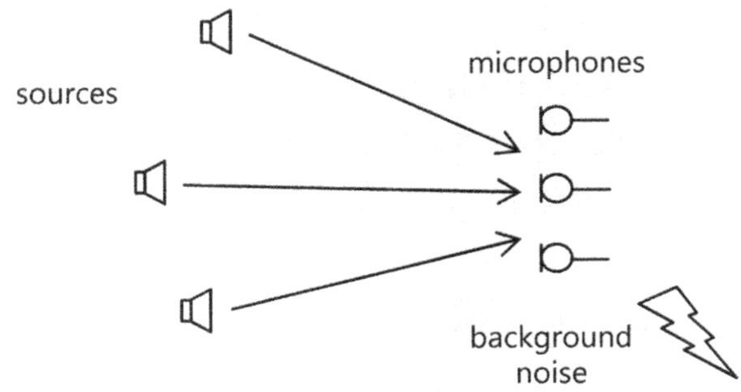

Fig. 11.1 Source separation and denoising we deal with in this paper

tiple microphones but also a single-channel approach using a single microphone [6–9]. A main advantage of the multichannel approach is that it can perform source separation and denoising with little or even no distortion in the desired source signal.

Especially, multichannel source separation and denoising based on source sparseness [10–20] have turned out to be highly effective and robust in the real world [16, 17, 19, 20]. Various signals including speech are known to have sparseness in the time-frequency domain: a small percentage of the time-frequency components of a signal capture a large percentage of its overall energy [10]. The source sparseness is often exploited by assuming that the observed signals are *dominated* by a single source signal or by background noise at each time-frequency point. We call this a *sparseness assumption*. The dominating source signal or background noise at each time-frequency point can be represented by *masks*. Once we have obtained these masks, we can estimate the source signals either by applying the masks directly to the observed signals (*masking*) [10–14, 16, 17, 19, 21] or by applying beamformers designed based on the masks [15, 18, 20, 22].

The key to the effectiveness of this approach is accurate estimation of the masks, which is usually performed based on either spatial information [10–20] or spectral information [21, 22]. We focus on the former, which employs source location features extracted from the observed signals, such as time and level differences between microphones. The sparseness assumption implies that the source location features form clusters, each of which corresponds to a source signal or the background noise. These clusters can be found by clustering the source location features to obtain the masks. This is typically done by fitting a mixture model to the features, where the appropriate design of the features and the mixture model is significant to mask estimation accuracy.

In this chapter, we introduce a recently proposed clustering method, *observation vector clustering*, which has attracted attention for its effectiveness [11, 13, 15–20]. This method has been employed in many evaluation campaigns successfully [17, 20]. We introduce algorithms for observation vector clustering based on a complex Watson mixture model (cWMM), a complex Bingham mixture model (cBMM), and a complex Gaussian mixture model (cGMM).

The rest of this chapter is organized as follows. Section 11.2 overviews source separation and denoising based on the observation vector clustering. Section 11.3

introduces algorithms for observation vector clustering based on the cWMM, the cBMM, and the cGMM. Section 11.4 describes experiments, and Sect. 11.5 concludes this chapter.

11.2 Source Separation and Denoising Based on Observation Vector Clustering

This section overviews source separation and denoising based on observation vector clustering. Figure 11.2 shows the overall processing flow of this method. In *mask estimation*, masks are estimated from the observed signals. In *source signal estimation*, source signals are estimated by masking or beamforming based on the estimated masks.

11.2.1 Mask Estimation

Figure 11.3 shows the processing flow of *mask estimation* in Fig. 11.2. In *feature extraction*, a source location feature vector is extracted from the observed signals. In *frequency-wise clustering*, clustering of the extracted feature vector is performed in each frequency bin. As a result, posterior probabilities are obtained, which indicate how much the individual clusters contribute to each time-frequency point. In *permutation alignment*, the masks are obtained from the posterior probabilities; the details will be explained later.

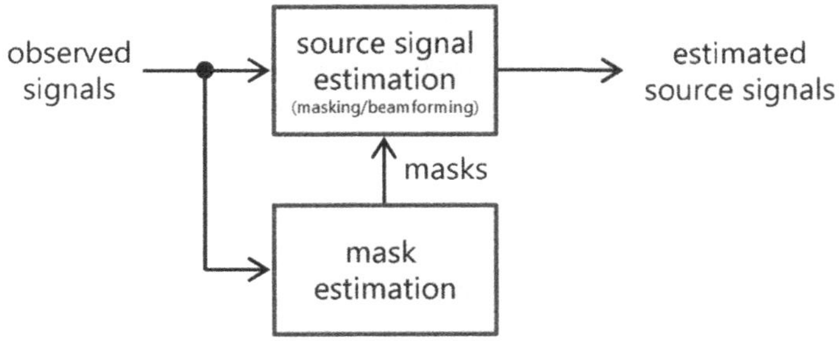

Fig. 11.2 Overall processing flow of source separation and denoising based on observation vector clustering

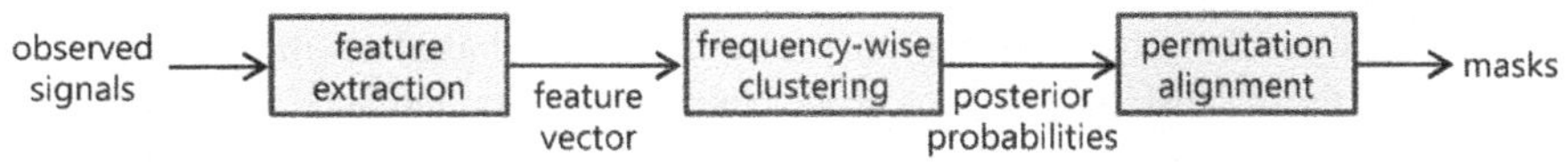

Fig. 11.3 Processing flow of *mask estimation* in Fig. 11.2

Feature Extraction

In *feature extraction* in Fig. 11.3, a source location feature vector is extracted at each time-frequency point. Conventionally, time and level differences between microphones were often employed as source location features. In contrast, in the observation vector clustering, we operate directly on an observation vector composed of multichannel complex spectra.

Let $y_{tf}^{(m)} \in \mathbb{C}$ denote the observed signal at the mth microphone in the short-time Fourier transform (STFT) domain. Here, $m \in \{1, \ldots, M\}$ denotes the microphone index; $t \in \{1, \ldots, T\}$ the frame index; $f \in \{1, \ldots, F\}$ the frequency bin index; M the number of microphones in the array; T the number of frames; F the number of frequency bins up to the Nyquist frequency. We define the *observation vector* by $\mathbf{y}_{tf} \triangleq \left[y_{tf}^{(1)}\ y_{tf}^{(2)}\ \cdots\ y_{tf}^{(M)}\right]^{\mathsf{T}} \in \mathbb{C}^M$, where the superscript ^{T} denotes transposition.

We employ the observation vector $\mathbf{y}_{tf}$ as the feature vector $\mathbf{z}_{tf}$:

$$\mathbf{z}_{tf} = \mathbf{y}_{tf}. \tag{11.1}$$

In this case, $\mathbf{z}_{tf}$ lies in the complex linear space $\mathbb{C}^M$. Alternatively, we can also employ a normalized observation vector $\frac{\mathbf{y}_{tf}}{\|\mathbf{y}_{tf}\|}$ as the feature vector $\mathbf{z}_{tf}$:

$$\mathbf{z}_{tf} = \frac{\mathbf{y}_{tf}}{\|\mathbf{y}_{tf}\|}, \tag{11.2}$$

where $\|\cdot\|$ denotes the Euclidean norm. In this case, $\mathbf{z}_{tf}$ lies on the unit hypersphere S^{M-1} in $\mathbb{C}^M$ centered at the origin, because $\|\mathbf{z}_{tf}\| = 1$ (see Fig. 11.4).

In the following, we describe our modeling of the observation vector $\mathbf{y}_{tf}$. We consider both noiseless and noisy cases.

First, we consider the noiseless case, where $N(\geq 2)$ source signals are recorded by the microphones without noise. The number of sources, N, is assumed to be given throughout this chapter. In this noiseless case, $\mathbf{y}_{tf}$ is modeled by $\mathbf{y}_{tf} = \sum_{n=1}^{N} s_{tf}^{(n)} \mathbf{h}_{tf}^{(n)}$. Here, $s_{tf}^{(n)}$ denotes the nth source signal in the STFT domain, and $\mathbf{h}_{tf}^{(n)}$ denotes the steering vector for the nth source. The steering vector $\mathbf{h}_{tf}^{(n)}$ represents the acoustic transfer characteristics from the nth source to the microphones. Under the sparseness assumption, the above model can be approximated by $\mathbf{y}_{tf} = s_{tf}^{(\nu)} \mathbf{h}_{tf}^{(\nu)}$, where $\nu = d_{tf}$ denotes the index of the source signal that dominates $\mathbf{y}_{tf}$ at the time-frequency point (t, f). Here, both $s_{tf}^{(n)}$ and $\mathbf{h}_{tf}^{(n)}$ are unknown.

Next, we consider the noisy case, where $N(\geq 1)$ source signal(s) are recorded by the microphones in the presence of background noise. In this case, $\mathbf{y}_{tf}$ is modeled

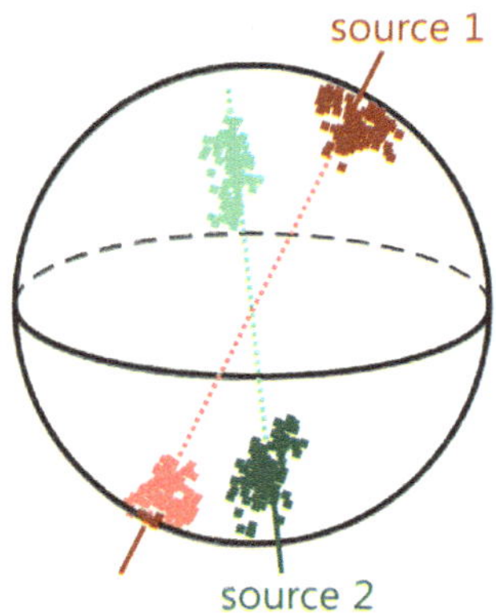

Fig. 11.4 Example of the source location feature vector for two sources. Here, $\mathbb{C}^M$ has been simplified to $\mathbb{R}^3$ for illustration

by $\mathbf{y}_{tf} = \sum_{n=1}^{N} s_{tf}^{(n)} \mathbf{h}_{tf}^{(n)} + \mathbf{v}_{tf}$, where $\mathbf{v}_{tf}$ denotes the contribution of the background noise to $\mathbf{y}_{tf}$. Under the sparseness assumption, this model can be approximated by

$$\mathbf{y}_{tf} = \begin{cases} s_{tf}^{(\nu)} \mathbf{h}_{tf}^{(\nu)} + \mathbf{v}_{tf}, & \text{if } d_{tf} = \nu \in \{1, \ldots, N\}, \\ \mathbf{v}_{tf}, & \text{if } d_{tf} = 0. \end{cases} \tag{11.3}$$

Here, d_{tf} denotes the index of the source signal or the background noise that dominates $\mathbf{y}_{tf}$ at the time-frequency point (t, f), where the case $d_{tf} = 0$ corresponds to the background noise and the cases $d_{tf} \in \{1, \ldots, N\}$ to the source signals. Note that the background noise $\mathbf{v}_{tf}$ is assumed to be contained in $\mathbf{y}_{tf}$ at all time-frequency points, because it is usually not sparse. $s_{tf}^{(n)}$, $\mathbf{h}_{tf}^{(n)}$, and $\mathbf{v}_{tf}$ are all unknown.

In both cases, our goal is to estimate $s_{tf}^{(n)}$ given $\mathbf{y}_{tf}$.

Frequency-Wise Clustering

In *frequency-wise clustering* in Fig. 11.3, clustering of the feature vector $\mathbf{z}_{tf}$ is performed in each frequency bin. As a result, the posterior probability $\tilde{\gamma}_{tf}^{(k)}$ is obtained for each cluster k, which indicates how much the kth cluster contributes to the time-frequency point (t, f).

The clustering can be performed by fitting a mixture model

$$p(\mathbf{z}_{tf} | \Theta_f) = \sum_k \alpha_f^{(k)} p\left(\mathbf{z}_{tf} \middle| \tilde{d}_{tf} = k, \Theta_f\right) \tag{11.4}$$

to $\mathbf{z}_{tf}$. Here, $\tilde{d}_{tf}$ denotes the index of the cluster that $\mathbf{z}_{tf}$ belongs to; $\alpha_f^{(k)} \triangleq P\left(\tilde{d}_{tf} = k \middle| \Theta_f\right)$ the prior probability of $\tilde{d}_{tf} = k$; $p\left(\mathbf{z}_{tf} \middle| \tilde{d}_{tf} = k, \Theta_f\right)$ the conditional probability density function of $\mathbf{z}_{tf}$ under $\tilde{d}_{tf} = k$; $\sum_k$ the sum over all possible values of k (i.e., $\sum_{k=1}^{K}$ for the noiseless case; $\sum_{k=0}^{K}$ for the noisy case); Θ_f the set of all model parameters in (11.4). $\alpha_f^{(k)}$ satisfies $\sum_k \alpha_f^{(k)} = 1$ and $\alpha_f^{(k)} \geq 0$.

Θ_f is estimated by the maximization of the log-likelihood function

$$L(\Theta_f) = \sum_{t=1}^{T} \ln p(\mathbf{z}_{tf}|\Theta_f), \tag{11.5}$$

which can be done by the expectation-maximization (EM) algorithm. Once Θ_f has been estimated, we obtain the posterior probability $\tilde{\gamma}_{tf}^{(k)}$ based on Bayes' theorem [23] as follows:

$$\tilde{\gamma}_{tf}^{(k)} \triangleq P(\tilde{d}_{tf} = k|\mathbf{z}_{tf}, \Theta_f) \tag{11.6}$$

$$= \frac{\alpha_f^{(k)} p(\mathbf{z}_{tf}|\tilde{d}_{tf} = k, \Theta_f)}{\sum_l \alpha_f^{(l)} p(\mathbf{z}_{tf}|\tilde{d}_{tf} = l, \Theta_f)}. \tag{11.7}$$

Here, $\tilde{\gamma}_{tf}^{(k)}$ satisfies $\sum_k \tilde{\gamma}_{tf}^{(k)} = 1$ and $\tilde{\gamma}_{tf}^{(k)} \geq 0$.

Permutation Alignment

In *permutation alignment* in Fig. 11.3, the masks are obtained by using the posterior probabilities $\tilde{\gamma}_{tf}^{(k)}$.

The index k of the clusters and the index n of the source signals and the background noise do not necessarily coincide, but there is permutation ambiguity between them. This implies that $\tilde{\gamma}_{tf}^{(k)}$ for the same k may correspond to different source signals at different frequencies. Therefore, we need to permute the cluster indexes k so that each k corresponds to the same source signal or background noise in all frequency bins, which is called permutation alignment. As a result of the permutation alignment, we obtain the masks $\gamma_{tf}^{(n)}$.

Many methods have been proposed for permutation alignment [16, 24–26]. Especially, Sawada et al. [16] has proposed an effective method based on correlation of posterior probabilities $\tilde{\gamma}_{tf}^{(k)}$ between frequencies.

11.2.2 Source Signal Estimation

In *source signal estimation* in Fig. 11.2, source signals are estimated by masking or beamforming based on the estimated masks.

Masking

When masking is employed, the source signals are estimated by multiplying an observed signal by the estimated masks $\gamma_{tf}^{(n)}$ as follows:

$$\hat{s}_{tf}^{(n)} = \gamma_{tf}^{(n)} y_{tf}^{(\mu)}. \tag{11.8}$$

Here, μ denotes the index of the reference microphone.

Beamforming

Here we consider the noisy case. Among many types of beamformers, we focus on the MVDR beamformer. The MVDR beamformer is especially suitable for the front end of ASR, because it can perform source separation and denoising without distorting the desired source signal.

The output of the MVDR beamformer is given by

$$\hat{s}_{tf}^{(n)} = \frac{\mathbf{h}_{tf}^{(n)\mathsf{H}} \left(\boldsymbol{\Phi}_f^{\mathbf{y}}\right)^{-1} \mathbf{y}_{tf}}{\mathbf{h}_{tf}^{(n)\mathsf{H}} \left(\boldsymbol{\Phi}_f^{\mathbf{y}}\right)^{-1} \mathbf{h}_{tf}^{(n)}}. \tag{11.9}$$

$\boldsymbol{\Phi}_f^{\mathbf{y}}$ denotes the covariance matrix of $\mathbf{y}_{tf}$, which can be estimated by

$$\hat{\boldsymbol{\Phi}}_f^{\mathbf{y}} = \frac{1}{T} \sum_{t=1}^{T} \mathbf{y}_{tf} \mathbf{y}_{tf}^{\mathsf{H}}. \tag{11.10}$$

In the MVDR beamformer, accurate estimation of the steering vector $\mathbf{h}_{tf}^{(n)}$ is crucial.

Conventionally, $\mathbf{h}_{tf}^{(n)}$ was estimated based on the assumptions of planewave propagation and a known array geometry. These assumptions are often violated in the real world, and lead to degraded performances of the MVDR beamformer and therefore ASR. Here we present mask-based steering vector estimation, which does not rely on these assumptions, and therefore is more robust in the real world.

First, a covariance matrix $\boldsymbol{\Psi}_f^{(n)}$ corresponding to the nth source signal plus the background noise is estimated by

$$\boldsymbol{\Psi}_f^{(n)} = \frac{\sum_{t=1}^{T} \gamma_{tf}^{(n)} \mathbf{y}_{tf} \mathbf{y}_{tf}^{\mathsf{H}}}{\sum_{t=1}^{T} \gamma_{tf}^{(n)}}, \tag{11.11}$$

and a noise covariance matrix $\boldsymbol{\Psi}_f^{(0)}$ is estimated by

$$\boldsymbol{\Psi}_f^{(0)} = \frac{\sum_{t=1}^{T} \gamma_{tf}^{(0)} \mathbf{y}_{tf} \mathbf{y}_{tf}^{\mathsf{H}}}{\sum_{t=1}^{T} \gamma_{tf}^{(0)}}. \tag{11.12}$$

The noise contribution to $\boldsymbol{\Psi}_f^{(n)}$ is reduced by subtracting $\boldsymbol{\Psi}_f^{(0)}$ from $\boldsymbol{\Psi}_f^{(n)}$. The steering vector $\mathbf{h}_{tf}^{(n)}$ is estimated as a principal eigenvector of the resultant matrix $\boldsymbol{\Psi}_f^{(n)} - \boldsymbol{\Psi}_f^{(0)}$.

11.3 Mask Estimation Based on Modeling Directional Statistics

Several mixture models for the feature vector $\mathbf{z}_{tf}$ have been proposed to estimate the masks accurately. These mixture models include the cWMM, the cBMM, and the cGMM, which are specific examples of the general mixture model (11.4).

11.3.1 Mask Estimation Based on Complex Watson Mixture Model (cWMM)

Sawada et al. [13, 16] and Tran Vu et al. [15] have proposed to estimate masks based on modeling the feature vector (11.2) by a *complex Watson mixture model* (*cWMM*). The cWMM is composed of *complex Watson distributions* of Mardia et al. [27], and the complex Watson distribution is an extension of a real Watson distribution of Watson [28].

The probability density function (PDF) of the cWMM is given by

$$p(\mathbf{z}_{tf};\Theta_{\mathrm{W}f}) = \sum_k \alpha_f^{(k)} p_{\mathrm{W}}\Big(\mathbf{z}_{tf};\mathbf{a}_f^{(k)},\kappa_f^{(k)}\Big), \tag{11.13}$$

where p_{W} denotes a complex Watson distribution

$$p_{\mathrm{W}}(\mathbf{z};\mathbf{a},\kappa) \triangleq \frac{(M-1)!}{2\pi^M \mathscr{K}(1,M;\kappa)} \exp\Big(\kappa \big|\mathbf{a}^{\mathsf{H}}\mathbf{z}\big|^2\Big). \tag{11.14}$$

Both the complex Watson distribution and the cWMM are defined on the unit hypersphere in $\mathbb{C}^M$:

$$S^{M-1} \triangleq \Big\{\mathbf{z} \in \mathbb{C}^M \,\Big|\, \|\mathbf{z}\| = 1\Big\}, \tag{11.15}$$

which is illustrated in Fig. 11.4. Each complex Watson distribution in (11.13) models the distribution of $\mathbf{z}_{tf}$ for a cluster. k denotes the cluster index.

$$\Theta_{\mathrm{W}f} \triangleq \Big\{\alpha_f^{(k)}, \mathbf{a}_f^{(k)}, \kappa_f^{(k)} \Big| \forall k\Big\} \tag{11.16}$$

denotes the set of all model parameters of the cWMM (11.13), where $\alpha_f^{(k)}$ satisfies

$$\alpha_f^{(k)} \geq 0, \tag{11.17}$$

$$\sum_k \alpha_f^{(k)} = 1, \tag{11.18}$$

$\mathbf{a}_f^{(k)}$ denotes a parameter representing the mean orientation of $\mathbf{z}_{tf}$ for the kth cluster satisfying

$$\left\|\mathbf{a}_f^{(k)}\right\| = 1, \tag{11.19}$$

and $\kappa \in \mathbb{R}$ denotes a parameter representing the concentration of the distribution of $\mathbf{z}_{tf}$ for the kth cluster. ^{H} denotes conjugate transposition; $\mathscr{K}$ the confluent hypergeometric function of the first kind, also known as the Kummer function, which is defined by the following power series:

$$\mathscr{K}(\xi, \eta; \kappa) \triangleq 1 + \frac{\xi}{\eta}\frac{\kappa}{1!} + \frac{\xi(\xi+1)}{\eta(\eta+1)}\frac{\kappa^2}{2!} + \dots . \tag{11.20}$$

To analyze the behavior of (11.14) as a function of $\mathbf{z}$, note that (11.14) depends on $\mathbf{z}$ through the term $\left|\mathbf{a}^{\mathsf{H}}\mathbf{z}\right|$ only and increases[decreases] monotonically as $\left|\mathbf{a}^{\mathsf{H}}\mathbf{z}\right|$ increases when $\kappa > 0[\kappa < 0]$. Note also that

$$0 \le \left|\mathbf{a}^{\mathsf{H}}\mathbf{z}\right| \le 1, \tag{11.21}$$

which follows from the Cauchy-Schwartz inequality and $\|\mathbf{z}\| = \|\mathbf{a}\| = 1$. Therefore, for $\kappa > 0[\kappa < 0]$, (11.14) has the global minima[maxima] at

$$\left\{\mathbf{z} \in S^{M-1} \middle| \left|\mathbf{a}^{\mathsf{H}}\mathbf{z}\right| = 0\right\}, \tag{11.22}$$

increases[decreases] monotonically as $\left|\mathbf{a}^{\mathsf{H}}\mathbf{z}\right|$ increases, and has the global maxima[minima] at

$$\left\{\mathbf{z} \in S^{M-1} \middle| \left|\mathbf{a}^{\mathsf{H}}\mathbf{z}\right| = 1\right\}. \tag{11.23}$$

Note that (11.22) equals

$$\{\mathbf{z} \in S^{M-1} | \mathbf{a}^{\mathsf{H}}\mathbf{z} = 0\}, \tag{11.24}$$

and (11.23) equals

$$\{\exp(\mathrm{j}\theta)\mathbf{a} | \theta \in [0, 2\pi)\}. \tag{11.25}$$

It is straightforward to see that, for $\kappa = 0$, (11.14) is constant (i.e., uniform distribution on S^{M-1}). Based on the above property, we impose a constraint

$$\kappa_f^{(k)} > 0, \tag{11.26}$$

which is appropriate for our application.

Once the model parameters $\Theta_{\mathrm{W},f}$ have been estimated, the posterior probability $\tilde{\gamma}_{tf}^{(k)}$ can be obtained based on Bayes' theorem [23] by

$$\tilde{\gamma}_{tf}^{(k)} \leftarrow \frac{\alpha_f^{(k)} p_{\mathrm{W}}\left(\mathbf{z}_{tf}; \mathbf{a}_f^{(k)}, \kappa_f^{(k)}\right)}{\sum_l \alpha_f^{(l)} p_{\mathrm{W}}\left(\mathbf{z}_{tf}; \mathbf{a}_f^{(l)}, \kappa_f^{(l)}\right)}. \tag{11.27}$$

To estimate the model parameters $\Theta_{\mathrm{W}f}$, the cWMM (11.13) is fitted to the feature vector $\mathbf{z}_{tf}$, e.g., based on the maximization of the log-likelihood function

$$\sum_{t=1}^{T} \ln p(\mathbf{z}_{tf}; \Theta_{\mathrm{W}f}). \tag{11.28}$$

This is realized by, e.g., an expectation-maximization (EM) algorithm [23], which consists in alternate iteration of an E-step and an M-step. The E-step consists in updating the posterior probability $\gamma_{tf}^{(k)}$ by (11.27) using current estimates of the model parameters $\Theta_{\mathrm{W}f}$. The M-step consists in updating the model parameters $\Theta_{\mathrm{W}f}$ using the posterior probability $\gamma_{tf}^{(k)}$, which is realized by applying the following update rules:

$$\alpha_f^{(k)} \leftarrow \frac{1}{T} \sum_{t=1}^{T} \tilde{\gamma}_{tf}^{(k)}, \tag{11.29}$$

$$\mathbf{R}_f^{(k)} \leftarrow \frac{\sum_{t=1}^{T} \tilde{\gamma}_{tf}^{(k)} \mathbf{z}_{tf} \mathbf{z}_{tf}^{\mathsf{H}}}{\sum_{t=1}^{T} \tilde{\gamma}_{tf}^{(k)}}, \tag{11.30}$$

$$\left(\lambda_f^{(k)}, \mathbf{a}_f^{(k)}\right) \leftarrow \text{the largest eigenvalue and a corresponding eigenvector of } \mathbf{R}_f^{(k)}, \tag{11.31}$$

$$\mathbf{a}_f^{(k)} \leftarrow \frac{\mathbf{a}_f^{(k)}}{\|\mathbf{a}_f^{(k)}\|}, \tag{11.32}$$

$$\kappa_f^{(k)} \leftarrow \frac{M\lambda_f^{(k)} - 1}{2\lambda_f^{(k)}\left(1 - \lambda_f^{(k)}\right)} \left[1 + \sqrt{1 + \frac{4(M+1)\lambda_f^{(k)}\left(1 - \lambda_f^{(k)}\right)}{M - 1}}\right]. \tag{11.33}$$

See Appendix 1 for derivation of this EM algorithm.

A major limitation of the cWMM lies in that the complex Watson distribution (11.14) can represent a distribution that is rotationally symmetric about the axis $\mathbf{a}$ (see Fig. 11.5). Indeed, as we have already noted, (11.14) is a function of $|\mathbf{a}^{\mathsf{H}}\mathbf{z}|$, which can be regarded as the cosine of the angle between $\mathbf{a}$ and $\mathbf{z}$. However, the distribution of the feature vector $\mathbf{z}_{tf}$ for each cluster is not necessarily rotationally symmetric, depending on various conditions such as the array geometry and acoustic transfer characteristics. The cWMM therefore has a limited ability to approximate the distribution of $\mathbf{z}_{tf}$, which results in degraded mask estimation accuracy and therefore degraded performance of source separation and denoising. This motivates us to consider more flexible distributions, which are described in Sects. 11.3.2 and 11.3.3.

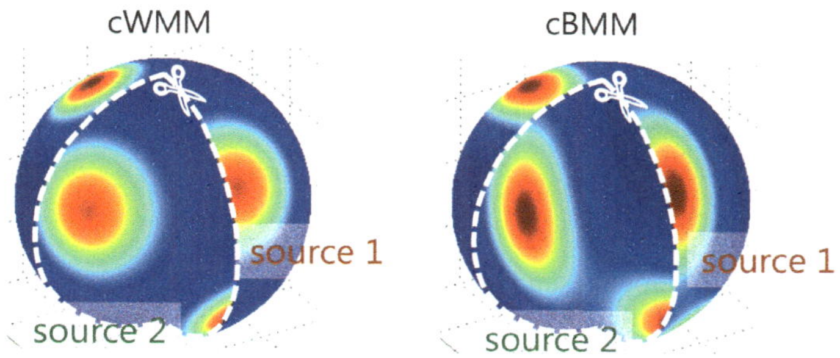

Fig. 11.5 Illustration of the cWMM and the cBMM for two sources

11.3.2 Mask Estimation Based on Complex Bingham Mixture Model (cBMM)

To overcome the above limitation of the cWMM, Ito et al. have proposed to estimate masks based on modeling the feature vector (11.2) by a *complex Bingham mixture model* (*cBMM*) [29]. The cBMM is composed of *complex Bingham distributions* of Kent [30], and the complex Bingham distribution is an extension of the real Bingham distribution of Bingham [31]. The complex Bingham distribution can represent not only rotationally symmetric but also elliptical distributions on the unit hypersphere (see Fig. 11.5), and can therefore better approximate the distribution of the feature vector $\mathbf{z}_{tf}$ than the complex Watson distribution. As a result, the cBMM can improve mask estimation accuracy and therefore source separation and denoising performance compared to the cWMM.

The PDF of the cBMM is given by

$$p\left(\mathbf{z}_{tf};\,\Theta_{\mathrm{B},f}\right) = \sum_{k} \alpha_f^{(k)} p_{\mathrm{B}}\left(\mathbf{z}_{tf};\,\mathbf{B}_f^{(k)}\right), \tag{11.34}$$

where p_{B} denotes a complex Bingham distribution

$$p_{\mathrm{B}}\left(\mathbf{z};\,\mathbf{B}\right) \triangleq c(\mathbf{B})^{-1} \exp\left(\mathbf{z}^{\mathsf{H}}\mathbf{B}\mathbf{z}\right). \tag{11.35}$$

Here, $c(\mathbf{B})$ denotes the following function defined for a Hermitian matrix $\mathbf{B}$.

$$c(\mathbf{B}) \triangleq \left[2\pi^{M} \sum_{m=1}^{M} \frac{\exp\left(\beta_m\right)}{\prod_{l\neq m}\left(\beta_m - \beta_l\right)}\right], \tag{11.36}$$

where β_m, $m = 1, \dots, M$, denote the eigenvalues of $\mathbf{B}$. Both the complex Bingham distribution and the cBMM are defined on the unit hypersphere S^{M-1}. Each complex Bingham distribution in (11.34) models the distribution of $\mathbf{z}_{tf}$ for a cluster.

$$\Theta_{\mathrm{B},f} \triangleq \left\{\alpha_f^{(k)}, \mathbf{B}_f^{(k)} \middle| \forall k\right\} \tag{11.37}$$

denotes the set of all model parameters of the cBMM, where $\mathbf{B}_f^{(k)}$ is a Hermitian parameter matrix, which represents not only the location and the concentration, but also the direction and the shape, of the complex Bingham distribution. Note that the expression for the normalization factor in (11.35) is valid only when the eigenvalues of $\mathbf{B}$ are all distinct, which is always satisfied in practice.

Once the model parameters $\Theta_{\mathrm{B},f}$ have been estimated, the posterior probability $\tilde{\gamma}_{tf}^{(k)}$ can be obtained by

$$\tilde{\gamma}_{tf}^{(k)} \leftarrow \frac{\alpha_f^{(k)} p_{\mathrm{B}}\left(\mathbf{z}_{tf}; \mathbf{B}_f^{(k)}\right)}{\sum_l \alpha_f^{(l)} p_{\mathrm{B}}\left(\mathbf{z}_{tf}; \mathbf{B}_f^{(l)}\right)}. \tag{11.38}$$

As in the cWMM case, $\Theta_{\mathrm{B},f}$ can be estimated by the maximum likelihood method based on the EM algorithm. The E-step consists in updating $\tilde{\gamma}_{tf}^{(k)}$ by (11.38) using the current $\Theta_{\mathrm{B},f}$ value. The M-step consists in updating $\Theta_{\mathrm{B},f}$ using $\tilde{\gamma}_{tf}^{(k)}$, which is realized by applying the following update rules:

$$\alpha_f^{(k)} \leftarrow \frac{1}{T}\sum_{t=1}^{T} \tilde{\gamma}_{tf}^{(k)}, \tag{11.39}$$

$$\mathbf{R}_f^{(k)} \leftarrow \frac{\sum_{t=1}^{T} \tilde{\gamma}_{tf}^{(k)} \mathbf{z}_{tf} \mathbf{z}_{tf}^{\mathsf{H}}}{\sum_{t=1}^{T} \tilde{\gamma}_{tf}^{(k)}}, \tag{11.40}$$

$$\left(\lambda_{fm}^{(k)}, \mathbf{a}_{fm}^{(k)}\right) \leftarrow \text{the } m\text{th largest eigenvalue and a corresponding eigenvector of } \mathbf{R}_f^{(k)}, \tag{11.41}$$

$$\mathbf{a}_{fm}^{(k)} \leftarrow \frac{\mathbf{a}_{fm}^{(k)}}{\left\|\mathbf{a}_{fm}^{(k)}\right\|}, \tag{11.42}$$

$$\mathbf{B}_f^{(k)} \leftarrow \sum_{m=1}^{M} \left(-\frac{1}{\lambda_{fm}^{(k)}} + \frac{1}{\lambda_{f1}^{(k)}}\right) \mathbf{a}_{fm}^{(k)} \mathbf{a}_{fm}^{(k)\mathsf{H}}. \tag{11.43}$$

See Appendix 2 for derivation of the above algorithm.

Note that the parameter matrix $\mathbf{B}_f^{(k)}$ has the following indeterminacy

$$p_{\text{B}}\left(\mathbf{z}_{tf};\mathbf{B}_f^{(k)}\right)=p_{\text{B}}\left(\mathbf{z}_{tf};\mathbf{B}_f^{(k)}+\xi\mathbf{I}\right),\ ^\forall\xi\in\mathbb{R},\tag{11.44}$$

which follows from $\|\mathbf{z}_{tf}\|=1$. Here, $\mathbf{I}$ denotes the $M\times M$ identity matrix. To remove this indeterminacy, in the above algorithm, ξ has been determined so that the largest eigenvalue of $\mathbf{B}_f^{(k)}$ equals zero.

11.3.3 Mask Estimation Based on Complex Gaussian Mixture Model (cGMM)

As an alternative method, Ito et al. have proposed to estimate masks based on modeling the feature vector (11.1) by a *complex (time-varying) Gaussian mixture model (cGMM)* [32], inspired by Duong et al. [33]. Note that the cGMM models the observation vector itself in (11.1), instead of its normalized version in (11.2). The cGMM is composed of complex Gaussian distributions, where the covariance matrices are parametrized by time-invariant spatial covariance matrices and time-variant power parameters.

The PDF of the cGMM is given by

$$p(\mathbf{z}_{tf};\Theta_{\text{G},f})=\sum_k\alpha_f^{(k)}p_{\text{G}}\left(\mathbf{z}_{tf};0,\phi_{tf}^{(k)}\mathbf{B}_f^{(k)}\right),\tag{11.45}$$

where p_{G} denotes a complex Gaussian distribution

$$p_{\text{G}}\left(\mathbf{z};\mathbf{g},\boldsymbol{\Sigma}\right)\triangleq\frac{1}{\pi^M\det\boldsymbol{\Sigma}}\exp[-(\mathbf{z}-\mathbf{g})^{\mathsf{H}}\boldsymbol{\Sigma}^{-1}(\mathbf{z}-\mathbf{g})],\tag{11.46}$$

with $\mathbf{g}$ being the mean and $\boldsymbol{\Sigma}$ the covariance matrix. Both the complex Gaussian distribution and the cGMM are defined in $\mathbb{C}^M$. Each complex Gaussian distribution in (11.45) models the distribution of $\mathbf{z}_{tf}$ for a cluster.

$$\Theta_{\text{G},f}\triangleq\left\{\alpha_f^{(k)},\mathbf{B}_f^{(k)}\middle|^\forall k\right\}\cup\left\{\phi_{tf}^{(k)}\middle|^\forall k,{}^\forall t\right\}\tag{11.47}$$

denotes the set of all model parameters of the cGMM, where $\mathbf{B}_f^{(k)}$ is a scaled covariance matrix modeling the direction of the observation vector in (11.1) (i.e., the normalized observation vector (11.2)), and $\phi_{tf}^{(k)}$ is a power parameter modeling the magnitude of the observation vector.

Once the model parameters $\Theta_{\text{G},f}$ have been estimated, the posterior probability $\tilde{\gamma}_{tf}^{(k)}$ can be obtained by

$$\tilde{\gamma}_{tf}^{(k)} \leftarrow \frac{\alpha_f^{(k)} p_{\mathrm{G}}\left(\mathbf{y}_{tf}; 0, \phi_{tf}^{(k)} \mathbf{B}_f^{(k)}\right)}{\sum_l \alpha_f^{(l)} p_{\mathrm{G}}\left(\mathbf{y}_{tf}; 0, \phi_{tf}^{(l)} \mathbf{B}_f^{(l)}\right)}. \tag{11.48}$$

As in the cWMM and the cBMM cases, $\Theta_{\mathrm{G},f}$ can be estimated by the maximum likelihood method based on the EM algorithm. The E-step consists in updating $\tilde{\gamma}_{tf}^{(k)}$ by (11.48) using the current $\Theta_{\mathrm{G},f}$ value. The M-step consists in updating $\Theta_{\mathrm{G},f}$ using $\tilde{\gamma}_{tf}^{(k)}$, which is realized by applying the following update rules:

$$\alpha_f^{(k)} \leftarrow \frac{1}{T} \sum_{t=1}^{T} \tilde{\gamma}_{tf}^{(k)}, \tag{11.49}$$

$$\mathbf{B}_f^{(k)} \leftarrow \frac{\sum_{t=1}^{T} \tilde{\gamma}_{tf}^{(k)} \mathbf{y}_{tf} \mathbf{y}_{tf}^{\mathsf{H}} / \phi_{tf}^{(k)}}{\sum_{t=1}^{T} \tilde{\gamma}_{tf}^{(k)}}, \tag{11.50}$$

$$\phi_{tf}^{(k)} \leftarrow \frac{1}{M} \mathbf{y}_{tf}^{\mathsf{H}} \left(\mathbf{B}_f^{(k)}\right)^{-1} \mathbf{y}_{tf}. \tag{11.51}$$

See Appendix 3 for derivation of the above algorithm.

11.4 Experimental Evaluation

We conducted source separation and denoising experiments to verify the effectiveness of observation vector clustering introduced in this chapter.

11.4.1 Source Separation

We first describe the source separation experiment. We assumed that the number of sources was known. We generated observed signals by convolving 8s-long English speech signals with room impulse responses measured in an experimental room (see Fig. 11.6). The sampling frequency of the observed signals was 8 kHz; the frame length 1024 points (128 ms); the frame shift 256 points (32 ms); the number of EM iterations 100. The permutation problem was resolved by Sawada's method [16]. Source signal estimates were obtained based on masking as in (11.8).

Figure 11.7 shows the signal-to-distortion ratio (SDR) [34] as a function of the reverberation time RT_{60}, and Fig. 11.8 shows an example of source separation results. The SDRs were averaged over 16 trials with eight combinations of speech signals and two distances between a loudspeaker and the array center. The azimuths of sources were 70° and 150° for $N = 2$, and 70°, 150°, and 245° for $N = 3$.

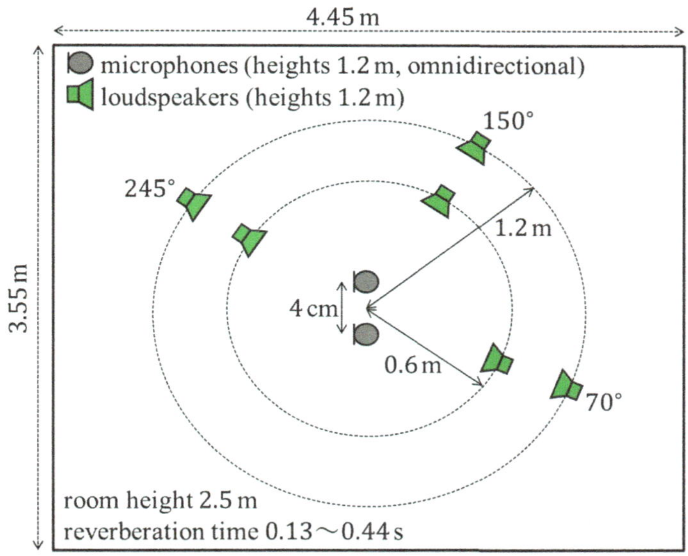

Fig. 11.6 Configurations in room impulse response measurement

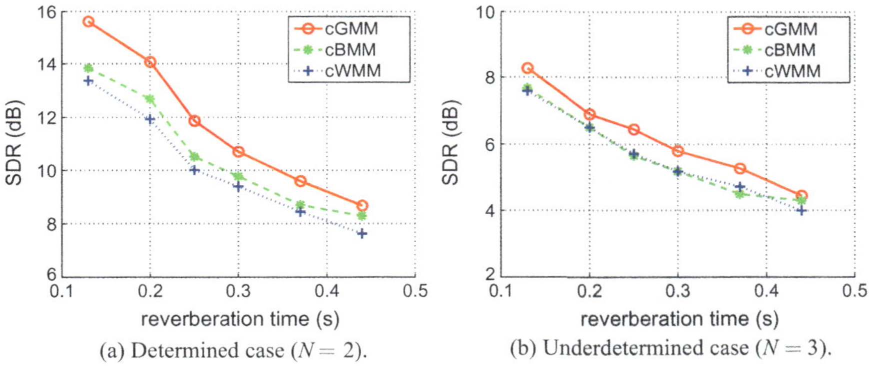

Fig. 11.7 Signal-to-distortion ratio (SDR) as a function of the reverberation time RT_{60}

11.4.2 Denoising

Now we move on to the denoising experiment. The performance was measured by the word error rate (WER) of ASR on the CHiME-3 task [35]. The CHiME-3 task consists in recognition of WSJ-5K prompts read from, and recorded by, a tablet device equipped with $M = 6$ microphones in four noisy public areas: on the bus (BUS), cafe (CAF), pedestrian area (PED), and street junction (STR). For further details about the data, we refer the readers to [35].

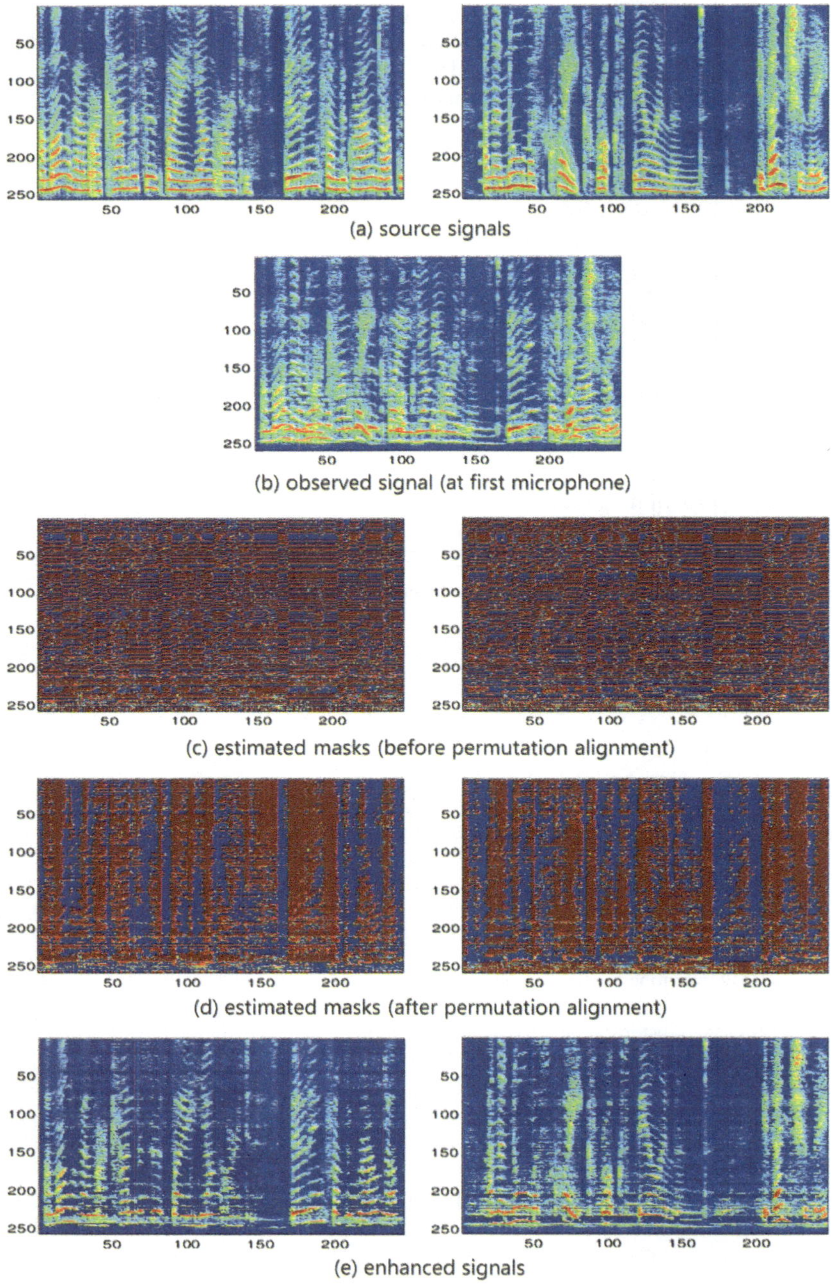

Fig. 11.8 Example of source separation results for $N = 2$ and $RT_{60} = 130\,\text{ms}$. The horizontal axis represents the time, and the vertical the frequency. To focus on low frequencies, which contain most speech energy, only the frequency range of 0 to 2 kHz is shown. The temporal range shown corresponds to 10 s

Denoising was performed by using MVDR beamformers designed using the estimated masks as in Sect. 11.2.2. Assuming that the background noise arrive from all directions equally (i.e., noise is diffuse), we set $\kappa_f^{(0)} = 0$ for the cWMM, $\mathbf{B}_f^{(0)} = 0$ for the cBMM, and $\mathbf{B}_f^{(0)} = \mathbf{I}$ for the cGMM. Permutation alignment was performed by the method proposed in [36], which is based on a common amplitude modulation property of speech. The frame length and the frame shift were 64 ms and 16 ms, respectively, and the window was hann.

ASR was performed by using a DNN-HMM-based acoustic model with a fully connected DNN (10 hidden layers) and an RNN-based language model. The acoustic model was trained on 18 hours of multicondition data.

The word error rate (WER) for the real data of the development set, averaged over all environments, was as follows:

- no denoising: 14.29 %,
- denoising with the cWMM: 10.2 %,
- denoising with the cBMM: 8.3 %,
- denoising with the cGMM: 9.3 %.

We see that the WER has been reduced significantly by mask-based MVDR beamforming.

11.5 Conclusions

In this chapter, we described multichannel source separation and denoising based on source sparseness. Particularly, we introduced recently proposed framework of observation vector clustering, which have been shown to be effective and robust in the real world. We also introduced specific algorithms for observation vector clustering, based on the cWMM, the cBMM, and the cGMM.

Appendix 1 Derivation of cWMM-Based Mask Estimation Algorithm

Here we derive the cWMM-based mask estimation algorithm in Sect. 11.3.1. The derivation of the E-step is straightforward and omitted. The update rules for the M-step is obtained by maximizing the following Q-function with respect to $\Theta_{\mathrm{W},f}$:

$$Q(\Theta_{\mathrm{W},f}) \triangleq \sum_{t=1}^{T} \sum_{k} \tilde{\gamma}_{tf}^{(k)} \ln\left[\alpha_f^{(k)} p_{\mathrm{W}}\left(\mathbf{z}_{tf}; \mathbf{a}_f^{(k)}, \kappa_f^{(k)}\right)\right] \tag{11.52}$$

$$= \sum_{k}\left(\sum_{t=1}^{T} \tilde{\gamma}_{tf}^{(k)}\right) \ln \alpha_f^{(k)} - \sum_{k}\left(\sum_{t=1}^{T} \tilde{\gamma}_{tf}^{(k)}\right) \ln \mathscr{K}\left(1, M; \kappa_f^{(k)}\right) \tag{11.53}$$

$$+\sum_k \kappa_f^{(k)} \mathbf{a}_f^{(k)\mathsf{H}} \left(\sum_{t=1}^T \tilde{\gamma}_{tf}^{(k)} \mathbf{z}_{tf} \mathbf{z}_{tf}^{\mathsf{H}}\right) \mathbf{a}_f^{(k)} + C$$

$$= \sum_k \left(\sum_{t=1}^T \tilde{\gamma}_{tf}^{(k)}\right) \left[\ln \alpha_f^{(k)} - \ln \mathscr{K}\left(1, M; \kappa_f^{(k)}\right) + \kappa_f^{(k)} \mathbf{a}_f^{(k)\mathsf{H}} \mathbf{R}_f^{(k)} \mathbf{a}_f^{(k)}\right] + C. \tag{11.54}$$

Here, $\mathbf{R}_f^{(k)}$ is defined by

$$\mathbf{R}_f^{(k)} \triangleq \frac{\sum_{t=1}^T \tilde{\gamma}_{tf}^{(k)} \mathbf{z}_{tf} \mathbf{z}_{tf}^{\mathsf{H}}}{\sum_{t=1}^T \tilde{\gamma}_{tf}^{(k)}}, \tag{11.55}$$

and C denotes a constant independent of $\Theta_{\mathrm{W},f}$.

The update rule for $\alpha_f^{(k)}$ is obvious: note the constraint (11.18) and apply the Lagrangian multiplier method.

The update rule for $\mathbf{a}_f^{(k)}$ is obtained by maximizing $Q(\Theta_{\mathrm{W},f})$ subject to (11.19). Noting (11.26), we see that this is equivalent to maximizing $\mathbf{a}_f^{(k)\mathsf{H}} \mathbf{R}_f^{(k)} \mathbf{a}_f^{(k)}$ subject to (11.19). From the linear algebra, $\mathbf{a}_f^{(k)}$ is therefore a unit-norm principal eigenvector of $\mathbf{R}_f^{(k)}$.

The update rule for $\kappa_f^{(k)}$ is obtained by maximizing

$$-\left(\sum_{t=1}^T \tilde{\gamma}_{tf}^{(k)}\right) \ln \mathscr{K}\left(1, M; \kappa_f^{(k)}\right) + \left(\sum_{t=1}^T \tilde{\gamma}_{tf}^{(k)}\right) \kappa_f^{(k)} \mathbf{a}_f^{(k)\mathsf{H}} \mathbf{R}_f^{(k)} \mathbf{a}_f^{(k)}. \tag{11.56}$$

Since $\mathbf{a}_f^{(k)}$ is a unit-norm principal eigenvector of $\mathbf{R}_f^{(k)}$, we have

$$\mathbf{a}_f^{(k)\mathsf{H}} \mathbf{R}_f^{(k)} \mathbf{a}_f^{(k)} = \lambda_f^{(k)}, \tag{11.57}$$

where $\lambda_f^{(k)}$ is the principal eigenvalue of $\mathbf{R}_f^{(k)}$. Therefore, we have the following nonlinear equation for $\kappa_f^{(k)}$:

$$\frac{\partial}{\partial \kappa_f^{(k)}} \mathscr{K}\left(1, M; \kappa_f^{(k)}\right) = \lambda_f^{(k)} \mathscr{K}\left(1, M; \kappa_f^{(k)}\right). \tag{11.58}$$

Using (3.8) in [37], (11.58) is approximately solved as follows:

$$\kappa_f^{(k)} = \frac{M \lambda_f^{(k)} - 1}{2\lambda_f^{(k)}\left(1 - \lambda_f^{(k)}\right)} \left[1 + \sqrt{1 + \frac{4(M+1)\lambda_f^{(k)}\left(1 - \lambda_f^{(k)}\right)}{M - 1}}\right]. \tag{11.59}$$

Appendix 2 Derivation of cBMM-Based Mask Estimation Algorithm

Here we derive the cBMM-based mask estimation algorithm in Sect. 11.3.2. The update rule for the E-step is obvious. The update rules for the M-step is obtained by maximizing the following Q-function with respect to $\Theta_{\mathrm{B},f}$:

$$Q(\Theta_{\mathrm{B},f}) \triangleq \sum_{t=1}^{T} \sum_{k} \tilde{\gamma}_{tf}^{(k)} \ln\left[\alpha_f^{(k)} p_{\mathrm{B}}\left(\mathbf{z}_{tf} ; \mathbf{B}_f^{(k)}\right)\right] \tag{11.60}$$

$$= \sum_{k}\left(\sum_{t=1}^{T} \tilde{\gamma}_{tf}^{(k)}\right) \ln \alpha_f^{(k)} - \sum_{k}\left(\sum_{t=1}^{T} \tilde{\gamma}_{tf}^{(k)}\right) \ln c\left(\mathbf{B}_f^{(k)}\right) \tag{11.61}$$

$$+ \sum_{k} \operatorname{tr}\left[\mathbf{B}_f^{(k)}\left(\sum_{t=1} \tilde{\gamma}_{tf}^{(k)} \mathbf{z}_{tf} \mathbf{z}_{tf}^{\mathsf{H}}\right)\right]$$

$$= \sum_{k}\left(\sum_{t=1}^{T} \gamma_{tf}^{(k)}\right)\left[\ln \alpha_f^{(k)} - \ln c\left(\mathbf{B}_f^{(k)}\right) + \operatorname{tr}\left(\mathbf{B}_f^{(k)} \mathbf{R}_f^{(k)}\right)\right]. \tag{11.62}$$

Here, $c(\mathbf{B})$ is defined by (11.36), and $\mathbf{R}_f^{(k)}$ by (11.55).

The update rule for $\alpha_f^{(k)}$ is obvious.

To derive the update rule for $\mathbf{B}_f^{(k)}$, let us denote the mth largest eigenvalue of $\mathbf{R}_f^{(k)}$ by $\lambda_{fm}^{(k)}$ and a corresponding unit-norm eigenvector by $\mathbf{v}_{fm}^{(k)}$. We assume that $\lambda_{fm}^{(k)}$, $m = 1, \ldots, M$, are all distinct and positive, which is always true in practice. $\mathbf{R}_f^{(k)}$ is represented as

$$\mathbf{R}_f^{(k)} = \sum_{m=1}^{M} \lambda_{fm}^{(k)} \mathbf{v}_{fm}^{(k)} \mathbf{v}_{fm}^{(k)\mathsf{H}}. \tag{11.63}$$

From a result in [38], $\mathbf{v}_{fm}^{(k)}$, $m = 1, \ldots, M$, are also the eigenvectors of $\mathbf{B}_f^{(k)}$. Hence, $\mathbf{B}_f^{(k)}$ is represented in the form

$$\mathbf{B}_f^{(k)} = \sum_{m=1}^{M} \beta_{fm}^{(k)} \mathbf{v}_{fm}^{(k)} \mathbf{v}_{fm}^{(k)\mathsf{H}}. \tag{11.64}$$

Substituting (11.63) and (11.64) into (11.62) and disregarding terms independent of $\beta_{fm}^{(k)}$, $m = 1, \ldots, M$, we have

$$\left(\sum_{t=1}^{T} \gamma_{tf}^{(k)}\right)\left[-\ln c\left(\mathbf{B}_f^{(k)}\right) + \sum_{m=1}^{M} \lambda_{fm}^{(k)} \beta_{fm}^{(k)}\right]. \tag{11.65}$$

Therefore, we have

$$\frac{\partial \ln c\left(\mathbf{B}_f^{(k)}\right)}{\partial \beta_{fm}^{(k)}} = \lambda_{fm}^{(k)}. \tag{11.66}$$

Using an approximation in [38], this nonlinear equation can be approximately solved as follows:

$$\beta_{fm}^{(k)} \sim -\frac{1}{\lambda_{fm}^{(k)}}. \tag{11.67}$$

Substituting (11.67) into (11.64) and adding a matrix of the form $\xi\mathbf{I}$ so that the largest eigenvalue of $\mathbf{B}_f^{(k)}$ is zero, we obtain the following update rule for $\mathbf{B}_f^{(k)}$:

$$\mathbf{B}_f^{(k)} \leftarrow \sum_{m=1}^{M}\left(-\frac{1}{\lambda_{fm}^{(k)}} + \frac{1}{\lambda_{f1}^{(k)}}\right)\mathbf{v}_{fm}^{(k)}\mathbf{v}_{fm}^{(k)\mathsf{H}}. \tag{11.68}$$

Appendix 3 Derivation of cGMM-Based Mask Estimation Algorithm

Here we derive the cGMM-based mask estimation algorithm in Sect. 11.3.3. The derivation of the E-step is straightforward and omitted. The update rules for the M-step is obtained by maximizing the following Q-function with respect to $\Theta_{\mathrm{G},f}$:

$$Q(\Theta_{\mathrm{G},f}) \triangleq \sum_{t=1}^{T}\sum_{k}\tilde{\gamma}_{tf}^{(k)} \ln\left[\alpha_f^{(k)} p_{\mathrm{G}}\left(\mathbf{y}_{tf}; 0, \phi_{tf}^{(k)}\mathbf{B}_f^{(k)}\right)\right] \tag{11.69}$$

$$= \sum_{k}\left(\sum_{t=1}^{T}\tilde{\gamma}_{tf}^{(k)}\right)\ln\alpha_f^{(k)} - M\sum_{t=1}^{T}\sum_{k}\tilde{\gamma}_{tf}^{(k)}\ln\phi_{tf}^{(k)} \tag{11.70}$$

$$-\sum_{k}\left(\sum_{t=1}^{T}\tilde{\gamma}_{tf}^{(k)}\right)\ln\det\mathbf{B}_f^{(k)} - \sum_{t=1}^{T}\sum_{k}\frac{\tilde{\gamma}_{tf}^{(k)}}{\phi_{tf}^{(k)}}\mathbf{y}_{tf}^{\mathsf{H}}\left(\mathbf{B}_f^{(k)}\right)^{-1}\mathbf{y}_{tf} + C$$

$$= \sum_{k}\left(\sum_{t=1}^{T}\tilde{\gamma}_{tf}^{(k)}\right)\ln\alpha_f^{(k)} - M\sum_{t=1}^{T}\sum_{k}\tilde{\gamma}_{tf}^{(k)}\ln\phi_{tf}^{(k)} \tag{11.71}$$

$$-\sum_{k}\left(\sum_{t=1}^{T}\tilde{\gamma}_{tf}^{(k)}\right)\ln\det\mathbf{B}_f^{(k)} - \sum_{k}\mathrm{tr}\left[\left(\mathbf{B}_f^{(k)}\right)^{-1}\left(\sum_{t=1}^{T}\frac{\tilde{\gamma}_{tf}^{(k)}}{\phi_{tf}^{(k)}}\mathbf{y}_{tf}\mathbf{y}_{tf}^{\mathsf{H}}\right)\right] + C.$$

Here, C denotes a constant independent of $\Theta_{\mathrm{G},f}$.

The update rule for $\alpha_f^{(k)}$ is obvious.

From (11.70), the update rule for $\phi_{tf}^{(k)}$ is given by

$$\phi_{tf}^{(k)} = \frac{1}{M}\mathbf{y}_{tf}^{\mathsf{H}}\left(\mathbf{B}_f^{(k)}\right)^{-1}\mathbf{y}_{tf}. \tag{11.72}$$

As for $\mathbf{B}_f^{(k)}$, it should satisfy

$$-\left(\sum_{t=1}^{T}\tilde{\gamma}_{tf}^{(k)}\right)\left(\mathbf{B}_f^{(k)}\right)^{-1} + \left(\mathbf{B}_f^{(k)}\right)^{-1}\left(\sum_{t=1}^{T}\frac{\tilde{\gamma}_{tf}^{(k)}}{\phi_{tf}^{(k)}}\mathbf{y}_{tf}\mathbf{y}_{tf}^{\mathsf{H}}\right)\left(\mathbf{B}_f^{(k)}\right)^{-1} = 0. \tag{11.73}$$

Therefore, the update rule for $\mathbf{B}_f^{(k)}$ is

$$\mathbf{B}_f^{(k)} = \frac{\sum_{t=1}^{T}\tilde{\gamma}_{tf}^{(k)}\mathbf{y}_{tf}\mathbf{y}_{tf}^{\mathsf{H}}/\phi_{tf}^{(k)}}{\sum_{t=1}^{T}\tilde{\gamma}_{tf}^{(k)}}. \tag{11.74}$$

References

1. P.A. Naylor, N.D. Gaubitch, *Speech Dereverberation.* (Springer, 2009)
2. M. Brandstein, D. Ward, *Microphone Arrays: Signal Processing Techniques and Applications.* (Springer, 2001)
3. R. Zelinski, A microphone array with adaptive post-filtering for noise reduction in reverberant rooms, in *Proceeding of ICASSP* (1988), pp. 2578–2581
4. S. Gannot, D. Burshtein, E. Weinstein, Signal enhancement using beamforming and nonstationarity with applications to speech. IEEE Trans. SP **49**(8), 1614–1626 (2001)
5. S. Doclo, M. Moonen, GSVD-based optimal filtering for single and multimicrophone speech enhancement. IEEE Trans. SP **50**(9), 2230–2244 (2002)
6. S.F. Boll, Suppression of acoustic noise in speech using spectral subtraction. IEEE Trans. ASSP **ASSP-27**(2), 113–120 (1979)
7. Y. Ephraim, D. Malah, Speech enhancement using a minimum-mean square error short-time spectral amplitude estimator. IEEE Trans. ASSP **32**(6), 1109–1121 (1984)
8. R. Miyazaki, H. Saruwatari, T. Inoue, Y. Takahashi, K. Shikano, K. Kondo, Musical-noise-free speech enhancement based on optimized iterative spectral subtraction. IEEE Trans. ASLP **20**(7), 2080–2094 (2012)
9. P. Smaragdis, Probabilistic decompositions of spectra for sound separation, in *Blind Speech Separation*, ed. by S. Makino, T.-W. Lee, H. Sawada (Springer, 2007), pp. 365–386
10. Ö. Yılmaz, S. Rickard, Blind separation of speech mixtures via time-frequency masking. IEEE Trans. SP **52**(7), 1830–1847 (2004)
11. S. Araki, H. Sawada, R. Mukai, S. Makino, Underdetermined blind sparse source separation for arbitrarily arranged multiple sensors. Signal Process. **87**(8), 1833–1847 (2007)
12. Y. Izumi, N. Ono, S. Sagayama, Sparseness-based 2ch BSS using the EM algorithm in reverberant environment, in *Proceeding of WASPAA* (2007), pp. 147–150
13. H. Sawada, S. Araki, S. Makino, A two-stage frequency-domain blind source separation method for underdetermined convolutive mixtures, in *Proceeding of WASPAA* (2007), pp. 139–142
14. M.I. Mandel, R.J. Weiss, D.P.W. Ellis, Model-based expectation-maximization source separation and localization. IEEE Trans. ASLP **18**(2), 382–394 (2010)
15. D.H. Tran Vu, R. Haeb-Umbach, Blind speech separation employing directional statistics in an expectation maximization framework, in *Proceeding of ICASSP* (2010), pp. 241–244
16. H. Sawada, S. Araki, S. Makino, Underdetermined convolutive blind source separation via frequency bin-wise clustering and permutation alignment. IEEE Trans. ASLP **19**(3), 516–527 (2011)

17. M. Delcroix, K. Kinoshita, T. Nakatani, S. Araki, A. Ogawa, T. Hori, S. Watanabe, M. Fujimoto, T. Yoshioka, T. Oba, Y. Kubo, M. Souden, S.-J. Hahm, A. Nakamura, Speech recognition in the presence of highly non-stationary noise based on spatial, spectral and temporal speech/noise modeling combined with dynamic variance adaptation, in *Proceeding of CHiME 2011 Workshop on Machine Listening in Multisource Environments* (2011), pp. 12–17
18. M. Souden, S. Araki, K. Kinoshita, T. Nakatani, H. Sawada, A multichannel MMSE-based framework for speech source separation and noise reduction. IEEE Trans. ASLP **21**(9), 1913–1928 (2013)
19. T. Nakatani, S. Araki, T. Yoshioka, M. Delcroix, M. Fujimoto, Dominance based integration of spatial and spectral features for speech enhancement. IEEE Trans. ASLP **21**(12), 2516–2531 (2013)
20. T. Yoshioka, N. Ito, M. Delcroix, A. Ogawa, K. Kinoshita, M. Fujimoto, C. Yu, W.J. Fabian, M. Espi, T. Higuchi, S. Araki, T. Nakatani, The NTT CHiME-3 system: Advances in speech enhancement and recognition for mobile multi-microphone devices, in *Proceeding of ASRU* (2015), pp. 436–443
21. Y. Wang, D. Wang, Towards scaling up classification-based speech separation. IEEE Trans. ASLP **21**(7), 1381–1390 (2013)
22. J. Heymann, L. Drude, R. Haeb-Umbach, Neural network based spectral mask estimation for acoustic beamforming, in *Proceeding of ICASSP* (2016), pp. 196–200
23. C. Bishop, *Pattern Recognition and Machine Learning*. (Springer, 2006)
24. N. Murata, S. Ikeda, A. Ziehe, An approach to blind source separation based on temporal structure of speech signals. Neurocomputing **41**(1–4), 1–24 (2001)
25. H. Sawada, R. Mukai, S. Araki, S. Makino, A robust and precise method for solving the permutation problem of frequency-domain blind source separation. IEEE Trans. SAP **12**(5), 530–538 (2004)
26. H. Sawada, S. Araki, S. Makino, Measuring dependence of bin-wise separated signals for permutation alignment in frequency-domain BSS, in *Proceeding of IEEE International Symposium on Circuits and Systems (ISCAS)* (2007), pp. 3247–3250
27. K.V. Mardia, I.L. Dryden, The complex Watson distribution and shape analysis. J. Roy. Stat. Soc.: Ser. B (Stat. Methodol.) **61**(4), 913–926 (1999)
28. G. Watson, Equatorial distributions on a sphere. Biometrika **52**, 193–201 (1965)
29. N. Ito, S. Araki, T. Nakatani, Modeling audio directional statistics using a complex Bingham mixture model for blind source extraction from diffuse noise, in *Proceeding of ICASSP* (2016), pp. 465–468
30. J.T. Kent, The complex Bingham distribution and shape analysis. J. Roy. Stat. Soc.: Ser. B (Methodol.) **56**(2), 285–299 (1994)
31. C. Bingham, An antipodally symmetric distribution on the sphere. Ann. Stat. **2**, 1201–1205 (1974)
32. N. Ito, S. Araki, T. Yoshioka, T. Nakatani, Relaxed disjointness based clustering for joint blind source separation and dereverberation, in *Proceeding of IWAENC* (2014), pp. 268–272
33. N.Q.K. Duong, E. Vincent, R. Gribonval, Under-determined reverberant audio source separation using a full-rank spatial covariance model. IEEE Trans. ASLP **18**(7), 1830–1840 (2010)
34. E. Vincent, R. Gribonval, C. Févotte, Performance measurement in blind audio source separation. IEEE Trans. ASLP **14**(4), 1462–1469 (2006)
35. J. Barker, R. Marxer, E. Vincent, S. Watanabe, The third 'CHiME' speech separation and recognition challenge: Dataset, task and baselines, in *Proceeding of ASRU* (2015), pp. 504–511
36. N. Ito, S. Araki, T. Nakatani, Permutation-free clustering of relative transfer function features for blind source separation, in *Proceeding of EUSIPCO* (2015), pp. 409–413
37. S. Sra, D. Karp, The multivariate Watson distribution: maximum-likelihood estimation and other aspects. J. Multivar. Anal. **114**, 256–269 (2013)
38. K.V. Mardia, P.E. Jupp, *Directional Statistics*. (Wiley, 1999)

Chapter 12
Multimicrophone MMSE-Based Speech Source Separation

Shmulik Markovich-Golan, Israel Cohen and Sharon Gannot

Abstract Beamforming methods using a microphone array successfully utilize spatial diversity for speech separation and noise reduction. Adaptive design of the beamformer based on various minimum mean squared error (MMSE) criteria significantly improves performance compared to fixed, and data-independent design. These criteria differ in their considerations to noise minimization and desired speech distortion. Three common data-dependent beamformers, namely, matched filter (MF), MWF and LCMV are presented and analyzed. Estimation methods for implementing the various beamformers are surveyed. Simple examples of applying the various beamformers to simulated narrowband signals in an anechoic environment and to speech signals in a real-life reverberant environment are presented and discussed.

12.1 Introduction

In this chapter we introduce multimicrophone methods for speech separation and noise reduction methods, which are based on *beamforming*. Traditionally, beamforming methods are adopted from classical array processing techniques, in which a *beam* of high response is *steered* towards the desired source, while suppressing other directions. These methods were mainly applied in communications and radar domains. They usually assume free-field propagation, i.e., the angle-of-arrival fully determines the source position, although several design methods take multi-path propagation into account.

S. Markovich-Golan (✉)
Communication and Devices Group, Intel Corporation, Petah Tikva, Israel
e-mail: shmulik.markovich-golan@intel.com

S. Markovich-Golan · S. Gannot
Faculty of Engineering, Bar Ilan University, 5290002 Ramat-Gan, IsraelS. Gannot
e-mail: sharon.gannot@biu.ac.il

I. Cohen
Department of Electrical Engineering, Technion, 32000 Haifa, Israel
e-mail: icohen@ee.technion.ac.il

S. Makino (ed.), *Audio Source Separation*, Signals and Communication Technology, https://doi.org/10.1007/978-3-319-73031-8_12

Statistically optimal beamformers are powerful multichannel filtering tools that optimize a certain design criteria while adapting to the received data, hence usually referred to as *data-dependent* approaches. A plethora of optimization criteria were proposed. The MVDR beamformer, also referred to as Capon beamformer [1], minimizes the noise power at the output of the beamformer subject to a unit gain constraint in the look direction. Frost [2] presented an adaptive implementation of the MVDR beamformer for wideband signals. Griffiths and Jim [3] proposed the GSC which is an efficient decomposition of the MVDR beamformer into two branches: one satisfying the constraint on the desired source, and the other for minimizing the noise (and interference).

Several researchers, e.g. [4] (see also Van Veen and Buckley [5]) have proposed modifications to the MVDR beamformer to deal with multiple linear constraints, denoted linearly constrained minimum variance (LCMV). Their work was motivated by the desire to apply further control to the array beampattern, beyond that of a steer-direction gain constraint. Hence, the LCMV can be applied to construct a beampattern satisfying certain constraints for a set of directions, while minimizing the array response in all other directions. Breed and Strauss [6] proved that the LCMV extension has also an equivalent GSC structure, which decouples the constraining and the minimization operations. The multichannel Wiener filter (MWF) is another important beamforming criterion, which minimizes the minimum mean squared error (MMSE) between the desired signal and the array output. It can be shown that the MWF decomposes into an MVDR beamformer followed by a single-channel Wiener post-filter [7]. A comprehensive analysis of beamformer criteria can be found in [5, 8].

Speech signals usually propagate in acoustic enclosures, e.g., rooms, and not in free-field. In the presence of obstacles, the sound wave is subject to diffractions and reflections, depending on its wavelength. Due to the typically small absorbtion coefficients of the obstacles, many successive wave reflections occur before their power decay. This induces multiple propagation paths between each source and each microphone, each with a different delay and attenuation factor. This phenomenon is often referred to as *reverberation*. The AIR (and its respective ATF) encompasses all these reflections and is usually a very long (few thousands taps) and time-varying filter.

Due to this intricate propagation regime, resorting to beampatterns as a function of the angle-of-arrival implies a reduction of a complex multi-parameter problem to an arbitrary single parameter problem. Classical beamformers, that construct their steering-vector under the assumption of free-field propagation, are often prone to performance degradation when applied in reverberant environments. It is therefore very important to take the reverberation effects into account while designing beamformers.

To circumvent the simplified free-field assumption, it was proposed [9, 10] to substitute the delay-only steering vector by the (normalized) ATFs relating the source and the microphones. This concept was later extended to the multiple sources scenario for extracting the desired source(s) from a mixture of desired and interference sources [11].

The MWF is also widely applied for speech enhancement, especially in its more flexible form, the speech distortion weighted-MWF (SDW-MWF) [12], which introduces a tradeoff factor that controls the amount of speech distortion versus the level of the residual noise.

A recent review paper [13] surveys many beamforming design criteria and their relation to BSS techniques.

12.2 Background

In this section we formulate the problem of speaker separation using spatial filtering methods. The signals and their propagation models are defined in Sect. 12.2.1. The following Sects. 12.2.2 and 12.2.3 are dedicated to defining spatial filters and criteria for evaluating their performances, respectively.

12.2.1 Generic Propagation Model

The speech sources are typically modeled in the STFT as quasi-stationary complex random processes with zero mean and time-varying variance, with stationarity time of the order of tens of milliseconds. Let us consider the case of J speech sources, denoted:

$$s_j(n,f) \sim \mathscr{N}\left(0, \phi_{s_j}(n,f)\right) \tag{12.1}$$

for $j = 1, \ldots, J$ where $\phi_{s_j}(n,f)$ denotes the time-varying signals spectra, and the indices $n = 0, 1, \ldots,$ and $f = 0, 1, \ldots, F-1$ stand for the time-frame and frequency-bin index, and F denotes the STFT window length.

Given a microphone array comprising M microphones, the received microphone signals are given in an $M \times 1$ vector notation by

$$\mathbf{x}(n,f) = \sum_{j=1}^{J} \mathbf{c}_j(n,f) + \mathbf{u}(n,f) \tag{12.2}$$

where $\mathbf{c}_j(n,f)$ for $j = 1, \ldots, J$ denotes the J vectors of speech sources as received by the microphone array and $\mathbf{u}(n,f)$ denotes the $M \times 1$ dimensional vector comprising the noise components received at the microphones. Modeling the speech sources as coherent point sources and modeling the AIR as time-invariant convolution system, the speech components at the microphones are modeled in the STFT domain as a simple multiplication

$$\mathbf{c}_j(n,f) \triangleq \mathbf{a}_j(f) s_j(n,f) \tag{12.3}$$

where

$$\mathbf{a}_j(f) = \left[a_{j1}(f), \cdots a_{jI}(f)\right]^T \tag{12.4}$$

denotes the $M \times 1$ dimensional vector of ATFs relating the j-th source and the microphone array. Note that we assume that the STFT length is longer than the *effective length* of the AIR, such that convolution in the time-domain can be approximated as multiplication in the STFT domain (theoreticaly, only cyclic-convolution in the time-domain transforms to multiplication in the STFT domain [14]). Note that we assume that the AIR are time-invariant, i.e., the sources and the enclosure are static. This assumption can be relaxed to slowly time-varying environments, in which case the separating algorithm needs to adapt faster than the the system variations. However, for brevity we consider here time-invariant systems. The covariance matrices of the sources are given by:

$$\boldsymbol{\Phi}_{c_j}(n, f) \triangleq \mathrm{E}\left[\mathbf{c}_j(n, f)\mathbf{c}_j^H(n, f)\right] = \mathbf{a}_j(f)\mathbf{a}_j^H(f)\phi_{s_j}(n, f). \tag{12.5}$$

The noise-field is also assumed stationary, and the covariance matrix of its components at the microphone array is defined as:

$$\boldsymbol{\Phi}_u(f) = \mathrm{E}\left[\mathbf{u}(n, f)\mathbf{u}^H(n, f)\right]. \tag{12.6}$$

Note that the noise-stationarity assumption can be relaxed to slowly time-varying statistics, however, for ease of notation and derivation we assume that the noise is stationary.

12.2.2 Spatial Filtering

The spatial filter which is designed to extract the j-th speech source is denoted by $\mathbf{w}_j(n, f)$. Its corresponding output is defined by:

$$y_j(n, f) \triangleq \mathbf{w}_j^H(n, f)\mathbf{x}(n, f). \tag{12.7}$$

Note that generally the spatial filter may vary over time. By substituting (12.2) into (12.7), the output of the j-th spatial filter is decomposed into different components:

$$y_j(n, f) = \sum_{j'=1}^{J} d_{j,j'}(n, f) + v_j(n, f) \tag{12.8}$$

where

$$d_{j,j'}(n, f) \triangleq \mathbf{w}_j^H(n, f)\mathbf{c}_{j'}(n, f) \tag{12.9}$$

is the component that corresponds to the j'-th source at the output of the j-th spatial filter and

$$v_j(n, f) \triangleq \mathbf{w}_j^H(n, f)\mathbf{u}(n, f) \tag{12.10}$$

is the noise component at the j-th output. The aim of the j-th spatial filter is to maintain the j-th speech source, i.e., $d_{j,j}(n, f) \approx s_j(n, f)$, attenuate the other speech sources, i.e., $d_{j,j'}(n, f) \approx 0$ for $j' \neq j$, and reduce the noise, i.e., $v_j(n, f) \approx 0$. Note that aiming to obtain the *dry* signal of the j-th source (the original source before the convolution with the AIR) is a cumbersome task, and that in many practical scenarios obtaining the desired source as picked up by one of the microphones, which is denoted the *reference* microphone, is sufficient. Let us assume that the first microphone is selected as the reference microphone, and therefore the desired source at the output of the j-th spatial filter is $d_{j,j}(n, f) = a_{j1}(f)s_j(n, f)$. The RTFs relating the received components of the j'-th source at all microphones with its component at the reference microphone is defined as [10]:

$$\tilde{\mathbf{a}}_{j'}(f) \triangleq \frac{\mathbf{a}_{j'}(f)}{a_{j'1}(f)} \tag{12.11}$$

for $j' = 1, \ldots, J$. In the following sections, for the sake of clarity, we present the derivations of the various spatial filtering criteria using the ATF vectors rather than the RTF vectors.

12.2.3 Second-Order Moments and Criteria

Let us consider the output of the j-th spatial filter which aims to extract the j-th source while reducing noise and other interfering speakers, and define criteria to evaluate its performance. The difference between the output of the spatial filter and the desired signal is denoted as the error signal. The variance of the error signal, also known as the mean squared error (MSE) which we denote as $\chi_j(n, f)$, can be decomposed to its various components:

$$\begin{aligned}\chi_j(n, f) &\triangleq \mathrm{E}\left[\left|y_j(n, f) - s_j(n, f)\right|^2\right] \\ &= \delta_j(n, f) + \sum_{j' \neq j} \psi_{d_{j,j'}}(n, f) + \psi_{vj}(n, f)\end{aligned} \tag{12.12}$$

where

$$\delta_j(n, f) \triangleq \mathrm{E}\left[\left|s_j(n, f) - d_{j,j}(n, f)\right|^2\right] = |1 - \mathbf{w}_j^H(n, f)\mathbf{a}_j(f)|^2 \phi_{s_j}(n, f) \tag{12.13}$$

is the distortion of the j-th source component,

$$\psi_{d_{j,j'}}(n,f) \triangleq \mathrm{E}\left[\left|d_{j,j'}(n,f)\right|^2\right] = \left|\mathbf{w}_j^H(n,f)\mathbf{a}_{j'}(f)\right|^2 \phi_{s_{j'}}(n,f) \tag{12.14}$$

is the variance of the residual j'-th signal component, for $j' \neq j$ and

$$\psi_{v_j}(n,f) \triangleq \mathrm{E}\left[\left|v_j(n,f)\right|^2\right] = \mathbf{w}_j^H(n,f)\boldsymbol{\Phi}_u(f)\mathbf{w}_j(n,f) \tag{12.15}$$

is the variance of the residual noise component.

For evaluating the distortion level at the enhanced j-th signal, we define the signal-to-distortion ratio (SDR) as the power ratio of the desired speech component and its distortion:

$$\begin{aligned}\mathrm{SDR}_{\mathrm{o},j}(n,f) &\triangleq \frac{\phi_{s_j}(n,f)}{\delta_j(n,f)} \\ &= \frac{1}{\left|1-\mathbf{w}_j^H(n,f)\mathbf{a}_j(f)\right|^2}.\end{aligned} \tag{12.16}$$

To evaluate the noise reduction of the spatial filter we define the signal-to-noise ratio (SNR) improvement, denoted $\Delta\mathrm{SNR}_j$, which is the ratio of the SNR at the output and at the input, denoted $\mathrm{SNR}_{\mathrm{o},j}$ and $\mathrm{SNR}_{\mathrm{i},j}$:

$$\mathrm{SNR}_{\mathrm{i},j}(n,f) \triangleq \frac{\mathrm{trace}\left(\boldsymbol{\Phi}_{c_j}(n,f)\right)}{\mathrm{trace}\left(\boldsymbol{\Phi}_u(f)\right)} \tag{12.17a}$$

$$\mathrm{SNR}_{\mathrm{o},j}(n,f) \triangleq \frac{\psi_{dj,j}(n,f)}{\psi_{vj}(f)} \tag{12.17b}$$

$$\begin{aligned}\Delta\mathrm{SNR}_j(n,f) &\triangleq \frac{\mathrm{SNR}_{\mathrm{o},j}(n,f)}{\mathrm{SNR}_{\mathrm{i},j}(n,f)} \\ &= \frac{\left|\mathbf{w}_j^H(n,f)\mathbf{a}_j(f)\right|^2/\mathbf{w}_j^H(n,f)\boldsymbol{\Phi}_u(f)\mathbf{w}_j(n,f)}{\|\mathbf{a}_j(f)\|^2/\mathrm{trace}\left(\boldsymbol{\Phi}_u(f)\right)}.\end{aligned} \tag{12.17c}$$

Note that the last expression of $\Delta\mathrm{SNR}_j$ is obtained by substituting the expressions from (12.5), (12.14), (12.15), (12.17a), and (12.17b).

The interfering speakers reduction is evaluated by using the SIR improvement, denoted as $\Delta\mathrm{SIR}_{jj'}$, defined for pairs of desired speaker and interfering speaker, denoted j and j' respectively, as the ratio the output SIR and the input SIR, denoted as $\mathrm{SIR}_{\mathrm{o},jj'}$ and $\mathrm{SIR}_{\mathrm{i},jj'}$:

$$\mathrm{SIR}_{\mathrm{i},jj'}(n,f) \triangleq \frac{\mathrm{trace}\left(\boldsymbol{\Phi}_{c_j}(n,f)\right)}{\mathrm{trace}\left(\boldsymbol{\Phi}_{c_{j'}}(n,f)\right)} \tag{12.18a}$$

$$\mathrm{SIR}_{\mathrm{o},jj'}(n,f) \triangleq \frac{\psi_{djj}(n,f)}{\psi_{djj'}(n,f)} \tag{12.18b}$$

$$\begin{aligned}\Delta\mathrm{SIR}_{jj'}(n,f) &\triangleq \frac{\mathrm{SIR}_{\mathrm{o},jj'}(n,f)}{\mathrm{SIR}_{\mathrm{i},jj'}(n,f)} \\ &= \frac{\left|\mathbf{w}_j^H(n,f)\mathbf{a}_j(f)\right|^2 / \left|\mathbf{w}_j^H(n,f)\mathbf{a}_{j'}(f)\right|^2}{\|\mathbf{a}_j(f)\|^2/\|\mathbf{a}_{j'}(f)\|^2}.\end{aligned} \tag{12.18c}$$

Note that the last expression of $\Delta\mathrm{SIR}_{jj'}$ is obtained by substituting the expressions from (12.5), (12.14), (12.18a) and (12.18b).

Finally, in order to evaluate the total interference and noise reduction, the signal-to-interference-and-noise ratio (SINR) improvement, denoted $\Delta\mathrm{SINR}_j$, is defined as the ratio of the SINR at the output and at the input, denoted $\mathrm{SINR}_{\mathrm{o},j}$ and $\mathrm{SINR}_{\mathrm{i},j}$:

$$\mathrm{SINR}_{\mathrm{i},j}(n,f) = \frac{\|\mathbf{a}_j(f)\|^2\phi_{s_j}(n,f)}{\sum_{j'\neq j}\|\mathbf{a}_{j'}(f)\|^2\phi_{s_{j'}}(n,f) + \mathrm{trace}\left(\boldsymbol{\Phi}_u(f)\right)} \tag{12.19a}$$

$$\mathrm{SINR}_{\mathrm{o},j}(n,f) = \frac{\psi_{dj,j}(n,f)}{\sum_{j'\neq j}\psi_{d_{j,j'}}(n,f) + \psi_{v_j}(f)} \tag{12.19b}$$

$$\begin{aligned}\Delta\mathrm{SINR}_j(n,f) = &\frac{\left|\mathbf{w}_j^H(n,f)\mathbf{a}_j(f)\right|^2}{\|\mathbf{a}_j(f)\|^2} \cdot \\ &\cdot \frac{\sum_{j'\neq j}\|\mathbf{a}_{j'}(f)\|^2\phi_{s_{j'}}(n,f) + \mathrm{trace}\left(\boldsymbol{\Phi}_u(f)\right)}{\sum_{j'\neq j}\left|\mathbf{w}_j^H(n,f)\mathbf{a}_{j'}(f)\right|^2\phi_{s_{j'}}(n,f) + \mathbf{w}_j^H(n,f)\boldsymbol{\Phi}_u(f)\mathbf{w}_j(n,f)}.\end{aligned} \tag{12.19c}$$

12.3 Matched Filter

In this section, the *matched filter* spatial filtering method is presented. Its design criterion is defined and explained in Sect. 12.3.1, and its performance is analyzed in Sect. 12.3.2. The MF based spatial filter was first introduced in [15], where it was implemented in the time domain.

12.3.1 Design

As suggested by its name, the matched filter is designed to match the ATFs of the desired source (here denoted as the j-th source). Formally, it is defined as:

$$\mathbf{w}_j^{\mathrm{MF}}(f) \triangleq \frac{\mathbf{a}_j(f)}{\|\mathbf{a}_j(f)\|^2} \tag{12.20}$$

where the scaling is designed to maintain a distortionless response towards the desired source, i.e.,

$$\left(\mathbf{w}_j^{\mathrm{MF}}(f)\right)^H \mathbf{a}_j(f) = 1 \tag{12.21}$$

and therefore $d_{jj'}(n, f) = s_j(n, f)$.

This criterion, can be shown optimal in the sense of maximizing the SNR at the output for the case of a single source contaminated by spatially white noise. The main advantage of this spatial filter lies in its simplicity as it is independent of the noise and interferences properties. In the special case of a desired source signal arriving from the far-field regime in an anechoic environment, the matched filter reduces to the well known delay-and-sum (DS) beamformer.

12.3.2 Performance

As stated in the previous section, the matched filter is designed to pass the j-th source undistorted. Hence, by substituting (12.21) in (12.13) we obtain that the distortion equals zero

$$\delta_j^{\mathrm{MF}}(n, f) = 0 \tag{12.22}$$

and by following (12.16) the SDR of the j-th source is infinite, i.e.:

$$\mathrm{SDR}_{\mathrm{o},j}^{\mathrm{MF}}(n, f) \to \infty. \tag{12.23}$$

Since the MF is designed independently of the noise and interference sound fields, the SIR and SNR improvements are accidental. The spectrogram of the j'-th interfering source at the output of the j-th output, $\psi_{d_{j,j'}}^{\mathrm{MF}}(n, f)$, and the corresponding SIR improvement of the j-th source with respect to the j'-th interfering source are:

$$\psi_{d_{j,j'}}^{\mathrm{MF}}(n, f) = \frac{\|\mathbf{a}_{j'}(f)\|^2}{\|\mathbf{a}_j(f)\|^2} \left|\rho_{jj'}(f)\right|^2 \phi_{s_{j'}}(n, f) \tag{12.24a}$$

$$\Delta\mathrm{SIR}_{jj'}^{\mathrm{MF}}(f) = \frac{1}{\left|\rho_{jj'}(f)\right|^2} \tag{12.24b}$$

where $\rho_{jj'}(f)$ is defined as the normalized projection of the desired source ATF onto the interfering source ATF (per frequency-bin):

$$\rho_{jj'}(f) \triangleq \frac{\mathbf{a}_j^H(f)\mathbf{a}_{j'}(f)}{\|\mathbf{a}_j(f)\| \cdot \|\mathbf{a}_{j'}(f)\|}. \tag{12.25}$$

Note that from the Cauchy-Schwarz inequality the reciprocal of the SIR improvement expression is bounded by $0 \leq |\rho_{jj'}(f)|^2 \leq 1$, therefore the SIR improvement is bounded by $1 \leq \Delta\text{SIR}_{jj'}^{\text{MF}}(f) < \infty$. The SIR improvement will reach its upper-bound with the j'-th source being nulled by the MF of the j-th source if their corresponding ATFs are orthogonal (i.e., $\rho_{jj'}(f) = 0$).

By substituting (12.20) in (12.15) and (12.17c) the spectrum of the noise at the j-th output, and the corresponding SNR improvement are given by:

$$\psi_{v_j}^{\text{MF}}(f) = \frac{\mathbf{a}_j^H(f)\boldsymbol{\Phi}_u(f)\mathbf{a}_j(f)}{\|\mathbf{a}_j(f)\|^4} \tag{12.26a}$$

$$\Delta\text{SNR}_j^{\text{MF}}(f) = \frac{\|\mathbf{a}_j(f)\|^2 \cdot \text{trace}\left(\boldsymbol{\Phi}_u(f)\right)}{\mathbf{a}_j^H(f)\boldsymbol{\Phi}_u(f)\mathbf{a}_j(f)}. \tag{12.26b}$$

12.4 Multichannel Wiener Filter

In this section we present the MWF and analyze its performance.

12.4.1 Design

Considering the problem of enhancing the j-th source, recall that the MSE of an arbitrary spatial filter $\mathbf{w}(f)$ is denoted $\chi_j(n, f)$ and is defined by (12.12). The MSE is comprised of the following components: (a) distortion (denoted $\delta_j(n, f)$); (b) residual interferers spectra (denoted $\psi_{d_{j,j'}}(n, f)$, for $j' \neq j$); (c) residual noise spectrum (denoted $\psi_{v_j}(n, f)$). The MWF is designed to minimize the MSE expression:

$$\begin{aligned}\mathbf{w}_j^{\text{WF}}(n, f) &\triangleq \text{argmin}_{\mathbf{w}} \chi_j(n, f) \\ &= \frac{\left(\sum_{j' \neq j} \boldsymbol{\Phi}_{c_{j'}}(n, f) + \boldsymbol{\Phi}_u(f)\right)^{-1} \mathbf{a}_j(f)}{\mathbf{a}_j^H(f)\left(\sum_{j' \neq j} \boldsymbol{\Phi}_{c_{j'}}(n, f) + \boldsymbol{\Phi}_u(f)\right)^{-1} \mathbf{a}_j(f) + 1/\phi_{s_j}(n, f)}.\end{aligned} \tag{12.27}$$

Note that computing the MWF in (12.27) requires knowledge of: (a) power spectral density (PSD) of the desired source (denoted $\phi_{s_j}(n, f)$); (b) ATFs of the desired source (denoted $\mathbf{a}_j(f)$); (c) PSD matrices of the interferes (denoted $\boldsymbol{\Phi}_{c_{j'}}(n, f)$ for $j' \neq j$); (d) PSD matrix of the noise (denoted $\Phi_u(f)$). The estimation of these parameters is discussed in details in Sect. 12.6.

Although, practical methods exist for estimating the required parameters and implementing the MWF for the single source case, some relaxation is required when considering the multiple sources case. The *long-term averaged SOS* MWF is defined similarly to (12.27) by replacing the instantaneous PSD and PSD matrices of the desired source and interferers, respectively, with long-term averages:

$$\overline{\mathbf{w}}_j^{\mathrm{WF}}(f) = \frac{\left(\sum_{j'\neq j} \overline{\boldsymbol{\Phi}}_{c_{j'}}(f) + \boldsymbol{\Phi}_u(f)\right)^{-1} \mathbf{a}_j(f)}{\mathbf{a}_j^H(f)\left(\sum_{j'\neq j} \overline{\boldsymbol{\Phi}}_{c_{j'}}(f) + \boldsymbol{\Phi}_u(f)\right)^{-1} \mathbf{a}_j(f) + 1/\overline{\phi}_{s_j}(f)} \tag{12.28}$$

where

$$\bar{\phi}_{s_j}(f) \triangleq \frac{1}{\sum_n 1_{s_j}(n, f)} \sum_n 1_{s_j}(n, f) \mathrm{E}\left[|s_j(n, f)|^2\right] \tag{12.29a}$$

$$\overline{\boldsymbol{\Phi}}_{c_{j'}} \triangleq \frac{1}{\sum_n 1_{s_{j'}}(n, f)} \sum_n 1_{s_{j'}}(n, f) \mathrm{E}\left[\mathbf{c}_{j'}(n, f)\mathbf{c}_{j'}^H(n, f)\right] \tag{12.29b}$$

and $1_{s_{j'}}(n, f)$ denotes an indicator function which equals 1 for time-frequency bins in which the j'-th source is active, for $j' \in \{1, \ldots, J\}$.

12.4.2 Performance

By substituting (12.27) in (12.19c), the SINR improvement of the MWF can be shown to be

$$\begin{aligned}\Delta \mathrm{SINR}_j^{\mathrm{WF}}(n, f) =& \frac{1}{\|\mathbf{a}_j(f)\|^2} \left(\sum_{j'\neq j} \|\mathbf{a}_{j'}(f)\|^2 \phi_{s_{j'}}(n, f) + \mathrm{trace}\left(\boldsymbol{\Phi}_u(f)\right)\right) \\ & \times \mathbf{a}_j^H(f) \left(\sum_{j'\neq j} \phi_{s_{j'}}(n, f)\mathbf{a}_{j'}(f)\mathbf{a}_{j'}^H(f) + \boldsymbol{\Phi}_u(f)\right)^{-1} \mathbf{a}_j(f).\end{aligned} \tag{12.30}$$

Note that the MWF allows to introduce distortion to the desired source, as long as it minimizes the variance of the total error between the desired source and the MWF output. The latter distortion may become high for example in low SIR cases when the number of interfering speech sources is larger than the number of microphones, i.e., $J - 1 > M$.

12.5 Multichannel LCMV

The criterion for designing the LCMV spatial filter is defined in Sect. 12.5.1, and its performance is analyzed in Sect. 12.5.2.

12.5.1 Design

Let us consider the design of the LCMV spatial filter which enhances the j-th speech source. The LCMV is designed to satisfy a set of J linear constraints, one for each speech source, that are defined by:

$$\mathbf{A}^H(f)\mathbf{w}_j^{\text{LCMV}}(f) = \mathbf{g}_j \tag{12.31}$$

where

$$\mathbf{A}(f) \triangleq \left[\mathbf{a}_1(f), \cdots, \mathbf{a}_J(f)\right] \tag{12.32}$$

is the source ATFs matrix and

$$\mathbf{g}_j \triangleq \left[\mathbf{0}_{1\times(j-1)}\ 1\ \mathbf{0}_{1\times(J-j)}\right]^T \tag{12.33}$$

is the desired response for each of the sources and $\mathbf{w}_j^{\text{LCMV}}(f)$ denotes the LCMV spatial filter at the f-th frequency-bin. Note that the desired response for the j-th source is $g_{j,j} = 1$, i.e., pass the j-th source undistorted, and the desired response for all other sources is $g_{j,j'} = 0$ for $j' \neq j$, i.e., null all other source.

The LCMV spatial filter is defined as the optimal solution of the following criterion:

$$\mathbf{w}_j^{\text{LCMV}}(f) \triangleq \operatorname{argmin}_{\mathbf{w}} \mathbf{w}^H \boldsymbol{\Phi}_u(f)\mathbf{w};\ \text{ s.t. } \mathbf{A}^H(f)\mathbf{w} = \mathbf{g}_j \tag{12.34}$$

which aims to minimize the power of the noise at the output of the spatial filter (defined by (12.15)) while satisfying the linear constraints set, defined in (12.31). The closed-form solution of the optimization problem in (12.34) is given by:

$$\mathbf{w}_j^{\text{LCMV}}(f) = \boldsymbol{\Phi}_u^{-1}(f)\mathbf{A}(f)\left(\mathbf{A}^H(f)\boldsymbol{\Phi}_u^{-1}(f)\mathbf{A}(f)\right)^{-1}\mathbf{g}_j. \tag{12.35}$$

An alternative form for implementing the LCMV, denoted GSC [3], conveniently separates the tasks of constraining the spatial filter and minimizing the noise variance. Additionally, the GSC can be efficiently implemented as a time-recursive procedure which tracks the noise statistics and, adapts to it, and converges to the optimal LCMV solution.

The performance and behavior of the LCMV are different than those of the MWF. On the one hand, the MWF gives equal weight to the three sources of error, i.e. distortion, interfering speakers and noise, when designing the spatial filter the sum of the three is minimized at the output. On the other hand, the LCMV maintains the desired signal undistorted and nulls (zero response) towards the interfering speech signals at the output. The remaining degrees of freedom (DoF) are designed to minimize the noise at the output. By doing so, conceptually, the LCMV gives significantly higher weights to the distortion and interfering speech components compared to the weight of the noise component. In [16] the multiple speech distortion weighted-MWF (MSDW-MWF) criterion which generalizes both MWF and LCMV criteria is defined. The latter enables component specific weights to each of the error sources at the output of the spatial filter. It extends the SDW-MWF to the multiple speakers case.

12.5.2 Performance

By design, the LCMV satisfies a set of J linear constraints, one per speech source. The constraint that corresponds to the j-th desired source is designed to maintain a distortionless response towards this source, and therefore the distortion equals zero

$$\delta_j^{\mathrm{LCMV}}(n, f) = 0 \tag{12.36}$$

and correspondingly the SDR is infinite

$$\mathrm{SDR}_{\mathrm{o},j}^{\mathrm{LCMV}}(n, f) \to \infty. \tag{12.37}$$

Similarly, as the rest of the $J - 1$ constraints are associated with interfering speech sources and are designed to null them out, their corresponding SIRs are infinite:

$$\Delta\mathrm{SIR}_{jj'}^{\mathrm{LCMV}}(f) \to \infty. \tag{12.38}$$

By substituting (12.35) in (12.15) the noise variance at the output of the LCMV and the corresponding SNR improvement are:

$$\psi_{vj}^{\mathrm{LCMV}}(f) = \mathbf{g}_j^H \left(\mathbf{A}^H(f)\boldsymbol{\Phi}_u^{-1}\mathbf{A}(f)\right)^{-1} \mathbf{g}_j \tag{12.39a}$$

$$\Delta\mathrm{SINR}_j^{\mathrm{LCMV}}(f) = \frac{\mathbf{g}_j^H \left(\mathbf{A}^H(f)\boldsymbol{\Phi}_u^{-1}\mathbf{A}(f)\right)^{-1} \mathbf{g}_j}{\sum_{j' \neq j} \|\mathbf{a}_{j'}(f)\|^2 \phi_{s_{j'}}(n, f) + \mathrm{trace}\left(\boldsymbol{\Phi}_u(f)\right)}. \tag{12.39b}$$

12.6 Parameters Estimation

12.6.1 Multichannel SPP Estimators

Speech presence probability (SPP) is a fundamental and crucial component of many speech enhancement algorithms, among them are the spatial filters described in the previous sections. In the latter, SPP governs the adaptation of various components which contribute to the calculation of the spatial filter. Specifically, it can be used to govern the estimation of noise and speech covariance matrices (see Sect. 12.6.2) and of RTFs (see Sect. 12.6.3).

The problem of estimating SPP is derived from the classic detection problem, also known as the radar problem, and its goal is to identify the temporal-spectral activity pattern of speech contaminated by noise. Explicitly, determining if a time-frequency bin contains a noisy speech component or just noise. Contrary to the VAD problem where low resolution is sufficient, high-resolution activity estimation in both time and frequency is required here for proper enhancement. Most single-channel SPP estimators are based on non-stationarity of speech as opposed to the stationarity of the noise. However, in low SNR cases the accuracy of the estimation degrades.

When utilized for controlling the gain in single-channel postfiltering, the estimated SPP is "tuned" to have a tendency towards speech. This relates to the single-channel processing tradeoff between speech distortion and noise reduction, and to the common understanding that speech distortion and artifacts are more detrimental for human listeners than increased noise level. In difference to its use for single-channel enhancement, where the effect of SPP errors (i.e., false-alarms and miss-detections) is short-term (in time), in spatial processing the consequences of such errors can be grave and spreads over a longer period. Miss-detections of speech, and its false classification as noise might lead to a major distortion, also known as the self cancellation phenomenon. On the other hand false-alarms, i.e., time-frequency bins containing noise which are mistakenly classified as desired speech, result in increased noise level at the output of the spatial filter, since it is designed to pass them through.

Several contributions extend SPP estimation to utilize spatial information when using an array of microphones. Here we present some of these methods. In [17] which is presented in Sect. 12.6.1.1, the single channel Gaussian signal model of both speech and noise is extended to multichannel input, yielding a multichannel SPP. In Sect. 12.6.1.2, the work of [18], suggesting to incorporate the spatial information embedded in the direct-to-reverberant ratio (DRR) into the speech a priori probability (SAP), is presented. Thereby utilizing the coherence property of the speech source, assuming diffuse noise. Multichannel SPP incorporating spatial diversity can be utilized to address complex scenarios of multiple speakers. In [19, 20], the authors extend the previous DRR based SAP and incorporate estimated speaker positions to distinguish between different speakers, see Sect. 12.6.1.3.

12.6.1.1 Multichannel Gaussian Variables Model Based SPP

All derivations in this section refer to a specific time-frequency bin (n, f) and are replicated for all time-frequency bins. For brevity, the time and frequency indexes are omitted in the rest of this section. The received microphone signals

$$\mathbf{x} = \mathbf{c} + \mathbf{u} \tag{12.40}$$

and the speech and noise components thereof, are modeled as Gaussian random variables:

$$\mathbf{c} \sim \mathcal{N}_c(\mathbf{0}, \boldsymbol{\Phi}_c) \tag{12.41a}$$
$$\mathbf{u} \sim \mathcal{N}_c(\mathbf{0}, \boldsymbol{\Phi}_u) \tag{12.41b}$$

where $\boldsymbol{\Phi}_c = \phi_s \mathbf{a}\mathbf{a}^H$ is the covariance matrix of the speech image at the microphone signals. Consequently, a multichannel Gaussian model is adopted for the noise only, and noisy speech hypothesis:

$$\mathbf{x}|\mathcal{H}_u \sim \mathcal{N}_c(\mathbf{0}, \boldsymbol{\Phi}_u) \tag{12.42a}$$
$$\mathbf{x}|\mathcal{H}_s \sim \mathcal{N}_c(\mathbf{0}, \boldsymbol{\Phi}_c + \boldsymbol{\Phi}_u). \tag{12.42b}$$

It can be shown [17] that the SPP, defined as:

$$p \triangleq P(\mathcal{H}_s|\mathbf{x}) \tag{12.43}$$

can be formulated as

$$p = \frac{\Lambda}{1 + \Lambda} \tag{12.44}$$

where Λ is the generalized likelihood ratio, which in our case equals

$$\Lambda = \frac{1-q}{q} \cdot \frac{1}{1 + \mathrm{tr}\left\{\boldsymbol{\Phi}_u^{-1}\boldsymbol{\Phi}_c\right\}} \cdot \exp\left\{\frac{\boldsymbol{x}^H\boldsymbol{\Phi}_u^{-1}\boldsymbol{\Phi}_c\boldsymbol{\Phi}_u^{-1}\boldsymbol{x}}{1 + \mathrm{tr}\left\{\boldsymbol{\Phi}_u^{-1}\boldsymbol{\Phi}_c\right\}}\right\}. \tag{12.45}$$

and q is the a priori speech absence probability. Define the multichannel SNR as

$$\xi \triangleq \mathrm{tr}\left\{\boldsymbol{\Phi}_u^{-1}\boldsymbol{\Phi}_c\right\} \tag{12.46}$$

and also define

$$\beta \triangleq \mathbf{x}^H\boldsymbol{\Phi}_u^{-1}\boldsymbol{\Phi}_c\boldsymbol{\Phi}_u^{-1}\mathbf{x}. \tag{12.47}$$

Substituting (12.45), (12.46) and (12.47) in (12.44) yields the multichannel SPP:

$$p = \left\{ 1 + \frac{q}{1-q} \cdot (1+\xi) \cdot \exp\left\{ -\frac{\beta}{1+\xi} \right\} \right\}^{-1}. \tag{12.48}$$

Note that the single-channel SPP (of the first microphone) can be derived as a special case of the multichannel SPP by substituting

$$\xi_1 = \frac{\Phi_{c,11}}{\Phi_{u,11}} \tag{12.49a}$$

$$\beta_1 = \gamma_1 \cdot \xi_1 \tag{12.49b}$$

with $\gamma_1 \triangleq \frac{|x_1|^2}{\Phi_{u,11}}$ defined as the posterior SNR and $\Phi_{c,11}$, $\Phi_{u,11}$ denote the speech and noise variances at the first microphone, respectively. The multichannel SPP can be interpreted as a single channel SPP applied to the output of an MVDR spatial filter designed to minimize the noise while maintaining a distortionless response towards the speech, with corresponding covariance matrices of $\boldsymbol{\Phi}_u$ and $\boldsymbol{\Phi}_c$, respectively.

The improvement of using the multichannel SPP depends on the spatial properties of the noise and of the speech. Two interesting special cases are the spatially white noise case and the coherent noise case. In the first case of a spatially white noise, the noise covariance matrix equals $\boldsymbol{\Phi}_u = \phi_u \mathbf{I}$ where $\mathbf{I}$ is the identity matrix. For this case the multichannel SNR equals $M \cdot \xi_1$ and is higher than the single-channel SNR by a factor of the number of microphones (assuming that the SNRs at all microphones are equal). In the second case of a coherent noise, the noise covariance matrix equals $\boldsymbol{\Phi}_u = \mathbf{a}_u \mathbf{a}_u^H \phi_{u,c} + \phi_{u,\text{nc}} \mathbf{I}$, where $\mathbf{a}_u$ and $\phi_{u,c}$ are the vector of ATFs relating the coherent interference and the microphone signals and its respective variance. It is further assumed that the microphones also contain spatially white noise components with variance $\phi_{u,\text{nc}}$. In this case, perfect speech detection is obtained, i.e., $p|\mathscr{H}_s \rightarrow 1$ and $p|\mathscr{H}_u \rightarrow 0$ regardless of the coherent noise power, assuming that the ATFs vectors of the speech and the coherent noise are not parallel and that the spatially white sensors noise power $\phi_{u,\text{nc}}$ is sufficiently low.

12.6.1.2 Coherence Based SAP

As presented in Sect. 12.6.1.1, computing the SPP requires the speech a priori probability (SAP), denoted q. The SAP can be either set to a constant [21] or derived from the received signals and updated adaptively according to past estimates of SPP and SNR [22, 23] (also known as the *decision-directed* approach). In [19] the multichannel generalization of the SPP (see Sect. 12.6.1.1 and [17]) is adopted and it is proposed to incorporate coherence information in the SAP.

Let us consider a scenario where a single desired speech component contaminated by a diffuse noise is received by a pair of omnidirectional microphones. The diffuse noise field can be modeled as an infinite number of equal power statistically independent interferences uniformly distributed over a sphere surrounding the microphone array. A well known result [24] is that the coherence of diffuse noise components received by a pair of microphones is

$$\gamma_{\text{diff}}(\ell, \lambda) = \text{sinc}\left(\frac{2\pi\ell}{\lambda}\right) \tag{12.50}$$

were λ is the wavelength and ℓ is the microphones spacing. The direct to diffuse ratio (DDR) is defined as the SNR in this case, i.e., the power ratio of the directional speech received by the microphone and the diffuse noise. Heuristically, high and low DDR values are transformed into low and high SAP, respectively. The estimation of the DDR is based on a sound field model where the sound pressure at any position and time-frequency bin is modelled as a superposition of a direct sound represented by a single monochromatic plane wave and an ideal diffuse field, for more details please refer to [19, 25]. The DDR is estimated by:

$$\Gamma = \text{Re}\left\{\frac{\gamma_{\text{diff}} - \hat{\gamma}}{\hat{\gamma} - \exp(\mathrm{j}\hat{\theta})}\right\} \tag{12.51}$$

where

$$\gamma \triangleq \frac{\mathrm{E}\left[x_1 x_2^*\right]}{\sqrt{\mathrm{E}\left[|x_1|^2\right] \cdot \mathrm{E}\left[|x_2|^2\right]}} \tag{12.52}$$

is the coherence between the microphone signals and $\theta \triangleq \angle\left(c_1 \cdot c_2^*\right)$ is the phase between the speech components received by the microphones. The coherence is computed from estimates of the auto-PSDs and cross-PSD of the microphones (see Sect. 12.6.2), and the phase θ is approximated from the phase of the cross-PSD by:

$$\hat{\theta} = \angle\left(\mathrm{E}\left[x_1 x_2^*\right]\right) \tag{12.53}$$

assuming that both SNR and DDR are high.

12.6.1.3 Multiple Speakers Position Based SPP

Consider the J speakers scenario in which the microphone signals can be formulated as:

$$\mathbf{x} = \sum_{j=1}^{J} \mathbf{c}_j + \mathbf{u}. \tag{12.54}$$

In [26], the authors propose to use a MWF for extracting a desired source from a multichannel convolutive mixture of sources. By incorporating position estimates into the SPP and classifying the dominant speaker per time-frequency point, the "interference" components' PSD matrix, comprising noise and interfering speakers, and the desired speaker components' PSD matrix are estimated and utilized for constructing a spatial filter. Speaker positions are derived by triangulation of DOA estimates obtained from distributed sub-arrays of microphones with known positions.

Individual sources SPPs are defined as:

$$p_j \triangleq p\left(\mathscr{H}_{s_j}|\mathbf{x}\right) = p\left(\mathscr{H}_{s_j}|\mathbf{x}, \mathscr{H}_s\right) p \tag{12.55}$$

where $\mathscr{H}_{s_j}$ denotes the hypothesis that the j-th speaker is active (per time-frequency point), and p is the previously defined SPP (for any speaker activity).

The conditional SPPs given the microphone signals are replaced by conditional SPPs given an estimate position of the dominant active speaker, denoted $\hat{\boldsymbol{\Theta}}$, i.e., it is assumed that:

$$p\left(\mathscr{H}_{s_j}|\hat{\boldsymbol{\Theta}}, \mathscr{H}_s\right) \approx p\left(\mathscr{H}_{s_j}|\mathbf{x}, \mathscr{H}_s\right). \tag{12.56}$$

The estimated position, given that a specific speaker is active, is modeled as a mixture of Gaussian variables centered at the sources' positions:

$$p\left(\hat{\boldsymbol{\Theta}}|\mathscr{H}_s\right) = \sum_{j=1}^{J} \pi_j \mathscr{N}\left(\hat{\boldsymbol{\Theta}}; \boldsymbol{\mu}_j, \boldsymbol{\Omega}_j\right) \tag{12.57}$$

where $\boldsymbol{\mu}_j$, $\boldsymbol{\Omega}_j$ and π_j are the mean, covariance and mixing coefficient of Gaussian vector distribution which corresponds to the estimated position of the j-th source, for $j = 1, \ldots, J$. The parameters of the distribution of $\hat{\boldsymbol{\Theta}}$ are estimated by an expectation maximization (EM) procedure given a batch of estimated positions. For a detailed explanation please refer to [26].

This work is further extended in [19], where a MWF is designed to extract sources arriving from a predefined "spot", i.e., a bounded area, while suppressing all other sources outside of the spot. This method is denoted by *spotforming*.

12.6.2 Covariance Matrix Estimators

The noise covariance matrix can be estimated by recursively averaging instantaneous covariance matrices weighted according to the SPP:

$$\begin{aligned}\widehat{\boldsymbol{\Phi}}_{\boldsymbol{u}}(n, f) =& \lambda'_u(n, f)\widehat{\boldsymbol{\Phi}}_{\boldsymbol{u}}(n-1, f) \\ &+ \left(1 - \lambda'_u(n, f)\right)\mathbf{x}(n, f)\mathbf{x}^H(n, f).\end{aligned} \tag{12.58}$$

where

$$\lambda'_u(n, f) \triangleq (1 - p(n, f))\,\lambda_u + p(n, f) \tag{12.59}$$

is a time-varying recursive averaging factor and λ_u is selected such that its corresponding estimation period ($\frac{1}{1-\lambda_u}$ frames) is shorter than the stationarity time of the noise. Alternatively, a hard binary weighting, obtained by applying a threshold to the SPP, can be used instead of the soft weighting.

The hypothesis that speaker j is present and the corresponding SPP are denoted in Sect. 12.6.1.1 as $\mathscr{H}_{s_j}(n, f)$ and $p_j(n, f)$, respectively. Similarly to (12.58), the covariance matrix of the spatial image of source j, denoted $\boldsymbol{\Phi}_{c_j}(n, f)$, can be estimated by

$$\begin{aligned}\widehat{\boldsymbol{\Phi}}_{c_j}(n, f) &= \lambda'_{c_j}(n, f)\widehat{\boldsymbol{\Phi}}_{c_j}(n - 1, f)\\ &+ (1 - \lambda'_{c_j}(n, f))(\mathbf{x}(n, f)\mathbf{x}^H(n, f) - \widehat{\boldsymbol{\Phi}}_u(n - 1, f))\end{aligned} \tag{12.60}$$

where

$$\lambda'_{c_j}(n, f) \triangleq \left(1 - p_j(n, f)\right)\lambda_c + p_j(n, f) \tag{12.61}$$

is a time-varying recursive-averaging factor, and λ_c is selected such that its corresponding estimation period ($\frac{1}{1-\lambda_c}$ frames) is shorter than the *coherence time* of the AIRs of speaker j, i.e. the time period over which the AIRs are assumed to be time-invariant. Note that: (1) usually the estimation period is longer than the speech nonstationarity time, therefore, although the spatial structure of $\boldsymbol{\Phi}_{c_j}(n, f)$ is maintained, the estimated variance is an average of the speech variances over multiple time periods, denoted $\bar{\phi}_{s_j}(n, f)$, rather than $\phi_{s_j}(n, f)$, the actual time-varying variance of the speaker ; (2) the estimate $\widehat{\boldsymbol{\Phi}}_{c_j}(n, f)$ keeps its past value when speaker j is absent.

12.6.3 Procedures for Semi-blind RTF Estimation

Two common approaches for RTF estimation are the covariance subtraction [27, 28] and the covariance whitening [11, 29] methods. Here, for brevity we assume a single speaker scenario. Both of these approaches rely on estimated noisy speech and noise-only covariance matrices, i.e. $\widehat{\boldsymbol{\Phi}}_{xj}(n, f)$ (where $\boldsymbol{\Phi}_{xj}(n, f) = \boldsymbol{\Phi}_{c_j}(n, f) + \boldsymbol{\Phi}_u(f)$) and $\widehat{\boldsymbol{\Phi}}_u(n, f)$. Given the estimated covariance matrices, covariance subtraction estimates the speaker RTF by

$$\tilde{\mathbf{a}}_{j,\mathrm{CS}}(f) \triangleq \frac{1}{\mathbf{i}_1^H(\widehat{\boldsymbol{\Phi}}_{c_j}(n, f) - \widehat{\boldsymbol{\Phi}}_u(n, f))\mathbf{i}_1}(\widehat{\boldsymbol{\Phi}}_{c_j}(n, f) - \widehat{\boldsymbol{\Phi}}_u(n, f))\mathbf{i}_1 \tag{12.62}$$

where $\mathbf{i}_1 = [\, 1\ \mathbf{0}_{1\times M-1}\,]^T$ is an $M \times 1$ selection vector for extracting the component of the reference microphone, here assumed to be the first micrphone.

The covariance whitening approach estimates the RTF by: (1) applying the generalized eigenvalue decomposition (GEVD) to $\widehat{\boldsymbol{\Phi}}_{xj}(n, f)$ with $\widehat{\boldsymbol{\Phi}}_{\boldsymbol{u}}(n, f)$ as the whitening matrix; (2) de-whitening the eigenvector corresponding to the strongest eigenvalue, denoted $\dot{\mathbf{a}}_j(f)$, namely $\widehat{\boldsymbol{\Phi}}_{\boldsymbol{u}}(n, f)\dot{\mathbf{a}}_j(f)$; (3) normalizing the de-whitened eigenvector by the reference microphone component. Explicitly:

$$\tilde{\mathbf{a}}_{j,\mathrm{CW}}(f) \triangleq (\mathbf{i}_1^H \widehat{\boldsymbol{\Phi}}_{\boldsymbol{u}}(n, f)\dot{\mathbf{a}}_j(f))^{-1} \widehat{\boldsymbol{\Phi}}_{\boldsymbol{u}}(n, f)\dot{\mathbf{a}}_j(f). \tag{12.63}$$

A preliminary analysis and comparison of the covariance subtraction and covariance whitening methods can be found in [30].

Other methods utilize the speech nonstationarity property, assuming that the noise has slow time-varying statistics. In [10], the problem of estimating the RTF of microphone i is formulated as a least squares (LS) problem where the l-th equation utilizes $\widehat{\phi}^l_{x_i x_1}$(f), the estimated cross-PSD of microphone i and the reference microphone in the l-th time segment. This cross-PSD satisfies:

$$\widehat{\phi}^l_{xj,i1}(f) = \tilde{a}_{j,i}(f)\widehat{\phi}^l_{xj,11}(f) + \widehat{\phi}_{\dot{u}i,xj1}(f) + \varepsilon^l_{j,i}(f) \tag{12.64}$$

where we use the relation $\mathbf{x}(n, f) = \tilde{\mathbf{a}}_j(f)x_1(n, f) + \dot{\mathbf{u}}(n, f)$. The unknowns are $\tilde{a}_{j,i}(f)$, i.e. the required RTF, and $\widehat{\phi}_{\dot{u}i,xj1}(f)$, which is a nuisance parameter. $\varepsilon^l_{j,i}(f)$ denotes the error term of the l-th equation. Multiple LS problems, one for each microphone, are solved for estimating the vector RTF. Note that, the latter method, also known as the *nonstationarity*-based RTF estimation, does not require a prior estimate of the noise covariance, since it simultaneously solves for RTF and the noise statistics. Similarly, a weighted least squares (WLS) problem with exponential weighting can be defined and implemented using a recursive least squares (RLS) algorithm [31]. Considering speech sparsity in the STFT domain, in [28] the SPPs were incorporated into the weights of the WLS problem, resulting in a more accurate solution.

12.7 Examples

In the following section some simple examples are used to present the behaviors and differences between the MF, MWF and LCMV spatial filters. The case of a narrowband signal in an anechoic environment is presented in Sect. 12.7.1, and the case of two speech sources in a reverberant environment is presented in Sect. 12.7.2.

12.7.1 Narrowband Signals at an Anechoic Environment

Consider the case of $J = 2$ narrowband sources occupying the f-th frequency-bin propagating in an anechoic environment and received by a uniform linear array (ULA) array comprising M microphones with microphone spacing ℓ. Define a spherical-coordinate system with the origin coincides with the center of the ULA, and rotated such that the ULA is placed along the elevation angle $\theta = \pm 90°$ and azimuth angle of $\phi = 0°$. The sources are positioned in the far-field at a large distance from the microphones and on the same plane as the microphones. The DOA of the sources with respect to the microphones array are denoted θ_j for $j = 1, 2$. By adopting the far-field free-space propagation model the ATF vectors of the sources are given by:

$$\mathbf{a}_j^0(f) = \left[1, \exp\left(-j2\pi \frac{\ell \sin(\theta_j)}{\lambda}\right), \cdots, \exp\left(-j2\pi \frac{(M-1)\ell \sin(\theta_j)}{\lambda}\right) \right]^T \quad (12.65)$$

for $j = 1, 2$ where λ is the wavelength corresponding to the f-th frequency-bin. The wavelength can be expressed as:

$$\lambda \triangleq \frac{\nu F}{f f_s} \quad (12.66)$$

with the continuous frequency which corresponds to the discrete frequency-bin f is $\frac{f f_s}{F}$ where f_s is the sample-rate, F is the length of STFT window and $\nu \approx 343$ m/s is the sound velocity. An additive white Gaussian noise with covariance matrix of

$$\boldsymbol{\Phi}_u^0(n, f) = \phi_u \mathbf{I}. \quad (12.67)$$

is contaminating the received microphone signals. The setup of the sources and microphones is depicted in Fig. 12.1. Note that the ATF vector is independent of the azimuth angle ϕ, and therefore the beampattern and all performance measures of any spatial filter in this case will have a cylindrical symmetry. Next, we compare the performance of various spatial filters that are applied in this problem, namely MF, MWF and LCMV. The performance criteria that we use are SNR, SIR, SINR and SDR which are evaluated empirically from the signals. For the MF spatial filter we derive simplified expressions for the performance criteria, whereas for the MWF and the LCMV spatial filters we use the previously defined generic scenario expressions.

Let us revisit the performance criteria of the MF for this case. By substituting the ATF vectors in (12.65), the scalar product of the j-th and j'-th ATF vectors, denoted by $\rho_{jj'}(f)$ in (12.25), can be expressed as:

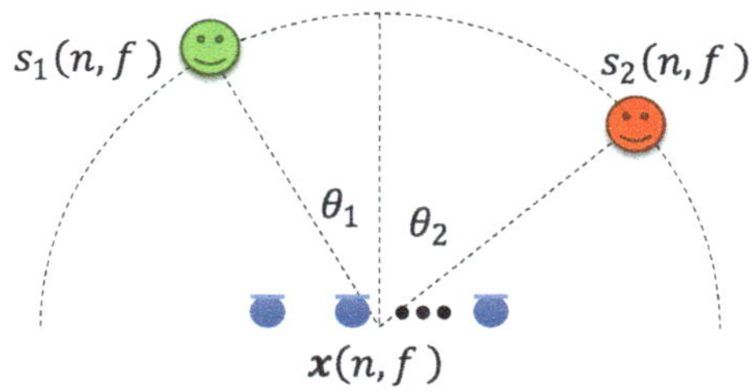

Fig. 12.1 Setup of the narrowband signal at an anechoic environment example

$$\rho^0_{jj'}(f) = \sum_{i=1}^{M} \exp\left(j2\pi \frac{(i-1)\,\ell\left(\sin\left(\theta_j\right) - \sin\left(\theta_{j'}\right)\right)}{\lambda}\right)$$
$$= M \cdot \text{diric}\left(2\pi \frac{\ell\left(\sin\left(\theta_j\right) - \sin\left(\theta_{j'}\right)\right)}{\lambda}\right) \cdot \exp\left(j\pi \frac{(M-1)\,\ell\left(\sin\left(\theta_j\right) - \sin\left(\theta_{j'}\right)\right)}{\lambda}\right) \tag{12.68}$$

where

$$\text{diric}\left(2\pi \frac{\ell\left(\sin\left(\theta_j\right) - \sin\left(\theta_{j'}\right)\right)}{\lambda}\right) \triangleq \frac{\sin\left(M\pi \frac{\ell(\sin(\theta_j) - \sin(\theta_{j'}))}{\lambda}\right)}{M \cdot \sin\left(\pi \frac{\ell(\sin(\theta_j) - \sin(\theta_{j'}))}{\lambda}\right)} \tag{12.69}$$

is the Dirichlet function which in general has a period of 4π. Note that $\left|\rho^0_{jj'}(f)\right|^2 = M^2$ for

$$\frac{\ell\left(\sin\left(\theta_j\right) - \sin\left(\theta_{j'}\right)\right)}{\lambda} = k \tag{12.70}$$

where $k = 0, \pm 1, \pm 2, \ldots$ is any integer number. Next, since the $\sin(\cdot)$ is bounded by $-1 \leq \sin(\cdot) \leq 1$ the left-hand side of (12.70) is bounded by:

$$-\frac{2\ell}{\lambda} \leq \frac{\ell\left(\sin\left(\theta_j\right) - \sin\left(\theta_{j'}\right)\right)}{\lambda} \leq \frac{2\ell}{\lambda}. \tag{12.71}$$

Hence, in order to avoid the spatial aliasing phenomenon, where undesired directions are passed through the spatial filter without any attenuation, the well-known constraint on the ratio between microphones spacing and the wavelength is given by:

$$\frac{\ell}{\lambda} < \frac{1}{2}. \tag{12.72}$$

Furthermore, note that for

$$M\frac{\ell\left(\sin\left(\theta_j\right)-\sin\left(\theta_{j'}\right)\right)}{\lambda}=k \tag{12.73}$$

for any integer k non-divisible by M with the a zero remainder, i.e. of the form $k \neq \iota M$ where ι is an integer, we obtain that $\left|\rho^0_{jj'}(f)\right|^2 = 0$. Explicitly, in the range of $-\frac{\pi}{2} \leq \theta_{j'} \leq \frac{\pi}{2}$ there are $M-1$ such DOAs that are perfectly attenuated by the MF, also referred to as nulls in the beampattern. By replacing $\rho_{jj'}(f)$ with $\rho^0_{jj'}(f)$ in the power of the residual j'-th interference at the j-th output, see (12.24a), and the corresponding SIR improvement of the MF, see (12.24b), the following simplified expressions are obtained:

$$\psi^{0,\mathrm{MF}}_{d_{j,j'}}(n,f) = \mathrm{diric}^2\left(2\pi\frac{\ell\left(\sin\left(\theta_j\right)-\sin\left(\theta_{j'}\right)\right)}{\lambda}\right)\phi_{s_{j'}}(n,f) \tag{12.74a}$$

$$\Delta\mathrm{SIR}^{0,\mathrm{MF}}_{jj'}(f) = \left(\mathrm{diric}^2\left(2\pi\frac{\ell\left(\sin\left(\theta_j\right)-\sin\left(\theta_{j'}\right)\right)}{\lambda}\right)\right)^{-1}. \tag{12.74b}$$

Considering the spatially non-correlated noise properties, see (12.67), and substituting it in the power of the noise at the output of the j-th source MF, see (12.26a), and the corresponding SNR improvement, see (12.26b), the latter can be expressed in this special case as:

$$\psi^{0,\mathrm{MF}}_{v_j}(f) = \frac{\phi_u(f)}{M} \tag{12.75a}$$

$$\Delta\mathrm{SNR}^{0,\mathrm{MF}}_{j}(f) = M. \tag{12.75b}$$

The corresponding criteria for the MWF and multichannel LCMV are more complicated, and their derivation is omitted.

We compare the spatial filters in a specific scenario of: (a) the microphone array comprises of $M = 4$ microphones with a microphone spacing of $\ell = 10\,\mathrm{cm}$; (b) the desired source is the first source which arrives from $\theta_1 = 0°$. In the following, we investigate the performance dependency on the parameters: (a) SNR and interference-to-noise ratio (INR); (b) interference DOA, θ_2; (c) frequency. For simplicity, we consider two subsets of the above mentioned parameters.

In the first parameters subset, the interference DOA and frequency are set to $\theta_2 = 10^o$ and $\frac{f \cdot f_s}{F} = 1715\,\mathrm{Hz}$, corresponding to $\frac{\ell}{\lambda} = \frac{1}{2}$ (in other figures we explore the performance depending on the frequency or wavelength). For this parameters selection the performance measures of the spatial filters are compared as a function of SNR and INR values that are selected within the range of $[-20\,\mathrm{dB}, 30\,\mathrm{dB}]$. The improvements in SNR, SIR and SINR and the SDR are depicted in Fig. 12.2a–d. We can observe in these figures the consequences of the different design criteria: (a) the MF is designed to maximized the SNR improvement with a spatially white noise (as in this example) and obtains the highest SNR improvement in Fig. 12.2a; (b) the LCMV is designed to null out the interfering sources and therefore obtains an infinite

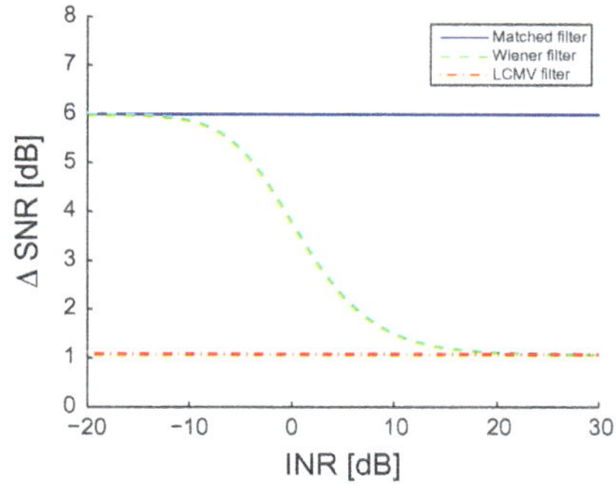

(a) SNR improvement.

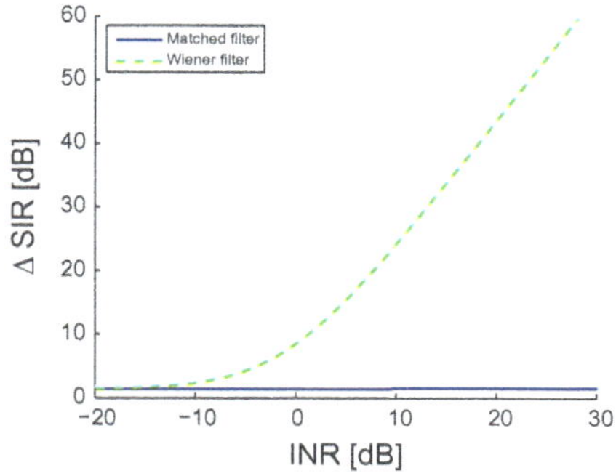

(b) SIR improvement.

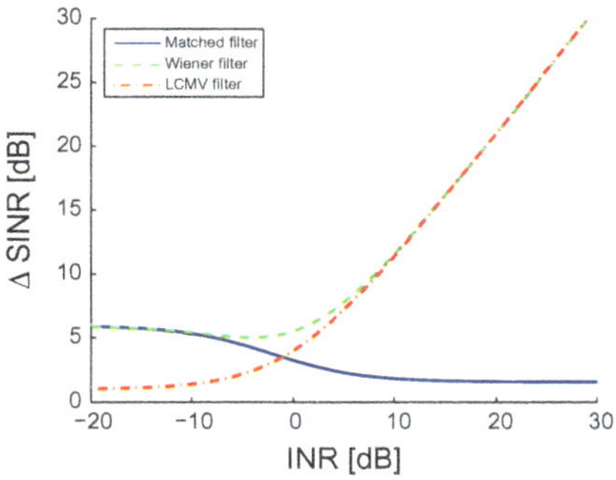

(c) SINR improvement.

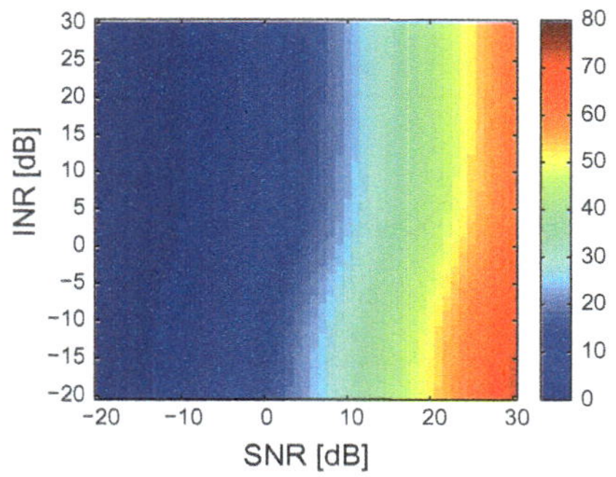

(d) SDR of the MWF.

Fig. 12.2 Performance comparison depending on input SNR and INR of various spatial filters in the narrowband case with 2 speech sources propagating in freespace and spatially white noise received by a ULA comprising 4 microphones

SIR improvement, which is of course higher than the finite SIR improvement of the other methods that are depicted in Fig. 12.2b; (c) the MWF is designed to maximize the SINR improvement and this is evidently seen in Fig. 12.2c. The MWF aims to maximize the SINR improvement and thus minimize the sum of interference and noise powers at its output. In the limit cases of INR [dB] $\rightarrow -\infty$ where the interference power is negligible and INR [dB] $\rightarrow \infty$ where the noise power is negligible, the MWF coincides with the MF and the LCMV, respectively. This can be clearly seen in Fig. 12.2a–c, where the performance of the MWF converges to that of the MF and LCMV for INR [dB] $\rightarrow -\infty$ and INR [dB] $\rightarrow \infty$, respectively. The MF and LCMV spatial filters are distortionless by design at any input SNR and INR levels, and therefore we do not depict their SDR. The SDR at the output of the MWF as a function of the input SNR and INR is depicted in Fig. 12.2d. The higher is the input SNR the higher is the relative weight of the distortion component compared to the interference and noise components in the MSE in (12.27), which is the MWF design criterion and correspondingly the higher is the SDR of the MWF.

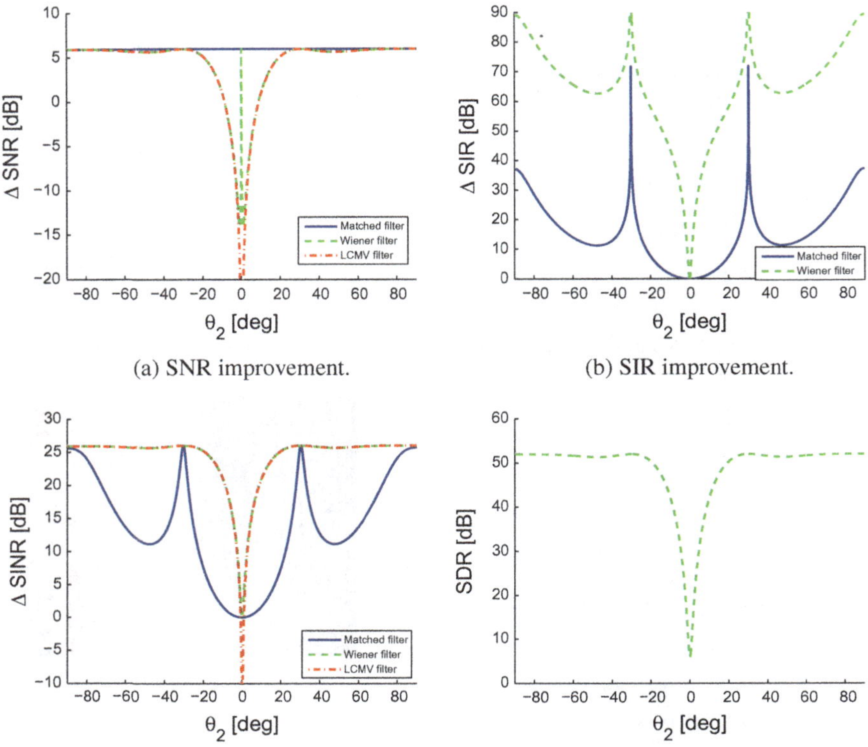

(a) SNR improvement.

(b) SIR improvement.

(c) SINR improvement.

(d) SDR of the MWF.

Fig. 12.3 Performance comparison of various spatial filters applied in the narrowband case (at frequency 1715 Hz), where 2 speech sources propagating in freespace and a spatially white additive noise are received by a ULA comprising 4 microphones. The desired source arrives from $\theta_1 = 0^\circ$ and the interfering source arrives from $\theta_2 \int [-90^\circ, 90^\circ]$

In the second parameters subset, the input SNR and INR are both set to 20 dB and the interference DOA and frequency are selected within the range of $[-90^\circ, 90^\circ]$ and [0 Hz, 8000 Hz], respectively. The SNR, SIR and SINR improvement as well as SDR for frequency 1715 Hz (for which $\frac{\ell}{\lambda} = 0.5$) depending the interference source DOA are depicted in Fig. 12.3 for the various spatial filters. As in the previous example and regardless of the DOA of the interference: (a) the MF is optimal in the sense of SNR improvement for a spatially white noise, see Fig. 12.3a; (b) the LCMV is optimal in the sense of SIR improvement, see Fig. 12.3b, as it completely nulls out the interference and obtains infinite SIR improvement, whereas for MF and MWF there is some residual interference at the output for almost all interference DOAs; (c) the MWF is optimal in the sense of SINR improvement, see Fig. 12.3c, although the SINR improvement of the LCMV is very similar for most interference DOAs.

The main difference between the LCMV and MWF can be observed when the interference DOA, θ_2, is close to that of the desired source, $\theta_1 = 0°$. The LCMV, which is designed to null the interference, "struggles" to satisfy its constraints as the interference and desired source DOAs become closer. As a result, the SNR improvement (which is a secondary objective for the LCMV) and correspondingly the SINR improvement are degraded (see Fig. 12.3a, c), and might even become negative (i.e. the spatial noise power at the output might become higher than the noise power at the input and in extreme cases might even become higher than the noise an interference power at the input). Furthermore, the LCMV is not defined for the singular case of $\theta_2 = \theta_1$. In this specific case, the MWF is not able to improve the SIR, however, it is able to improve the SNR. However, note in Fig. 12.3 that as the interference and desired source DOAs become close the SDR degrades. This is because the MWF converges in this case to the MF scaled by a single channel Wiener filter, which introduces more distortion as interference and noise power increases.

Another interesting observation in the SIR improvement (see Fig. 12.3b) is that for some DOAs ($\theta_2 \approx \pm 30°$) the SIR improvement of the MWF and MF also converge to infinity (as the optimal LCMV). The reason for that is that for these DOAs the interfering and desired ATF vectors are orthogonal (i.e. $\rho_{12}^0(f) = 0$, see (12.68)) and the corresponding SIR improvement is also infinite.

The SINR improvement of the MF and MWF depending on interference DOA and frequency are depicted in Figs. 12.4a, b. Clearly the SINR improvement of the MWF outperforms that of the MF. For brevity we omit the SINR improvement of the LCMV as it is similar to the improvement of the MWF almost always, except for when the interference DOA approaches the desired source DOA. The red regions in the SINR improvement of the MF in Fig. 12.4a, similarly to the peaks in the SINR improvement of the MF in Fig. 12.3c, correspond to cases where the desired source and interference ATF vectors are orthogonal. Note that positions of these peaks vary over frequency. The blue regions in the SINR improvement of the MF in Fig. 12.4a

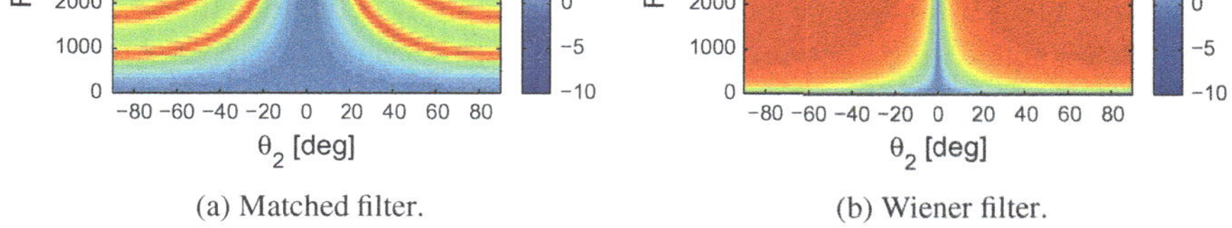

(a) Matched filter.

(b) Wiener filter.

Fig. 12.4 SINR improvement depending on interference DOA and frequency of various spatial filters in the narrowband case with 2 speech sources propagating in freespace and spatially white noise received by a ULA comprising 4 microphones (except for the one around $\theta_2 \approx \theta_1$) correspond to the spatial aliasing phenomenon. When the interference arrives from the DOA of the desired source or from a DOA which corresponds to a grating lobe of the spatial filter, it cannot be attenuated without degrading the desired source. Hence, the SINR improvement at these DOA is close to zero (blue color). Note that the positions of the grating lobes vary over frequencies. A similar phenomenon can be seen when observing the SINR improvement of the MWF in Fig. 12.4b. For the MWF, however, the areas in which the SINR is close to zero are narrower than in the MF, and the areas of high SINR improvement cover almost the entire interference DOA and frequency ranges.

12.7.2 Speech Signals at a Reverberant Environment

In this section we compare the performance of the various spatial filters in a scenario simulated by convolving recorded speech signals from the WSJCAM0 database [32] with AIRs drawn from a database collected in reverberant enclosures [33]. A ULA comprising $M = 4$ microphones with spacing of $\ell = 8$ cm is picking up signals of $J = 2$ speakers, a female and a male, located at a distance of 1 m from the array at DOAs of $-90°$ and $75°$, respectively, as well as diffuse noise that is generated using a diffuse noise simulator [34]. The SIR is set to 0 dB and the SNR is set to 15 dB.

The signals are transformed to the STFT domain, where MF, MWF and LCMV are designed to enhance the first speaker. Speech-free time-segments and single-talk time-segments of each of the speech sources are used as training segments from which the required parameters for the various spatial filters are estimated: (a) RTFs vector $\tilde{\mathbf{a}}_1(f)$ of the first source for the MF; (b) RTFs vector $\tilde{\mathbf{a}}_1(f)$ and spectrum $\overline{\phi}_{s1}(f)$ of the first source, covariance matrix of the second source $\boldsymbol{\Phi}_{c2}(f)$ and covariance matrix of the noise $\boldsymbol{\Phi}_u(f)$ for the MWF; (c) RTF vectors of both sources $\tilde{\mathbf{a}}_1(f)$,$\tilde{\mathbf{a}}_2(f)$ and covariance matrix of the noise $\boldsymbol{\Phi}_u(f)$ for the LCMV. The output of the spatial filter is transformed back to the time-domain, yielding the enhanced signal. A reference microphone and outputs of the various spatial filters decomposed to their various components (desired speech, interfering speech and noise) are depicted in Fig. 12.5. The corresponding spectrograms of the reference microphone and the outputs of the spatial filters are depicted in Fig. 12.6. The performance measures of each of the spatial filters per frequency-bin in terms of SNR, SIR and SINR improvement as well as SDR are depicted in Fig. 12.7. Considering the SIR and SINR improvements, it is clear from Figs. 12.5 and 12.7b, c, that the MWF is slightly better than the LCMV and that both are significantly better than the MF. While the MWF is expected to obtain the maximal SINR improvement, it is surprising that it outperforms the LCMV in terms of SIR improvement as well. The reason for that lies in the fact that LCMV designates a single constraint for nulling the interfering source, thus assuming a rank-1 model for the interference, while the MWF utilizes the complete covariance matrix of the interfering source, thus allowing to reduce interferences with higher ranks. Although, theoretically, the covariance matrix of coherent point sources is rank-1, in practice, finite window lengths and variations in the AIR (AIR might vary even

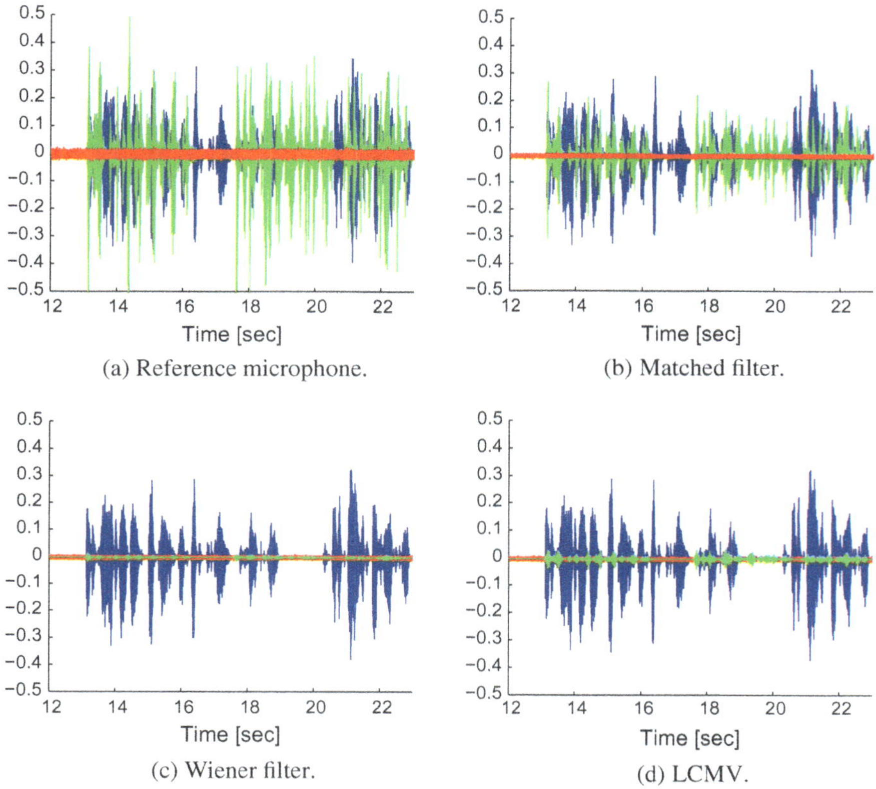

(a) Reference microphone.

(b) Matched filter.

(c) Wiener filter.

(d) LCMV.

Fig. 12.5 Input and output signals of various spatial filters in a simulated scenario with $J = 2$ speech signals contaminated by diffuse noise and received by a $M = 4$ microphones array in a reverberant environment. The signals are decomposed to their components: (1) desired speaker (blue); (2) interfering speaker (green); and (3) noise (red)

when the source is static due to slight variations in the enclosure) increase the matrix rank. Considering the SNR improvement, note that the MWF and LCMV are better than the MF in frequencies lower than 1000 Hz, and that for higher frequencies the MF is better than the MWF and LCMV. This result is attributed to the diffuse noise properties. In low frequencies, where $\frac{\ell}{\lambda} < \frac{1}{2}$ the diffuse noise has a strong coherent component which the data-dependent filters, MWF and LCMV, reduce efficiently. In higher frequencies the diffuse noise becomes spatially uncorrelated, in which case the MF is optimal and outperforms the MWF and LCMV which utilize their DoF to reduce the interfering speech.

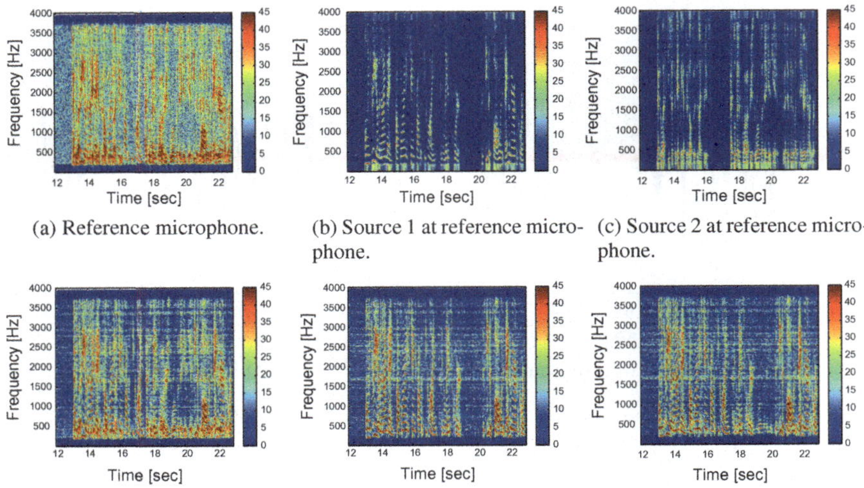

(a) Reference microphone. (b) Source 1 at reference microphone. (c) Source 2 at reference microphone.

(d) MF output. (e) MWF output. (f) LCMV output.

Fig. 12.6 Input and output spectrograms of various spatial filters aiming to enhance the first source in a simulated scenario with $J = 2$ speech signals contaminated by diffuse noise and received by a $M = 4$ microphones array in a reverberant environment. Input spectrogram of the reference signal and its speech components are respectively depicted in **a**, **b**, **c** and the outputs of the MF, MWF and LCMV spatial filters are respectively depicted in **d**, **e**, **f**

12.8 Summary

MMSE based criteria for designing beamformers, also referred to as spatial-filters, can be used in noise reduction and speech separation tasks. The following methods were presented and analyzed: (1) the MF, which maximizes the SNR at the output without distorting the speech signal, assuming a spatially white noise; (2) the MWF, which minimizes the MSE between the output signal and the desired speech signal, and assigns equal weights to the desired speech distortion, the variance of the interfering speakers at the output, and the noise variance at the output; and (3) the LCMV, which minimizes the noise variance at the output while satisfying a set of constraints designed to maintain the desired speech undistorted and to null out the interfering speakers. Estimation methods for implementing the various beamformers are surveyed. Specifically, methods for estimating the RTFs of speakers and for estimating the spatial covariance matrices of the noise and of the various speaker components were presented. The estimation methods are governed by the multichannel SPP, which was also presented. Some simple examples of applying the various beamformers to simulated narrowband signals in an anechoic environment and to speech signals in a real reverberant environment were presented and discussed.

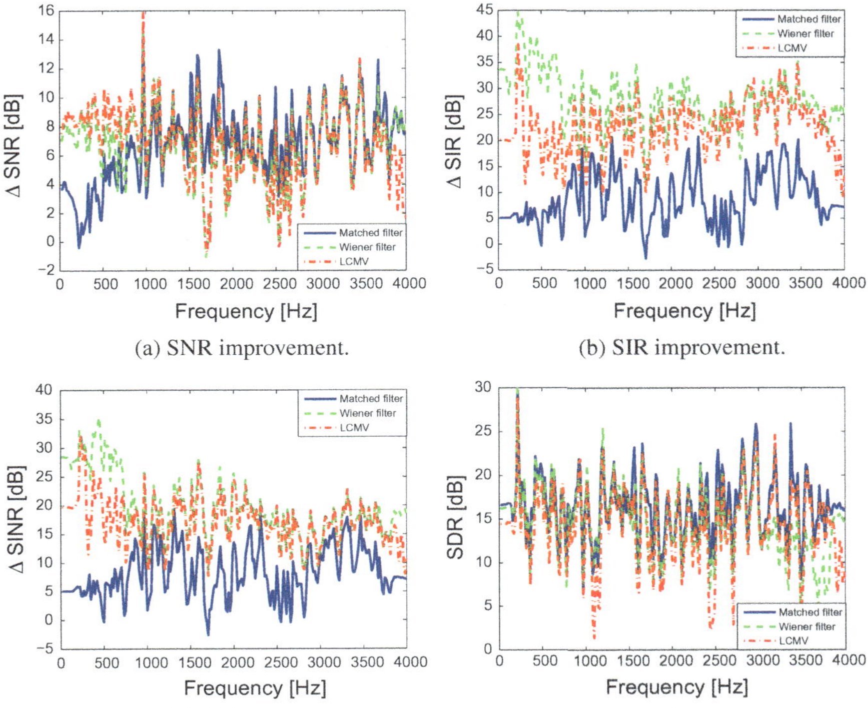

(a) SNR improvement.

(b) SIR improvement.

(c) SINR improvement.

(d) SDR.

Fig. 12.7 Performance criteria per frequency-bin of various spatial filters in a simulated scenario with $J = 2$ speech signals contaminated by diffuse noise and received by a $M = 4$ microphones array in a reverberant environment

References

1. J. Capon, High-resolution frequency-wavenumber spectrum analysis. Proc. IEEE **57**(8), 1408–1418 (1969)
2. O.L. Frost III, An algorithm for linearly constrained adaptive array processing. Proc. IEEE **60**(8), 926–935 (1972)
3. L.J. Griffiths, C.W. Jim, An alternative approach to linearly constrained adaptive beamforming. IEEE Trans. Antennas Propag. **30**(1), 27–34 (1982)
4. M. Er, A. Cantoni, Derivative constraints for broad-band element space antenna array processors. IEEE Trans. Acoust. Speech Signal Process. **31**(6), 1378–1393 (1983)
5. B.D. Van Veen, K.M. Buckley, Beamforming: a versatile approach to spatial filtering. IEEE Acoust. Speech Signal Process. Mag., 4–24 (1988)
6. B.R. Breed, J. Strauss, A short proof of the equivalence of LCMV and GSC beamforming. IEEE Signal Process. Lett. **9**(6), 168–169 (2002)
7. H.L. Van Trees, *Optimum Array Processing: Estimation Detection, Modulation Theory*, vol. IV (Wiley, New York, 2002)
8. H. Cox, R. Zeskind, M. Owen, Robust adaptive beamforming. IEEE Trans. Acoust. Speech Signal Process. **35**(10), 1365–1376 (1987)

9. S. Affes, Y. Grenier, A signal subspace tracking algorithm for microphone array processing of speech. IEEE Trans. Speech Audio Process. **5**(5), 425–437 (1997)
10. S. Gannot, D. Burshtein, E. Weinstein, Signal enhancement using beamforming and nonstationarity with applications to speech. IEEE Trans. Signal Process. **49**(8), 1614–1626 (2001)
11. S. Markovich, S. Gannot, I. Cohen, Multichannel eigenspace beamforming in a reverberant noisy environment with multiple interfering speech signals. IEEE Trans. Audio Speech Lang. Process. **17**(6), 1071–1086 (2009)
12. S. Doclo, A. Spriet, J. Wouters, M. Moonen, Speech distortion weighted multichannel Wiener filtering techniques for noise reduction, in *Speech Enhancement*. Signals and Communication Technology series (Springer, Berlin, 2005), pp. 199–228
13. S. Gannot, E. Vincent, S. Markovich-Golan, A. Ozerov, A consolidated perspective on multimicrophone speech enhancement and source separation. IEEE/ACM Trans. Audio Speech Lang. Process. **25**(4), 692–730 (2017)
14. Y. Avargel, I. Cohen, On multiplicative transfer function approximation in the short-time Fourier transform domain. IEEE Signal Process. Lett. **14**(5), 337–340 (2007)
15. J.L. Flanagan, A.C. Surendran, E.-E. Jan, Spatially selective sound capture for speech and audio processing. Speech Commun. **13**(1–2), 207–222 (1993)
16. S. Markovich-Golan, S. Gannot, I. Cohen, A weighted multichannel Wiener filter for multiple sources scenarios, in *Proceedings of the IEEE Convention of Electrical and Electronics Engineers in Israel (IEEEI)* (Eilat, Israel, 2012)
17. M. Souden, J. Chen, J. Benesty, S. Affes, Gaussian model-based multichannel speech presence probability. IEEE Trans. Audio Speech Lang. Process. **18**(5), 1072–1077 (2010)
18. M. Taseska, E.A. Habets, MMSE-based blind source extraction in diffuse noise fields using a complex coherence-based a priori sap estimator, in *Proceedings of the International Workshop Acoustic Signal Enhancement (IWAENC)* (VDE, 2012), pp. 1–4
19. M. Taseska, E. Habets, Spotforming using distributed microphone arrays, in *Proceedings of the IEEE Workshop on Applications of Signal Processing to Audio and Acoustics (WASPAA)*, 2013, pp. 1–4
20. M. Taseska, E. Habets, Informed spatial filtering for sound extraction using distributed microphone arrays. IEEE/ACM Trans. Audio Speech Lang. Proc. **22**(7), 1195–1207 (2014)
21. T. Gerkmann, C. Breithaupt, R. Martin, Improved a posteriori speech presence probability estimation based on a likelihood ratio with fixed priors. IEEE Trans. Audio Speech Lang. Proc. **16**(5), 910–919 (2008)
22. Y. Ephraim, D. Malah, Speech enhancement using a minimum-mean square error short-time spectral amplitude estimator. IEEE Trans. Acoust. Speech Signal Process. **32**(6), 1109–1121 (1984)
23. I. Cohen, B. Berdugo, Speech enhancement for non-stationary noise environments. Signal Process. **81**(11), 2403–2418 (2001)
24. M. Schroeder, Frequency correlation functions of frequency responses in rooms. J. Acoust. Soc. Am. **34**(12), 1819–1823 (1962)
25. O. Thiergart, G.D. Galdo, E.A. Habets, Signal-to-reverberant ratio estimation based on the complex spatial coherence between omnidirectional microphones, in *Proceedings IEEE International Conference on Acoustics, Speech and Signal Processing (ICASSP)*, 2012, pp. 309–312
26. M. Taseska, E.A. Habets, MMSE-based source extraction using position-based posterior probabilities, in *Proceedings IEEE International Conference on Acoustics, Speech and Signal Processing (ICASSP)* (IEEE, 2013), pp. 664–668
27. S. Doclo, M. Moonen, Multimicrophone noise reduction using recursive GSVD-based optimal filtering with ANC postprocessing stage. IEEE Trans. Speech Audio Process. **13**(1), 53–69 (2005)
28. I. Cohen, Relative transfer function identification using speech signals. IEEE Trans. Speech Audio Process. **12**(5), 451–459 (2004)
29. A. Bertrand, M. Moonen, Distributed node-specific LCMV beamforming in wireless sensor networks. IEEE Trans. Signal Process. **60**, 233–246 (2012)

30. S. Markovich-Golan, S. Gannot, Performance analysis of the covariance subtraction method for relative transfer function estimation and comparison to the covariance whitening method, in *Proceedings of the IEEE International Conference on Acoustics, Speech and Signal Processing (ICASSP)* (Brisbane, Australia, 2015)
31. T. Dvorkind, S. Gannot, Time difference of arrival estimation of speech source in a noisy and reverberant environment. Signal Process. **85**(1), 177–204 (2005)
32. T. Robinson, J. Fransen, D. Pye, J. Foote, S. Renals, WSJCAMO: a British English speech corpus for large vocabulary continuous speech recognition, in *Proceedings of the IEEE International Conference on Acoustics, Speech and Signal Proceesing (ICASSP)*, vol. 1, 1995, pp. 81–84
33. E. Hadad, F. Heese, P. Vary, S. Gannot, Multichannel audio database in various acoustic environments, in *Proceedings of the International Workshop Acoustic Signal Enhancement (IWAENC)* (IEEE, 2014), pp. 313–317
34. E. Habets, S. Gannot, Generating sensor signals in isotropic noise fields. J. Acoust. Soc. Am. **122**(6), 3464–3470 (2007)

Chapter 13
Musical-Noise-Free Blind Speech Extraction Based on Higher-Order Statistics Analysis

Hiroshi Saruwatari and Ryoichi Miyazaki

Abstract In this chapter, we introduce a musical-noise-free blind speech extraction method using a microphone array for application to nonstationary noise. In the recent noise reduction study, it was found that optimized iterative spectral subtraction (SS) results in speech enhancement with almost no musical noise generation, but this method is valid only for stationary noise. The method presented in this chapter consists of iterative blind dynamic noise estimation by, e.g., independent component analysis (ICA) or multichannel Wiener filtering, and musical-noise-free speech extraction by modified iterative SS, where multiple iterative SS is applied to each channel while maintaining the multichannel property reused for the dynamic noise estimators. Also, in relation to the method, we discuss the justification of applying ICA to signals nonlinearly distorted by SS. From objective and subjective evaluations simulating a real-world hands-free speech communication system, we reveal that the method outperforms the conventional speech enhancement methods.

13.1 Introduction

In the past few decades, many applications of speech communication systems have been investigated, but it is well known that these systems always suffer from the deterioration of speech quality under adverse noise conditions. In a study of speech enhancement, many types of statistical signal estimation methods have been proposed, e.g., the maximum likelihood estimator of short-time spectral amplitude (spectral subtraction (SS) [1–4]), the minimum mean-square error estimator of the complex-valued spectrum (Wiener filtering (WF) [5]), the Bayesian estimator of short-time spectral amplitude (the minimum mean-square error short-time spectral amplitude (MMSE-STSA) estimator [6] and theminimum mean-square error

H. Saruwatari (✉)
The University of Tokyo, Tokyo 113-8656, Japan
e-mail: hiroshi_saruwatari@ipc.i.u-tokyo.ac.jp

R. Miyazaki
Department of Computer Science and Electronic Engineering,
National Institute of Technology, Tokuyama College, Gakuendai,
Shunan, Yamaguchi 745-8585, Japan

S. Makino (ed.), *Audio Source Separation*, Signals and Communication Technology, https://doi.org/10.1007/978-3-319-73031-8_13

log-spectral amplitude estimator (MMSE-LSA) [7]), and the MAP estimator [8]. SS is the commonly used noise reduction method that has high noise reduction performance with low computational complexity. However, in this method, artificial distortion, referred to as *musical noise*, arises owing to nonlinear signal processing, leading to a serious deterioration of sound quality [9, 10]. Therefore, to assess and control the generation of musical noise, several studies were conducted using higher-order statistics [11–13].

To achieve high-quality noise reduction with low musical noise, an iterative SS method has been proposed [14–16]. This method is performed through signal processing, in which weak SS processes are iteratively applied to the input signal. Also, Inoue et al. have reported the very interesting phenomenon that this method with appropriate parameters gives equilibrium behavior in the growth of higher-order statistics with increasing number of iterations [17]. This means that almost no musical noise is generated even with high noise reduction, which is one of the most desirable properties of single-channel nonlinear noise reduction methods. Following this finding, Miyazaki et al. have derived the optimal parameters satisfying the no-musical-noise-generation condition by analysis based on higher-order statistics [18]. We have defined this method as *musical-noise-free* speech enhancement, where no musical noise is generated even for a high signal-to-noise ratio (SNR) in iterative SS. In this chapter, firstly, we explain the overview of musical-noise-free iterative SS.

In conventional iterative SS, however, it is assumed that the input noise signal is stationary, meaning that we can estimate the expectation of noise power spectral density from a time-frequency period of a signal that contains only noise. In contrast, under real-world acoustical environments, such as a nonstationary noise field, although it is necessary to dynamically estimate noise, this is very difficult. Therefore, in this chapter, secondly, we describe an advanced iterative signal extraction method using a microphone array that can be applied to nonstationary noise [19]. This method consists of iterative blind dynamic noise estimation by independent component analysis (ICA) [20–23] and musical-noise-free speech extraction by modified iterative SS, where multiple iterative SS is applied to each channel while maintaining the multichannel property reused for ICA.

Thirdly, in relation to the above-mentioned method, we discuss the justification of applying ICA to signals nonlinearly distorted by SS. We theoretically clarify that the degradation in ICA-based noise estimation obeys an amplitude variation in room transfer functions between the target user and microphones. Next, to reduce speech distortion, we introduce a channel selection strategy into ICA, where we automatically choose less varied inputs to maintain the high accuracy of noise estimation. Furthermore, we introduce a time-variant noise power spectral density (PSD) estimator [24] instead of ICA to improve the noise estimation accuracy. From objective and subjective evaluations, it is revealed that the presented method outperforms various types of the conventional methods.

Note that there exist many investigations for musical noise assessment using higher-order statistics [25–29] and the study on musical-noise-free speech enhancement was carried out for several methods except for iterative SS, namely, iterative WF [30], the iterative MMSE-STSA estimator [31] and the iterative generalized

MMSE-STSA estimator [32]. In this chapter, however, only SS-based method is dealt with because of ease in the mathematical derivations and readers' understanding. Also, the theoretical analysis and results in ICA-based noise estimation are valid for other independent linear factor analysis algorithms, e.g., independent vector analysis [33–35] and independent low-rank matrix analysis [36–39]. However, we focus our attention on only ICA in this chapter owing to its simpleness.

13.2 Single-Channel Speech Enhancement with Musical-Noise-Free Properties

13.2.1 Conventional Non-iterative Spectral Subtraction

We apply a short-time discrete Fourier transform (DFT) to the observed signal, which is a mixture of target speech and noise, to obtain the time-frequency signal. We formulate conventional *non-iterative SS* [1] in the time-frequency domain as follows:

$$Y(f,\tau) = \begin{cases} \sqrt{|X(f,\tau)|^2 - \beta \mathrm{E}[|N|^2]}\exp(j\arg(X(f,\tau))) \\ \quad (\text{if} \quad |X(f,\tau)|^2 > \beta \mathrm{E}[|N|^2]), \\ \eta X(f,\tau) \qquad (\text{otherwise}), \end{cases} \tag{13.1}$$

where $Y(f,\tau)$ is the enhanced target speech signal, $X(f,\tau)$ is the observed signal, f denotes the frequency subband, τ is the frame index, β is the oversubtraction parameter, and η is the flooring parameter. Here, $\mathrm{E}[|N|^2]$ is the expectation of the random variable $|N|^2$ corresponding to the noise power spectra. In practice, we can approximate $\mathrm{E}[|N|^2]$ by averaging the observed noise power spectra $|N(f,\tau)|^2$ in the first K-sample frames, where we assume the absence of speech in this period and noise stationarity. However, this often requires high-accuracy voice activity detection. In addition, many methods for dynamic estimation of the expectation of the noise PSD have been proposed [4], but always suffered from difficulty in rapidly changing nonstationary noise.

Generally speaking, conventional spectral subtraction suffers from the inherent problem of musical noise generation. For example, a large oversubtraction parameter affords a large noise reduction but considerable musical noise is also generated. To reduce the amount of musical noise generated, we often increase the flooring parameter, but this decreases noise reduction; thus, there exists a trade-off between noise reduction and musical noise generation.

13.2.2 Iterative Spectral Subtraction

In an attempt to achieve high-quality noise reduction with low musical noise, an improved method based on iterative SS was proposed in previous studies [14–16].

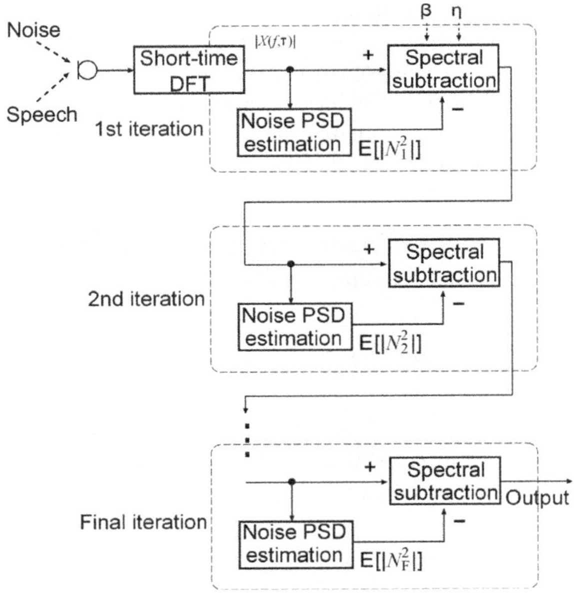

Fig. 13.1 Block diagram of iterative SS

This method is performed through signal processing, in which the following *weak* SS processes are recursively applied to the noise signal (see Fig. 13.1). (I) The average power spectrum of the input noise is estimated. (II) The estimated noise prototype is then subtracted from the input with the parameters specifically set for weak subtraction, e.g., a large flooring parameter η and a small subtraction parameter β. (III) We then return to step (I) and substitute the resultant output (partially noise reduced signal) for the input signal.

13.2.3 Modeling of Input Signal

In this chapter, we assume that the input signal X in the power spectral domain is modeled using the gamma distribution as

$$P(x) = \frac{x^{\alpha-1}}{\Gamma(\alpha)\theta^{\alpha}} \exp(-x/\theta), \tag{13.2}$$

where $x \geq 0$, $\alpha > 0$, and $\theta > 0$. Here, α is the shape parameter, θ is the scale parameter, and $\Gamma(\alpha)$ is the gamma function, defined as

$$\Gamma(\alpha) = \int_0^{\infty} t^{\alpha-1} \exp(-t)dt. \tag{13.3}$$

If the input signal is Gaussian noise, its complex-valued DFT coefficients also have the Gaussian distributions in the real and imaginary parts. Therefore, the p.d.f. of its power spectra obeys the chi-square distribution with two degrees of freedom, which corresponds to the gamma distribution with $\alpha = 1$. Also, if the input signal is super-Gaussian noise, the p.d.f. of its power spectra obeys the gamma distribution with $\alpha < 1$. We make assumption here that θ is assumed to be the deterministically known noise PSD and estimation artifacts of the noise PSD are not taken into account in this chapter. Also, the estimation of α for real-world (short-term) data is explained in, e.g., Ref. [40].

13.2.4 Metric of Musical Noise Generation: Kurtosis Ratio

We speculate that the amount of musical noise is highly correlated with the number of isolated power spectral components and their level of isolation (see Fig. 13.2). In this chapter, we call these isolated components *tonal components*. Since such tonal components have relatively high power, they are strongly related to the weight of the tail of their probability density function (p.d.f.). Therefore, quantifying the tail of the p.d.f. makes it possible to measure the number of tonal components. Thus, we adopt kurtosis, one of the most commonly used higher-order statistics, to evaluate the percentage of tonal components among all components. A larger kurtosis value indicates a signal with a heavy tail, meaning that the signal has many tonal components. Kurtosis is defined as

$$\text{kurt} = \frac{\mu_4}{\mu_2^2}, \tag{13.4}$$

where "kurt" is the kurtosis and μ_m is the mth-order moment, given by

$$\mu_m = \int_0^\infty x^m P(x) dx, \tag{13.5}$$

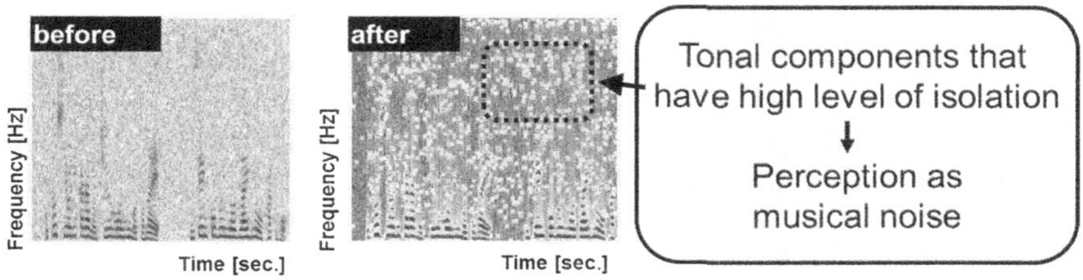

Fig. 13.2 Example of generation of tonal component after signal processing, where input signal is speech with white Gaussian noise and output is processed signal by SS

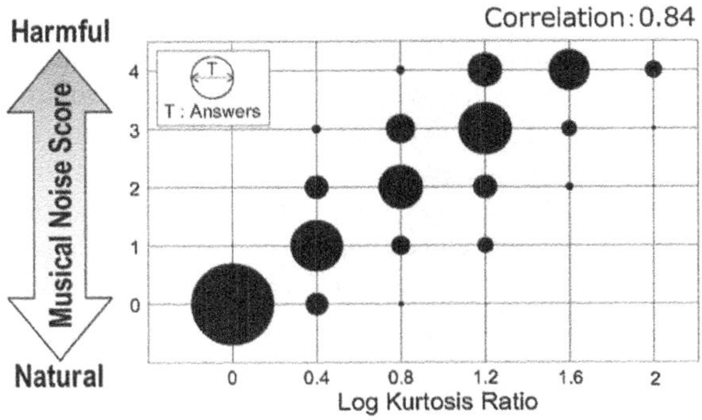

Fig. 13.3 Relation between kurtsis ratio (in log scale) and human-perceptual score of degree of musical noise generation [11]

where $P(x)$ is the p.d.f. of the random variable X. Note that μ_m is not a central moment but a raw moment. Thus, (13.4) is not kurtosis in the mathematically strict definition but a modified version; however, we still refer to (13.4) as kurtosis in this chapter.

In this study, we apply such a kurtosis-based analysis to a time-frequency period of subject signals for the assessment of musical noise. Thus, this analysis should be conducted during, for example, periods of silence in speech when we evaluate the degree of musical noise arising in remaining noise. This is because we aim to quantify the tonal components arising in the noise-only part, which is the main cause of musical noise perception, and not in the target-speech-dominant part.

Although kurtosis can be used to measure the number of tonal components, note that the kurtosis itself is not sufficient to measure the amount of musical noise. This is obvious since the kurtosis of some unprocessed noise signals, such as an interfering speech signal, is also high, but we do not recognize speech as musical noise. Hence, we turn our attention to the change in kurtosis between before and after signal processing to identify only the musical-noise components. Thus, we adopt the *kurtosis ratio* as a measure to assess musical noise [11–13]. This measure is defined as

$$\text{kurtosis ratio} = \frac{\text{kurt}_{\text{proc}}}{\text{kurt}_{\text{org}}}, \tag{13.6}$$

where $\text{kurt}_{\text{proc}}$ is the kurtosis of the processed signal and kurt_{org} is the kurtosis of the original (unprocessed) signal. This measure increases as the amount of generated musical noise increases. In Ref. [11], it was reported that the kurtosis ratio is strongly correlated with the human perception of musical noise. Figure 13.3 shows an example of the relation between the kurtsis ratio (in log scale) and a human-perceptual score of degree of musical noise generation, where we can confirm the strong correlation.

13.2.5 Musical Noise Generation in Non-iterative Spectral Subtraction

In conventional non-iterative spectral subtraction, the long-term-averaged power spectrum of a noise signal is utilized as the estimated noise power spectrum. Then, the estimated noise power spectrum multiplied by the oversubtraction parameter β is subtracted from the observed power spectrum. When a gamma distribution is used to model the noise signal, its mean is $\alpha_{\mathrm{n}}\theta_{\mathrm{n}}$, where α_{n} and θ_{n} are the shape and scale parameters of noise, respectively (the subscript "n" indicates that the parameters belong to noise). Thus, the amount of subtraction is $\beta\alpha_{\mathrm{n}}\theta_{\mathrm{n}}$. The subtraction of the estimated noise power spectrum in each frequency band can be considered as a shift of the p.d.f. in the zero-power direction, given by

$$\frac{1}{\theta_{\mathrm{n}}^{\alpha_{\mathrm{n}}}\Gamma(\alpha_{\mathrm{n}})}(z+\beta\alpha_{\mathrm{n}}\theta_{\mathrm{n}})^{\alpha_{\mathrm{n}}-1}\exp\left\{-\frac{z+\beta\alpha_{\mathrm{n}}\theta_{\mathrm{n}}}{\theta_{\mathrm{n}}}\right\}, \tag{13.7}$$

where z is the random variable of the p.d.f. after spectral subtraction.

As a result, negative-power components with nonzero probability arise. To avoid this, such negative components are replaced by observations that are multiplied by a positive value η (flooring parameter). This means that the region corresponding to the probability of the negative components, which forms a section cut from the original gamma distribution, is compressed by the effect of the flooring, resulting in

$$\frac{1}{(\eta^2\theta_{\mathrm{n}})^{\alpha_{\mathrm{n}}}\Gamma(\alpha_{\mathrm{n}})}z^{\alpha_{\mathrm{n}}-1}\exp\left\{-\frac{z}{\eta^2\theta_{\mathrm{n}}}\right\}. \tag{13.8}$$

Note that the flooring parameter η is squared in the p.d.f. because the multiplication of η is conducted in the amplitude spectrum domain (see the second branch in (13.1)) but we now consider its effect in the power spectrum domain.

Finally, the floored components are superimposed on the laterally shifted p.d.f. Thus, the resultant p.d.f. after spectral subtraction, $P_{\mathrm{SS}}(z)$, can be written as

$$P_{\mathrm{SS}}(z) = \begin{cases} \frac{1}{\theta_{\mathrm{n}}^{\alpha_{\mathrm{n}}}\Gamma(\alpha_{\mathrm{n}})}(z+\beta\alpha_{\mathrm{n}}\theta_{\mathrm{n}})^{\alpha_{\mathrm{n}}-1}\exp\left\{-\frac{z+\beta\alpha_{\mathrm{n}}\theta_{\mathrm{n}}}{\theta_{\mathrm{n}}}\right\} & \\ \quad + \frac{1}{(\eta^2\theta_{\mathrm{n}})^{\alpha_{\mathrm{n}}}\Gamma(\alpha_{\mathrm{n}})}z^{\alpha_{\mathrm{n}}-1}\exp\left\{-\frac{z}{\eta^2\theta_{\mathrm{n}}}\right\} & (0 \le z < \beta\alpha_{\mathrm{n}}\eta^2\theta_{\mathrm{n}}), \\ \frac{1}{\theta_{\mathrm{n}}^{\alpha_{\mathrm{n}}}\Gamma(\alpha_{\mathrm{n}})}(z+\beta\alpha_{\mathrm{n}}\theta_{\mathrm{n}})^{\alpha_{\mathrm{n}}-1}\exp\left\{-\frac{z+\beta\alpha_{\mathrm{n}}\theta_{\mathrm{n}}}{\theta_{\mathrm{n}}}\right\} & (\beta\alpha_{\mathrm{n}}\eta^2\theta_{\mathrm{n}} \le z), \end{cases} \tag{13.9}$$

To characterize non-iterative spectral subtraction, the mth-order moment of z is required. For $P_{\mathrm{SS}}(z)$, the mth-order moment is given by

$$\mu_m^{\mathrm{SS}} = \int_0^\infty z^m \cdot P_{\mathrm{SS}}(z) dz$$
$$= \int_0^\infty z^m \frac{1}{\theta_{\mathrm{n}}^{\alpha_{\mathrm{n}}} \Gamma(\alpha_{\mathrm{n}})} (z + \beta\alpha_{\mathrm{n}}\theta_{\mathrm{n}})^{\alpha_{\mathrm{n}}-1} \exp\left\{-\frac{z + \beta\alpha_{\mathrm{n}}\theta_{\mathrm{n}}}{\theta_{\mathrm{n}}}\right\} dz$$
$$+ \int_0^{\beta\alpha_{\mathrm{n}}\eta^2\theta_{\mathrm{n}}} z^m \frac{1}{(\eta^2\theta_{\mathrm{n}})^{\alpha_{\mathrm{n}}} \Gamma(\alpha_{\mathrm{n}})} z^{\alpha_{\mathrm{n}}-1} \exp\left\{-\frac{z}{\eta^2\theta_{\mathrm{n}}}\right\} dz, \tag{13.10}$$

where z is the random variable of the p.d.f. after spectral subtraction. We now expand the first term of the right-hand side of (13.10). Here, let $t = (z + \beta\alpha_{\mathrm{n}}\theta_{\mathrm{n}})/\theta_{\mathrm{n}}$, then $\theta_{\mathrm{n}} dt = dz$ and $z = \theta_{\mathrm{n}}(t - \beta\alpha_{\mathrm{n}})$. Consequently,

$$\int_0^\infty z^m \frac{1}{\theta_{\mathrm{n}}^{\alpha_{\mathrm{n}}} \Gamma(\alpha_{\mathrm{n}})} (z + \beta\alpha_{\mathrm{n}}\theta_{\mathrm{n}})^{\alpha_{\mathrm{n}}-1} \exp\left\{-\frac{z + \beta\alpha_{\mathrm{n}}\theta_{\mathrm{n}}}{\theta_{\mathrm{n}}}\right\} dz$$
$$= \int_{\beta\alpha_{\mathrm{n}}}^\infty \theta_{\mathrm{n}}^m (t - \beta\alpha_{\mathrm{n}})^m \frac{1}{\theta_{\mathrm{n}}^{\alpha_{\mathrm{n}}} \Gamma(\alpha_{\mathrm{n}})} (\theta_{\mathrm{n}} t)^{\alpha_{\mathrm{n}}-1} \exp\{-t\} \theta_{\mathrm{n}} dt$$
$$= \frac{\theta_{\mathrm{n}}^m}{\Gamma(\alpha_{\mathrm{n}})} \int_{\beta\alpha_{\mathrm{n}}}^\infty \sum_{l=0}^m (-\beta\alpha_{\mathrm{n}})^l \frac{\Gamma(m+1)}{\Gamma(l+1)\Gamma(m-l+1)} t^{m-l} t^{\alpha_{\mathrm{n}}-1} \exp\{-t\} dt$$
$$= \frac{\theta_{\mathrm{n}}^m}{\Gamma(\alpha_{\mathrm{n}})} \sum_{l=0}^m (-\beta\alpha_{\mathrm{n}})^l \frac{\Gamma(m+1)}{\Gamma(l+1)\Gamma(m-l+1)} \Gamma(\alpha_{\mathrm{n}} + m - l, \beta\alpha_{\mathrm{n}}), \tag{13.11}$$

where we use the binomial theorem given by

$$(t + a)^m = \sum_{l=0}^m a^l \frac{\Gamma(m+1)}{\Gamma(l+1)\Gamma(m-l+1)} t^{m-l}, \tag{13.12}$$

and $\Gamma(a, b)$ is the upper incomplete gamma function defined as

$$\Gamma(a, b) = \int_b^\infty t^{a-1} \exp\{-t\} dt. \tag{13.13}$$

Next we consider the second term of the right-hand side of (13.10). Here, let $t = z/(\eta^2\theta_{\mathrm{n}})$, then $\eta^2\theta_{\mathrm{n}} dt = dz$. Thus,

$$\int_0^{\beta\alpha_{\mathrm{n}}\eta^2\theta_{\mathrm{n}}} z^m \frac{1}{(\eta^2\theta_{\mathrm{n}})^{\alpha_{\mathrm{n}}} \Gamma(\alpha_{\mathrm{n}})} z^{\alpha_{\mathrm{n}}-1} \exp\left\{-\frac{z}{\eta^2\theta_{\mathrm{n}}}\right\} dz$$
$$= \int_0^{\beta\alpha_{\mathrm{n}}} (\eta^2\theta_{\mathrm{n}} t)^m \frac{1}{(\eta^2\theta_{\mathrm{n}})^{\alpha_{\mathrm{n}}} \Gamma(\alpha_{\mathrm{n}})} (\eta^2\theta_{\mathrm{n}} t)^{\alpha_{\mathrm{n}}-1} \exp\{-t\} \eta^2\theta_{\mathrm{n}} dt$$

$$= \frac{\eta^{2m} \theta_{\mathrm{n}}^{m}}{\Gamma(\alpha_{\mathrm{n}})} \int_{0}^{\beta\alpha_{\mathrm{n}}} t^{\alpha_{\mathrm{n}}-1+m} \exp\{-t\}\, dt$$

$$= \frac{\eta^{2m} \theta_{\mathrm{n}}^{m}}{\Gamma(\alpha_{\mathrm{n}})} \gamma(\alpha_{\mathrm{n}} + m, \beta\alpha_{\mathrm{n}}), \tag{13.14}$$

where $\gamma(a, b)$ is the lower incomplete gamma function defined as

$$\gamma(a, b) = \int_{0}^{b} t^{a-1} \exp\{-t\} dt. \tag{13.15}$$

As a result, the mth-order moment after spectral subtraction, μ_m^{SS}, is a composite of (13.11) and (13.14), and is given by [17]

$$\mu_m^{\mathrm{SS}} = \theta_{\mathrm{n}}^{m} \mathscr{M}(\alpha_{\mathrm{n}}, \beta, \eta, m), \tag{13.16}$$

where

$$\mathscr{M}(\alpha_{\mathrm{n}}, \beta, \eta, m) = \mathscr{S}(\alpha_{\mathrm{n}}, \beta, \eta) + \eta^{2m} \mathscr{F}(\alpha_{\mathrm{n}}, \beta, \eta), \tag{13.17}$$

$$\mathscr{S}(\alpha_{\mathrm{n}}, \beta, m) = \sum_{l=0}^{m} (-\beta\alpha_{\mathrm{n}})^{l} \frac{\Gamma(m+1)\Gamma(\alpha_{\mathrm{n}}+m-l, \beta\alpha_{\mathrm{n}})}{\Gamma(\alpha_{\mathrm{n}})\Gamma(l+1)\Gamma(m-l+1)}, \tag{13.18}$$

$$\mathscr{F}(\alpha_{\mathrm{n}}, \beta, m) = \frac{\gamma(\alpha_{\mathrm{n}}+m, \beta\alpha_{\mathrm{n}})}{\Gamma(\alpha_{\mathrm{n}})}. \tag{13.19}$$

From (13.4), (13.16), and (13.17), the kurtosis after SS can be expressed as

$$\mathrm{kurt} = \frac{\mathscr{M}(\alpha_{\mathrm{n}}, \beta, \eta, 4)}{\mathscr{M}^{2}(\alpha_{\mathrm{n}}, \beta, \eta, 2)}. \tag{13.20}$$

Using (13.6) and (13.20), we also express the kurtosis ratio as

$$\text{kurtosis ratio} = \frac{\mathscr{M}(\alpha_{\mathrm{n}}, \beta, \eta, 4)/\mathscr{M}^{2}(\alpha_{\mathrm{n}}, \beta, \eta, 2)}{\mathscr{M}(\alpha_{\mathrm{n}}, 0, 0, 4)/\mathscr{M}^{2}(\alpha_{\mathrm{n}}, 0, 0, 2)}. \tag{13.21}$$

Also, as a measure of the noise reduction performance, the noise reduction rate (NRR) [41], the output SNR minus the input SNR in dB, can be given in terms of a 1st-order moment as [17]

$$\mathrm{NRR} = 10 \log_{10} \frac{\alpha_{\mathrm{n}}}{\mathscr{M}(\alpha_{\mathrm{n}}, \beta, \eta, 1)}. \tag{13.22}$$

13.2.6 Musical-Noise-Free Speech Enhancement

In [18], Miyazaki et al. proposed musical-noise-free noise reduction, where no musical noise is generated even for a high SNR in iterative SS. In the study, some of the authors discovered an interesting phenomenon that the kurtosis ratio sometimes does not change even after SS via mathematical analysis based on (13.21) [17]. This indicates that the kurtosis ratio can be maintained at unity even after iteratively applying SS to improve the NRR, and thus no musical noise is generated owing to the *domino-toppling* phenomenon. Following this finding, the authors derived the optimal parameters satisfying the musical-noise-free condition [18] by finding a fixed-point status in the kurtosis ratio, i.e., by solving

$$\frac{\mathscr{M}(\alpha_{\rm n}, 0, 0, 4)}{\mathscr{M}^2(\alpha_{\rm n}, 0, 0, 2)} = \frac{\mathscr{M}(\alpha_{\rm n}, \beta, \eta, 4)}{\mathscr{M}^2(\alpha_{\rm n}, \beta, \eta, 2)}. \tag{13.23}$$

The inductive result is that the kurtosis ratio never changes even at a large number of (ideally "infinite") iterations. In this situation, sufficient noise reduction can be gained if the NRR improvement in each iteration is even small but positive. This corresponds to musical-noise-free noise reduction. In summary, we can formulate a new theorem on musical-noise-free conditions as follows.
(I) Fixed-point kurtosis condition: The kurtosis should be equal before and after spectral subtraction in each iteration. This corresponds to a fixed point for the 2nd- and 4th-order moments.
(II) NRR growth condition: The amount of noise reduction should be larger than 0 dB in each iteration, relating to a change in the 1st-order moment.

Although the parameters to be optimized are η and β, we hereafter derive the optimal η given a fixed β for ease of closed-form analysis. First, we change (13.20) for

$$\mathrm{kurt}(\alpha_{\rm n}, \beta, \eta) = \frac{\mathscr{S}(\alpha_{\rm n}, \beta, 4) + \eta^8 \mathscr{F}(\alpha_{\rm n}, \beta, 4)}{\left(\mathscr{S}(\alpha_{\rm n}, \beta, 2) + \eta^4 \mathscr{F}(\alpha_{\rm n}, \beta, 2)\right)^2}. \tag{13.24}$$

Next, the fixed-point kurtosis condition corresponds to the kurtosis being equal before and after spectral subtraction, thus

$$\frac{\mathscr{S}(\alpha_{\rm n}, \beta, 4) + \eta^8 \mathscr{F}(\alpha_{\rm n}, \beta, 4)}{\left(\mathscr{S}(\alpha_{\rm n}, \beta, 2) + \eta^4 \mathscr{F}(\alpha_{\rm n}, \beta, 2)\right)^2} = \frac{(\alpha_{\rm n}+3)(\alpha_{\rm n}+2)}{(\alpha_{\rm n}+1)\alpha_{\rm n}}. \tag{13.25}$$

Let $\mathscr{H} = \eta^4$, and (13.25) yields the following quadratic equation in $\mathscr{H}$.

$$\begin{aligned}
&\left(\mathscr{F}(\alpha_{\rm n}, \beta, 4)(\alpha_{\rm n}+1)\alpha_{\rm n} - \mathscr{F}^2(\alpha_{\rm n}, \beta, 2)(\alpha_{\rm n}+3)(\alpha_{\rm n}+2)\right)\mathscr{H}^2 \\
&-2\mathscr{S}(\alpha_{\rm n}, \beta, 2)\mathscr{F}(\alpha_{\rm n}, \beta, 2)(\alpha_{\rm n}+3)(\alpha_{\rm n}+2)\mathscr{H} \\
&+\mathscr{S}(\alpha_{\rm n}, \beta, 4)(\alpha_{\rm n}+1)\alpha_{\rm n} - \mathscr{S}^2(\alpha_{\rm n}, \beta, 2)(\alpha_{\rm n}+3)(\alpha_{\rm n}+2) = 0.
\end{aligned} \tag{13.26}$$

Thus, we can derive a closed-form estimate of $\mathscr{H}$ from the given noise shape parameter $\alpha_{\rm n}$ and oversubtraction parameter β as

$$\begin{aligned}\mathscr{H} =&\{\mathscr{F}(\alpha_{\rm n},\beta,4)(\alpha_{\rm n}+1)\alpha_{\rm n}-\mathscr{F}^2(\alpha_{\rm n},\beta,2)(\alpha_{\rm n}+3)(\alpha_{\rm n}+2)\}^{-1}\\ &\Big[\mathscr{S}(\alpha_{\rm n},\beta,2)\mathscr{F}(\alpha_{\rm n},\beta,2)(\alpha_{\rm n}+3)(\alpha_{\rm n}+2)\\ &\pm\Big[\{\mathscr{S}(\alpha_{\rm n},\beta,2)\mathscr{F}(\alpha_{\rm n},\beta,2)(\alpha_{\rm n}+3)(\alpha_{\rm n}+2)\}^2\\ &-\big\{\mathscr{F}(\alpha_{\rm n},\beta,4)(\alpha_{\rm n}+1)\alpha_{\rm n}-\mathscr{F}^2(\alpha_{\rm n},\beta,2)(\alpha_{\rm n}+3)(\alpha_{\rm n}+2)\big\}\\ &\big\{\mathscr{S}(\alpha_{\rm n},\beta,4)(\alpha_{\rm n}+1)\alpha_{\rm n}-\mathscr{S}^2(\alpha_{\rm n},\beta,2)(\alpha_{\rm n}+3)(\alpha_{\rm n}+2)\big\}\Big]^{\frac{1}{2}}\Big]. \end{aligned} \tag{13.27}$$

Finally, $\eta = \mathscr{H}^{1/4}$ is the resultant flooring parameter that satisfies the fixed-point kurtosis condition.

From (13.22), the NRR growth condition is expressed as

$$\mathrm{NRR} = 10\log_{10}\frac{\alpha_{\rm n}}{\mathscr{S}(\alpha_{\rm n},\beta,1)+\eta^2\mathscr{F}(\alpha_{\rm n},\beta,1)} > 0. \tag{13.28}$$

Here, since $\eta > 0$, we can solve the inequality as

$$0 < \eta < \sqrt{\frac{\alpha_{\rm n} - \mathscr{S}(\alpha_{\rm n},\beta,1)}{\mathscr{F}(\alpha_{\rm n},\beta,1)}}. \tag{13.29}$$

In summary, we can choose the parameters simultaneously satisfying the fixed kurtosis point condition and NRR growth condition using (13.27) and (13.29). Figure 13.4 shows an example of the kurtosis ratio in optimized iterative SS, where Gaussian noise is assumed. We can confirm the flat trace of the kurtosis, indicating no musical noise generation.

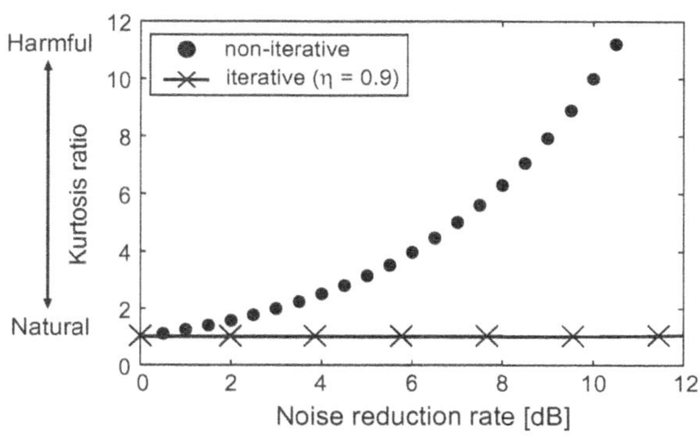

Fig. 13.4 Relation between NRR and kurtosis ratio obtained from theoretical analysis for case of Gaussian noise

13.3 Extension to Multichannel Blind Signal Processing

13.3.1 Blind Spatial Subtraction Array

In the previous section, we assumed that the input noise signal is stationary, meaning that we can estimate the expectation of a noise signal from a time-frequency period of a signal that contains only noise, i.e., speech absence. However, in actual environments, such as a nonstationary noise field, it is necessary to dynamically estimate the noise PSD.

To solve this problem, Takahashi et al. previously proposed blind spatial subtraction array (BSSA) [42], which involves accurate noise estimation by ICA followed by a speech extraction procedure based on SS (see Fig. 13.5). BSSA improves the noise reduction performance, particularly in the presence of both diffuse and nonstationary noises; thus, almost all the environmental noise can be dealt with. However, BSSA always suffers from musical noise owing to SS. In addition, the output signal of BSSA degenerates to a *monaural* (not multichannel) signal, meaning that ICA cannot be reapplied; thus, we cannot iteratively estimate the noise power spectra. Therefore, it is impossible to directly apply iterative SS to the conventional BSSA.

13.3.2 Iterative Blind Spatial Subtraction Array

In this section, we introduce a multi-iterative blind signal extraction method integrating iterative blind noise estimation by ICA and iterative noise reduction by SS. As mentioned previously, the conventional BSSA cannot iteratively and accurately estimate noise by ICA because the conventional BSSA performs a delay and sum (DS) operation before SS. To solve this problem, Takahashi et al. have proposed an improved BSSA structure that performs multiple independent SS in each channel before DS; we call this structure *channelwise SS* [43–45]. Using this structure, we can equalize the number of channels of the observed signal to that of the signals after

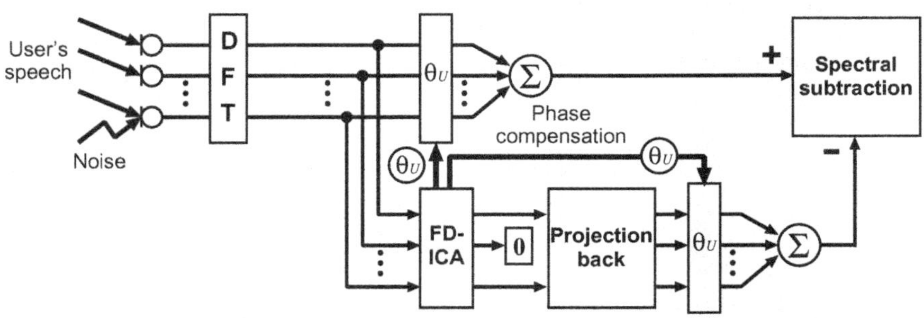

Fig. 13.5 Block diagram of BSSA [42]

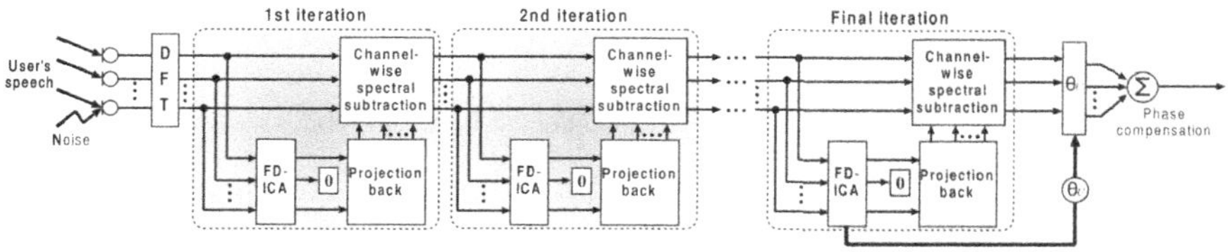

Fig. 13.6 Block diagram of iterative BSSA [19]

channelwise SS. Therefore, we can iteratively apply noise estimation by ICA and speech extraction by SS (see Fig. 13.6). Also, the advantage of the structure is that ICA has the possibility of adaptively estimating the *distorted wavefront* of a speech signal to some extent even after SS, because ICA is a blind signal identification method that does not require knowledge of the target signal direction. Details of this issue will be discussed in Sect. 13.3.3. Hereafter, we refer to this type of BSSA as *iterative BSSA*.

We conduct iterative BSSA in the following manner, where the superscript $[i]$ represents the value in the ith iteration of SS (initially $i = 0$).

(I) The observed signal vector of the K-channel array in the time-frequency domain, $\mathbf{X}^{[0]}(f, \tau)$, is given by

$$\mathbf{X}^{[0]}(f, \tau) = \mathbf{H}(f)S(f, \tau) + \mathbf{N}(f, \tau), \tag{13.30}$$

where $\mathbf{H}(f) = [H_1(f), H_2(f) \ldots, H_K(f)]^{\mathrm{T}}$ is a column vector of the transfer functions from the target signal position to each microphone, $S(f, \tau)$ is the target speech signal, and $\mathbf{N}(f, \tau)$ is a column vector of the additive noise.

(II) Next, we perform signal separation using ICA as [20]

$$\mathbf{O}^{[i]}(f, \tau) = \mathbf{W}_{\mathrm{ICA}}^{[i]}(f)\mathbf{X}^{[i]}(f, \tau), \tag{13.31}$$

$$\mathbf{W}_{\mathrm{ICA}}^{[i][p+1]}(f) = \mu[\mathbf{I} - \langle\varphi(\mathbf{O}^{[i]}(f, \tau))(\mathbf{O}^{[i]}(f, \tau))^{\mathrm{H}}\rangle_\tau] \cdot \mathbf{W}_{\mathrm{ICA}}^{[i][p]}(f) + \mathbf{W}_{\mathrm{ICA}}^{[i][p]}(f), \tag{13.32}$$

where $\mathbf{W}_{\mathrm{ICA}}^{[i][p]}(f)$ is a demixing matrix, μ is the step-size parameter, $[p]$ is used to express the value of the pth step in the ICA iterations, $\mathbf{I}$ is the identity matrix, $\langle\cdot\rangle_\tau$ denotes a time-averaging operator, and $\varphi(\cdot)$ is an appropriate nonlinear vector function. Then, we construct a *noise-only vector*,

$$\mathbf{O}_{\mathrm{noise}}^{[i]}(f, \tau) = [O_1^{[i]}(f, \tau), \ldots, O_{U-1}^{[i]}, 0, O_{U+1}^{[i]}(f, \tau), \ldots, O_K^{[i]}(f, \tau)]^{\mathrm{T}}, \tag{13.33}$$

where U is the signal number for speech, and we apply the projection back operation to remove the ambiguity of the amplitude and construct the estimated noise signal, $\mathbf{Z}^{[i]}(f, \tau)$, as

$$\mathbf{Z}^{[i]}(f, \tau) = \mathbf{W}_{\mathrm{ICA}}^{[i]}(f)^{-1}\mathbf{O}_{\mathrm{noise}}^{[i]}(f, \tau). \tag{13.34}$$

(III) Next, we perform SS independently in each input channel and derive the multiple target-speech-enhanced signals. This procedure can be given by

$$X_k^{[i+1]}(f, \tau) = \begin{cases} \sqrt{|X_k^{[i]}(f, \tau)|^2 - \beta|Z_k^{[i]}(f, \tau)|^2} \exp(j \arg(X_k^{[i]}(f, \tau))) \\ (\text{if} \quad |X_k^{[i]}(f, \tau)|^2 > \beta|Z_k^{[i]}(f, \tau)|^2), \\ \eta X_k^{[i]}(f, \tau) \quad (\text{otherwise}), \end{cases} \tag{13.35}$$

where $X_k^{[i+1]}(f, \tau)$ is the target-speech-enhanced signal obtained by SS at a specific channel k. Then we return to step (II) with $\mathbf{X}^{[i+1]}(f, \tau)$. When we obtain sufficient noise reduction performance, we proceed to step (IV).

(IV) Finally, we obtain the resultant target-speech-enhanced signal by applying DS to $\mathbf{X}^{[*]}(f, \tau)$, where $*$ is the number of iterations after which sufficient noise reduction performance is obtained. This procedure can be expressed by

$$Y(f, \tau) = \mathbf{W}_{\mathrm{DS}}^{\mathrm{T}}(f)\mathbf{X}^{[*]}(f, \tau), \tag{13.36}$$

$$\mathbf{W}_{\mathrm{DS}}(f) = [W_1^{(\mathrm{DS})}(f), \ldots, W_K^{(\mathrm{DS})}(f)], \tag{13.37}$$

$$W_k^{(\mathrm{DS})}(f) = \frac{1}{K} \exp(-2\pi j (f/M) f_{\mathrm{s}} d_k \sin\theta_U / c), \tag{13.38}$$

$$\theta_U = \sin^{-1} \frac{\arg\left(\dfrac{\left[\mathbf{W}_{\mathrm{ICA}}^{[*]}(f)^{-1}\right]_{kU}}{\left[\mathbf{W}_{\mathrm{ICA}}^{[*]}(f)^{-1}\right]_{k'U}} \right)}{2\pi f_{\mathrm{s}} c^{-1} (d_k - d_{k'})}, \tag{13.39}$$

where $Y(f, \tau)$ is the final output signal of iterative BSSA, $\mathbf{w}_{\mathrm{DS}}$ is the filter coefficient vector of DS, M is the DFT size, f_{s} is the sampling frequency, d_k is the microphone position, c is the sound velocity, and θ_U is the estimated direction of arrival of the target speech obtained by ICA's demixing matrix [46]. Moreover, $[\mathbf{A}]_{lj}$ represents the entry in the lth row and jth column of $\mathbf{A}$.

13.3.3 Accuracy of Wavefront Estimated by Independent Component Analysis After Spectral Subtraction

In this subsection, we discuss the accuracy of the estimated noise signal in each iteration of iterative BSSA. In actual environments, not only point-source noise but also non-point-source (e.g., diffuse) noise often exists. It is known that ICA is proficient in noise estimation rather than speech estimation under such a noise condition [42, 47, 48]. This is because the target speech can be regarded as a point-source signal (thus, the wavefront is static in each subband) and ICA acts as an effective blocking filter of the speech wavefront even in a time-invariant manner, resulting in good noise estimation. However, in iterative BSSA, we should address the inherent question of whether the distorted speech wavefront after nonlinear noise reduction such as SS can be blocked by ICA or not; thus, we determine whether the speech component after channelwise SS can become a point source again.

Hereafter, we quantify the degree of point-source-likeness for SS-applied speech signals. For convenience of discussion, a simple two-channel array model is assumed. First, we define the speech component in each channel after channelwise SS as

$$\hat{S}_1(f,\tau) = H_1(f)S(f,\tau) + \Delta S_1(f,\tau), \tag{13.40}$$

$$\hat{S}_2(f,\tau) = H_2(f)S(f,\tau) + \Delta S_2(f,\tau), \tag{13.41}$$

where $S(f,\tau)$ is the original point-source speech signal, $\hat{S}_k(f,\tau)$ is the speech component after channelwise SS at the kth channel, and $\Delta S_k(f,\tau)$ is the speech component distorted by channelwise SS. Also, we assume that $S(f,\tau)$, $\Delta S_1(f,\tau)$, and $\Delta S_2(f,\tau)$ are uncorrelated with each other. Obviously, $\hat{S}_1(f,\tau)$ and $\hat{S}_2(f,\tau)$ can be regarded as being generated by a point source if $\Delta S_1(f,\tau)$ and $\Delta S_2(f,\tau)$ are zero, i.e., a valid static blocking filter can be obtained by ICA as

$$\begin{aligned}
&[\mathbf{W}_{\mathrm{ICA}}(f)]_{11}\hat{S}_1(f,\tau) + [\mathbf{W}_{\mathrm{ICA}}(f)]_{12}\hat{S}_2(f,\tau) \\
&= ([\mathbf{W}_{\mathrm{ICA}}(f)]_{11}H_1(f) + [\mathbf{W}_{\mathrm{ICA}}(f)]_{12}H_2(f))S(f,\tau) \\
&= 0,
\end{aligned} \tag{13.42}$$

where we assume $U = 1$ and, e.g., $[\mathbf{W}_{\mathrm{ICA}}(f)]_{11} = H_2(f)$ and $[\mathbf{W}_{\mathrm{ICA}}(f)]_{12} = -H_1(f)$. However, if $\Delta S_1(f,\tau)$ and $\Delta S_2(f,\tau)$ become nonzero as a result of SS, ICA does not have a valid speech-blocking filter with a static (time-invariant) form.

Second, the cosine distance between speech power spectra $|\hat{S}_1(f,\tau)|^2$ and $|\hat{S}_2(f,\tau)|^2$ is introduced in each frequency subband to indicate the degree of point-source-likeness as

$$\mathrm{COS}(f) = \frac{\sum_\tau |\hat{S}_1(f,\tau)|^2 |\hat{S}_2(f,\tau)|^2}{\sqrt{\sum_\tau |\hat{S}_1(f,\tau)|^4}\sqrt{\sum_\tau |\hat{S}_2(f,\tau)|^4}}. \tag{13.43}$$

From (13.43), the cosine distance reaches its maximum value of unity if and only if $\Delta S_1(f,\tau) = \Delta S_2(f,\tau) = 0$, regardless of the values of $H_1(f)$ and $H_2(f)$, meaning that the SS-applied speech signals $\hat{S}_1(f,\tau)$ and $\hat{S}_2(f,\tau)$ can be assumed to be produced by the point source. The value of $\mathrm{COS}(f)$ decreases with increasing magnitudes of $\Delta S_1(f,\tau)$ and $\Delta S_2(f,\tau)$ as well as with increasing difference between $H_1(f)$ and $H_2(f)$; this indicates the non-point-source state.

Third, we evaluate the degree of point-source-likeness in each iteration of iterative BSSA by using $\mathrm{COS}(f)$. We statistically estimate the distorted speech component of the enhanced signal in each iteration. Here, we assume that the original speech power spectrum $|S(f,\tau)|^2$ obeys a gamma distribution with a shape parameter of 0.1 (this is a typical value for speech [49–54]) as

$$|S(f,\tau)|^2 \sim \frac{x^{-0.9}}{\Gamma(0.1)\theta_{\mathrm{s}}^{0.1}} \exp(-x/\theta_{\mathrm{s}}), \tag{13.44}$$

where θ_{s} is the speech scale parameter. Regarding the amount of noise to be subtracted, the 1st-order moment of the noise power spectra is equal to $\theta_{\mathrm{n}}\alpha_{\mathrm{n}}$ when the number of iterations, i, equals zero. Also, the value of α_{n} does not change in each iteration when we use the specific parameters β and η that satisfy the musical-noise-free condition because the kurtosis ratio does not change in each iteration. If we perform SS only once, the rate of noise decrease is given by

$$\mathscr{M}(\alpha_{\mathrm{n}}, \beta, \eta, 1)/\alpha_{\mathrm{n}}, \tag{13.45}$$

and thus, the amount of residual noise after the ith iteration is given by

$$\begin{aligned}\mu_1^{[i]} &= \theta_{\mathrm{n}}\alpha_{\mathrm{n}} \left\{\mathscr{M}(\alpha_{\mathrm{n}}, \beta, \eta, 1)/\alpha_{\mathrm{n}}\right\}^i \\ &= \theta_{\mathrm{n}}\mathscr{M}^i(\alpha_{\mathrm{n}}, \beta, \eta, 1)\alpha_{\mathrm{n}}^{1-i}.\end{aligned} \tag{13.46}$$

Next, we assume that the speech and noise are disjoint, i.e., there are no overlaps in the time-frequency domain, and that speech distortion is caused by subtracting the average noise from the pure speech component. Thus, the speech component $|\hat{S}_k^{[i+1]}(f,\tau)|^2$ at the kth channel after the ith iteration is represented by subtracting the amount of residual noise (13.46) as

$$\begin{aligned}&|\hat{S}_k^{[i+1]}(f,\tau)|^2 = \\ &\begin{cases} |\hat{S}_k^{[i]}(f,\tau)|^2 - \beta\theta_{\mathrm{n}}\mathscr{M}^i(\alpha_{\mathrm{n}}, \beta, \eta, 1)\alpha_{\mathrm{n}}^{1-i} \\ (\text{if} \quad |\hat{S}_k^{[i]}(f,\tau)|^2 > \beta\theta_{\mathrm{n}}\mathscr{M}^i(\alpha_{\mathrm{n}}, \beta, \eta, 1)\alpha_{\mathrm{n}}^{1-i}), \\ \eta^2|\hat{S}_k^{[i]}(f,\tau)|^2 \quad (\text{otherwise}). \end{cases}\end{aligned} \tag{13.47}$$

Here, we define the input SNR as the average of both channel SNRs,

$$\begin{aligned}\mathrm{ISNR}(f) &= \frac{1}{2}\left(\frac{0.1|H_1(f)|^2\theta_{\mathrm{s}}}{\alpha_{\mathrm{n}}\theta_{\mathrm{n}}} + \frac{0.1|H_2(f)|^2\theta_{\mathrm{s}}}{\alpha_{\mathrm{n}}\theta_{\mathrm{n}}}\right) \\ &= \frac{0.1\theta_{\mathrm{s}}}{2\alpha_{\mathrm{n}}\theta_{\mathrm{n}}}(|H_1(f)|^2 + |H_2(f)|^2).\end{aligned} \tag{13.48}$$

If we normalize the speech scale parameter θ_{s} to unity, from (13.48), the noise scale parameter θ_{n} is given by

$$\theta_{\mathrm{n}} = \frac{0.1(|H_1(f)|^2 + |H_2(f)|^2)}{2\alpha_{\mathrm{n}}\mathrm{ISNR}(f)}, \tag{13.49}$$

and using (13.49), we can reformulate (13.47) as

$$|\hat{S}_k^{[i+1]}(f,\tau)|^2 = \begin{cases} |\hat{S}_k^{[i]}(f,\tau)|^2 - \beta\frac{0.1(|H_1(f)|^2+|H_2(f)|^2)}{2\mathrm{ISNR}(f)}\mathscr{M}^i(\alpha_{\mathrm{n}},\beta,\eta,1)\alpha_{\mathrm{n}}^{-i} \\ (\text{if } |\hat{S}_k^{[i]}(f,\tau)|^2 > \beta\frac{0.1(|H_1(f)|^2+|H_2(f)|^2)}{2\mathrm{ISNR}(f)}\mathscr{M}^i(\alpha_{\mathrm{n}},\beta,\eta,1)\alpha_{\mathrm{n}}^{-i}), \\ \eta^2|\hat{S}_k^{[i]}(f,\tau)|^2 \quad (\text{otherwise}). \end{cases} \tag{13.50}$$

Furthermore, we define the transfer function ratio (TFR) as

$$\mathrm{TFR}(f) = |H_1(f)/H_2(f)|^2, \tag{13.51}$$

and if we normalize $|H_1(f)|^2$ to unity in each frequency subband, $|H_1(f)|^2 + |H_2(f)|^2$ becomes $1 + 1/\mathrm{TFR}(f)$. Finally, we express (13.50) in terms of the input SNR $\mathrm{ISNR}(f)$ and the transfer function ratio $\mathrm{TFR}(f)$ as

$$|\hat{S}_k^{[i+1]}(f,\tau)|^2 = \begin{cases} |\hat{S}_k^{[i]}(f,\tau)|^2 - \beta\frac{0.1(1+1/\mathrm{TFR}(f))}{2\mathrm{ISNR}(f)}\mathscr{M}^i(\alpha_{\mathrm{n}},\beta,\eta,1)\alpha_{\mathrm{n}}^{-i} \\ (\text{if } \ |\hat{S}_k^{[i]}(f,\tau)|^2 > \beta\frac{0.1(1+1/\mathrm{TFR}(f))}{2\mathrm{ISNR}(f)}\mathscr{M}^i(\alpha_{\mathrm{n}},\beta,\eta,1)\alpha_{\mathrm{n}}^{-i}), \\ \eta^2|\hat{S}_k^{[i]}(f,\tau)|^2 \quad (\text{otherwise}). \end{cases} \tag{13.52}$$

As can be seen, the speech component is subjected to greater subtraction and distortion as $\mathrm{ISNR}(f)$ and/or $\mathrm{TFR}(f)$ decrease.

Figure 13.7 shows the relation between the TFR and the corresponding value of $\mathrm{COS}(f)$ calculated by (13.43) and (13.52). In Fig. 13.7, we plot the average of $\mathrm{COS}(f)$ over whole frequency subbands. The noise shape parameter α_{n} is set to 0.2 with the assumption of super-Gaussian noise (this corresponds to the real noises used in Sect. 13.5), the input SNR is set to 10, 5, or 0 dB, and the noise scale parameter θ_{n} is uniquely determined by (13.49) and the previous parameter settings. The TFR is set from 0.4 to 1.0 ($|h_1(f)|$ is fixed to 1.0). Note that the TFR is highly correlated to the room reverberation and the interelement spacing of the microphone array; we determined the range of the TFR by simulating a typical moderately reverberant room and the array with 2.15 cm interelement spacing used in Sect. 13.5 (see the example of the TFR in Fig. 13.8). For the internal parameters used in iterative BSSA in this simulation, β and η are 8.5 and 0.9, respectively, which satisfy the musical-noise-free condition. In addition, the smallest value on the horizontal axis is 3 dB in Fig. 13.7 because DS is still performed even when $i = 0$.

From Figs. 13.7a and b, which correspond to relatively high input SNRs, we can confirm that the degree of point-source-likeness, i.e., $\mathrm{COS}(f)$, is almost maintained when the TFR is close to 1 even if the speech components are distorted by iterative BSSA. Also, it is worth mentioning that the degree of point-source-likeness is still

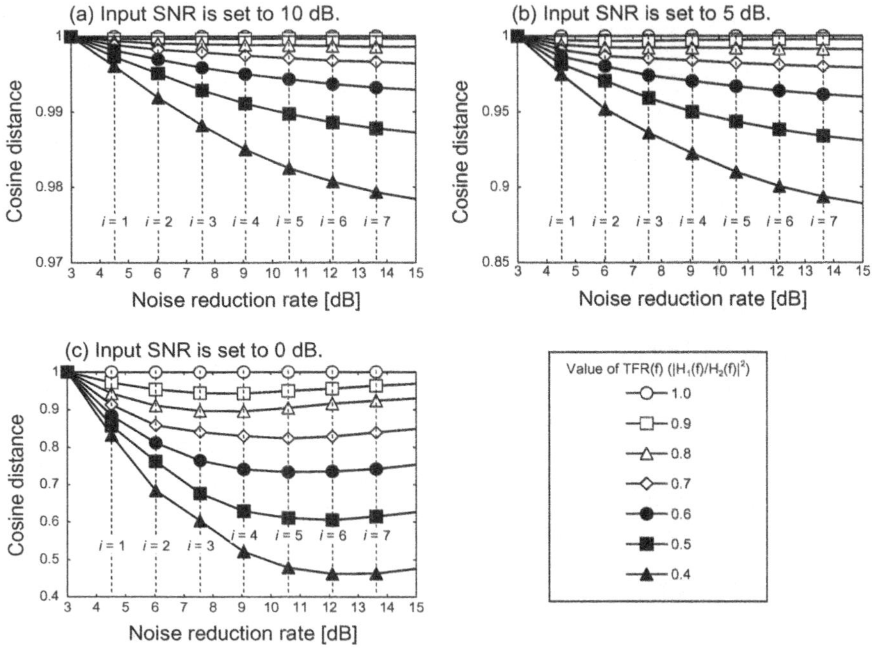

Fig. 13.7 Relation between number of iterations of iterative BSSA and cosine distance. Input SNR is **a** 10 dB, **b** 5 dB, and **c** 0 dB

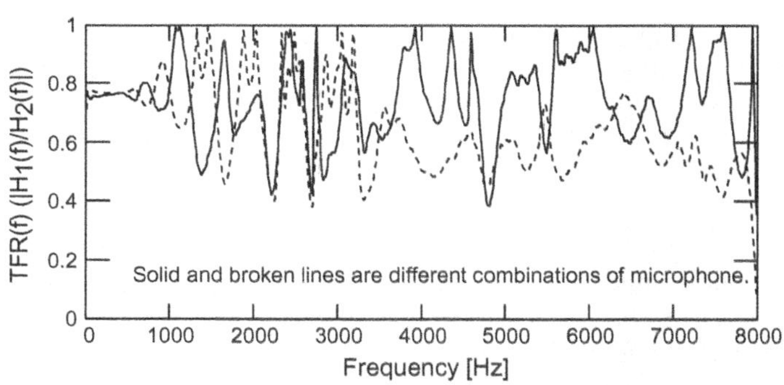

Fig. 13.8 Typical examples of TFR(f) $(|H_1(f)/H_2(f)|^2)$ in each frequency subband

above 0.9 even when the TFR is decreased to 0.4 and i is increased to 6. This means that almost 90% of the speech components can be regarded as a point source and thus can be blocked by ICA. In contrast, from Fig. 13.7c, which shows the case of a low input SNR, when the TFR is dropped to 0.4 and i is more than 3, the degree of point-source-likeness is lower than 0.6. Thus, less than 60% of the speech components can be regarded as a point source, and this leads to poor noise estimation.

13.4 Improvement Scheme for Poor Noise Estimation

13.4.1 Channel Selection in Independent Component Analysis

In this subsection, we introduce a channel selection strategy in ICA for achieving high accuracy of noise estimation. As mentioned previously, speech distortion is subjected to $\mathrm{ISNR}(f)$ and $\mathrm{TFR}(f)$, and the accuracy of noise estimation is degraded along with speech distortion. Figure 13.8 shows typical examples of the TFR. From Fig. 13.8, we can confirm that the TFRs in different combinations of microphones are not the same in each frequency subband; at a specific frequency, one microphone pair has higher $\mathrm{TFR}(f)$ than another pair, and vice versa at another frequency. Thus, we are able to select an appropriate combination of microphones to obtain a higher TFR.

Therefore, we introduce the channel selection method into ICA in each frequency subband, where we automatically choose less varied inputs to maintain high accuracy of noise estimation. Hereafter, we describe the detail of the channel selection method. First, we calculate the average power of the observed signal $X_k(f,\tau)$ at the kth channel as

$$\mathrm{E}_\tau[|X_k(f,\tau)|^2]=\mathrm{E}_\tau[|S(f,\tau)|^2]|H_k(f)|^2+\mathrm{E}_\tau[|N_k(f,\tau)|^2]. \qquad (13.53)$$

Here, $\mathrm{E}_\tau[|S(f,\tau)|^2]$ is a constant, and if we assume a diffuse noise field, $\mathrm{E}_\tau[|N_k(f,\tau)|^2]$ is also a constant. Thus, we can estimate the relative order of $|H_k(f)|^2$ by comparing (13.53) for every k.

Next, we sort $\mathrm{E}_\tau[|X_k(f,\tau)|^2]$ in descending order and select the channels corresponding to a high amplitude of $|H_k(f)|^2$ satisfying the following condition:

$$\max_k \mathrm{E}_\tau[|X_k(f,\tau)|^2]\cdot\xi \le \mathrm{E}_\tau[|X_k(f,\tau)|^2], \qquad (13.54)$$

where $\xi(<1)$ is the threshold for the selection.

Finally, we perform noise estimation based on ICA using the selected channels in each frequency subband, and we apply the projection back operation to remove the ambiguity of the amplitude and construct the estimated noise signal.

13.4.2 Time-Variant Noise Power Spectral Density Estimator

In the previous section, we revealed that the speech components cannot be regarded as a point source, and this leads to poor noise estimation in iterative BSSA. To solve this problem, we introduce a time-variant noise PSD estimator [24] instead of ICA to improve the noise estimation accuracy. This method has been developed for future high-end binaural hearing aids and performs a prediction of the left noisy signal from the right noisy signal via the Wiener filter, followed by an auto-PSD of the difference between the left noisy signal and the prediction. By applying the noise PSD estimated from this estimator to (13.35), we can perform speech extraction. The procedure of this noise PSD estimator is described in Appendix.

13.5 Experiments in Real World

13.5.1 Experimental Conditions

We conducted objective and subjective evaluation experiments to confirm the validity of iterative BSSA under the diffuse and nonstationary noise condition. The size of the experimental room was $4.2 \times 3.5 \times 3.0$ m^3 and the reverberation time was approximately 200 ms. We used a two-, three-, or four-element microphone array with an interelement spacing of 2.15 cm, and the direction of the target speech was set to be normal to the array. All the signals used in this experiment were sampled at 16 kHz with 16-bit accuracy. The DFT size was 1024, and the frame shift length was 256. We used 5 male and 5 female speakers (one utterance per speaker) as sources of the original target speech signal. The input SNR was -5, 0, 5, and 10 dB.

13.5.2 Objective Evaluation

We conducted an objective experimental evaluation under the same NRR condition. First, Figs. 13.9, 13.10, 13.11, and 13.12 show the kurtosis ratio and cepstral distortion obtained from the experiments with real traffic noise and railway station noise, where we evaluate 10-dB NRR (i.e., output SNRs = 5, 10, 15, and 20 dB) signals processed by five conventional methods, namely, the MMSE-STSA estimator, the Log MMSE estimator incorporating speech-presence uncertainty [55], single-channel musical-noise-free iterative spectral subtraction, the multichannel speech enhancement method integrating the minimum variance beamformer and the Log MMSE estimator for postfiltering, and BSSA, in addition to several types of iterative BSSAs (using ICA or a time-variant noise estimator with/without channel selection). Here, we did not apply the channel selection method to the two-microphone case because ICA or time-variant noise estimation requires at least two-channel signals. Also, we applied a minimum statistics noise PSD estimator [4] to the MMSE STSA estimator and musical-noise-free iterative spectral subtraction, and we use the decision-directed approach for a priori SNR estimation in the MMSE STSA estimator and the log MMSE estimator. From Figs. 13.9 and 13.11, we can confirm that the iterative BSSA methods outperform the MMSE STSA estimator, the Log MMSE estimator, and the conventional BSSA in terms of kurtosis ratio. In particular, the

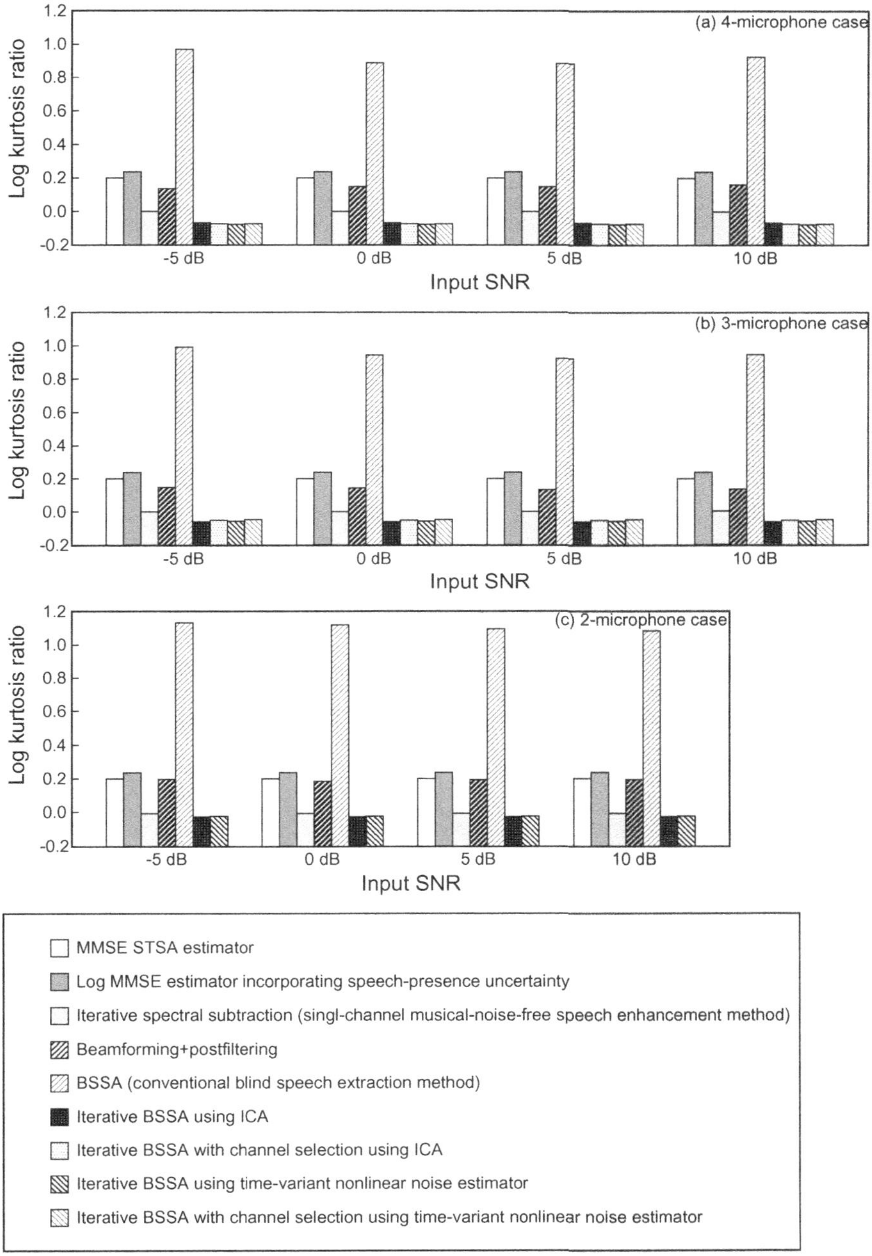

Fig. 13.9 Kurtosis ratio obtained from experiment for traffic noise under 10-dB NRR condition

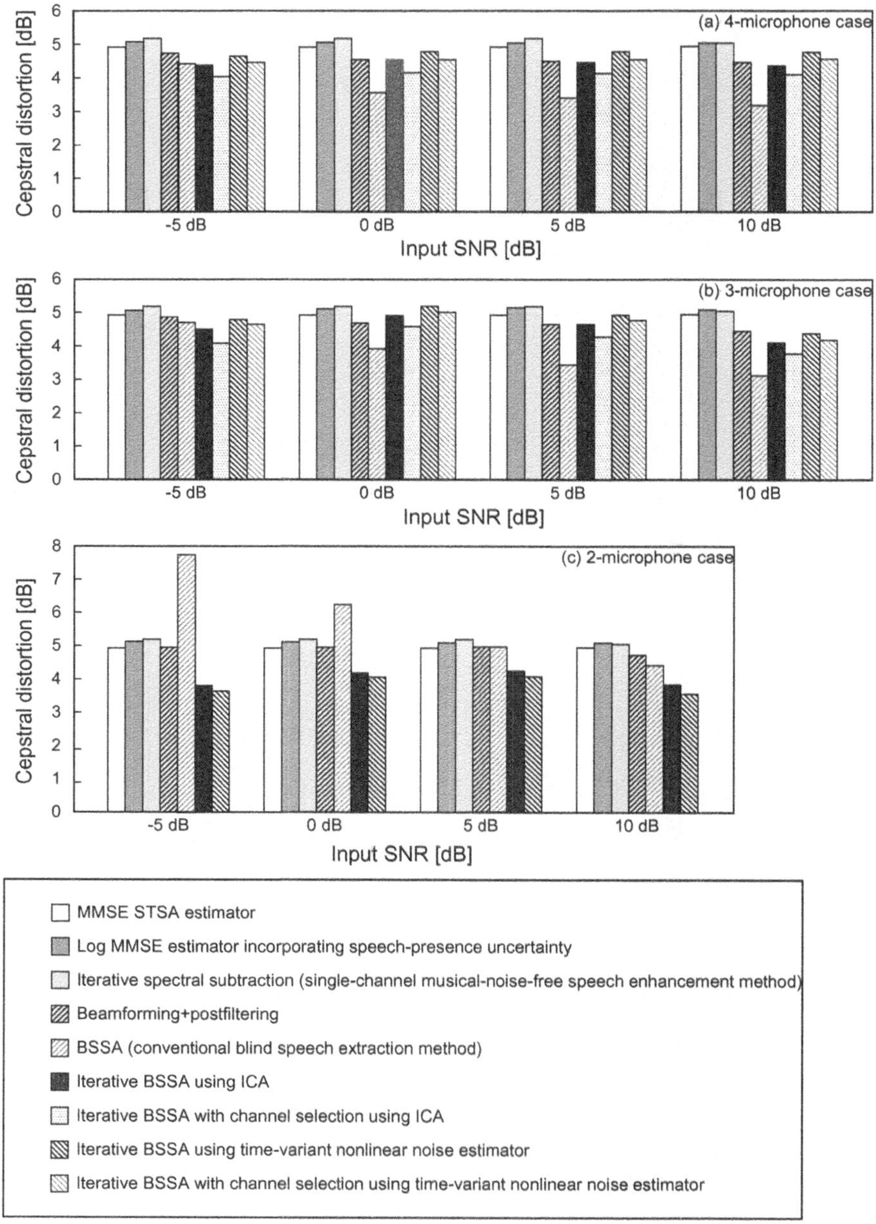

Fig. 13.10 Cepstral distortion obtained from experiment for traffic noise under 10-dB NRR condition

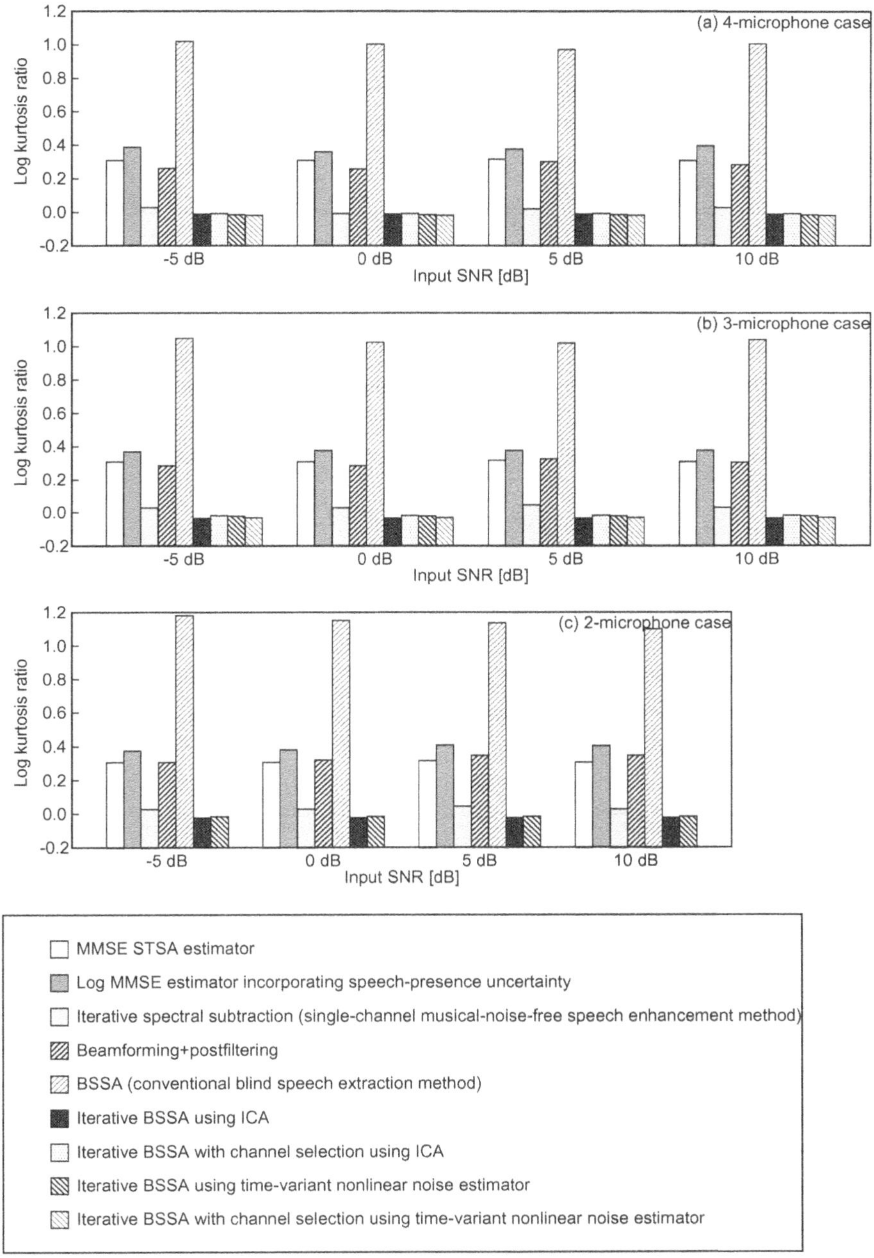

Fig. 13.11 Kurtosis ratio obtained from experiment for railway station noise under 10-dB NRR condition

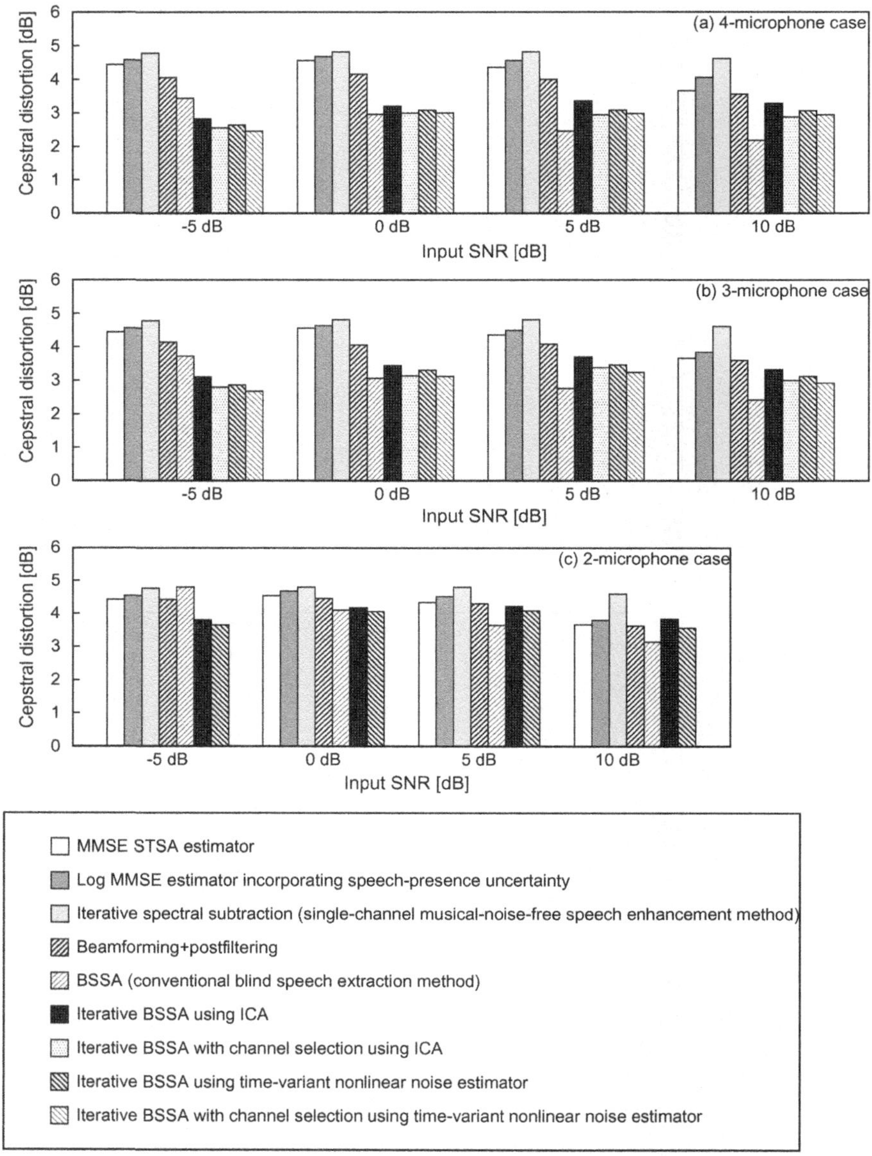

Fig. 13.12 Cepstral distortion obtained from experiment for railway station noise under 10-dB NRR condition

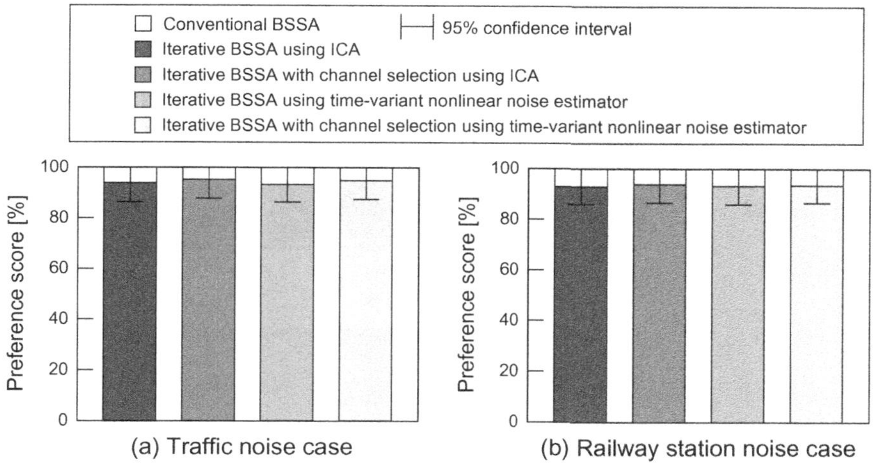

Fig. 13.13 Subjective evaluation results for **a** traffic noise and **b** railway station noise

kurtosis ratios of the iterative BSSA methods are mostly close to 1.0. This means that the iterative methods did not generate any musical noise. However, the iterative BSSA methods lead to greater speech distortion compared with the conventional BSSA (see Figs. 13.10 and 13.12). Therefore, a trade-off exists between the amount of musical noise generation and speech distortion in the conventional BSSA and iterative BSSA methods. This result implies the disadvantage of iterative BSSA, i.e., large speech distortion, which has been theoretically predicted in Sect. 13.3.3. However, since the speech distortion of the proposed iterative BSSA with channel selection is lower than that of the original iterative BSSA, we can confirm the validity of the channel selection method.

13.5.3 Subjective Evaluation

Since we found the above-mentioned trade-off, we next conducted a subjective evaluation for setting the performance competition. In the evaluation, we presented a pair of 10-dB NRR signals processed by the conventional BSSA and four of iterative BSSAs (using ICA or a time-variant noise estimator with/without channel selection) in random order to 10 examinees, who selected which signal they preferred from the viewpoint of total sound quality, e.g., less musical noise, less speech distortion, and so forth.

The result of this experiment is shown in Fig. 13.13 for (a) traffic noise and (b) railway station noise. It is found that the output signals of some iterative BSSAs are preferred to that of the conventional BSSA, indicating the higher sound quality of the iterative methods in terms of human perception. This result is plausible because

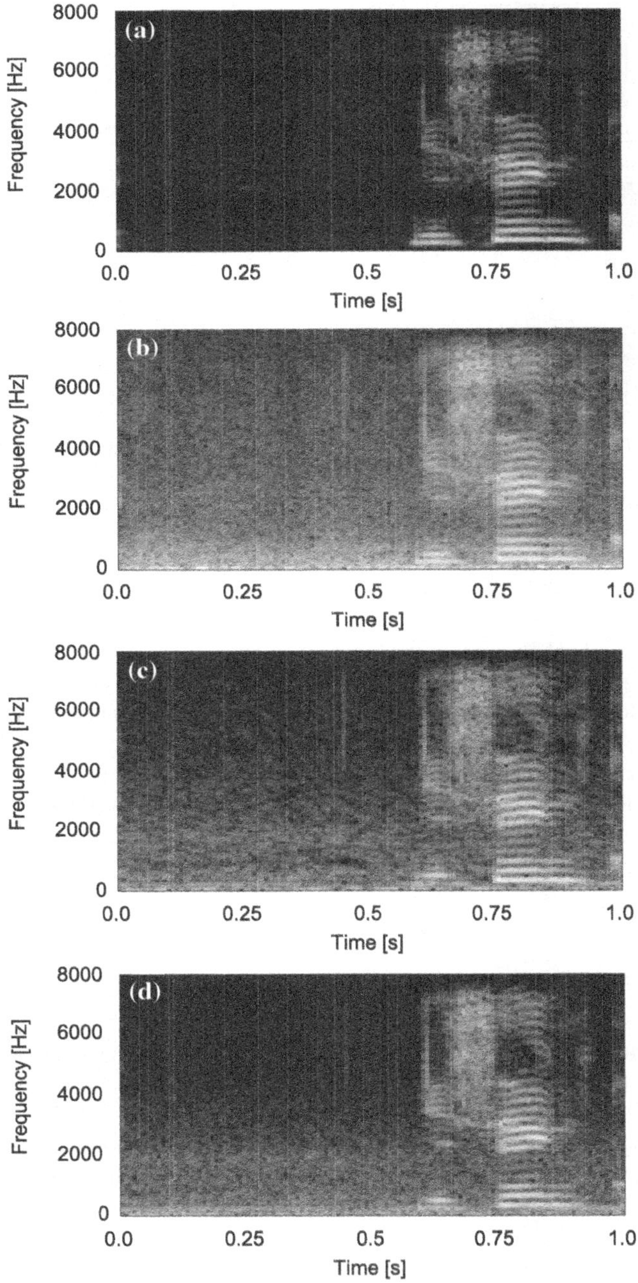

Fig. 13.14 Spectrogram for **a** clean signal, **b** observed signal, **c** signal extracted by BSSA, and **d** signal extracted by iterative BSSA

humans are often more sensitive to musical noise than to speech distortion as indicated in past studies, e.g., [12].

To visually understand the above-mentioned result, Fig. 13.14 shows part of the spectrograms of (a) a clean signal, (b) an observed signal, (c) a signal extracted by BSSA, and (d) a signal extracted by iterative BSSA, where the input SNR is set to 5 dB with real traffic noise and the NRR is 10 dB. From Fig. 13.14, it is confirmed that iterative BSSA reduces the number of isolated components in time-frequency domain sequences, which is a factor contributing to musical noise, compared with BSSA. Also, there are no major differences in the speech components of the clean signal, the signal processed by BSSA, and the signal processed by iterative BSSA; thus, we can conclude that the intelligibility of iterative BSSA is no less than that of BSSA.

13.6 Conclusions and Remarks

In this chapter, we addressed a musical-noise-free blind speech extraction method using a microphone array that can be applied to nonstationary noise. Firstly, we introduced iterative BSSA using a new BSSA structure, which generates almost no musical noise even with increasing noise reduction performance.

Secondly, in relation to the method, we discussed the justification of applying ICA to signals nonlinearly distorted by SS. We theoretically clarified that the degradation in ICA-based noise estimation obeys an amplitude variation in room transfer functions between the target user and microphones. Therefore, we gave the introduction of a channel selection strategy in ICA and a time-variant noise PSD estimator to improve the noise estimation accuracy.

Finally, from the objective evaluation experiments, we confirmed a trade-off between the amount of musical noise generation and speech distortion in the conventional and iterative BSSA. However, in a subjective preference test, iterative BSSA obtained a higher preference score than the conventional BSSA. Thus, iterative BSSA is advantageous to the conventional BSSA in terms of sound quality.

Implementation on a small hardware still receives much attention in industrial applications. Due to the limitation of space, however, the authors skip the discussion on this issue. Instead, several studies [56–60] have dealt with the issue of real-time implementation of ICA and BSSA, which would be helpful for the readers.

Acknowledgements This work was partially supported by SECOM Science and Technology Foundation.

Appendix

This appendix provides a brief review of the time-variant nonlinear noise estimator. For more detailed information, Ref. [24] is available.

Let $X_1(f,\tau)$ and $X_2(f,\tau)$ be noisy signals received at the microphones in the time-frequency domain, defined as

$$X_1(f,\tau) = H_1(f)S(f,\tau) + N_1(f,\tau), \tag{13.55}$$

$$X_2(f,\tau) = H_2(f)S(f,\tau) + N_2(f,\tau), \tag{13.56}$$

where $H_1(f)$ and $H_2(f)$ are the transfer functions from the target signal position to each microphone. Next, the auto-power PSDs in each microphone, $\Gamma_{11}(f)$ and $\Gamma_{22}(f)$, can be expressed as follows:

$$\Gamma_{11}(f,\tau) = |H_1(f)|^2\Gamma_{\mathrm{SS}}(f,\tau) + \Gamma_{\mathrm{NN}}(f,\tau), \tag{13.57}$$

$$\Gamma_{22}(f,\tau) = |H_2(f)|^2\Gamma_{\mathrm{SS}}(f,\tau) + \Gamma_{\mathrm{NN}}(f,\tau), \tag{13.58}$$

where $\Gamma_{\mathrm{SS}}(f,\tau)$ is the PSD of the target speech signal and $\Gamma_{\mathrm{NN}}(f,\tau)$ is the PSD of the noise signal. In this chapter, we assume that the left and right noise PSDs are approximately the same, i.e., $\Gamma_{\mathrm{N_1N_1}}(f,\tau) \simeq \Gamma_{\mathrm{N_2N_2}}(f,\tau) \simeq \Gamma_{\mathrm{NN}}(f,\tau)$.

Next, we consider the Wiener solution between the left and right transfer functions, which is defined as

$$H_{\mathrm{W}}(f,\tau) = \frac{\Gamma_{12}(f,\tau)}{\Gamma_{22}(f,\tau)}, \tag{13.59}$$

where $\Gamma_{12}(f)$ is the cross-PSD between the left and right noisy signals. The cross-PSD expression then becomes

$$\Gamma_{12}(f,\tau) = \Gamma_{\mathrm{SS}}(f,\tau)H_1(f)H_2^*(f). \tag{13.60}$$

Therefore, substituting (13.60) into (13.59) yields

$$H_{\mathrm{W}}(f,\tau) = \frac{\Gamma_{\mathrm{SS}}(f,\tau)H_1(f)H_2^*(f)}{\Gamma_{22}(f,\tau)}. \tag{13.61}$$

Furthermore, using (13.57) and (13.58), the squared magnitude response of the Wiener solution in (13.61) can also be expressed as

$$|H_{\mathrm{W}}(f,\tau)|^2 = \frac{(\Gamma_{11}(f,\tau) - \Gamma_{\mathrm{NN}}(f,\tau))(\Gamma_{22}(f,\tau) - \Gamma_{\mathrm{NN}}(f,\tau))}{\Gamma_{22}^2(f,\tau)}. \tag{13.62}$$

Equation (13.62) is rearranged into the following quadratic equation:

$$\Gamma_{\mathrm{NN}}^2(f,\tau) - \Gamma_{\mathrm{NN}}(f,\tau)\,(\Gamma_{11}(f,\tau) + \Gamma_{22}(f,\tau)) + \Gamma_{\mathrm{EE}}(f,\tau)\Gamma_{22}(f,\tau) = 0, \tag{13.63}$$

where

$$\Gamma_{\mathrm{EE}}(f,\tau) = \Gamma_{11}(f,\tau) - \Gamma_{22}(f,\tau)|H_{\mathrm{W}}(f)|^2. \tag{13.64}$$

Consequently, the noise PSD $\Gamma_{\mathrm{NN}}(f)$ can be estimated by solving the quadratic equation in (13.63) as follows:

$$\Gamma_{\mathrm{NN}}(f,\tau) = \frac{1}{2}(\Gamma_{11}(f,\tau) + \Gamma_{22}(f,\tau)) - \Gamma_{\mathrm{avg}}(f,\tau), \tag{13.65}$$

$$\Gamma_{\mathrm{avg}}(f,\tau) = \frac{1}{2}\{(\Gamma_{11}(f,\tau) + \Gamma_{22}(f,\tau))^2 - 4\Gamma_{\mathrm{EE}}(f,\tau)\Gamma_{22}(f,\tau)\}^{0.5}. \tag{13.66}$$

References

1. S.F. Boll, Suppression of acoustic noise in speech using spectral subtraction. IEEE Trans. Acoust. Speech Signal Process. **27**(2), 113–120 (1979)
2. M. Berouti, R. Schwartz, J. Makhoul, Enhancement of speech corrupted by acoustic noise, in *Proceeding of ICASSP* (1979), pp. 208–211
3. R. McAulay, M. Malpass, Speech enhancement using a soft-decision noise suppression filter. IEEE Trans. Acoust. Speech Signal Process. **28**(2), 137–145 (1980)
4. R. Martin, Spectral subtraction based on minimum statistics, in *Proceeding of EUSIPCO* (1994), pp. 1182–1185
5. P.C. Loizou, *Speech Enhancement Theory and Practice* (CRC Press, Taylor & Francis Group FL, 2007)
6. Y. Ephraim, D. Malah, Speech enhancement using a minimum mean-square error short-time spectral amplitude estimator. IEEE Trans. Acoust. Speech Signal Process. **32**(6), 1109–1121 (1984)
7. Y. Ephraim, D. Malah, Speech enhancement using a minimum mean square error log-spectral amplitude estimator. IEEE Trans. Acoust. Speech Signal Process. **33**(2), 443–445 (1985)
8. T. Lotter, P. Vary, Speech enhancement by MAP spectral amplitude estimation using a super-Gaussian speech model. EURASIP J. Appl. Signal Process. **2005**, 1110–1126 (2005)
9. O. Cappe, Elimination of the musical noise phenomenon with the Ephraim and Malah noise suppressor. IEEE Trans. Speech Audio Process. **2**(2), 345–349 (1994)
10. Z. Goh, K.-C. Tan, B. Tan, Postprocessing method for suppressing musical noise generated by spectral subtraction. IEEE Trans. Speech Audio Process. **6**(3), 287–292 (1998)
11. Y. Uemura, Y. Takahashi, H. Saruwatari, K. Shikano, K. Kondo, Automatic optimization scheme of spectral subtraction based on musical noise assessment via higher-order statistics, in *Proceeding of IWAENC* (2008)
12. Y. Uemura, Y. Takahashi, H. Saruwatari, K. Shikano, K. Kondo, Musical noise generation analysis for noise reduction methods based on spectral subtraction and MMSE STSA estimation, in *Proceeding of ICASSP* (2009), pp. 4433–4436
13. Y. Takahashi, R. Miyazaki, H. Saruwatari, K. Kondo, Theoretical analysis of musical noise in nonlinear noise reduction based on higher-order statistics, in *Proceeding of APSIPA Annual Summit and Conference* (2012)

14. K. Yamashita, S. Ogata, T. Shimamura, Spectral subtraction iterated with weighting factors, in *Proceeding of IEEE Speech Coding Workshop* (2002), pp. 138–140
15. M.R. Khan, T. Hansen, Iterative noise power subtraction technique for improved speech quality, in *Proceeding of ICECE* (2008), pp. 391–394
16. S. Li, J.-Q. Wang, M. Niu, X.-J. Jing, T. Liu, Iterative spectral subtraction method for millimeter-wave conducted speech enhancement. J. Biomed. Sci. Eng. **2010**(3), 187–192 (2010)
17. T. Inoue, H. Saruwatari, Y. Takahashi, K. Shikano, K. Kondo, Theoretical analysis of iterative weak spectral subtraction via higher-order statistics, in *Proceeding of IEEE International Workshop on Machine Learning for Signal Processing* (2010), pp. 220–225
18. R. Miyazaki, H. Saruwatari, T. Inoue, Y. Takahashi, K. Shikano, K. Kondo, Musical-noise-free speech enhancement based on optimized iterative spectral subtraction. IEEE Trans. Audio Speech Lang. Process. **20**(7), 2080–2094 (2012)
19. R. Miyazaki, H. Saruwatari, S. Nakamura, K. Shikano, K. Kondo, J. Blanchette, M. Bouchard, Musical-noise-free blind speech extraction integrating microphone array and iterative spectral subtraction. Signal Process. (Elsevier) **102**, 226–239 (2014)
20. P. Comon, Independent component analysis, a new concept? Signal Process. (Elsevier) **36**, 287–314 (1994)
21. S. Araki, R. Mukai, S. Makino, T. Nishikawa, H. Saruwatari, The fundamental limitation of frequency domain blind source separation for convolutive mixtures of speech. IEEE Trans. Speech Audio Process. **11**(2), 109–116 (2003)
22. H. Sawada, R. Mukai, S. Araki, S. Makino, A robust and precise method for solving the permutation problem of frequency-domain blind source separation. IEEE Trans. Speech Audio Process. **12**(5), 530–538 (2004)
23. H. Saruwatari, T. Kawamura, T. Nishikawa, A. Lee, K. Shikano, Blind source separation based on a fast-convergence algorithm combining ICA and beamforming. IEEE Trans. Audio Speech Lang. Process. **14**(2), 666–678 (2006)
24. A. Homayoun, M. Bouchard, Improved noise power spectrum density estimation for binaural hearing aids operating in a diffuse noise field environment. IEEE Trans. Audio Speech Lang. Process. **17**(4), 521–533 (2009)
25. T. Inoue, H. Saruwatari, Y. Takahashi, K. Shikano, K. Kondo, Theoretical analysis of musical noise in generalized spectral subtraction based on higher order statistics. IEEE Trans. Audio Speech Lang. Process. **19**(6), 1770–1779 (2011)
26. H. Yu, T. Fingscheidt, A figure of merit for instrumental optimization of noise reduction algorithms, in *Proceeding of DSP in Vehicles* (2011)
27. H. Yu, T. Fingscheidt, Black box measurement of musical tones produced by noise reduction systems, in *Proceeding of ICASSP* (2012), pp. 4573–4576
28. S. Kanehara, H. Saruwatari, R. Miyazaki, K. Shikano, K. Kondo, Theoretical analysis of musical noise generation in noise reduction methods with decision-directed a priori SNR estimator, in *Proceeding of IWAENC* (2012)
29. S. Kanehara, H. Saruwatari, R. Miyazaki, K. Shikano, K. Kondo, Comparative study on various noise reduction methods with decision-directed a priori SNR estimator via higher-order statistics, in *Proceeding of APSIPA Annual Summit and Conference* (2012)
30. R. Miyazaki, H. Saruwatari, K. Shikano, K. Kondo, Musical-noise-free speech enhancement based on iterative Wiener filtering, in *Proceeding of IEEE International Symposium on Signal Processing and Information Technology* (2012)
31. S. Nakai, H. Saruwatari, R. Miyazaki, S. Nakamura, K. Kondo, Theoretical analysis of biased MMSE short-time spectral amplitude estimator and its extension to musical-noise-free speech enhancement, in *Proceeding of Hands-Free Speech Communication and Microphone Arrays* (2014)
32. H. Saruwatari, Statistical-model-based speech enhancement with musical-noise-free properties, in *Proceeding of IEEE International Conference on Digital Signal Processing* (2015), pp. 1201–1205
33. A. Hiroe, Solution of permutation problem in frequency domain ICA using multivariate probability density functions, in *Proceeding of ICA* (2006), pp. 601–608

34. T. Kim, H.T. Attias, S.-Y. Lee, T.-W. Lee, Blind source separation exploiting higher-order frequency dependencies. IEEE Trans. Audio Speech Lang. Process. **15**(1), 70–79 (2007)
35. N. Ono, Stable and fast update rules for independent vector analysis based on auxiliary function technique, in *Proceeding of WASPAA* (2011), pp. 189–192
36. D. Kitamura, N. Ono, H. Sawada, H. Kameoka, H. Saruwatari, Relaxation of rank-1 spatial constraint in overdetermined blind source separation, in *Proceeding of EUSIPCO* (2015), pp. 1271–1275
37. D. Kitamura, N. Ono, H. Sawada, H. Kameoka, H. Saruwatari, Determined blind source separation unifying independent vector analysis and nonnegative matrix factorization. IEEE/ACM Trans. Audio Speech Lang. Process. **24**(9), 1626–1641 (2016)
38. Y. Mitsui, D. Kitamura, S. Takamichi, N. Ono, H. Saruwatari, Blind source separation based on independent low-rank matrix analysis with sparse regularization for time-series activity, in *Proceeding of ICASSP* (2017), pp. 21–25
39. S. Mogami, D. Kitamura, Y. Mitsui, N. Takamune, H. Saruwatari, N. Ono, Independent low-rank matrix analysis based on complex Student's t-distribution for blind audio source separation, in *Proceeding of IEEE International Workshop on Machine Learning for Signal Processing* (2017)
40. F.D. Aprilyanti, J. Even, H. Saruwatari, K. Shikano, S. Nakamura, T. Takatani, Suppression of noise and late reverberation based on blind signal extraction and Wiener filtering. Acoust. Sci. Technol. **36**(4), 302–313 (2015)
41. H. Saruwatari, S. Kurita, K. Takeda, F. Itakura, T. Nishikawa, Blind source separation combining independent component analysis and beamforming. EURASIP J. Appl. Signal Process. **2003**, 1135–1146 (2003)
42. Y. Takahashi, T. Takatani, K. Osako, H. Saruwatari, K. Shikano, Blind spatial subtraction array for speech enhancement in noisy environment. IEEE Trans. Audio Speech Lang. Process. **17**(4), 650–664 (2009)
43. Y. Takahashi, H. Saruwatari, K. Shikano, K. Kondo, Musical-noise analysis in methods of integrating microphone array and spectral subtraction based on higher-order statistics. EURASIP J. Adv. Signal Process. **2010**(431347), 25 pages (2010)
44. H. Saruwatari, Y. Ishikawa, Y. Takahashi, T. Inoue, K. Shikano, K. Kondo, Musical noise controllable algorithm of channelwise spectral subtraction and adaptive beamforming based on higher-order statistics. IEEE Trans. Audio Speech Lang. Process. **19**(6), 1457–1466 (2011)
45. R. Miyazaki, H. Saruwatari, K. Shikano, Theoretical analysis of amounts of musical noise and speech distortion in structure-generalized parametric spatial subtraction array. IEICE Trans. Fundam. **95-A**(2), 586–590 (2012)
46. S. Kurita, H. Saruwatari, S. Kajita, K. Takeda, F. Itakura, Evaluation of blind signal separation method using directivity pattern under reverberant conditions, in *Proceeding of ICASSP*, vol. 5 (2000), pp. 3140–3143
47. J. Even, H. Saruwatari, K. Shikano, T. Takatani, Speech enhancement in presence of diffuse background noise: Why using blind signal extraction? in *Proceeding of ICASSP* (2010), pp. 4770–4773
48. J. Even, C. Ishi, H. Saruwatari, N. Hagita, Close speaker cancellation for suppression of non-stationary background noise for hands-free speech interface, in *Proceeding of INTERSPEECH* (2010), pp. 977–980
49. R. Prasad, H. Saruwatari, K. Shikano, Probability distribution of time-series of speech spectral components, IEICE Trans. Fundam. **E87-A**(3), 584–597 (2004)
50. R. Prasad, H. Saruwatari, K. Shikano, Estimation of shape parameter of GGD function by negentropy matching. Neural Process. Lett. **22**, 377–389 (2005)
51. T.H. Dat, K. Takeda, F. Itakura, Generalized gamma modeling of speech and its online estimation for speech enhancement, in *Proceeding of ICASSP*, vol. 4 (2005), pp. 181–184
52. I. Andrianakis, P.R. White, MMSE speech spectral amplitude estimators with chi and gamma speech priors, in *Proceeding of ICASSP* (2006), pp. III-1068–III-1071
53. R. Wakisaka, H. Saruwatari, K. Shikano, T. Takatani, Speech prior estimation for generalized minimum mean-square error short-time spectral amplitude estimator. IEICE Trans. Fundam. **95-A**(2), 591–595 (2012)

54. R. Wakisaka, H. Saruwatari, K. Shikano, T. Takatani, Speech kurtosis estimation from observed noisy signal based on generalized Gaussian distribution prior and additivity of cumulants, in *Proceeding of ICASSP* (2012), pp. 4049–4052
55. I. Cohen, Optimal speech enhancement under signal presence uncertainty using log-spectra amplitude estimator. IEEE Signal Process. Lett. **9**(4), 113–116 (2002)
56. H. Buchner, R. Aichner, W. Kellermann, A generalization of blind source separation algorithms for convolutive mixtures based on second-order statistics. IEEE Trans. Speech Audio Process. **13**(1), 120–134 (2005)
57. Y. Mori, H. Saruwatari, T. Takatani, S. Ukai, K. Shikano, T. Hiekata, Y. Ikeda, H. Hashimoto, T. Morita, Blind separation of acoustic signals combining SIMO-model-based independent component analysis and binary masking. EURASIP J. Appl. Signal Process. **2006**(34970), 17 pages (2006)
58. T. Hiekata, Y. Ikeda, T. Yamashita, T. Morita, R. Zhang, Y. Mori, H. Saruwatari, K. Shikano, Development and evaluation of pocket-size real-time blind source separation microphone. Acoust. Sci. Technol. **30**(4), 297–304 (2009)
59. Y. Omura, H. Kamado, H. Saruwatari, K. Shikano, Real-time semi-blind speech extraction with speaker direction tracking on Kinect, in *Proceeding of APSIPA Annual Summit and Conference* (2012)
60. Y. Bando, H. Saruwatari, N. Ono, S. Makino, K. Itoyama, D. Kitamura, M. Ishimura, M. Takakusaki, N. Mae, K. Yamaoka, Y. Matsui, Y.i Ambe, M. Konyo, S. Tadokoro, K. Yoshii, H.G. Okuno, Low-latency and high-quality two-stage human-voice-enhancement system for a hose-shaped rescue robot. J. Robot. Mechatron. **29**(1), 198–212 (2017)

Chapter 14
Audio-Visual Source Separation with Alternating Diffusion Maps

David Dov, Ronen Talmon and Israel Cohen

Abstract In this chapter we consider the separation of multiple sound sources of different types including multiple speakers and transients, which are measured by a single microphone and by a video camera. We address the problem of separating a particular sound source from all other sources focusing specifically on obtaining an underlying representation of it while attenuating all other sources. By pointing the video camera merely to the desired sound source, the problem becomes equivalent to extracting the common source to the audio and the video modalities while ignoring the other sources. We use a kernel-based method, which is particularly designed for this task, providing an underlying representation of the common source. We demonstrate the usefulness of the obtained representation for the activity detection of the common source and discuss how it may be further used for source separation.

14.1 Introduction

We consider a complex sound scene with multiple sound sources, which are highly non stationary and are of different types including speech and transients. The sources may share similar characteristics, e.g., multiple speakers with similar voices may be present in the scene. The sound sources are measured by a single microphone, so that no spatial information of the sources is available. The objective in this chapter is to extract a particular sound source from all other sources in the scene by providing an underlying representation of the source, in which all other *interfering sources* are

---

This research was supported by the Israel Science Foundation (grant no. 576/16).

D. Dov (✉) · R. Talmon · I. Cohen
Andrew and Erna Viterbi Faculty of Electrical Engineering,
The Technion-Israel Institute of Technology, 32000 Haifa, Israel
e-mail: davidd@tx.technion.ac.il

R. Talmon
e-mail: ronen@ee.technion.ac.il

I. Cohen
e-mail: icohen@ee.technion.ac.il

S. Makino (ed.), *Audio Source Separation*, Signals and Communication Technology, https://doi.org/10.1007/978-3-319-73031-8_14

attenuated. We address this goal by incorporating a video signal pointed merely at the particular sound source, so that the objective transforms to the problem of extracting the *common source* to the audio and the video signals. We use a data driven method termed "alternating diffusion maps" [1], which is particularly designed for extracting the common source via the fusion between the audio and the video signals.

Alternating diffusion maps belong to a class of kernel based geometric methods, which are originally designed for dimensionality reduction of single modal signals, i.e., those measured by a single sensor [2–6]. These methods are based on constructing an affinity kernel, which aggregates affinities (similarities) between samples of the signal, and they provide low dimensional representations through the eigenvalue decomposition of the affinity kernel. The kernel methods were recently extended to the processing of multimodal signals in [1, 7–18]. The multi-modal kernel methods are mostly based on constructing separate affinity kernels for each modality and fusing (incorporating) the modalities by a combination of the affinity kernels.

The fusion of audio visual signals is often addressed in previous studies for the analysis of speech signals. These studies focus on modeling the shape and the movement of the mouth by the design of video-based features such as key-points at the mouth area [19–21]; width, height and intensity levels of the mouth [22–24]; and motion vectors [25, 26]. Then, features of the audio and the video signals are concatenated into a single vector—a fusion approach often referred to as early fusion [27, 28]. The new vectors are then processed as data that is measured in a single sensor. Another common approach for the fusion of audio visual signals, often designed for specific tasks such as classification, is based on the combinations of measure functions constructed separately from each modality. For example in [29], we presented a method for voice activity detection, which is based on fusing two voice activity measures, constructed separately and similarly in specifically designed domains of the audio and the video signals.

In this chapter, we present a sound source separation approach based on the alternating diffusion maps method, in which audio and video signals are fused via the product between affinity kernels. Alternating diffusion maps differs from the common approaches for audio-visual fusion since it allows to consider complex relations between the modalities. The method has an interpretation of iterating between two diffusion steps, one for each modality such that the diffusion step in one modality attenuates factors, which are specific to the other modality, i.e., the interfering sources in our case. Therefore, the eigenvalue decomposition of the kernel product allows us to obtain an underlying representation of the signal, in which the interfering sources are attenuated.

We demonstrate the usefulness of the obtained representation for the activity detection of the common source by presenting a simple unsupervised detection algorithm. We apply the algorithm for the detection of different common sources. When the video camera is pointed at the face of a speaker, and the interfering sources are acoustic noises and transients, which are short term environmental sounds such as door-knocks [30–32], the algorithm operates as a voice activity detector [33]. We further demonstrate that a useful representation is obtained even when the common source is of the same type as the interfering sources such that they share similar acoustic characteristics. Specifically, the proposed approach provides good distinction between different speakers implying that the representation of the common source may be further used for the problem of speaker diarization (who spoke when). We note that the objective in this chapter of obtaining an underlying representation, in which the common source is separated from other sources, deviates from the classical problem of source separation. Yet, we show in this chapter the usefulness of the representation of the common source for applications, directly related to source separation, and discuss how this representation may be further used for source separation.

The remainder of this chapter is organized as follows. In Sect. 14.2, we formulate the problem. In Sect. 14.3, we describe the use of alternating diffusion maps for building a representation of the common source, in which the other sources are attenuated. Based on the new representation, we present an algorithm for the activity detection of the common source. Finally, we show experimental results in Sect. 14.4 demonstrating the usefulness of the representation of common sources for their activity detection, and discuss how to separate them from the interfering sources.

14.2 Problem Formulation

Consider an audio-visual signal measured by a single microphone and by a video camera. A common practice in the processing of audio signals is dividing them into consecutive time frames with a certain overlap. Here we consider N consecutive pairs of frames of the audio and the video signals:

$$(\mathbf{v}_1, \mathbf{w}_1), (\mathbf{v}_2, \mathbf{w}_2), \ldots, (\mathbf{v}_N, \mathbf{w}_N), \tag{14.1}$$

where $\mathbf{v}_n \in \mathbb{R}^{L_v}$ and $\mathbf{w}_n \in \mathbb{R}^{L_w}$ are feature representations of the nth audio and the nth video frame, and L_v and L_w are the corresponding number of the features, respectively. We note that throughout this chapter we follow the notations in [33], where v and w correspond to two different views, i.e., two different modalities—audio and video in our case. Specifically, the pair $(\mathbf{v}_n, \mathbf{w}_n)$ is assumed aligned. The particular selection of the features is described in Sect. 14.4.

The audio-visual signal is used to measure a certain sound source, i.e., the common source, which we denote by x, in the presence of other interfering sources. For simplicity of the formulation, we assume a single interfering sound source, which we denote by y, and accordingly, the audio signal is given by a certain function f of these sources:

$$\mathbf{v}_n = f(x, y). \tag{14.2}$$

Similarly, the video signal is given by a function g of the common source and of an interfering video source z:

$$\mathbf{w}_n = g(x, z).$$

We do not focus in this chapter on the interfering video sources and use in the experimental results relatively clean video signals. Yet, interferences such as non-speech mouth movements of a speaker may be naturally present in the recordings. The goal is to obtain a new representation of the measured signal according to the common source x while attenuating the interfering sources y and z:

$$\boldsymbol{\phi}_n = h\left(x\right),$$

where $\boldsymbol{\phi}_n \in \mathbb{R}^L$, is the new representation of the nth frame, and h is a function that maps the common source to the new representation. We note that throughout this chapter we do not define explicitly the functions f, g and h nor the sources x, y and z. These are merely used for an analysis that shows how to construct an underlying representation of the measured signal according to the common source x, i.e., $\boldsymbol{\phi}_n$, using the alternating diffusion maps method.

14.3 Separation of the Common Source via Alternating Diffusion Maps

14.3.1 Alternating Diffusion Maps

Alternating Diffusion maps [6] is a kernel based geometric method for multimodal data fusion, which is based on the construction of two affinity kernels, $\mathbf{K}^v \in \mathbb{R}^{N\times N}$ and $\mathbf{K}^w \in \mathbb{R}^{N\times N}$, from the audio and the video signal, respectively. The affinity kernels are real-valued symmetric matrices whose entries are in the range of $[0, 1]$. Here we describe the construction of $\mathbf{K}^v$; the construction of $\mathbf{K}^w$ is similar. We use a Gaussian kernel whose (n, m)th entry is denoted by $K^v_{n,m}$, and is given by:

$$K^v_{n,m} = \exp\left[-\frac{\|\mathbf{v}_n - \mathbf{v}_m\|^2}{\sigma_v^2}\right], \tag{14.3}$$

where $\|\cdot\|$ is the l_2 norm, and σ_v is a scaling parameter, which is also called the kernel bandwidth. The term $K^v_{n,m}$ defines a measure of affinity (similarity) between frames $\mathbf{v}_n, \mathbf{v}_m \in \mathbb{R}^{L_v}$, which are viewed as data points in a (high) L_v dimensional domain. The closer the data points are in the high dimensional domain, the higher the affinity between them. The kernel bandwidth σ_v defines a measure of locality between the data points such that when $\|\mathbf{v}_n - \mathbf{v}_m\|^2 \gg \sigma_v^2$ the affinity between frames $\mathbf{v}_n$ and $\mathbf{v}_m$ approaches zero. The affinity kernel $\mathbf{K}^v$ aggregates local relations (affinities) between each pair of the N audio frames, which is also often referred to as the geometry of the data. In addition, the affinity kernel has an interpretation from graph theory such that the frames $\mathbf{v}_1, \mathbf{v}_2, ..., \mathbf{v}_N$ are viewed as nodes of a graph and the weight of the edge between frames $\mathbf{v}_n$ and $\mathbf{v}_m$ is given by $K^v_{n,m}$.

We note that the choice of the affinity kernel and in particular the metric function between the frames has an important implication on the obtained representation. In [34] we suggested using a modified version of the Mahalanobis distance based on short term temporal statistics of the measured signal, instead of the l_2 norm used here [35]. To analyze the metric based on the Mahalanobis distance, we suggested an underlying model of speech and transient sound sources, and showed that this metric improves the separation between them.

Next, we construct a normalized version of the affinity kernel, $\mathbf{M}^v \in \mathbb{R}^{N \times N}$, by normalizing the rows of $\mathbf{K}^v$:

$$\mathbf{M}^v = \left(\mathbf{D}^v\right)^{-1} \mathbf{K}^v, \tag{14.4}$$

where $\mathbf{D}^v \in \mathbb{R}^{N \times N}$ is a diagonal matrix whose nth element on the diagonal, denoted by $D^v_{n,n}$, is given by:

$$D^v_{n,n} = \sum_{m=1}^{N} K^v_{n,m}. \tag{14.5}$$

The matrix $\mathbf{M}^v$ is a row stochastic matrix and it has an interpretation of a transition probabilities matrix of a random walker. Specifically, $M^v_{n,m}$ is viewed as the probability of transitioning from node n to node m on the graph.

Similarly to $\mathbf{M}^v$, let $\mathbf{M}^w$ be a row normalized affinity kernel obtained from the video data according to (14.3)–(14.5). The audio and the video data are fused by constructing a unified affinity kernel $\mathbf{M} \in \mathbb{R}^{N \times N}$ via the product of the (normalized) affinity kernels:

$$\mathbf{M} = \mathbf{M}^v \cdot \mathbf{M}^w. \tag{14.6}$$

The matrix $\mathbf{M}$ is also row stochastic, and it aggregates the affinities between the frames of the signal in the two modalities. Specifically, its (n, m)th entry is given by the rule of matrix product by:

$$M_{n,m} = \sum_{l=1}^{N} M^v_{n,l} M^w_{l,m}.$$

The matrix $\mathbf{M}$ may be viewed as defining a random walk on a graph whose nodes are the pairs of frames in (14.1). Accordingly, the probability of transitioning from node n to node m on this graph is given by the probability of transitioning from node n to any other point l in the graph constructed from the audio signal in one step, and then transitioning from node l to node m in the video-related graph in the second step.

14.3.2 Separation of the Common Source

The motivation to fuse the audio and the video signals via the product of affinity kernels in (14.6) for source separation lies in the analysis of the continuous counterpart of $\mathbf{M}$, which is considered as a diffusion operator [6]. Lederman and Talmon [1] showed that the continuous counterpart of $\mathbf{M}$ is an operator whose application is equivalent to applying two diffusion operators, one for each modality. They further showed that the application of a diffusion operator in one modality integrates out factors, which are specific to the other modality, i.e., the interfering sources in our case. This implies that while the unified affinity kernel $\mathbf{M}$ is constructed from the two modalities, i.e., from the sources x, y and z, it is equivalent to a diffusion operator, which is effectively a function of merely the common source x.

A representation of the common source is obtained through the eigenvalue decomposition of $\mathbf{M}$. Let $\lambda_0, \lambda_1, ...\lambda_L$ be the $L+1$ largest eigenvectors of $\mathbf{M}$ arranged in a decreasing order and let $\boldsymbol{\theta}_0, \boldsymbol{\theta}_1, ..., \boldsymbol{\theta}_L \in \mathbb{R}^N$ be their corresponding eigenvectors. Since the matrix $\mathbf{M}$ is row stochastic, the first eigenvector $\boldsymbol{\theta}_0$ is an all ones vector, so it is discarded along with the corresponding eigenvalue since they do not contain information [6]. Using the other L eigenvalues and eigenvectors we construct a matrix $\boldsymbol{\Theta} \in \mathbb{R}^{N \times L}$:

$$\boldsymbol{\Theta} = (\lambda_1 \boldsymbol{\theta}_1, \lambda_2 \boldsymbol{\theta}_2, ..., \lambda_L \boldsymbol{\theta}_L). \tag{14.7}$$

Note that the nth entry of each eigenvector corresponds to the nth row of $\mathbf{M}$, and as a result to the nth pair of frames $(\mathbf{v}_n, \mathbf{w}_n)$. The columns of the matrix $\boldsymbol{\Theta}$ define a new coordinate system on the data such the new representation of the nth frame, $\boldsymbol{\phi}_n$, is given by the nth row of $\boldsymbol{\Theta}$:

$$\boldsymbol{\phi}_n = \left[\Theta_{n,1}, \Theta_{n,2}, ..., \Theta_{n,L}\right]^T, \tag{14.8}$$

where ^{T} denotes transpose. Since $\boldsymbol{\phi}_n$ is constructed from the unified affinity kernel $\mathbf{M}$, it is a representation of the measured signal according to the common source x, in which the interfering sources y and z are attenuated, i.e., $\boldsymbol{\phi}_n = h(x)$. We summarize the alternating diffusion maps method in Algorithm 1.

Algorithm 1 Alternating Diffusion Maps

1: Obtain N consecutive pairs of frames $\{\mathbf{v}_n, \mathbf{w}_n\}_{n=1}^N$ in (14.1)
2: Calculate the affinity kernels of the audio and the video modalities, $\mathbf{K}^v$ and $\mathbf{K}^w$, respectively, according to (14.3)
3: Using (14.4)-(14.5), calculate the normalized affinity kernels $\mathbf{M}^v$ and $\mathbf{M}^w$
4: Obtain the unified kernel $\mathbf{M}$ by fusing the modalities via the product of kernels in (14.6)
5: Obtain the first L eigenvalues $\lambda_1, \lambda_2, ..., \lambda_L$ and their corresponding eigenvectors $\boldsymbol{\theta}_1, \boldsymbol{\theta}_2, ..., \boldsymbol{\theta}_L$
6: Construct the matrix $\boldsymbol{\Theta}$ in (14.7)
7: Use the nth row in $\boldsymbol{\Theta}$ as the new representation $\boldsymbol{\phi}_n$ of frame n

14.3.3 Online Extension

The mapping of the audio-visual signal $(\mathbf{v}_n, \mathbf{w}_n) \rightarrow \boldsymbol{\phi}_n$ provides a representation separating the common source from all other sources. However, it is based on batch processing of N time frames, which has to be available in advance. Although this requirement poses a significant limitation in online applications, it has a relatively simple solution based on the Nyström method, which is used for the extension of eigenvectors to new entries [36]. First, N consecutive frames are collected in a batch manner and the corresponding eigenvectors $\boldsymbol{\theta}_1, \boldsymbol{\theta}_2, \ldots, \boldsymbol{\theta}_L$ are constructed. These N frames may be viewed as a calibration dataset. Then, given a new incoming point (frame) m, the new entry of the lth eigenvector is given by:

$$\theta_l(m) = \frac{1}{\lambda_l} \sum_{n=1}^{N} M_{m,n} \theta_l(n). \tag{14.9}$$

This type of extension may be interpreted as a weighted nearest neighbor interpolation such that the weight of point n in the interpolation is proportional to the probability of transitioning from the new point m to point n in the calibration dataset. Intuitively, a higher weight will be assigned to points in the interpolation, to which the random walker will jump from node m with high probability. According to (14.8) and (14.9), the new representation of the frame m may be efficiently calculated via the product:

$$\boldsymbol{\phi}_m^T = \left[M_{m,1}, M_{m,2}, \ldots, M_{m,N}\right] \cdot \boldsymbol{\Theta}. \tag{14.10}$$

14.3.4 Source Activity Detection

As an application of the obtained representation $\boldsymbol{\phi}_n$, in which the common source is separated from all other sources, we consider in Sect. 14.4 the task of the activity detection of the common source. Let $\mathcal{H}_0^x$ and $\mathcal{H}_1^x$ be two hypotheses of the absence and the presence of the common source x, respectively, and let $\mathbb{1}_m^x$ be an indicator of the presence of source x in frame m, given by:

$$\mathbb{1}_m^x = \begin{Bmatrix} 1 \,; \, m \in \mathcal{H}_1^x \\ 0 \,; \, m \in \mathcal{H}_0^x \end{Bmatrix}.$$

The goal in the task of source activity detection is to estimate $\mathbb{1}_m^x$ for each frame m, i.e., to cluster the frames according to the activity of the common source.

As we show in Sect. 14.4, this task may be successfully addressed using the new representation; specifically, we propose using the first coordinate of $\boldsymbol{\phi}_m$ for clustering the frames:

$$\hat{\mathbb{1}}_m^x = \begin{Bmatrix} 1; \; \phi_m(1) \geq \tau \\ 0; \; \text{otherwise} \end{Bmatrix}, \tag{14.11}$$

where $\phi_m(1)$ is the first coordinate in $\boldsymbol{\phi}_m$, and τ is a threshold value selected according to a particular application at hand. In fact, $\hat{\mathbb{1}}_m^x$ is obtained based on the leading eigenvector $\boldsymbol{\theta}_1$ according to (14.7) and (14.8), which solves the well-known normalized cut problem, and is widely used for clustering [37]. Specifically, in [37], the sign of the eigenvector is used for binary clustering, while here we use the leading eigenvector to construct a continuous measure. In addition, here the eigenvector is constructed from the unified affinity kernel such that it represents the measured signal according to the common source, while the effect of the other sources is attenuated.

We note that in an online setting, the algorithm for the activity detection may be viewed as an unsupervised learning algorithm such that first N pairs of frames in the sequence are used as a training set without labels to construct the source activity measure for each training frame, i.e., $\phi_1(1), \phi_2(1), ..., \phi_N(1)$. Then, given a new incoming frame m, the level of the activity of the common source in the new frame, i.e., $\phi_m(1)$, is calculated using the training set and compared to a threshold for source activity detection. We summarize the algorithm for source activity detection in anonline setting in Algorithm 2.

Algorithm 2 Source Activity Detection

1: Obtain N consecutive pairs of frames $\{\mathbf{v}_n, \mathbf{w}_n\}_{n=1}^N$ in (14.1)
2: Using Algorithm 1, calculate the new representation $\{\boldsymbol{\phi}_1, \boldsymbol{\phi}_2, ..., \boldsymbol{\phi}_N\}$ of the signals, in which the common source is separated from the interfering sources

Given a new incoming frame m:
3: Calculate the interpolation weights $\{M_{m,1}, M_{m,2}, ..., M_{m,N}\}$
4: Using online extension, obtain the new representation of frame m, $\boldsymbol{\phi}_m$, according to (14.10)
5: **if** $\phi_m(1) > \tau$ **then**
  $\hat{\mathbb{1}}_m = 1$
else
  $\hat{\mathbb{1}}_n = 0$
end if

14.4 Experimental Results

14.4.1 Experimental Setup

We simulate a sound scene, in which multiple sound sources including speech, background noises and transients are active simultaneously. Specifically, the speech signals are taken from the audio-visual database we presented in [29], and the transients are taken from a free online corpus [38]. The speech signals and the transients are rescaled to have the same maximal amplitude. This is a common normalization used instead of, e.g., the ratio between the energies of the speech signal and the transients. Due to their short duration, transients often have low energy levels, which do not reflect properly the dominant effect of their presence in sound scenes. All signals are resampled to 8 kHz and are synthetically added to simulate the complex sound scene. The signals are divided into sequences of $\approx$60 s; in each sequence, there exists a single sound source which is also recorded by a video camera. The video signal has the resolution of 640×480 pixels and it is processed in $\approx$25 fps; to align between the video and the audio frames, we use audio frames comprising 634 samples.

We demonstrate the use of the obtained representation, in which the common source is separated from all other sources, for the activity detection of the common source. To evaluate the performance of the activity detection algorithm, we set the ground truth for the true activity of the common source using its clean recordings such that frames whose energy value is above 1% of the maximal energy value in the sequence are considered active. We note that this type of ground truth setting has a fine resolution of few tens of milliseconds, so it allows for evaluating the ability of the algorithm to detect very short active sequences with 50% overlap.

For the feature representation of the video signal, we use motion vectors, which are widely used for the representation of video recordings of speech [25, 26]. The motion vectors are calculated using the Lucas-Kanade method [39, 40] over blocks of 10×10 pixels. The feature representation of the nth video frame, $\mathbf{w}_n \in \mathbb{R}^{L_w}$, is given by a concatenation into a column vector of the absolute values of the velocities in each block of pixels in frames $n-1,\ n,\ n+1$. In this context, we note that the alternating diffusion maps method used to separate the common source from the other sources is a data driven method in the sense that the frames of the signal are viewed as high dimensional data points. As a result, it does not rely on a temporal information between the frames; to incorporate the temporal information, we use data from three consecutive frames $n-1,\ n,\ n+1$ for the representation of the nth frame. For the representation of the audio signal we use the perceptually meaningful Mel-Frequency Cepstral Coefficients (MFCC), which are widely used for speech and music processing applications [41–43]. The MFCCs are constructed from the responses of a filter bank with filters whose bandwidths increase proportionally to their central frequency imitating the human auditory system. Similarly to the video features, $\mathbf{v}_n \in \mathbb{R}^{L_w}$ is constructed by concatenating the MFCCs of frames $n-1,\ n,\ n+1$.

14.4.2 Activity Detection of the Common Source

In the first experiment, we consider two sound sources—speech of a single speaker and a single type of transient. The video camera is pointed at the face of the speaker such that speech is the common source x and the transient is the interfering source y in (14.2). An example of a typical video frame is presented in Fig. 14.1. In this setting, the algorithm for source activity detection performs as a voice detector. An example of the obtained detection of the common source is presented in Fig. 14.2, in which we refer to the source activity detection algorithm presented in Sect. 14.3 as the "proposed algorithm". We compare the proposed algorithm to an alternative kernel approach based on a different combination between the affinity kernels of

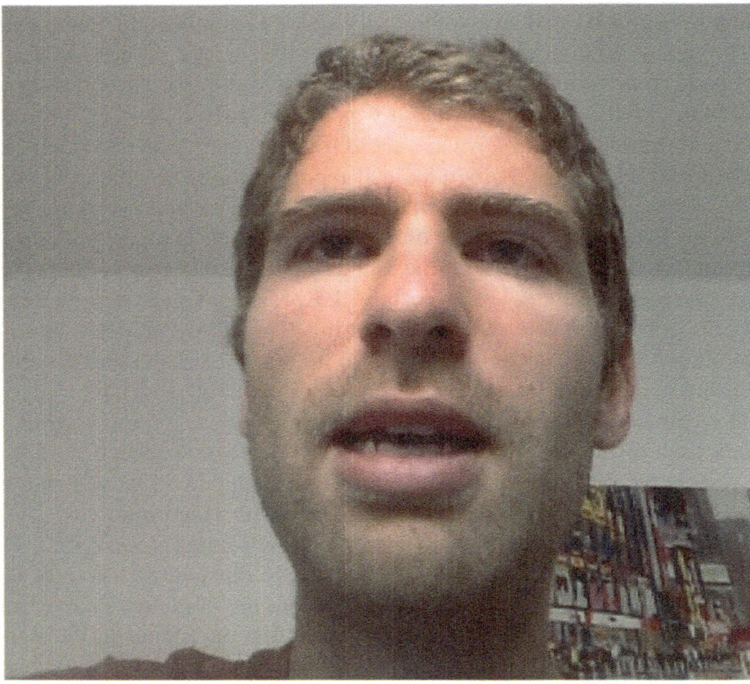

Fig. 14.1 An example of a video frame

each modality. Specifically, we consider the combination of the kernels using the Hadamard product, i.e., $\mathbf{M}^v \circ \mathbf{M}^w$, where $\circ$ denotes pointwise multiplication. In this context, we note that for simplicity of comparison, the signals are processed in a batch manner for both methods. For both algorithms, we set a threshold to 90% correct detection rate and compare their false alarms. Although Hadamard differs from the proposed approach only in the type of combination between the affinity kernels, it provides significantly more false alarms and it detects most transients as speech.

In Figs. 14.3, 14.4 and 14.5 we present a quantitative evaluation of the source activity detection in the form of receiver operating characteristic (ROC) curves, which are plots of probability of detection versus the probability of false alarm. The ROC curves are obtained by sweeping the value of the threshold τ in (14.11) from the maximal to the minimal values of the source activity measure such that the lower the threshold value the higher both the probability of detection and the probability of false alarm. We compare the proposed algorithm for source activity detection, in addition to Hadamard, to its single modal versions, which are based on the eigenvalue decomposition of the single modal kernels $\mathbf{M}^v$ and $\mathbf{M}^w$ rather than of the unified kernel $\mathbf{M}$. In addition, we compare the proposed algorithm to

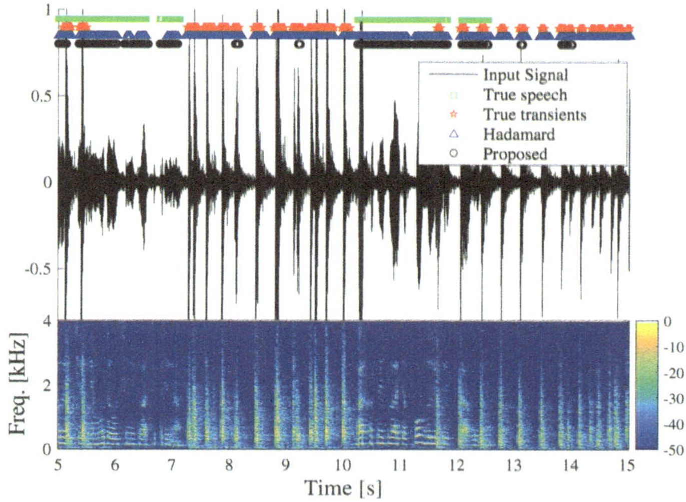

Fig. 14.2 Qualitative assessment of the source activity detection algorithm. Common source: speech. Interfering source: hammering. (Top) Time domain, input signal—black solid line, true speech—green squares, true hammering—red stars, "Hadamard"—blue triangles, proposed algorithm—black circles. (Bottom) Spectrogram of the input audio signal

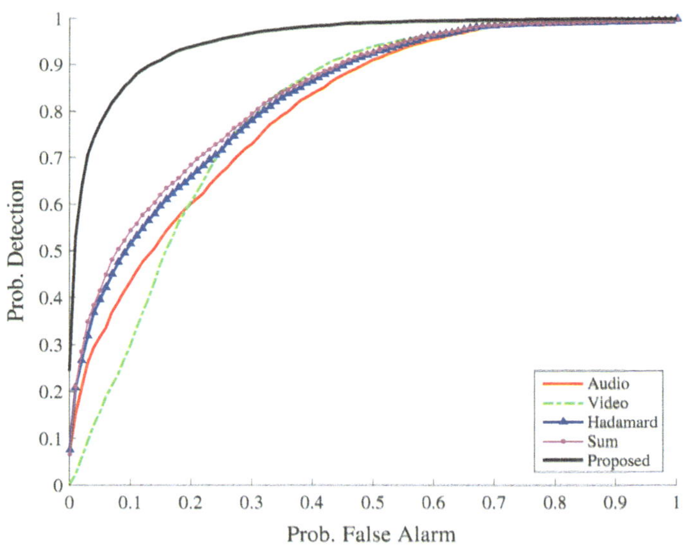

Fig. 14.3 Probability of the detection versus probability of false alarm. Common source: speech. Interfering source: hammering

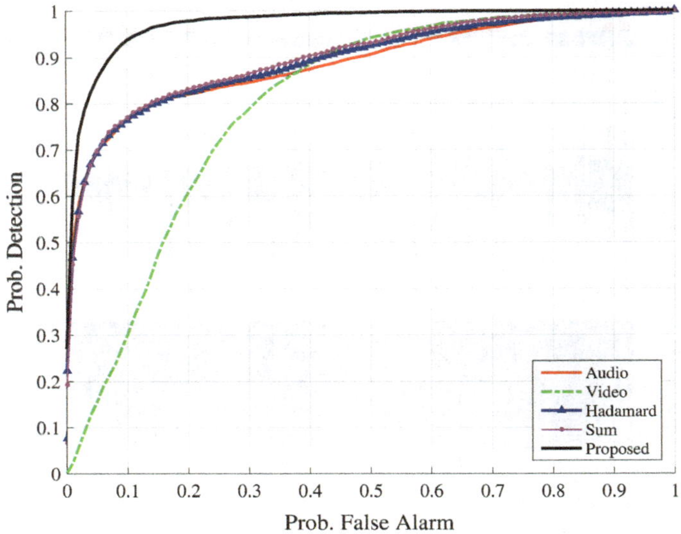

Fig. 14.4 Probability of the detection versus probability of false alarm. Common source: speech. Interfering source: door-knocks

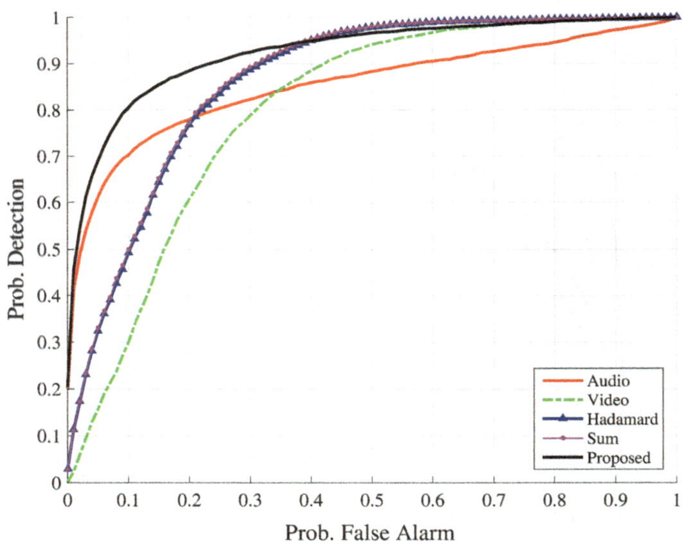

Fig. 14.5 Probability of the detection versus probability of false alarm. Common source: speech. Interfering source: microwave

another multimodal approach, which is obtained using a simple sum between the kernel $\mathbf{M}^v + \mathbf{M}^w$, and is termed "Sum" in the plots. Figures 14.3, 14.4 and 14.5 demonstrate that the single modal approaches perform inferior to the multi-modal approaches. In the single modal approaches, the representation obtained through the eigenvalue decomposition of the single modal kernel is called "diffusion maps" and it is widely used for clustering [6]. The poor results of the approach based only on the audio signal implies that the different sound sources speech and transients share similar characteristics, so they are not separated (clustered) properly. Combining the two modalities using the alternative fusion approaches "Hadamard" and "Sum" only slightly improves the activity detection of the common source, and the proposed algorithm outperform all other methods. The improved performance of the proposed algorithm implies that alternating diffusion maps indeed provide a representation of the audio-visual signal, in which the common source is separated from the other sources.

In the next experiment, we consider a setting, in which multiple sources are present including transients, background noise and speech from two speakers. The video camera is pointed to one of the speakers, which is considered as the common source while the other speech source is considered interference. This setting is particularly challenging not only due to the presence of multiple interferences, but specifically due to the presence of an interfering speech which has similar characteristics to the common source. In Fig. 14.6 we use relatively low levels of noise, and present an example of how alternating diffusion maps extracts the common speech source while ignoring the interfering speaker. It can be seen in Fig. 14.6 that the proposed source activity detection algorithm successfully tracks the common source, while providing only few false alarms in the presence of the interfering source. In contrast, Hadamard provides significantly more false alarms wrongly detecting the interfering speech. Figures 14.7 and 14.8 demonstrate the improved performance of the proposed method in the activity detection of the common source compared to all other methods. These results imply that the representation obtained by the alternating diffusion maps method successfully separates the common source even in the presence of challenging interferences with similar characteristics to the common source. We note that the ability to detect the activity of a particular speaker while ignoring the activity of the other speaker may be particularly useful for the task of source diarization (who spoke when). In this task, speech activity of each speaker in a sound scene has to be detected while ignoring the activity of the other speakers and noises. For example in a teleconference room, one approach to address this task would be to use multiple video cameras—each pointed at a different speaker, and to apply the source activity detection algorithm separately for each one of them.

14.4.3 Discussion—Sound Source Separation

The alternating diffusion maps method provides a representation of the audio-visual signal, by mapping the signal into a new coordinate system constructed from the

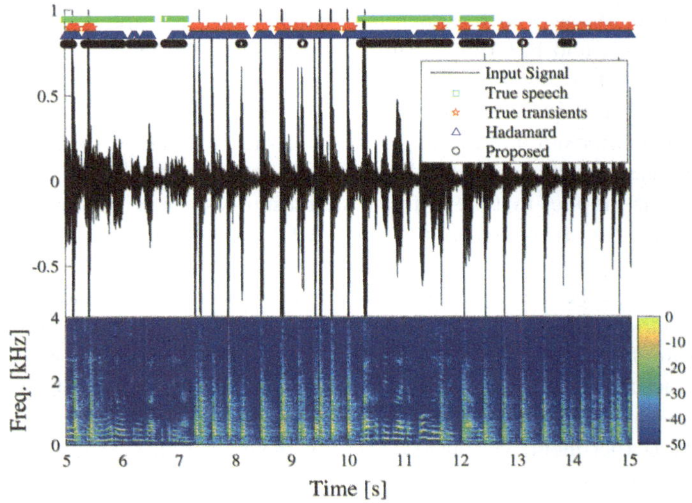

Fig. 14.6 Qualitative assessment of the source activity detection algorithm. Common source: speech. Interfering sources: speech, babble noise with 20 dB SNR. (Top) Time domain, input signal—black solid line, true common source (speech)—green squares, true interfering speech source—red stars, "Hadamard"—blue triangles, proposed algorithm—black circles. (Bottom) Spectrogram of the input signal

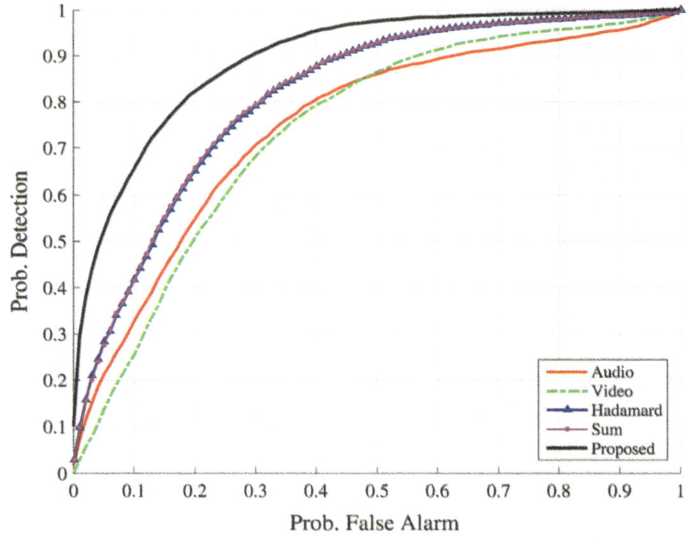

Fig. 14.7 Probability of the detection versus probability of false alarm. Common source: speech. Interfering sources: speech, babble noise with 20 dB SNR

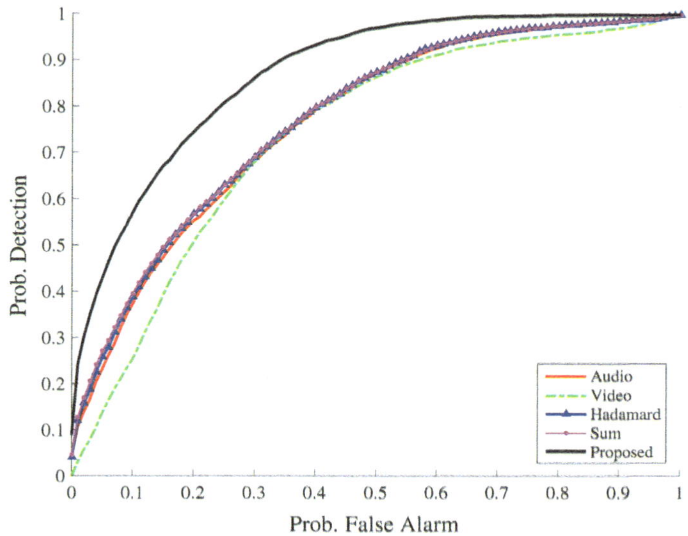

Fig. 14.8 Probability of the detection versus probability of false alarm. Common source: speech. Interfering sources: speech, door-knocks transients and white Gaussian noise with 15 dB SNR

eigenvalue decomposition of the unified affinity kernel **M**. In the new representation the common source is separated from all other sources, so it may be used, e.g., for the activity detection of the common source as we show in this chapter. However, the mapping to the new coordinate system is non-linear so that the question of how to reproduce the original sound produced by the common source from its new representation still requires further research. Based on the study presented in [44], one possible kernel-based solution for this question is the incorporation of the underlying representation of the common source in a non-local filtering technique. The main idea is to construct non-local filters for the estimation of the spectral variance of each source, and then incorporate the estimates in a spectral gain filter for source separation.

The spectral variance of the common source in frame n may be obtained by a weighted average of the spectral variance of the measured audio signal in frames $k \in 1, 2, \ldots, N$, where the selection of proper weights, denoted by $w(n, k)$, is the key for a good estimation. Assuming that the activity of the common source is known in advance, the authors in [44] proposed a statistical model showing that assigning $w(n, k) = 1$ for pairs of frames (n, k) in which the common source is active, and zero otherwise, indeed leads to the estimate of the common source. Intuitively, large weights are selected only for frames that share similar content with frame n in terms of the common source. In such a case, the content of the common source remains unchanged, while the other parts of the spectral variance are averaged out. We note that for a proper estimation of the spectral variance of the common source, in practice, spectral subtraction is further applied to reduce an additive error term related to the spectral variance of the interfering sources; for more details, we refer the reader to [44].

One approach for the selection of the weights $w(n, k)$ is to use the unified affinity kernel setting $w(n, k) = M_{n,k}$. As described in Sect. 14.3, the unified affinity kernel $\mathbf{M}$ in (14.6) represents relations between the frames according to the common source such that when both frames n and k are similar to each other in term of the common source $M_{n,k}$ has a high value. Another approach would be to construct the new representation from the common source $\boldsymbol{\phi}_1, \boldsymbol{\phi}_2, \ldots, \boldsymbol{\phi}_N$, and then to use the Euclidean distance between the frames in the new representation setting the coefficients, for example, to $w(n, k) = \exp\left[-\left\|\boldsymbol{\phi}_n - \boldsymbol{\phi}_k\right\|^2\right]$. The motivation to use the Euclidean distance in the domain of the new representation is that it approximates the meaningful diffusion distance [6]. Talmon, Cohen, and Gannot [44] have studied these two approaches for the separation of speech from transients in a single modal setting, and they indeed found that the second approach provides better performance.

For source separation, each source is enhanced using the optimally modified log spectral amplitude (OM-LSA) algorithm [45], in which the spectral gain function provides an optimal solution minimizing the quadratic error of the log of the spectral amplitude of a particular source assuming that the spectral variances of all sources are known. In practice, the estimates of the spectral variances of the sources may be obtained via the non-local filtering technique using multiple video cameras, each pointed at a different sound source.

14.5 Conclusions

We have addressed the problem of extracting a particular sound source measured in a complex sound scene by a single microphone and by a video camera. By pointing the video camera to this particular sound source, the problem becomes equivalent to extracting the common source to the video and the audio modalities, and as we showed, it can be addressed using the alternating diffusion maps method. The latter provides an underlying representation of the common source while attenuating the interfering sources as we demonstrated for the activity detection of the common source. The alternating diffusion maps is applied in a data driven manner to the measured signal, so that no specific assumptions on the types of the source are used. Therefore even though we demonstrated its use for extracting speech sources, it has the potential to be used for other sources. For example, an audio-visual recording of keyboard taps may be used for extracting the taps in the presence of speech and other transients. In addition, although we have focused in this chapter on obtaining an underlying representation of the common source, the alternating diffusion maps method has the potential to be further used for the separation of the common source via non-local filtering as shown in previous studies in a single modal setting.

References

1. R.R. Lederman, R. Talmon, Learning the geometry of common latent variables using alternating-diffusion. Appl. Comput. Harmon. Anal. (2015)
2. S.T. Roweis, L.K. Saul, Nonlinear dimensionality reduction by locally linear embedding. Science **290**(5500), 2323–2326 (2000)
3. M. Balasubramanian, E.L. Schwartz, J.B. Tenenbaum, V. de Silva, J.C. Langford, The isomap algorithm and topological stability. Science **295**(5552), 7–7 (2002)
4. M. Belkin, P. Niyogi, Laplacian eigenmaps for dimensionality reduction and data representation. Neural Comput. **15**(6), 1373–1396 (2003)
5. D.L. Donoho, C. Grimes, Hessian eigenmaps: locally linear embedding techniques for high-dimensional data. Proc. Nat. Acad. Sci. **100**(10), 5591–5596 (2003)
6. R. Coifman, S. Lafon, Diffusion maps. Appl. Comput. Harmon. Anal. **21**(1), 5–30 (2006)
7. D. Zhou, C.J.C. Burges, Spectral clustering and transductive learning with multiple views, in *Proceedings of the 24th International Conference on Machine Learning, Corvallis, OR, USA* (2007), pp. 1159–1166
8. M.B. Blaschko, C.H. Lampert, Correlational spectral clustering, in *Proceedings of the IEEE Conference on Computer Vision and Pattern Recognition (CVPR), Anchorage, AK* (2008), pp. 1–8
9. V.R. De Sa, P.W. Gallagher, J.M. Lewis, V.L. Malave, Multi-view kernel construction. Mach. Learn. **79**(1–2), 47–71 (2010)
10. A. Kumar, P. Rai, H. Daume, Co-regularized multi-view spectral clustering, Adv. Neural Inf. Process. Syst., 1413–1421 (2011)
11. A. Kumar, H. Daumé, A co-training approach for multi-view spectral clustering, in *Proceedings of the 28th International Conference on Machine Learning (ICML), Bellevue, Washington, USA* (2011), pp. 393–400
12. Y.Y. Lin, T.L. Liu, C.S. Fuh, Multiple kernel learning for dimensionality reduction. IEEE Trans. Pattern Anal. Mach. Intell. **33**(6), 1147–1160 (2011)
13. B. Wang, J. Jiang, W. Wang, Z.H. Zhou, Z. Tu, Unsupervised metric fusion by cross diffusion, in *Proceedings of the IEEE Conference on Computer Vision and Pattern Recognition (CVPR), Providence, RI* (2012), pp. 2997–3004
14. H.C. Huang, Y.Y. Chuang, C.S. Chen, Affinity aggregation for spectral clustering, in *Proceedings IEEE Conference on Computer Vision and Pattern Recognition (CVPR), Providence, RI* (2012), pp. 773–780
15. B. Boots, G. Gordon, Two-manifold problems with applications to nonlinear system identification, in *Proceedings of the 29th International Conference on Machine Learning (ICML), Edinburgh, Scotland, GB* (2012), pp. 623–630
16. M.M. Bronstein, K. Glashoff, T.A. Loring, Making laplacians commute (2013), arXiv:1307.6549
17. O. Lindenbaum, A. Yeredor, M. Salhov, A. Averbuch, Multiview diffusion maps (2015), arXiv preprint arXiv:1508.05550
18. T. Michaeli, W. Wang, T. Livescu, Nonparametric canonical correlation analysis, in *Proceedings of the International Conference on Machine Learning (ICML), New York, USA* (2016)
19. A. Aubrey, B. Rivet, Y. Hicks, L. Girin, J. Chambers, C. Jutten, Two novel visual voice activity detectors based on appearance models and retinal filltering, *Proceedings of the 15th European Signal Processing Conference (EUSIPCO)* (2007), pp. 2409–2413
20. E. Ong, R. Bowden, Robust lip-tracking using rigid flocks of selected linear predictors, *Proceedings of the 8th IEEE International Conference on Automatic Face and Gesture Recognition* (2008)
21. Q. Liu, W. Wang, P. Jackson, A visual voice activity detection method with adaboosting, in *Proceedings of the Sensor Signal Processing for Defence (SSPD)* (IET, 2011), pp. 1–5
22. D. Sodoyer, B. Rivet, L. Girin, J. Schwartz, C. Jutten, An analysis of visual speech information applied to voice activity detection, *Proceedings of the 31st IEEE International Conference on Acoustics, Speech and Signal Processing (ICASSP)*, vol. 1 (2006)

23. D. Sodoyer, B. Rivet, L. Girin, C. Savariaux, J. Schwartz, C. Jutten, A study of lip movements during spontaneous dialog and its application to voice activity detection. J. Acoust. Soc. Am. **125**, 1184 (2009)
24. S. Siatras, N. Nikolaidis, M. Krinidis, I. Pitas, Visual lip activity detection and speaker detection using mouth region intensities. IEEE Trans. Circuits Syst. Video Technol. **19**(1), 133–137 (2009)
25. A. Aubrey, Y. Hicks, J. Chambers, Visual voice activity detection with optical flow. IET Image Proc. **4**(6), 463–472 (2010)
26. P. Tiawongsombat, M. Jeong, J. Yun, B. You, S. Oh, Robust visual speakingness detection using bi-level HMM. Pattern Recogn. **45**(2), 783–793 (2012)
27. P. Atrey, M. Hossain, A. El Saddik, M. Kankanhalli, Multimodal fusion for multimedia analysis: a survey. Multimed. Syst. **16**(6), 345–379 (2010)
28. S. Tamura, M. Ishikawa, T. Hashiba, S. Takeuchi, S. Hayamizu, A robust audio-visual speech recognition using audio-visual voice activity detection, in *Proceedings of the Annual Conference of International Speech Communication Association (INTERSPEECH)* (2010), pp. 2694–2697
29. D. Dov, R. Talmon, I. Cohen, Audio-visual voice activity detection using diffusion maps. IEEE/ACM Trans. Audio Speech Lang. Process. **23**(4), 732–745 (2015)
30. R. Talmon, I. Cohen, S. Gannot, R.R. Coifman, Supervised graph-based processing for sequential transient interference suppression. IEEE Trans. Audio Speech Lang. Process. **20**(9), 2528–2538 (2012)
31. A. Hirszhorn, D. Dov, R. Talmon, I. Cohen, Transient interference suppression in speech signals based on the OM-LSA algorithm, *Proceedings of the International Workshop on Acoustic Signal Enhancement (IWAENC)* (2012), pp. 1–4
32. R. Talmon, I. Cohen, S. Gannot, Clustering and suppression of transient noise in speech signals using diffusion maps, in *Proceedings of the 36th IEEE International Conference on Acoustics, Speech and Signal Processing (ICASSP)* (IEEE, 2011), pp. 5084–5087
33. D. Dov, R. Talmon, I. Cohen, Kernel-based sensor fusion with application to audio-visual voice activity detection. IEEE Trans. Signal Process. **64**(24), 6406–6416 (2016)
34. D. Dov, R. Talmon, I. Cohen, Kernel method for voice activity detection in the presence of transients. IEEE/ACM Trans. Audio Speech Lang. Process. **24**(12), 2313–2326 (2016)
35. P.C. Mahalanobis, On the generalized distance in statistics. Proc. Nat. Inst. Sci. (Calcutta) **2**, 49–55 (1936)
36. C. Fowlkes, S. Belongie, F. Chung, J. Malik, Spectral grouping using the Nyström method. IEEE Trans. Pattern Anal. Mach. Intell. **26**(2), 214–225 (2004)
37. J. Shi, J. Malik, Normalized cuts and image segmentation. IEEE Trans. Pattern Anal. Mach. Intell. **22**(8), 888–905 (2000)
38. http://www.freesound.org
39. J. Barron, D. Fleet, S. Beauchemin, Performance of optical flow techniques. Int. J. Comput. Vis. **12**(1), 43–77 (1994)
40. A. Bruhn, J. Weickert, C. Schnörr, Lucas/Kanade meets Horn/Schunck: combining local and global optic flow methods. Int. J. Comput. Vis. **61**(3), 211–231 (2005)
41. S.B. Davis, P. Mermelstein, Comparison of parametric representations for monosyllabic word recognition in continuously spoken sentences. IEEE Trans. Acoust. Speech Signal Process. **28**(4), 357–366 (1980)
42. H. Hirsch, D. Pearce, The aurora experimental framework for the performance evaluation of speech recognition systems under noisy conditions, *ASR2000-Automatic Speech Recognition: Challenges for the New Millenium ISCA Tutorial and Research Workshop (ITRW)* (2000)
43. B. Logan, Mel frequency cepstral coefficients for music modeling, *Proceedings of the 1st International Conference on Music Information Retrieval (ISMIR)* (2000)
44. R. Talmon, I. Cohen, S. Gannot, Single-channel transient interference suppression with diffusion maps. IEEE Trans. Audio Speech Lang. Process. **21**(1), 132–144 (2013)
45. I. Cohen, B. Berdugo, Speech enhancement for non-stationary noise environments. Signal Process. **81**(11), 2403–2418 (2001)

Index

S. Makino (ed.), *Audio Source Separation*, Signals and Communication Technology, https://doi.org/10.1007/978-3-319-73031-8

The manufacturer's authorised representative in the EU is Springer Nature Customer Service Centre GmbH, Europaplatz 3, 69115 Heidelberg, Germany. If you have any concerns regarding our products, please contact ProductSafety@springernature.com

Printed and bound by CPI Group (UK) Ltd, Croydon, CR0 4YY
30/09/2026
02225230-0014